Lecture Notes in Computer Science 9348

More information about this series at http://www.springer.com/series/7407

Wolfram Kahl · Michael Winter
José N. Oliveira (Eds.)

Relational and Algebraic Methods in Computer Science

15th International Conference, RAMiCS 2015
Braga, Portugal, September 28 – October 1, 2015
Proceedings

Springer

Editors
Wolfram Kahl
McMaster University
Hamilton, Ontario
Canada

José N. Oliveira
Universidade do Minho
Braga
Portugal

Michael Winter
Brock University
St. Catharines, Ontario
Canada

ISSN 0302-9743 ISSN 1611-3349 (electronic)
Lecture Notes in Computer Science
ISBN 978-3-319-24703-8 ISBN 978-3-319-24704-5 (eBook)
DOI 10.1007/978-3-319-24704-5

Library of Congress Control Number: 2015949476

LNCS Sublibrary: SL1 – Theoretical Computer Science and General Issues

Springer Cham Heidelberg New York Dordrecht London

Printed on acid-free paper

Springer International Publishing AG Switzerland is part of Springer Science+Business Media
(www.springer.com)

Preface

Relations and formal languages are omnipresent in computer science and in software design. While quantifier-oriented (first- or higher-)order logics can be used to specify and reason about relations, this "element-level style" often obfuscates the structure of specifications and makes reasoning harder. A useful analogy is to consider how element-level reasoning gives way to matrix-level calculations in linear algebra. Similarly, *relation algebra* allows for calculational, largely quantifier-free reasoning about relations, and shares a large subtheory with *Kleene algebra*, the mathematical theory of the regular expressions used for the specification of certain formal languages.

An international collaboration to establish a conference series as a forum for the use of relational methods in computer science, RelMiCS, was initiated during the "38th Banach Semester on Algebraic Methods in Logic and their Computer Science Application" in Warsaw, Poland, September and October 1991. Adapting essentially a one-and-a-half year rhythm, the first 11 RelMiCS conferences were held from 1994 to 2009 on all inhabited continents except Australia. Starting with RelMiCS 7, these were were held as joint events with "Applications of Kleene Algebras" (AKA) conferences. At RelMiCS 11 / AKA 6 in Doha, Qatar, it was decided to continue the series under the unifying name "Relational and Algebraic Methods in Computer Science (RAMiCS)." The next events, RAMiCS 12–14, were then held in Rotterdam, The Netherlands, in 2011, Cambridge, UK, in 2012 and Marienstatt, Germany, in 2014.

This volume contains the proceedings of the 15th International Conference on Relational and Algebraic Methods in Computer Science (RAMiCS 2015), held in Braga, Portugal, from September 28 to October 1, 2015, exactly 24 years after the Banach Semester that resulted in founding this conference series.

The call for papers invited submissions about the theory of relation algebras and Kleene algebras, process algebras, fixed point calculi, idempotent semirings, quantales, allegories, and dynamic algebras, and cylindric algebras, and about their applications in areas such as verification, analysis and development of programs and algorithms, algebraic approaches to logics of programs, modal and dynamic logics, interval and temporal logics, etc.

We were fortunate to be able to invite Gheorghe Stefanescu and Ian Hodkinson who, with their presentations on "A Quest for Kleene Algebra in 2 Dimensions" and "Connections Between Relation Algebras and Cylindric Algebras", nicely emphasized the two traditional theoretical pillars of the RAMiCS conferences, and Ernst-Erich Doberkat, whose presentation "Towards a Probabilistic Interpretation of Game Logic," opened up new opportunities related to modal logic.

The body of this volume is made up of invited papers accompanying these three invited talks, and of 20 contributions by researchers from around the world

The papers have been arranged into three groups:

Theoretical Foundations
: Including studies of relation-algebraic theories ranging from nominal Kleene algebra to allegories and covering a range of relation concepts, including multirelations, n-ary relations, and relational resource semantics

Reasoning About Computations and Programs
: With contributions addressing refinement, type checking, and verified relation- and Kleene-algebraic programming

Applications of Relational and Algebraic Methods
: Including to fuzzy databases, rough set theory, preferences, optimization, and text categorization

The contributed papers were selected by the Program Committee from 25 relevant submissions. Each submission was reviewed by at least three Program Committee members; the Program Committee did not meet in person, but had over one week of intense electronic discussions.

We are very grateful to the members of the Program Committee and the subreviewers for their care and diligence in reviewing the submitted papers. We would like to thank the members of the RAMiCS Steering Committee for their support and advice especially in the early phases of the conference organization. We are grateful to INESC TEC and the University of Minho for generously providing administrative support, and we gratefully appreciate the excellent facilities offered by the EasyChair conference administration system. Last but not least, we thank FCT (Fundação para a Ciência e a Tecnologia, Portugal) for their financial support.

July 2015

Wolfram Kahl
Michael Winter
José N. Oliveira

Organization

Organizing Committee

Conference Chair

José N. Oliveira	University of Minho, Portugal

Program Co-chairs

Wolfram Kahl	McMaster University, Canada
Michael Winter	Brock University, Canada

Local Organizers

Luís S. Barbosa	University of Minho, Portugal
Manuel A. Cunha	University of Minho, Portugal
António N. Ribeiro	University of Minho, Portugal

Program Committee

Rudolf Berghammer	Christian-Albrechts-Universität zu Kiel, Germany
Jules Desharnais	Université Laval, Canada
Marcelo Frias	University of Buenos Aires, Argentina
Hitoshi Furusawa	Kagoshima University, Japan
Steven Givant	Mills College, USA
Timothy G. Griffin	University of Cambridge, UK
Walter Guttmann	University of Canterbury, New Zealand
Robin Hirsch	University College of London, UK
Peter Höfner	NICTA Ltd., Australia
Ali Jaoua	Qatar University, Qatar
Peter Jipsen	Chapman University, USA
Wolfram Kahl	McMaster University, Canada
Roger Maddux	Iowa State University, USA
Ali Mili	Tunis, Tunisia; NJIT, USA
Bernhard Möller	Universität Augsburg, Germany
Martin E. Müller	Universität Augsburg, Germany
José N. Oliveira	Universidade do Minho, Portugal

Ewa Orłowska	National Institute of Telecommunications, Poland
Agnieszka Rusinowska	Université Paris 1, France
Gunther Schmidt	Universität der Bundeswehr München, Germany
Renate Schmidt	University of Manchester, UK
Isar Stubbe	Université du Littoral-Côte-d'Opale, France
Michael Winter	Brock University, Canada

Steering Committee

Rudolf Berghammer	Christian-Albrechts-Universität zu Kiel, Germany
Jules Desharnais	Université Laval, Canada
Ali Jaoua	Qatar University, Qatar
Peter Jipsen	Chapman University, USA
Bernhard Möller	Universität Augsburg, Germany
José N. Oliveira	Universidade do Minho, Portugal
Ewa Orłowska	National Institute of Telecommunications, Poland
Gunther Schmidt	Universität der Bundeswehr München, Germany
Michael Winter	Brock University, Canada

Additional Reviewers

Ernst-Erich Doberkat
Alexander Kurz
Annabelle McIver
Koki Nishizawa
Patrick Roocks
Agnieszka Rusinowska
Alberto Simões
John Stell
Insa Stucke
Toshinori Takai
Norihiro Tsumagari

Sponsoring Institutions

INESC TEC
Universidade do Minho
FCT (Fundação para a Ciência e a Tecnologia, Portugal)

Contents

Invited Papers

Theoretical Foundations

Reasoning about Computations and Programs

Applications of Relational and Algebraic Methods

Invited Papers

A Quest for Kleene Algebra in 2 Dimensions

Gheorghe Stefanescu

Department of Computer Science, University of Bucharest, Romania
gheorghe.stefanescu@fmi.unibuc.ro

Abstract. The term Kleene algebra refers to a certain algebraic structure, built up using sequential composition, its iterated version, and union. The key operation is composition: on strings, it connects the final point of the first string to the initial point of the second string.

The quest for Kleene algebra in 2 dimensions starts with the clarification of the notions of word and composition in 2 dimensions. A 2-dimensional word is an arbitrary shape area, consisting of unit square cells, and filled with letters. Word composition puts two words together, without overlapping, and controls the contact elements of the contours of these words. This method actually defines a family of composition operations, indexed by the restrictions used to control the words' contact parts.

Finite automata and regular expressions are extended to 2 dimensions. The former is relatively easy and it reduces to tiling. For the latter, a few recently introduced classes of regular expressions $n2RE$ and $x2RE$ are presented. The formalism is completed with a mechanism to specify and solve recursive systems o equations for generating languages in 2 dimensions.

Finite automata and regular expressions are equivalent and Kleene algebra provides a beautiful algebraic setting to formalize this result. A section on the limits of our current understanding on lifting this result to 2 dimensions is included.

Finally, we briefly show that, enriched with spatial and temporal data attached to tiles, the formalism leads to a natural model for interactive, distributed programs.

Keywords: finite automata, regular expressions, Kleene algebra, self-assembling tile systems, 2-dimensional languages, 2-dimensional regular expressions, recursive specifications, interactive systems, scenarios, relational semantics.

1 Introduction

The term *Kleene algebra* is used for a certain algebraic structure, defined using sequential composition, its iterated version, union, zero, and identities.

Words and Composition in 2 Dimensions. A key operation in Kleene algebra is composition. The definition of composition in the model of Kleene algebra on strings is simple. A string is a totally ordered structure and, beside its contents (letters), it has 2 end points: an initial point and a final point. Composition puts

W. Kahl et al. (Eds.): RAMiCS 2015, LNCS 9348, pp. 3–26, 2015.
DOI: 10.1007/978-3-319-24704-5_1

two strings together connecting the final point of the first string to the initial point of the second string.

To extend this composition to 2-dimensional words, first we need to specify the class of 2-dimensional words we are working with. For us, a 2-dimensional word is an arbitrary shape area, consisting of unit square cells, placed with corners in points with integer coordinates, and filled with letters. Beside its contents, such an area has a contour. A natural definition of 2-dimensional word composition is: *put the words together, without overlapping, and control the contact elements of the contours of these words in the composed word.* While in the case of strings we have a unique composition operation, now we have a family of composition operations, indexed by the restrictions used to control these contour contact parts.

Finite Automata, in 2 Dimensions. Finite automata provide a low-level formalism for sequential computation. Technically, (sequential) composition is modelled by attaching state labels to the transitions and connect them using these labels: namely, for two transitions a and b, with labels $a : p \to p'$ and $b : q \to q'$, the sequential composition $a \cdot b : p \to q'$ is possible if the contact labels are equal, i.e., $p' = q$.

A natural way to extend transitions in 2 dimensions is by considering two orthogonal (west-east and north-south) finite automata transitions with the same label, say $a : p \to p'$ and $a : q \to q'$. This means, we are attaching 4 labels to each transition, say $a : (p, q) \to (q, q')$, getting a *tile*, i.e., a cell letter with labels on each west/north/east/south border. However, in 2 dimensions the linear order of sequentiality is lost, hence in the abstract tile models we make no distinction between the border labels – nevertheless, in applications modelling distributed, interactive programs the direction from west-north to east-south is preserved. As in the case of automata, tile composition produces scenarios (2-dimensional paths) and requires to have the same label on common borders. To conclude, path composition of transitions in finite automata becomes scenario composition of tiles in 2 dimensions.

Except for transitions, we need to consider initial and final labels. The resulting 2-dimensional extension of finite automata is called *self-assembling tile system* (SATS). We want to stress that the accepted words of a SATS are defined in terms of scenarios built up from tiles, having *specified labels* on each west/north/east/south *external border*. This last condition is needed for having a compositional model.

Regular Expressions and Recursive Definitions, in 2 Dimensions. Regular expressions are equivalent with finite automata, but they offer a quite different specification mechanism. Rather than starting with 1-letter transitions, they use sets of large chunks of transitions ("events", as Kleene called them [25]) and compose them using union, sequential composition, and its iterated version. Fortunately, the labels used by accepted paths in finite automata can be eliminated and equivalent label-free specifications using regular expressions can be found.

Extending regular expressions in 2 dimensions is not an easy task. Somehow, on has to mimic scenario composition of SATSs, where two complicate shapes can be glueing together whenever the labels of their contact parts are the same.

Recently two classes of 2-dimensional regular expressions have been proposed: *n2RE* [4] and *x2RE* [5]. The former uses constraints on 2-dimensional word contours, defined by a boolean logic built up from comparison formulas described in terms of corners and edges. In addition, the latter uses the information that certain cells are extreme (they have at most one neighbour in the word). Whether one of this logics is powerful enough to capture all the patterns of labels occurring at the contact parts of SATS scenario compositions is not yet known. Hence, currently we do not have a class of 2-dimensional regular expressions equivalent to SATSs.

SATS with 2-color Border Tiles. We have investigated a few hundreds of SATSs consisting of 2-color border tiles. We hope the analysis of this particular class of SATSs will gather enough information to identify a good regular expressions candidate for all SATSs.

Adding Data to Get Programs, in 2 Dimensions. The development of this fundamental investigation is parallel with an attempt to develop rigours and powerful programming formalisms for open, interactive, large-scale, distributed systems. Extending a classical slogan that "program = control + data", the new slogan is:

interactive, distributed program = (control & interaction) + (spatial & temporal) data

Our interest in SATSs or 2-dimensional regular expressions mainly comes from a basic fact that these models are used to specify the contol & interaction part of distributed systems. Understanding the structure of the languages defined by these models is of fundamental importance for understanding distributed programs. A structured programming language Agapia [15], [37] has been introduced, using regular expressions over rectangular words. Arbitrary shape words may be useful for extensions including spatial and temporal pointers [20].

Structure of the Paper: We start with a brief recall of basic Kleene algebra results in 1 dimension. Then, we define words and languages in 2 dimensions. In Section 4 we identify a Language Product Problem to be used for specifying 2-dimensional languages as product of two 1-dimensional languages. Next, self-assembling tile systems (SATSs), as a 2-dimensional version of finite automata, are presented. A section is devoted to 2-dimensional word composition, a key operation in getting a compositional approach to 2-dimensional tiling. The next section, Section 7, describes regular expressions and recursive systems of equations in 2 dimensions. In Section 8 we present a few partial results regarding languages accepted by SATSs using 2-colors border tiles. Our current limitations in understating SATS languages, applications to interactive systems, and comments on related works conclude the paper.

2 A Reference Point: Kleene Algebra in 1 Dimension

A well-known result says that finite automata and regular expressions are equivalent. There is a beautiful algebraic formalization of this result using Kleene algebras. We sketch the result below, following Chapter 8 of [44].

Abstract Automata: Let $(S,+,\cdot,0,1)$ be a semiring [44] and "$*$" an operation on (square) *matrices* over S. A *matricial presentation* of a regular language is a matrix $M=\begin{pmatrix} A & B \\ C & D\end{pmatrix}$ over $(S,+,\cdot,0,1)$ of type $(1+n)\times(1+n)$, where: $1,\ldots,n$ represent the states of the automaton; A represents the direct input-output connection (if any); $B=(b_j)_{1\leq j\leq n}$ specifies the input states; $C=(c_i)_{1\leq i\leq n}$ specifies the final states; and $D=(d_{ij})_{1\leq i,j\leq n}$ specifies the transitions.

The *language* specified by a matricial presentation, defined with the above notation, is $\mathcal{L}(M)=A+B\cdot D^*\cdot C$, where "$+$" and "$\cdot$" are the usual operations on matrices induced by the operations in S.

Simulation: Two matricial presentations $M_i=\begin{pmatrix} A_i & B_i \\ C_i & D_i\end{pmatrix}$, with n_i states, for $i\in\{1,2\}$, are *similar* via a relation $\rho\subseteq\{1,\ldots,n_1\}\times\{1,\ldots,n_2\}$, denoted $M_1\to_\rho M_2$, if: $A_1=A_2$, $B_1\rho=B_2$, $C_1=\rho C_2$, and $D_1\rho=\rho D_2$.

The importance of simulation relation comes from the following facts:

Soundness: *Simulation preserves the language.*

Completeness: *Two nondeterministic finite automata accept the same language if and only if they may be connected*[1] *by a chain of simulations.*

Conway and Kleene Algebras. Let $\mathbf{M}=(\mathcal{M}(m,n)_{m,n},+,\cdot,*,0_{m,n},\mathsf{I}_n)$ be a doubly-ranked family, enriched with operations:

$0_{m,n}\in\mathcal{M}(m,n)$; $\mathsf{I}_n\in\mathcal{M}(n,n)$; $+:\mathcal{M}(m,n)\times\mathcal{M}(m,n)\to\mathcal{M}(m,n)$; $\cdot:\mathcal{M}(m,n)\times\mathcal{M}(n,p)\to\mathcal{M}(m,p)$; $*:\mathcal{M}(n,n)\to\mathcal{M}(n,n)$.

Suppose $(\mathcal{M}(m,n)_{m,n},+,\cdot,0_{m,n},\mathsf{I}_n)$ is a *semiring of matrices* [44]. We also consider the following *axioms for star*:

(I) $(\mathsf{I}_n)^*=\mathsf{I}_n$
(S) $(a+b)^*=(a^*\cdot b)^*\cdot a^*$
(P) $(a\cdot b)^*=\mathsf{I}_n+a\cdot(b\cdot a)^*\cdot b$
(Inv) $a\cdot\rho=\rho\cdot b\Rightarrow a^*\cdot\rho=\rho\cdot b^*$, where ρ is a matrix over 0,1.

All these axioms define *Kleene theories.* Without (Inv) one gets the axioms for *Conway theories.* They are *idempotent* if $a+a=a$.

---

[1] For example, for two automata $\mathcal{A}_1$ and $\mathcal{A}_2$, accepting the same language, the relation

$$\mathcal{A}_1\overset{rel}{\longleftarrow}D(\mathcal{A}_1)\overset{inj}{\longleftarrow}\cdot\overset{sur}{\longrightarrow}D(\mathcal{A}_1)_{min}=D(\mathcal{A}_2)_{min}\overset{sur}{\longleftarrow}\cdot\overset{inj}{\longrightarrow}D(\mathcal{A}_2)\overset{rel}{\longrightarrow}\mathcal{A}_2$$

holds, where: D denotes the deterministic automaton obtained by classical powerset construction; $(\ldots)_{min}$ denotes the minimal deterministic automaton; and $\overset{rel}{\longrightarrow}/\overset{inj}{\longrightarrow}/\overset{sur}{\longrightarrow}$ denotes simulation via relations/injective-functions/surjective-functions, respectively.

Fundamental Result (Theorem 8.26 in [44]): *The idempotent Kleene theory axioms give a correct (i.e., sound and complete) axiomatization for the algebra of regular languages.*

The definitions of accepted language and simulation are given in terms of matrices. For completeness, a key technical result is a method to compute the star of a matrix in terms of star of smaller matrices.

Star of Matrices (Theorem 8.21 in [44]): In an idempotent Conway theory the following identity is valid

$$\begin{bmatrix} a & b \\ c & d \end{bmatrix}^* = \begin{bmatrix} a^* + a^*bwca^* & a^*bw \\ wca^* & w \end{bmatrix}$$

where $w = (ca^*b + d)^*$.

Repeatedly applied, this identity leads to a procedure to compute the star of a matrix in terms of star of its elements. Actually, this is another, more formal, presentation of Kleene theorem. It can be proved in the weaker setting of Conway theories. As a final remark, we mention that the Conway theory axioms themselves follow from the weaker version where they are required only on elements, not on matrices [28].

3 Words and Languages, in 2 Dimensions

Words. A *2-dimensional word* is a finite area of unit cells, placed with their corners in points with integer coordinates, and labelled with letters from a specified alphabet. A *2-dimensional letter* is a 2-dimensional word consisting of a unique cell. A *2-dimensional language* is a set of 2-dimensional words.

By convention, these 2-dimensional words are invariant with respect to translation by integer offsets, but not with respect to mirror or rotation.

A word may have several disconnected components. A *component* (respectively, a *horizontally-vertically connected component*) is a maximal set of cells, connected using horizontal, vertical and diagonal directions (respectively, using only horizontal and vertical directions). Tiles, to be introduced in Section 5, constrain the letters of horizontal and vertical neighbouring cells. Consequently, our focus will be on the structure of horizontally-vertically connected components, shortly called *hv-components.*

Examples of words are presented in Fig. 1: A rectangular word is presented in (a); Then, in (b), we present a word of arbitrary shape, with no holes and 1 component; Finally, a word with 1 hole and 2 components is shown in (c). The words in Fig. 1(a), (b), (c) have 1, 3, 7 hv-components, respectively.

Formal Representation. Formally, a 2-dimensional word is represented as (c, v), where c specifies its *contour* and v is *a listing of its letters*, contained in the internal area of the contour and sorted by rows. For example, the word in Fig. 1(b) is represented by the pair

$$(rdrrdrdlluldlurulu, acedab)$$

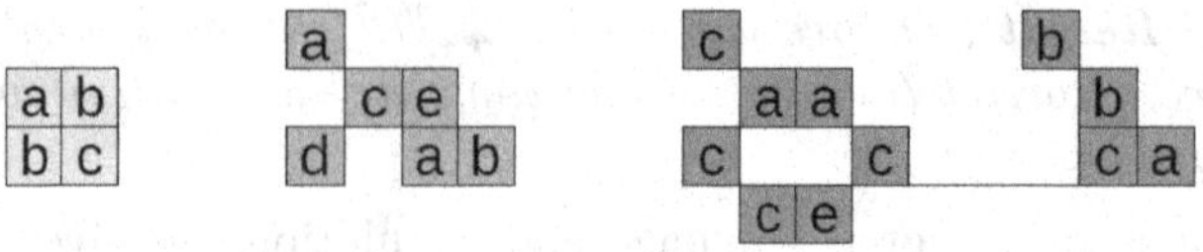

(a) (b) (c)

Fig. 1. Examples of 2-dimensional words

where $u/d/r/l$ stands for "up" / "down" / "right" / "left", respectively, and $acedab$ is the listing of the letters in the word, sorted by rows.

This word representation is unique up to "identities" and the placement of the starting point used in the representation of its contour; here, by definition, an *identity* is a word with a contour surrounding an empty internal area. For instance, the word representation $(rdlu, a)$ is equivalent to $(lurEd, a)$, where E is the identity $rurdrluldl$.

Parts of Contours. In the sequel, we will refer to several parts of word contours. A few such *elements of interest on word contours* are:

side borders: `w` (west), `e` (east), `n` (north), and `s` (south). For instance, a unit edge of the contour is an `e`-edge if it is a vertical edge and the cell on the left is inside the word.

land corners: `nw`, `ne`, `sw`, and `sw` - these are corner points seen from inside the word. For instance, a point on the contour is a `nw`-corner if the cell at the bottom-right of the point is inside the word, while the other 3 cells around the point are not inside the word.

golf corners: `nw'`, `ne'`, `sw'`, and `sw'` - these are corner points seen from outside the word. For instance, a point on the contour is a `nw'`-corner if the cell at the bottom-right of the point is outside the word, while those at the top-right and bottom-left of the point are inside the word.

These elements are used to define restrictions controlling word composition.

4 LPP: A Language Product Problem

4.1 The Problem

(LPP) *Given two string languages Lrow and Lcol, over the same alphabet V,* ***find*** *their product $Lrow \otimes Lcol$, seen as a 2-dimensional language over V.*
By definition, the *product $Lrow \otimes Lcol$* consists of all 2-dimensional words, of arbitrary shape, satisfying the following properties:

1. for any row, any maximal continuous sequence of the word, lying in this row and read in the left-to-right order, defines a string in *Lrow*;
2. for any column, any maximal continuous sequence of the word, lying in this column and read in the top-to-bottom order, defines a string in *Lcol*.

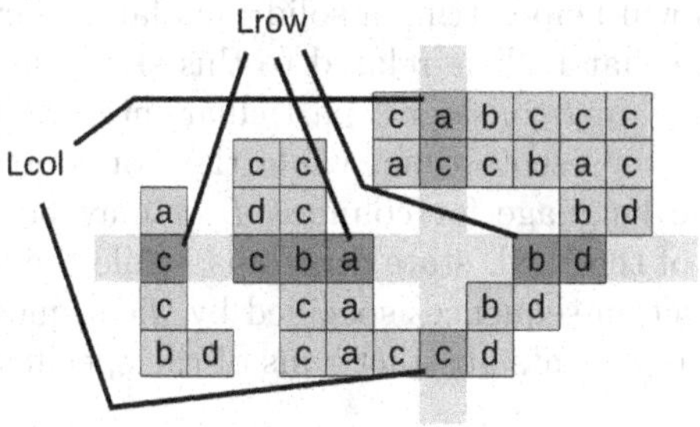

Fig. 2. The $\otimes$-product: $Lrow \otimes Lcol$

An illustration of the $\otimes$-product is presented in Fig. 2. On the emphasized row, the strings c, cba, bd are from $Lrow$, while on the emphasized column, the strings ac, c are from $Lcol$.

A key term in this problem statement is the word "find". We actually ask for

> *a constructive description of this language as a recursive tiling procedure*[2] *generating all 2-dimensional words in* $Lrow \otimes Lcol$.

This characterization would provide a better understanding of the structure of the words in $Lrow \otimes Lcol$.

We can restrict ourselves to the study of those 2-dimensional words consisting of one *hv*-component, as different *hv*-components can be independently processed.

4.2 A Motivation: Combining UML State and Sequence Diagrams

UML [39] is a popular and powerful specification language used to cope with the complexity of modern software systems. It allows the user to specify different views of the system (s)he wants to develop. Two particularly useful views are described using state diagrams and sequence diagrams. The former allows the user to look at the running of the processes, ignoring their interactions, while the latter allows to focus on chains of process interactions, ignoring their state evolution. After providing these independent specifications, the user has to figure out how the system will evolve, provided both specifications hold.

> *Question:* Is it difficult to combine state diagrams and sequence diagrams?

We think ***it is a difficult problem*** and its study will make a worthwhile contribution aiming to help UML users with results and tools useful to understand the result of their state and sequence diagram combination. At the practical

[2] The interest in tiling comes from a basic fact that, in 2 dimensions, word composition (catenation) is defined as a family of restricted tiling operators.

level, combined state and sequence diagrams are already in use, hence the LPP formalism and results will help getting a solid foundation for these approaches.

It is not hard to see that LPP is related to this state and sequence diagram integration. Indeed, one may consider a particular interpretation of LPP when the vertical dimension represents time, while the horizontal dimension represents space. Then, the language for columns *Lcol* may be taken to represent the control sequences of the UML state diagrams, while the row language *Lrow* to represent the interaction sequences specified by the sequence diagram. Their product $Lrow \otimes Lcol$ represents abstract runs of the specified system.

4.3 An Example: The Dutch-Roof Language

This example is taken from [48] - more details can be found there, including a recursive tiling specification. It uses four letters $0, 2, a, c$. The language for rows is specified by the regular expression[3] $Lrow = a^*c(0+2a^*c)^*$. Here, a typical word has the following structure: take a c followed by an arbitrary number of sequences $2c$; then, insert between 2 and c an arbitrary number of a's and between c and 2 an arbitrary number of 0's; and finally, put any number of a's in front of the word and any number of 0's at the end of the word. For columns, the language is specified by the simpler regular expression $Lcol = c(0+2+a)^*$. A typical word for columns consists of a top letter c, followed by an arbitrary number of occurrences of 0's, 2's, or a's.

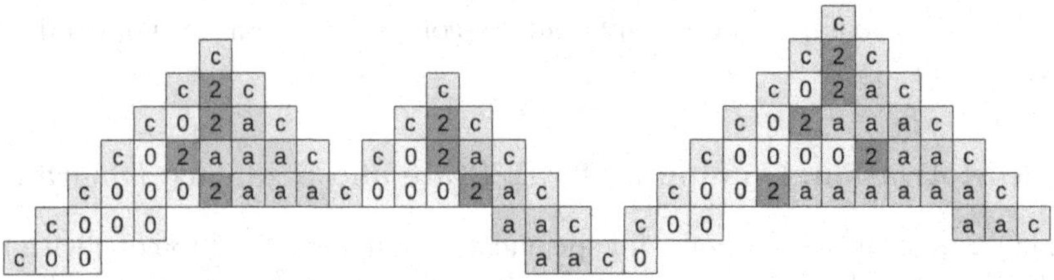

Fig. 3. Dutch-roof language [48]: the language obtained as the product $Lrow \otimes Lcol$ of $Lrow = a^*c(0+2a^*c)^*$ and $Lcol = c(0+2+a)^*$

The words in the language $Lrow \otimes Lcol$ have a structure resembling the structure of Dutch roofs. An example is presented in Fig. 3. A *hv*-component in a Dutch-roof word can be obtained by tiling using the following procedure:

(1. typical case) (i) Build a continuous sequences of c's, going along the diagonals, from left to right, and alternating the directions in a roof-like style: up-down-up-down-... (each sequence can start/end with either "up" or "down"). (ii) From a peak c, there is a kind of separation line of 2's going down, each 2 having one c at the left and one c at the right (on its row),

[3] Equivalently, the language *Lrow* can be defined by a simple finite automaton consisting of: states $\{0,1\}$, transitions $\{0 \xrightarrow{0} 0, 0 \xrightarrow{2} 1, 1 \xrightarrow{c} 0, 1 \xrightarrow{a} 1\}$, initial state 1, and final state 0.

but such that no more than one 2 is on each row between two consecutive c's. (iii) Connect in a stair-like style each bottom 2 with a bottom c on the left[4] and fill the area between c's, 2's, and this line with 0's. (iv) Do a similar construction on the right, but, finally, fill the area with a's instead of 0's.

(2. roof's ends) (i) If the construction starts with a "down" direction for c's, then add the following step: draw a stair-like line from the 1st c (a peak element) to a bottom c on the right and fill the area with a's. (ii) If the construction ends with an "up" direction for c's, then do a similar construction at the right end of the roof, filling the area with 0's.

(3. linking bottom 2's with further bottom c's) The bottom c's, selected to be chosen in step 1.(iii) or 2.(ii), need not be the nearest bottom c's on the left. However, if a more distant c is chosen, than each bottom c between the chosen one and the current 2 should have a 2 on the row, on the left, such that the full area on the left of that ignored bottom c is filled (no holes are allowed). A similar observation holds with regards to the right direction and the areas filled with a's.

4.4 The Pitfalls of Renaming Operator

In concurrent systems modelling, it is quite common to use independent state evolution and interaction evolution views and to combine them. For instance, this method is used in the definition of classical 2-dimensional regular expressions (for rectangular words) [18],[31], of regular expressions for Petri nets [17], or of regular expressions for timed automata [2]. Technically, these formalisms use the renaming operator as follows:

- first, a renaming is used to make all occurrences of the letters be different;
- then, an independent characterisation of evolution on each dimension is provided, say Lr for rows and Lc for columns[5];
- next, one combines these two views using, say, a notation $Lr \times Lc$;
- finally, use a renaming operator, say ρ, to come back to the original letters.

Therefore, the result is a characterisation of the system behaviours presented as $\rho(Lr \times Lc)$.

It is not hard to see that a key step is the third step, a step similar to LPP. However, a structural characterization of the resulting language is left open in these formalisms.

The point we want to discuss here is on the *mismatch resulting from the combination of renaming ρ and $\otimes$-product.* Formally, *the following relation is* ***not*** *always true*

$$\rho(Lcol \otimes Lrow) = \rho(Lcol) \otimes \rho(Lrow)$$

showing this kind of language characterisations, based on renaming and product, critically depends on the used letters.

[4] This line connects the south-west corner of 2 to the south-east corner of c.

[5] For Petri nets or timed automata the result is in terms of 1-dimensional traces, hence an extra flattening operator mapping 2-dimensional words to (set of) 1-dimensional words, is used. An example of flattening operator is presented in [46].

A particular consequence of this negative result is that one can not safely use even letter-to-letter substitutions. For example, if we rename all letters $0, 2, a, c$ as x in the example on the Dutch-roof language (presented in Fig. 3), then the resulted language $[x^*x(x + xx^*x)^*] \otimes [x(x + x + x)^*]$ consists of all rectangles of x's having at least 1 row and 1 column. This language is different form what one gets by replacing $0, 2, a, c$ with x directly in the Dutch-roof language.

4.5 Tiling Specifications for the LPP Results

The LPP explicitly asks to find an intrinsic characterisation of the words of a $\otimes$-product in terms of word composition, defined as a family of restricted tiling operators. Such a characterisation depends on the structure of the words and it is independent of the used letters. The approach can be safely combined with a letter-to-letter renaming. Even general letter-to-word or letter-to-language substitutions may be safely used in this framework, lifting to 2 dimensions a key feature of string languages.

5 Finite Automata in 2 Dimensions as Self-assembling Tile Systems

Tiles and Scenarios. A *tile* is a letter enriched with additional information on each border. Abstractly, this information is represented as an element from a finite set[6] and is called *border label.* The role of border labels is to impose local glueing constraints on self-assembling tiles: two neighbouring cells, sharing a horizontal or a vertical border, should agree on the label on that border. A *scenario* is similar to a 2-dimensional word, but: (1) each letter is replaced by a tile; and (2) horizontal or vertical neighbouring cells have the same label on the common border. Examples of tiles and scenarios are presented in Fig. 4(a).

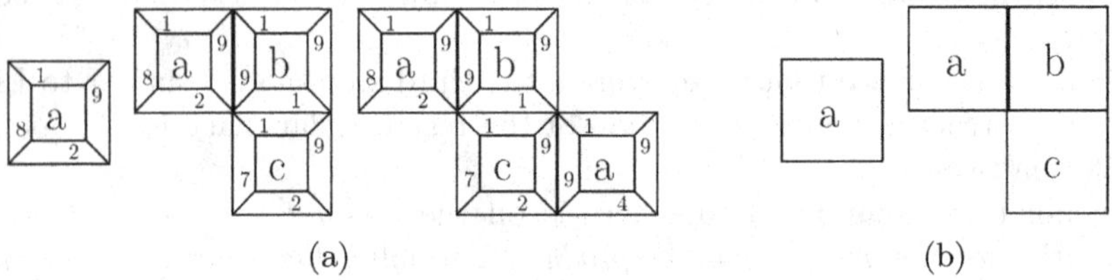

Fig. 4. Scenarios and accepted words

[6] Often, we use sets of numbers or sets of colors.

Self-assembling Tile Systems. A *self-assembling tile system* (shortly, *SATS*) is defined by a finite set of tiles, together with a specification of what border labels are to be used on the west/north/east/south external borders. An *accepting scenario* of a SATS F is a scenario obtained by self-assembling tiles form F and having the specified labels on the external borders. Finally, *the 2-dimensional language accepted by a SATS F*, denoted $\mathcal{L}(\mathcal{F})$, is the set of 2-dimensional words obtained from the accepting scenarios of F, dropping the border labels.

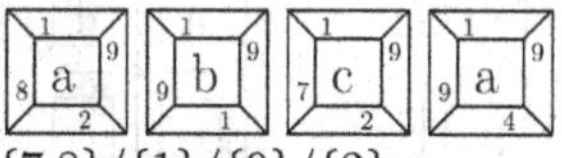

Fig. 5. A example of SATS

Example. Let F be the SATS defined by the specification in Fig. 5. All scenarios in Fig. 4(a) are correct scenarios of F. The first two are accepting scenarios, while the last one is not (there is a label 4, different of 2, on the south border). By dropping the border labels in the accepting scenarios in Fig. 4(a) one gets the accepted words in Fig. 4(b).

6 Composition

It is not at all obvious how to define 2-dimensional word composition, extending usual 1-dimensional word composition (catenation). We start with the simpler definition of scenario composition, then we try to capture the scenario composition effect, on the associated words, into the definition of word composition.

6.1 Scenario Composition

Scenario Composition. For two scenarios v and w, the *scenario composite* $v.w$ consists of all valid scenarios resulting from putting v and w together, without overlapping. This actually means if v and w share some borders in a particular placement, then the labels on the shared borders should be the same.

Example. We consider two scenarios v and w, presented in Fig. 6(a), (b). The composite $v.w$ has 3 results, sharing at least one cell border; they are presented in Fig. 6(c), (d), (e).

6.2 Word Composition

Word composition is defines as a family of restricted word composition operators, based on constraints using relevant information of word contours (see [4], [5], and [6]).

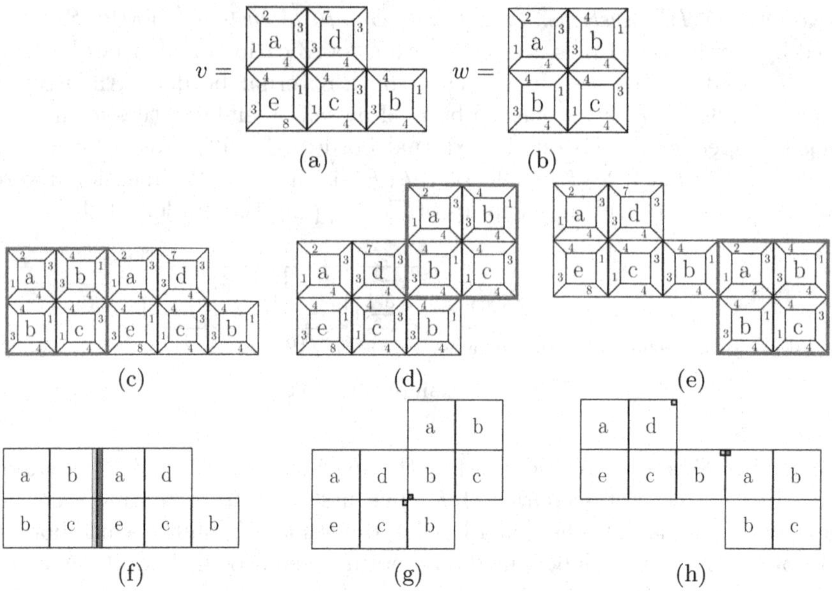

Fig. 6. Scenario composition and (restricted) word compositions

General Restricted Word Composition. Take two words v and w, with selected features:

1. a selected point p on the contour of v
2. a selected point q on the contour of w
3. a subset Y of elements of the contour of v
4. a subset G of elements of the contour of w
5. the subset B of the contact elements of both contours after composing v with w by identifying p and q,

and a ternary relation $R(U, V, W)$. Then, define

$$v_p\ R(Y,G,B)\ w_q = \begin{cases} \text{– the word obtained from } v \text{ and } w \text{ by identifying } p \text{ and } q & \text{, if } R(Y,G,B) \text{ is true} \\ \text{– undefined} & \text{, otherwise.} \end{cases}$$

The *restricted composition operator* $\_R(U, V, W)\_$ is

$$v\ R(U, V, W)\ w = \{v_p\ R(Y, G, B)\ w_q :\ \text{for all } p, q, Y, G, B\}.$$

Its *iterated version* is denoted by $\_{*}R(U, V, W)\_$ (put a "$*$" in front).

Example. Fig. 7 illustrates this definition. The selected composition points p and q are indicated by the little arrows; Y consists of the emphasized elements on the contour of v; the elements of G are emphasized on the contour of w; finally, B consists of the emphasized elements in the composed picture described in (c). For this example, a relation R making the restricted composition valid is:

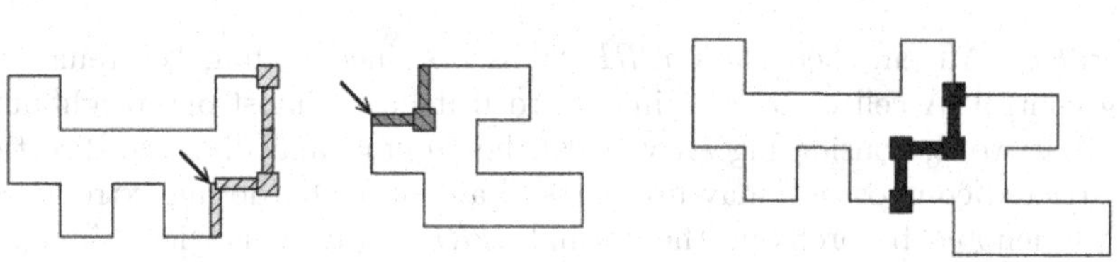

(a) v (b) w (c) $v\ R(Y,G,B)\ w$

Fig. 7. General restricted word composition

$G \subseteq Y \wedge G \subseteq B$ (after composition, all the emphasized elements on the contour of w are on the common border and included in the set of emphasized elements of v).

Example: Scenario Composition vs. Word Composition. To get the words corresponding to the scenario composition in Fig. 6(c,d,e), we can use the compositions illustrated in Fig. 6(f,g,h). They are obtained with the following restrictions:

1. v`(w=e)`w - the west border of v is equal to the east border of w; this composition is shown in Fig. 6(f) and yields a similar result as in Fig. 6(c);
2. v`(sw'=sw)`w - the south-west golf corners of v (there is only one) are identified with the south-west land corners of w (there is only one); composition with this restriction is shown in Fig. 6(g) and is similar to Fig. 6(d);
3. v`(ne>nw)`w - the north-east corners of v (there are two) includes the north-west corners of w (there is only one); this is shown in Fig. 6(h) and is similar to Fig. 6(e).

7 Regular Expressions

7.1 Regular Expressions

A new approach for defining classes of 2-dimensional regular expressions has been introduced in [4], [5], [6]. This approach uses arbitrary shape words and classes of restricted composition operators.

Class n2RE. Besides union, zero, and identities, the basic class *n2RE* uses *compositions and iterated compositions* corresponding to the following *restrictions*:

1. the selected elements of the word contours are: *side borders*, *land corners*, and *golf corners*;
2. the atomic comparison operators are: *equal-to* '=', *included-in* '<', *non-empty intersection* '#';
3. the general comparison formulas are *boolean formulas* built up from the atomic formulas defined in item 2.

(see [5] for more details and examples).

Class x2RE. An enriched class *x2RE* [5] is obtained adding "extreme cells" glueing control. A cell is *extreme* in a word if it has at most one neighbouring cell in that word, considering all vertical, horizontal, and diagonal directions. The restricted composition may use parts of a contour bordering extreme cells. They are denoted by prefixing the normal *n2RE* restrictions with 'x'; e.g., `xw`, `xse`, `xnw'`, etc. For instance, v`(e>xw)`w is true if the west borders of the extreme cells in w are included in the east borders of v.

7.2 Systems of Recursive Equations

Recursive Definitions. A system of recursive equations is defined using variables representing sets of 2-dimensional words and regular expressions. Formally, a *system of recursive equations* is defined by a set of equations

$$\begin{cases} X_1 = \sum_{i_1=1,k_1} E_{1i_1}(X_1,\ldots,X_n) \\ \ldots \\ X_n = \sum_{i_n=1,k_n} E_{ni_n}(X_1,\ldots,X_n) \end{cases}$$

where X_i are variables (denoting sets of 2-dimensional words) and E_{ij} are regular expressions from a specified class, extended with occurrences of variables X_i.

If nothing else is said, the default class is *n2RE*. Moreover, we can use $*$-free regular expressions, as $*$-operation can be defined by recursion.

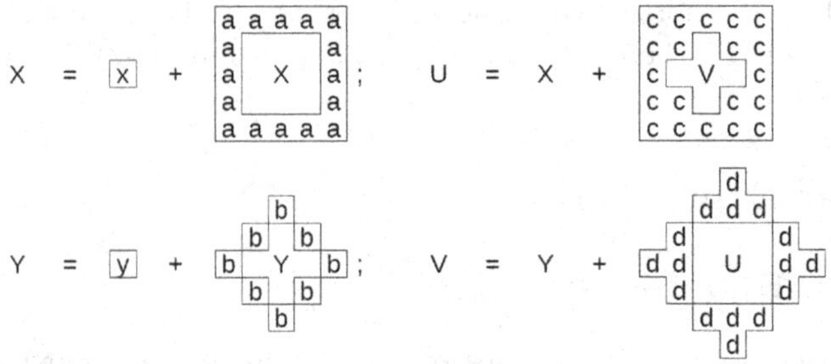

Fig. 8. Recursive equations

Examples: 1. The language consisting of square words filled with a, except for the center which contains x, may be represented by the equation (this equation is illustrated in Fig. 8):

$$(*) \qquad X = x + E(X)$$

where:

$$E_r = ((a*\texttt{(e=w)})\texttt{(se=ne)}(a*\texttt{(s=n)}))$$
$$\texttt{(sw=ne)}\ ((a*\texttt{(e=w)})\texttt{(nw=sw)}(a*\texttt{(s=n)}))$$
$$E_{rect} = (E_r(\texttt{(nw>ne)\&(nw>sw)})a)\ (\texttt{(se>ne)\&(se>sw)})\ a$$
$$E(X) = X(\texttt{(n<s)\&(e<w)\&(s<n)\&(w<e)})E_{rect}$$

The expression E_r generates a rectangle without 2 corners, E_{rect} a rectangle, and $E(X)$ a rectangle with an X inside. The recursive procedure (*) starts with a square x, so we get precisely the required square words.

2. The example above can be adapted to produce square diamonds of b's, having y in the center (Y in Fig. 8). One may use the se=nw and sw=ne restrictions to produce diagonal bars, then use extreme cells to locate the corners in the heads of the bars to be connected. The result is an expression in $x2RE$, not in $n2RE$.

3. Finally, U and V in Fig. 8 describe a mutually recursive definition built on top of the languages in items 1 and 2.

More examples of regular expressions and recursive specifications may be found in [5], [3], [48]. A recursive specification for the Dutch-roof language, following the tiling procedure informally presented in Subsection 3.3, is presented in [48].

7.3   Comparing SATS, LPP, and Recursive Specifications

Basic fact: *If S is a SATS, consisting of tiles labelled with distinct letters, and Lr and Lc are the languages accepted by the finite automata resulting projecting S on the horizontal and vertical dimension, then $\mathcal{L}(S) = Lr \otimes Lc$.*

However, this may be false if S has several tiles labelled with the same letter. The workaround this problem is to find a recursive specification over some class of 2-dimensional regular expressions. This recursive specification may be safely used even when several tiles have the same letter.

$$P = Lr \otimes Lc \quad \xrightarrow[Lr,Lc]{=} \quad \begin{array}{c} \mathcal{L}(S) \\ (S \text{ is a SATS, with} \\ \text{distinct-letters}) \end{array} \quad \xrightarrow{=} \quad \begin{array}{c} E \\ (2RE+\text{Recursion}) \end{array}$$

$$\searrow \rho(P) \qquad\qquad \rho^{-1}\uparrow \qquad\qquad\qquad\qquad \downarrow \rho$$

$$\rho(Lr) \otimes \rho(Lc) \quad \xrightarrow[\rho(Lr),\rho(Lc)]{?=} \quad \begin{array}{c} \mathcal{L}(\rho(S)) \\ (\rho(S) \text{ is a SATS,} \\ \text{with} \\ \text{nondistinct-letters}) \end{array} \quad \xrightarrow{=} \quad \begin{array}{c} \rho(E) \\ (2RE+\text{Recursion}) \end{array}$$

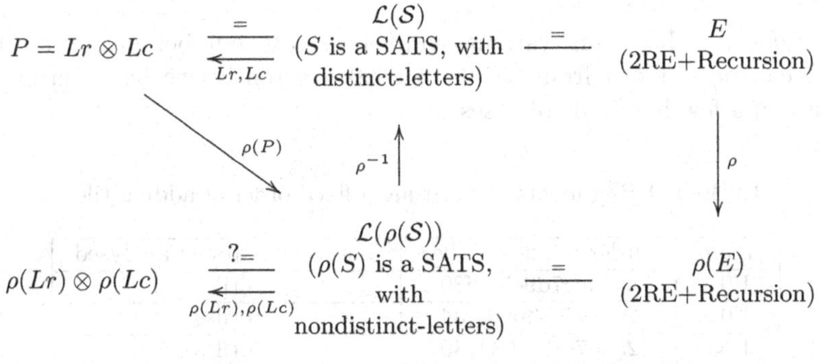

Fig. 9. SATS, LLP, and recursive specifications

The role of the LPP problem is to get information on the structure of the language associated to a SATS with distinct letters and to find a recursive characterization of this language. Later on, the method may be applied to any SATS: first, get distinct tile labels, i.e., choose an S with distinct labels and a renaming ρ such that our system is of the form $\rho(S)$; then, apply the previous step; and finally, rename the label in the recursive specification according to the initial labelling. Shortly, the left part of the diagram, which is not stable under substitution, is replaced by the right part.

8 Languages Generated by 2-Colors Border Tiles

Tiles. A non-trivial SATS has at least 2 labels for each vertical and horizontal dimension. Up to a bijective representation, the tiles of SATSs using 2 labels on each dimension can be identified with subsets of elements in the following set

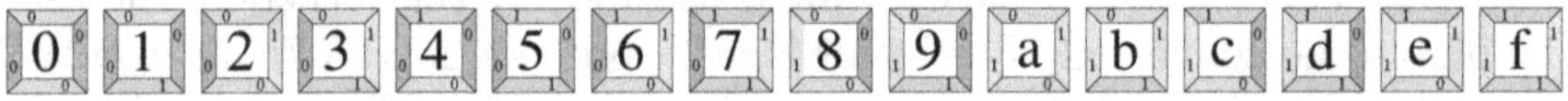

In this representation, the labels for the vertical and the horizontal borders are 0 and 1. The letter associated to a tile is the hexadecimal number obtained from the binary representation of the sequence of its west-north-east-south 0/1 digits, in this order. All together, there are 65536 distinct subsets; a subset is denoted by $Ft_1t_2 \dots t_k$, where $t_1, t_2, \dots, t_k$ are the tiles of the subset.

External Labels. For simplicity, we consider SATSs with *only one* label for each west/north/east/south external border. There are 16 possibilities, each one denoted by a hexadecimal number representing the sequence of the west-north-east-south 0/1 labels used for the external borders.

SATS Notation. The SATSs to be investigated are represented as $Ft_1t_2 \dots t_k.z$, where $t_1, t_2, \dots, t_k$ are the tiles and the binary digits of z specify the labels used for the external borders. As an example, the SATS used in the Dutch-roof language is $F02ac.c$, consisting of tiles 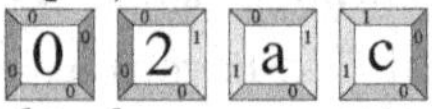 and labels 1/1/0/0 used for external west/north/east/south borders, respectively.

State-of-the-Art. By using various symmetries, the number of subsets to be studied can be reduced from 65536 to 2890. Currently, we have clarified the structure of a few hundreds of cases.

Table 1. UB students work, using a fixed order of adding tiles

Team	Order of adding tiles	The longest prefix analysed
E01	a614c27d9e5b3f80	5 tiles
E02	2ab8157df096ce34	6 tiles
E03	2a357bd6c89f1e40	5 tiles
E05	fb0259134adec786	5 tiles
E06	0ac3b17f64d5e892	4 tiles

Part of the work has been done by the University of Bucharest (UB) students. In the 2015 spring semester, the UB master students participating to the Design of Interactive Applications course were exposed to a research project. They were grouped in up to 5 member teams. Each team selected a permutation of the sixteen tiles $0, \dots, f$, specifying the order of adding new tiles. For each prefix analysed they considered all 16 SATSs obtained taking all combinations of labels for the external borders. 5 participating teams performed pretty well and their results are presented in Table 1. As a general remark, we mention that the tiling becomes more complicate when the automaton for *Lrow* and the automaton for *Lcol* have paths in both directions $0 \to 1$ and $1 \to 0$. We hope to have a published version of these results available soon.

9 A Nebulous Point: Kleene Algebra in 2 Dimensions

Kleene Theorem, in 2 Dimensions. In 1 dimension, Kleene theorem relates two rather different methods to generate regular languages: finite automata (FA) and regular expressions (RE). Finite automata process one-action-at-a-time and use state labels to handle continuation. By contrast, regular expressions combine large chunks of computation, leading to a more structural and scalable notation. Among others, finite automata are used in low-level models and in flowchart programs, while regular expressions are more suited for high-level models and for structured programs.

Self-assembling tile systems (SATS), our 2-dimensional extension of finite automata, inherit finite automata properties: they use labels for continuation and process one-action-at-a-time.

SATS languages have a very rich structure, mostly yet waiting to be revealed. While we have a mechanism to define classes of regular expressions in 2 dimensions, what is the appropriate one for SATS languages is still unclear. Find a Kleene theorem in 2 dimensions, to relate SATS and a specific class of regular expressions in 2 dimensions, depends on the understanding of the structure of SATS languages and the ability to identify equivalent regular expression based representations for these languages.

Kleene Algebra, in 2 Dimensions. Even solving the above problems, going to an algebraic setting, similar to 1-dimensional Kleene algebra, is still more challenging. For completeness, we need to formally capture SATS equivalence, perhaps as simulation captures FA equivalence. But SATS equivalence is undecidable [30], [42]. Restricted to rectangular words, even the emptiness problem is undecidable [30]. In the current setting of arbitrary shape words, we expect even the membership problem is undecidable; more precisely, we conjecture the model of SATSs over arbitrary shape 2-dimensional words is universal.

Rewards. If there are so many obstacles, why does one try to follow this quest? The effort deserves to be done, we think: it is a good intellectual challenge and even partial results may have a huge impact in active research areas as image processing (computer vision) and modelling large scale distributed computing systems.

10 Interactive Programs

Reading this section requires some familiarity with the register-voice interactive systems model (rv-IS) – a few pointers to the literature are indicated at the end of the section. For the reader unfamiliar with the rv-IS model the message is the following: the passing from rectangular words to arbitrary shape words supports interactive programming models where spatial and temporal resources are directly managed, for example by using spatial and temporal pointers.

From Self-assembling Tile Systems to Interactive, Distributed Programs. A classical slogan says that "program = control + data". The control part for simple sequential programs is provided by finite automata. We can extend this slogan saying that

"**interactive, distributed program** **= (control & interaction) + (spatial & temporal) data**"

The control & interaction part may be specified by SATSs. Spatial and temporal data can be added to SATSs to get completely specified interactive, distributed programs. A basic step is to enrich tiles with data associated to their border labels to get specific basic blocks, called *interactive modules*. Our convention is that the data on the north and south borders represent spatial data (memory states), the data on the west and east borders represent temporal data (communication messages), and the cell itself has an associated relational transformation connecting these data. SATSs may be used to produce scenarios describing system runs, built up out of these interactive modules.

A model along these lines, the *register-voice interactive systems (rv-IS)* model, accompanied by a kernel programming language and a few specification and verification techniques has been introduced in 2004 [46] and gradually developed till now - the most notable development is the introduction of Agapia in 2007 [15], a structured programming language based on the rv-IS model.

Words and Scenarios: Rectangular or Arbitrary Shapes? The control & interaction part of distributed programs may also be specified by particular classes of 2-dimensional regular expressions, producing more structured programs. As regular expressions do not use labels, the result is a programming languages setting without *"go-to"* statements (neither control, nor interaction can use go-to labels). As a side effect of lacking "go-to" labels on the interaction part, one gets a setting with name-free processes.

Both published versions of Agapia (v0.1 [15] and v0.2 [37]) have semantics defined in terms of rectangular words. They use horizontal, vertical, and diagonal compositions and their iterated versions. The v0.2 version allows to define recursive programs. A recent compiler, developed by Paduraru [35] and producing either MPI or OpenMP executables, has shown good execution time for programs specified in Agapia high-level setting compared with programs directly written in MPI or OpenMP.

In a recent study regarding process synchronization [20], the authors have introduced temporal pointers. We claim that arbitrary shape scenarios provide a good formalism to deal with spatial and temporal pointers. One example is presented in Fig. 10. If we want a synchronization by temporal pointers between C and Y (to be run at the some time interval) and to have both B and Y

running on the same process (to coordinate their memory allocations), then one can use a simple composition as in Fig. 10(b). Using rectangular scenarios, one may use a scenario as in Fig. 10(a), which preserve the functionality, but not the spatio-temporal constraints induced by the pointers.

The scenario in Fig. 10(a) uses particular cells, interpreted as simple wiring constants: $0 =$, I = , − = , ⌞ = , $\neg =$, $+ =$; for instance, the 1st row is $AB0\neg 0$I. They represent empty cell, vertical identity, horizontal identity, space-to-time converter (speaker), time-to-space converter (recorder), and cross identity, respectively. They can be used to migrate a process (as from C to Y), or to delay a communication (as from B to X).

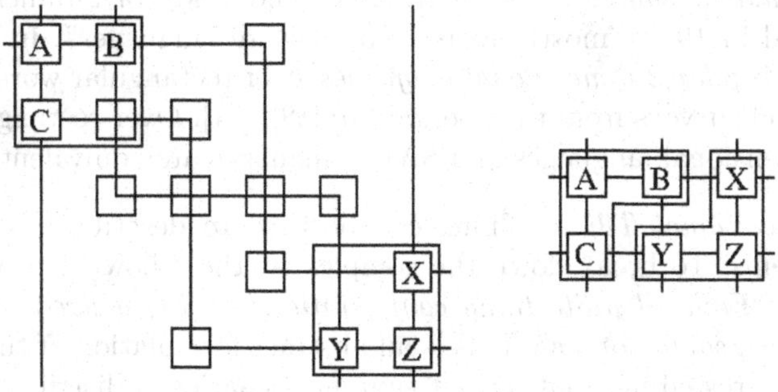

(a) (b)

Fig. 10. Rectangular vs. arbitrary shape scenarios in the rv-IS model

Interactive Computation. Interactive computation [50] is an important component of the software-intensive infrastructures of our society. Classical models for process interaction include, among many other models, process algebra models [8], Petri nets [38], dataflow networks [11], etc. In these models, process synchronization is a key feature. For instance, in process algebra models, synchronization is achieved by handshake communications, while in Petri nets and dataflow models, explicit transitions and dataflow nodes are respectively used. These models treats interaction as a primary feature, considering sequential computation to be either derived from communication or implicitly included in the dataflow node behaviour. A more recent proposal falling into this class is the ORC programming language [33], based on name-free processes and structured interaction.

Our rv-IS interaction model is based on scenarios and 2-dimensional traces. *Traces* represent running scenarios modulo graph-isomorphism and projected on classes and states. In this interpretation one focuses on data stored in or flowing through the cells of the interactive system. A notion of *trace-based refinement* for (structured) interactive systems has been recently presented in [14].

11 Related Work

Tiling. Tiling [21] is an old and popular subject. We include a few pointers to the literature below, trying to classify the type of problems studied.

Tiling, Infinite Behaviour. An interesting tiling problem was proposed by Wang in 1961: *"Given a set of tiles, is there a tiling of the whole 2-dimensional plane?"* Wang has conjectured that only regular periodic tilings are possible, but this is not true: the smallest set for which an aperiodic tiling does exist consists in 11 tiles and it uses 4 colors [23].

Cellular Automata. Cellular automata studies are also focusing on infinite behaviours. They were introduced in 1940s and studied, among others, in connection with self-reproducing systems [49], trying to get a new perspective on natural sciences [51], or looking for applications on adaptive systems [32].

2-Dimensional Languages, Classical Results. The study of 2-dimensional languages started in 1960s, mostly related to "picture languages". In 1990s a robust class[7] of *regular 2-dimensional languages*, over rectangular words, has been identified; good surveys from that period are [18], [31]. Over rectangular words, regular 2-dimensional languages and SATS languages are equivalent.

Finite, Compositional Tiling. Tiling can be used to describe large scale distributed systems. To break down the complexity, the following problem needs to be solved: *"Find all finite tiling configurations having a fixed label on each west/north/east/south border".* This is an abstract formulation of the basic fact that we are interested in input-output running scenarios of distributed systems, corresponding to tiling configurations, which start from initial states (on north), initial interaction classes (on west) and, in a finite number of steps, reach final states (on south) and final interaction classes (on east). This is the problem we have addressed in this paper; some additional references are included in the "space-time duality" paragraph.

Self-assembling Tiling with Unique Outcome. A more recent model using tiles is the Winfrees abstract Tile Assembly Model (see [36], for a survey). The problem of interest here is: *"Find a set of tiles such that any tiling yields the same specified final configuration."* Coming from practical considerations, the aim is to design self-assembling systems for quick and error-free production of complex substances.

Self-assembling in Computing. We finish with a brief comment on the use of the "self-assembling" term in our computing setting. According to some conventions, a (chemical) self-assembling system is an assembling system with the following distinctive features related to the *order*, the *interactions*, and the used *building blocks*: (1) usually, the resulting configurations have higher order; (2) they use "weak interactions" for coordination; and (3) larger or heterogeneous building blocks may be used. Properties (1) and (3) are clearly related to scenario composition and 2-dimensional regular expressions. Property (2) is more related to physical systems,

[7] One particularly interesting equivalent characterization of regular 2-dimensional languages use (existential) monadic second order logic over 2 successors [19]. The role of 2nd order variables is similar to that played by the variables used in our recursive specifications.

but a similar one may be considered in our computing setting by distinguishing computing activities (on the same machine) and coordinating activities. In short, there are strong enough reasons to use the "self-assembling" term for constructing scenarios/languages in our distributed computing formalism.

Kleene Algebra: From 1 to 2 Dimensions. Sequential computation is a well-established research domain. A witness of its maturity is the rich collection of algebraic theories based on regular expressions and the associated regular algebra, see, e.g., [25], [40], [12], [29], [43], [26], [9], [10]. Recent extensions of regular algebra to network algebra [44] shows deep connections with classical mathematics, especially via the "trace monoidal category" structure [24], [41]. Kleene algebra with tests [27] has been recently applied to real networks modelling [1].

The approach presented in this paper is in line with other attempts to formally describe distributed, interactive programs. An early paper on 2-dimensional connectors is [34] focusing on VLSI design. Two formalisms for parallel and/or distributed computation, directly based on regular expressions, are regular expressions for Petri nets [17] and for timed automata [2]. Other related formalisms include the tile logic specification model [16], the BIP rigorous approach to component-based system design [7], or the concurrent Kleene algebra [22].

Space-Time Duality. Our work on this subject started fifteen years ago with the exploration of *space-time duality* and its role in organizing the space of interactive computation; see Chapter 12 (Section 12.5) in [44]. Till now, we have advanced a few steps in this direction:

1. A 2-dimensional version of finite automata has been presented in [45], [46]. (It was called "finite interactive system" and is equivalent with SATS, but it was used only over rectangular 2-dimensional words.)
2. A space-time invariant[8] model extending *flowcharts* was shown in [46], [47].
3. A space-time invariant extension of (structured) *while programs* was presented in [15].
4. An enriched version of 3. supporting *recursion* was presented in [37].
5. Space-time invariant (2-dimensional) *regular expressions* are presented in [4], [5], [6].
6. Verification methods lifting *Floyd and Hoare logics* to 2-dimensions have been described, too.
7. A notion of *trace-based refinement*, for this space-time invariant model, was introduced in [13], [14].
8. Finally, a space-time invariant extended model, including spatial and temporal pointers[9], has been recently discovered [20], [35].

[8] The expression "space-time invariant" is a shortened version of "invariant with respect to a formally defined space-time duality transformation" (see, e.g., [47]).

[9] To be similar with spatial pointers, temporal pointers need the capacity to refer to both past and future moments in time. The former is captured by recording the past information, while the latter by speculation (guess the forthcoming value and continue the running; keep only the branch(s) with a consistent guess).

12 Conclusions

In this paper we have described a roadmap to extend finite automata, regular expressions, and Klenee algebra in 2 dimensions. The paper has been focused more on defining the concepts and presenting examples and basic properties, rather then on technical details. As a last, more personal, conclusion we think that the topics is very interesting, with high potential of applications, but it needs a collective effort to be properly done.

Acknowledgements. The manuscript had been completed during the author's 2015 summer visit at the University of Illinois at Urbana-Champaign (UIUC) and he wants to thank Grigore Rosu and the Computer Science Department at UIUC for the excellent working conditions provided.

References

1. Anderson, C., Foster, N., Guha, A., Jeannin, J., Kozen, D., Schlesinger, C., Walker, D.: Netkat: semantic foundations for networks. In: POPL 2014, pp. 113–126 (2014)
2. Asarin, E., Caspi, P., Maler, O.: Timed regular expressions. Journal of the ACM 49, 172–206 (2002)
3. Banu-Demergian, I.T.: The study of interaction in computing systems. PhD thesis, University of Bucharest (2014)
4. Banu-Demergian, I.T., Paduraru, C.I., Stefanescu, G.: A new representation of two-dimensional patterns and applications to interactive programming. In: Arbab, F., Sirjani, M. (eds.) FSEN 2013. LNCS, vol. 8161, pp. 183–198. Springer, Heidelberg (2013)
5. Banu-Demergian, I.T., Stefanescu, G.: Towards a formal representation of interactive systems. Fundamenata Informaticae 131, 313–336 (2014)
6. Banu-Demergian, I.T., Stefanescu, G.: On the contour representation of two-dimensional patterns. Carpathian Journal Mathematics (2016), Also: Arxiv, CoRR abs/1405.3791 (to appear)
7. Basu, A., Bensalem, S., Bozga, M., Combaz, J., Jaber, M., Nguyen, T., Sifakis, J.: Rigorous component-based system design using the BIP framework. IEEE Software 28, 41–48 (2011)
8. Bergstra, J., Ponse, A., Smolka, S. (eds.): Handbook of Process Algebra. Elsevier (2001)
9. Bloom, S.L., Esik, Z.: Equational axioms for regular sets. Mathematical Structures in Computer Science 3, 1–24 (1993)
10. Bloom, S.L., Esik, Z.: Iteration Theories: The Equational Logic of Iterative Processes. Springer, Berlin (1993)
11. Broy, M., Olderog, E.: Trace-oriented models of concurrency. In: [8], pp. 101–196. Elsevier (2001)
12. Conway, J.H.: Regular Algebra and Finite Machines. Chapman and Hall (1971)
13. Diaconescu, D., Leustean, I., Petre, L., Sere, K., Stefanescu, G.: Refinement-preserving translation from Event-B to register-voice interactive systems. In: Derrick, J., Gnesi, S., Latella, D., Treharne, H. (eds.) IFM 2012. LNCS, vol. 7321, pp. 221–236. Springer, Heidelberg (2012)

14. Diaconescu, D., Petre, L., Sere, K., Stefanescu, G.: Refinement of structured interactive systems. In: Ciobanu, G., Méry, D. (eds.) ICTAC 2014. LNCS, vol. 8687, pp. 133–150. Springer, Heidelberg (2014)
15. Dragoi, C., Stefanescu, G.: Agapia v0.1: A programming language for interactive systems and its typing system. Electronic Notes in Theoretical Computer Science 203, 69–94 (2008)
16. Gadducci, F., Montanari, U.: The tile model. In: Proof, Language, and Interaction: Essays in Honour of Robin Milner, pp. 133–166. The MIT Press (2000)
17. Garg, V., Ragunath, M.: Concurrent regular expressions and their relationship to Petri nets. Theoretical Computer Science 96, 285–304 (1992)
18. Giammarresi, D., Restivo, A.: Two-dimensional languages. In: Handbook of Formal Languages, pp. 215–267. Springer (1997)
19. Giammarresi, D., Restivo, A., Seibert, S., Thomas, W.: Monadic second-order logic over rectangular pictures and recognizability by tiling systems. Information and Computation 125, 32–45 (1996)
20. Gramatovici, R., Petre, L., Sere, K., Stefanescu, A., Stefanescu, G.: Synchronization in timed interactive systems. Technical Report 1047, TUCS (2012)
21. Grunbaum, B., Shephard, G.: Tilings and Patterns. W.H. Freeman and Co. (1987)
22. Hoare, T., van Staden, S., Möller, B., Struth, G., Villard, J., Zhu, H., O'Hearn, P.: Developments in Concurrent Kleene Algebra. In: Höfner, P., Jipsen, P., Kahl, W., Müller, M.E. (eds.) RAMiCS 2014. LNCS, vol. 8428, pp. 1–18. Springer, Heidelberg (2014)
23. Jeandel, E., Rao, M.: An aperiodic set of 11 Wang tiles. CoRR, abs/1506.06492 (2015)
24. Joyal, A., Street, R., Verity, D.: Traced monoidal categories. Mathematical Proceedings of the Cambridge Philosophical Society 119, 447–468 (1996)
25. Kleene, S.C.: Representation of events in nerve nets and finite automata. Automata Studies, Princeton University Press, pp. 3–41 (1956)
26. Kozen, D.: A completeness theorem for Kleene algebras and the algebra of regular events. In: LICS 1991, pp. 214–225. IEEE (1991)
27. Kozen, D.: Kleene algebra with tests. ACM Trans. Program. Lang. Syst. 19, 427–443 (1997)
28. Krob, D.: Matrix versions of a aperiodic K-rational identities. Theoretical Informatics and Applications 25, 423–444 (1991)
29. Kuich, W., Salomaa, A.: Semirings, automata and languages. Springer, Berlin (1985)
30. Latteux, M., Simplot, D.: Context-sensitive string languages and recognizable picture languages. Information and Computation 138, 160–169 (1997)
31. Lindgren, K., Moore, C., Nordahl, M.: Complexity of two-dimensional patterns. Journal of Statistical Physics 91, 909–951 (1998)
32. Miller, J., Page, S.: Complex adaptive systems: an introduction to computational models of social life. Princeton University Press (2009)
33. Misra, J., Cook, W.: Computation orchestration. Software and System Modelling 6, 83–110 (2007)
34. Molitor, P.: Free net algebras in VLSI-Theory. Fundamenta Informaticae 11, 117–142 (1988)
35. Paduraru, C.I.: Research on AGAPIA language, compiler and applications. PhD thesis, University of Bucharest (2015)
36. Patitz, M.: An introduction to tile-based self-assembly and a survey of recent results. Natural Computing 13, 195–224 (2014)

37. Popa, A., Sofronia, A., Stefanescu, G.: High-level structured interactive programs with registers and voices. J. UCS 13, 1722–1754 (2007)
38. Reisig, W.: Petri nets: An introduction. Springer Science & Business Media (2012)
39. Rumbaugh, J., Jacobson, I., Booch, G.: Unified Modeling Language Reference Manual. Pearson Higher Education (2004)
40. Salomaa, A.: Two complete axiom systems for the algebra of regular events. Journal of the ACM (JACM) 13, 158–169 (1966)
41. Selinger, P.: A survey of graphical languages for monoidal categories. Lecture Notes in Physics, vol. 813, pp. 289–355. Springer (2011)
42. Sofronia, A., Popa, A., Stefanescu, G.: Undecidability results for finite interactive systems. Romanian Journal of Information Science and Technology 12, 265–279 (2009), Also: Arxiv, CoRR abs/1001.0143
43. Stefanescu, G.: On flowchart theories: Part II. The nondeterministic case. Theoretical Computer Science 52, 307–340 (1987)
44. Stefanescu, G.: Network algebra. Springer (2000)
45. Stefanescu, G.: Interactive Systems: From Folklore to Mathematics. In: de Swart, H. (ed.) RelMiCS 2001. LNCS, vol. 2561, pp. 197–211. Springer, Heidelberg (2002)
46. Stefanescu, G.: Interactive systems with registers and voices. Draft, National University of Singapore (2004)
47. Stefanescu, G.: Interactive systems with registers and voices. Fundamenta Informaticae 73, 285–305 (2006)
48. Stefanescu, G.: Self-assembling interactive modules: A research programme. CoRR, abs/1506.05499 (2015)
49. von Neumann, J., Burks, A: Theory of self-reproducing automata. University of Illinois Press (1966)
50. Wegner, P.: Interactive foundations of computing. Theoretical Computer Science 192, 315–351 (1998)
51. Wolfram, S.: A new kind of science. Wolfram Media Champaign (2002)

Connections between Relation Algebras and Cylindric Algebras

Ian Hodkinson

Department of Computing, Imperial College London

Abstract. We give an informal description of a recursive representability-preserving reduction of relation algebras to cylindric algebras.

1 Introduction

Relation algebras form one of the principal algebraic approaches to binary relations. Introduced by Tarski in 1941 [9], their history actually goes back much further, to work of Peirce, Schröder, De Morgan, and even Boole. One of the key algebraic approaches to relations of higher arity is *cylindric algebras,* introduced by Tarski and his students Louise Chin and Frederick Thompson in the late 1940s.

Finding connections between relation algebras and cylindric algebras has been a prickly problem for a long time. There are a number of reasons why the problem might be of interest.

1. Algebraic logic generally seems to comprise a large number of formally different kinds of algebra — relation algebras, cylindric algebras, diagonal-free algebras, substitution algebras, polyadic (equality) algebras, and so on. Once one has proved a result for one kind of algebra, one is under some scientific obligation to try to prove it again for others.
   Sometimes, doing this involves substantial technical innovation. But often, it can seem like merely copying out the old proof with minor modifications to take account of the different type of algebra. The core argument, often combinatorial in nature, remains the same. This leads in the direction of off-putting repetitive papers, allegations of 'salami slicing', and unpleasant subjective debates about how incremental a paper is.
   In both cases, it would be valuable to have some reasonably general 'transfer theorems' allowing direct export of results from one kind of algebra to another. Indeed, such theorems might be more illuminating than just reformulating the same argument in slightly different terms.
2. A case in point is the 'negative' result that *there is no algorithm to decide whether a finite relation algebra is representable* [1, 2]. The proof was complicated. Redoing it for cylindric algebras is even more complicated. A 'transfer result' would be very helpful here. It would snatch a 'positive' result from the jaws of negativity. (It is not my fault that problems are undecidable, but as one distinguished logician on the RAMiCS programme committee once told me, 'People get fed up with negative results.')

W. Kahl et al. (Eds.): RAMiCS 2015, LNCS 9348, pp. 27–42, 2015.
DOI: 10.1007/978-3-319-24704-5_2

3. The question of connections between algebras of different arities (and perhaps varying in other features too) is of interest in its own right. It has a distinguished history, including work of Monk [8] and Maddux [5–7]. It raises intricate technical challenges.
4. Unlike first-order logic, algebraic logic is rather picky about arities. Usually there is a separate class of algebras for each arity (though some such as Craig have defined algebras comprising relations of multiple arities). So it would be nice to shed light on this separation and perhaps show it is less strict than appears.
5. Both relation algebras and cylindric algebras are listed on `http://ramics2015.di.uminho.pt` as in the scope of RAMiCS — so why not study their connections?

In this short note, we will attempt to give a gentle introduction to some work in this area. As a case study, we will focus on the problem mentioned in point 2 above: it is known to be undecidable whether a finite relation algebra is representable; *can we use this result to show the same for finite n-dimensional cylindric algebras,* for each finite $n \geq 3$?[1]

The obvious approach is to find a *recursive reduction* of the first problem to the second. That is, we find a recursive function f that, given a finite relation algebra $\mathcal{A}$, returns a finite n-dimensional cylindric algebra $f(\mathcal{A})$ that is representable when and only when $\mathcal{A}$ is representable.[2] It would then of course follow that no algorithm could decide representability of n-dimensional cylindric algebras. For such an algorithm could be coupled with f to provide an algorithm to decide representability of finite relation algebras, something that [1] assures us does not exist.

In section 3 below, we will recall briefly some earlier work on connections between relation algebras and cylindric algebras, and discuss the prospects for using it to construct such an f. Then, in section 4, we explain informally a simplified form of the construction of f from [3].

Since the full proofs are already in print, in this short note we will give only informal descriptions, not proofs. For clarity, we will make some simplifying assumptions. So: we will mostly restrict attention to the case of *finite simple algebras.* We will refrain from considering other algebras such as polyadic algebras and diagonal-free algebras. We will consider cylindric algebras only of finite dimensions $n \geq 3$. (The case $n < 3$ is easy — representability is decidable in this case — and infinite-dimensional cylindric algebras take us in some sense outside the realm of finite algebras.)

[1] We caution the reader that just because it appears 'harder' to represent cylindric algebras than relation algebras, it doesn't follow that the question of deciding *whether* a finite cylindric algebra is representable is harder than the corresponding question for relation algebras.

[2] In passing, we mention a converse problem: is there a recursive function g that, given a finite n-dimensional cylindric algebra $\mathcal{C}$, returns a finite relation algebra $g(\mathcal{C})$ that is representable iff $\mathcal{C}$ is representable?

2 Definitions

We recall the necessary basics. We adopt the standard convention that denotes the domain of an algebra $\mathcal{A}$ by A.

2.1 Relation Algebras

A *relation algebra* is an algebra

$$\mathcal{A} = (A, +, -, 0, 1, 1', \breve{}, ;),$$

where $(A, +, -, 0, 1)$ is a boolean algebra, called the *boolean reduct* of $\mathcal{A}$, $(A, ;, 1')$ is a monoid, $\breve{}$ is a unary function on A, and $\mathcal{A}$ satisfies the Peircean law: $(a ; b) \cdot c \neq 0 \iff (\breve{a} ; c) \cdot b \neq 0 \iff (c ; \breve{b}) \cdot a \neq 0$, for all $a, b, c \in A$, where $a \cdot b = -(-a + -b)$ (we will not use these properties in detail here).

We say that $\mathcal{A}$ is *simple* if $1 ; a ; 1 = 1$ for each non-zero $a \in A$, and *finite* if A is finite. Mostly we will consider only finite simple relation algebras here. We warn the reader that for arbitrary relation algebras, some definitions and results below need to be modified, or may even fail.

Representations. A *(square) representation* of $\mathcal{A}$ is a one-one map $h : A \to \wp(U^2)$, for some 'base' set U, such that for all $a, b \in A$,

1. $h(a + b) = h(a) \cup h(b)$
2. $h(-a) = U^2 \setminus h(a)$
3. $h(0) = \emptyset$
4. $h(1) = U^2$
5. $h(1') = \{(x, x) : x \in U\}$
6. $h(\breve{a}) = \{(x, y) \in U^2 : (y, x) \in h(a)\}$
7. $h(a ; b) = \{(x, y) \in U^2 : \exists z((x, z) \in h(a) \wedge (z, y) \in h(b))\}$.

So h *represents* each $a \in A$ as a binary relation on U, and the algebraic operations correspond via h to 'concrete' operations on binary relations. Not every finite simple relation algebra is *representable* (i.e., has a representation). By [2, theorem 18.13], it is undecidable whether a finite simple relation algebra is representable.

Atoms, Atomic Relation Algebras. We can define a standard 'boolean' partial ordering $\leq$ on A by $a \leq b$ iff $a + b = b$. An *atom* of $\mathcal{A}$ is a $\leq$-minimal non-zero element of A. We write $At\,\mathcal{A}$ for the set of atoms of $\mathcal{A}$. We say that $\mathcal{A}$ is *atomic* if every non-zero element of A lies $\leq$-above an atom. Every finite relation algebra is atomic.

Atom structures and Complex Algebras. By standard duality, an atomic relation algebra has an associated *atom structure:* a relational structure $\mathrm{At}\,\mathcal{A} = (At\,\mathcal{A}, R_{1'}, R_{\breve{}}, R_{;})$, where $R_{1'} = \{a \in At\,\mathcal{A} : a \leq 1'\}$, $R_{\breve{}} = \{(a, b) \in (At\,\mathcal{A})^2 : b \leq \breve{a}\}$, and $R_{;} = \{(a, b, c) \in (At\,\mathcal{A})^3 : c \leq a ; b\}$.

If $\mathcal{A}$ is finite, it is completely determined by At $\mathcal{A}$ up to isomorphism. Indeed, given any structure $\mathcal{S} = (S, R_{1'}, R_{\smile}, R_{;})$ of the signature of relation algebra atom structures, we can define its *complex algebra* $\mathfrak{Cm}\,\mathcal{S}$:

$$\mathfrak{Cm}\,\mathcal{S} = (\wp(S), \cup, \setminus, \emptyset, S, R_{1'}, \breve{}, ;),$$

where, for $a, b \subseteq S$, $\breve{a} = \{s \in S : \exists t \in a(R_{\smile}(t, s))\}$ and $a \,;b = \{s \in S : \exists t \in a \exists u \in b(R_{;}(t, u, s))\}$. Given suitable conditions on $\mathcal{S}$, this algebra $\mathfrak{Cm}\,\mathcal{S}$ will be a relation algebra, and $\mathcal{S} \cong \mathrm{At}\,\mathfrak{Cm}\,\mathcal{S}$. For finite relation algebras $\mathcal{A}$, we have $\mathcal{A} \cong \mathfrak{Cm}\,\mathrm{At}\,\mathcal{A}$.

The atom structure of a finite relation algebra $\mathcal{A}$ is a good way to handle $\mathcal{A}$, since it is exponentially smaller. We can define particular finite relation algebras by specifying their atom structures.

Networks. Another important concept to do with atoms and atom structures is that of *network*.

Given a finite simple relation algebra $\mathcal{A}$, an *$\mathcal{A}$-network (over N_1)* is a pair $N = (N_1, N_2)$, where N_1 is a set of 'nodes', and $N_2 : N_1 \times N_1 \to At\,\mathcal{A}$ is a 'labelling function' satisfying, for all $x, y, z \in N_1$,

1. $N_2(x, x) \leq 1'$,
2. $N_2(x, y) = N_2(y, x)^{\smile}$ (we note that in relation algebras, $\smile$ takes atoms to atoms),
3. $N_2(x, y) \leq N_2(x, z) \,; N_2(z, y)$.

Frequently we drop the indices 1, 2, deducing them by context.

Networks Arise from Parts of representations. The key observation here is that if $h : A \to \wp(U^2)$ is a representation of the (finite simple) relation algebra $\mathcal{A}$, then for each $u, v \in U$ there is a unique atom $a \in A$ such that $(u, v) \in h(a)$ (it is an exercise to show that this atom exists and is unique). Writing this atom as $\lambda(u, v)$, so that $\lambda : U^2 \to At\,\mathcal{A}$ is a function, it can be checked that for each $X \subseteq U$, the pair

$$(X, \lambda \restriction X^2) \tag{1}$$

is an $\mathcal{A}$-network. The 'whole' network $N = (U, \lambda)$ satisfies an additional 'saturation' condition

4. for each $x, y \in N$ and atoms $a, b \in A$, if $N(x, y) \leq a \,; b$, then there exists $z \in N$ with $N(x, z) = a$ and $N(z, y) = b$.

Conversely, any 'saturated' $\mathcal{A}$-network N satisfying condition 4 can be viewed as a representation h of $\mathcal{A}$ via $h(a) = \{(x, y) \in N : N(x, y) \leq a\}$, for each $a \in A$. Or almost. This h need not respect 1', since $N'(x, y) \leq 1'$ does not imply $x = y$. We say that h is a *loose representation* of $\mathcal{A}$. To get a pukka representation, we need to factor out by the equivalence relation $h(1')$ on N.

We end with some notation that will be useful. For $\mathcal{A}$-networks $N = (N_1, N_2)$ and $M = (M_1, M_2)$, and any objects $i_1, \ldots, i_k$, we write

$$N =_{i_1, \ldots, i_k} M$$

if $N_1 \setminus \{i_1, \ldots, i_k\} = M_1 \setminus \{i_1, \ldots, i_k\} = I$, say, and $N(i,j) = M(i,j)$ for all $i, j \in I$. That is, M and N *agree off of* $\{i_1, \ldots, i_k\}$.

2.2 Cylindric Algebras

Just as relation algebras 'algebraise' binary relations, so cylindric algebras algebraise relations of higher arities. From now on, fix some finite dimension (or arity) $n \geq 3$. An *n-dimensional cylindric algebra* is an algebra

$$\mathcal{C} = (C, +, -, 0, 1, \mathsf{d}_{ij}, \mathsf{c}_i : i, j < n),$$

where $(C, +, -, 0, 1)$ is a boolean algebra as before, the d_{ij} are constants, and the c_i are unary functions on C, satisfying certain equations not needed here. The algebra $\mathcal{C}$ is said to be *finite* if C is finite, and *simple* if $\mathsf{c}_0\mathsf{c}_1 \cdots \mathsf{c}_{n-1}a = 1$ for each non-zero $a \in C$.

A *(square) representation* of $\mathcal{C}$ is a map $h : C \to \wp(U^n)$, for some base set U, respecting the boolean operations as before, and with

1. $h(\mathsf{d}_{ij}) = \{(x_0, \ldots, x_{n-1}) \in U^n : x_i = x_j\}$
2. $h(\mathsf{c}_i a) = \{(x_0, \ldots, x_{n-1}) \in U^n : \exists (y_0, \ldots, y_{n-1}) \in h(a)(x_j = y_j$ for each $j \in n \setminus \{i\})\}$

for each $i, j < n$ (we identify n with $\{0, 1, \ldots, n-1\}$) and each $a \in C$. So this time, each element of the algebra is 'represented' as an n-ary relation on U. The elements of C are like first-order formulas written with variables $x_0, \ldots, x_{n-1}$; d_{ij} is like $x_i = x_j$, and $\mathsf{c}_i a$ is like $\exists x_i a$. Again, not every n-dimensional finite simple cylindric algebra ($n \geq 3$) is representable.

Atoms and atomic cylindric algebras are defined as for relation algebras. The *atom structure* of an atomic n-dimensional cylindric algebra $\mathcal{C}$ as above is the structure

$$\operatorname{At} \mathcal{C} = (At\, \mathcal{C}, R_{\mathsf{d}_{ij}}, R_{\mathsf{c}_i} : i, j < n),$$

where $R_{\mathsf{d}_{ij}} = \{a \in At\, \mathcal{C} : a \leq \mathsf{d}_{ij}\}$ and $R_{\mathsf{c}_i} = \{(a, b) \in (At\, \mathcal{C})^2 : b \leq \mathsf{c}_i a\}$. A structure $\mathcal{S} = (S, R_{\mathsf{d}_{ij}}, R_{\mathsf{c}_i} : i, j < n)$ in this signature is called a *n-dimensional cylindric-type atom structure*. Again, we can form its complex algebra:

$$\mathfrak{Cm}\, \mathcal{S} = (\wp(S), \cup, \setminus, \emptyset, S, R_{\mathsf{d}_{ij}}, \mathsf{c}_i : i, j < n),$$

where $\mathsf{c}_i X = \{y \in S : R_{\mathsf{c}_i}(x, y)$ for some $x \in X\}$, for each $i < n$ and $X \subseteq S$. Under favourable conditions, $\mathfrak{Cm}\, \mathcal{S}$ will be an n-dimensional cylindric algebra, and again we have $\mathcal{S} \cong \operatorname{At} \mathfrak{Cm}\, \mathcal{S}$, and $\mathcal{C} \cong \mathfrak{Cm} \operatorname{At} \mathcal{C}$ for each finite cylindric algebra $\mathcal{C}$.

One can also define networks for cylindric algebras, analogously to the relation algebra case.

3 Earlier Work

Now let us review some earlier work connecting relation algebras and cylindric algebras. We confine ourselves to the most relevant topics. For a far more thorough survey, see [7].

3.1 Monk

In [8], Monk gave a method of turning an arbitrary relation algebra $\mathcal{A}$ into a 3-dimensional cylindric algebra $\mathcal{C}$, preserving representability both ways — that is, $\mathcal{A}$ is representable iff $\mathcal{C}$ is representable.

[8, p.63] states that the idea is due to Lyndon. [8, p.81] adds that 'This description occurs in a letter from Lyndon to Thompson dated May, 1949... in this letter he restricts himself to the case of proper relation algebras.' For our purposes, we can take a *proper* relation algebra to be a representable one; so by extending the construction to arbitrary relation algebras, Monk made a considerable advance. [8, p.81] also states that reference to the embedding has occurred in several places. The earliest of them is [4].

The construction is important, but rather complicated, and I'm reluctant to summarise it for fear of misrepresentation. However, a related idea is to regard the elements of a relation algebra $\mathcal{A}$ as binary relation symbols, consider the set of all first-order formulas using these symbols and written with only the variables x_0, x_1, x_2, and quotient it out by a certain equivalence relation (actually a congruence) suggested by the relation algebra operations. For example, $\exists x_2(a(x_0, x_2) \wedge b(x_2, x_1))$ (where $a, b \in A$) would be equivalent to $(a\,;b)(x_0, x_1)$. If done properly, the congruence classes form a 3-dimensional cylindric algebra that is representable just when $\mathcal{A}$ is representable. For more on this, and much else, see [10].

For finite relation algebras, Monk's construction is recursive, and it follows by Turing reduction that *representability of finite 3-dimensional cylindric algebras is undecidable.*

Monk does not give any construction in dimensions higher than 3. For this, we need to pass to work of Maddux.

3.2 Maddux: Cylindric Bases

In a number of publications, including notably [5, 6] and the survey [7], Maddux gave a new way of constructing cylindric algebras of any dimension from atomic relation algebras, using sets of networks called *cylindric bases.* We will continue to simplify things by restricting consideration to finite simple relation algebras. For these, in dimension 3, Maddux's construction reproduces Monk's construction up to isomorphism. We will go into some detail about it, since we need it later.

Idea. Let us try to motivate the idea of cylindric basis. Suppose we are given a representation $h : A \to \wp(U^2)$ of a (finite simple) relation algebra $\mathcal{A}$. Recall from formula (1) in section 2.1 that a subset of the base set U can be viewed as an $\mathcal{A}$-network. We can make this a little tighter by considering maps instead of subsets.

Definition 1. *Let $N = (N_1, N_2)$ be an $\mathcal{A}$-network, and $h : A \to \wp(U^2)$ a representation of $\mathcal{A}$ with base U.*

1. *A partial map* $f : N_1 \to U$ *is said to be a* partial embedding of N into h *if* $(f(x), f(y)) \in h(N_2(x, y))$ *for all* $x, y \in \mathrm{dom}(f)$.
2. *We say that* f *is a* total embedding, *or just an* embedding, *if* $\mathrm{dom}(f) = N_1$.
3. *We also say that* N embeds homogeneously into h *if every partial embedding of* N *into* h *extends to a total one.*

Recall that $n \geq 3$ is our fixed finite dimension. For any n-tuple $(u_0, \ldots, u_{n-1}) \in U^n$, we can form an $\mathcal{A}$-network

$$N_{(u_0,\ldots,u_{n-1})} = (n, \nu)$$

whose set of nodes is $n = \{0, 1, \ldots, n-1\}$, where for each $i, j < n$, the label $\nu(i, j)$ is the unique atom a of $\mathcal{A}$ with $(u_i, u_j) \in h(a)$ — that is, $\nu(i, j) = \lambda(u_i, u_j)$ in our earlier notation. Manifestly, the map $(i \mapsto u_i)_{i<n}$ is an embedding of $N_{(u_0,\ldots,u_{n-1})}$ into h. In model-theoretic terms, $N_{(u_0,\ldots,u_{n-1})}$ describes the *atomic type* of the tuple $(u_0, \ldots, u_{n-1})$ in the representation.

Let $\mathcal{N}_n(\mathcal{A})$ denote the set of all $\mathcal{A}$-networks whose set of nodes is n. Each network $N \in \mathcal{N}_n(\mathcal{A})$ defines a (possibly empty) *n-ary relation* on U, namely

$$\{(u_0, \ldots, u_{n-1}) \in U^n : N_{(u_0,\ldots,u_{n-1})} = N\}.$$

This is the set of n-tuples onto which we can embed N. See figure 1 in the case $n = 3$.

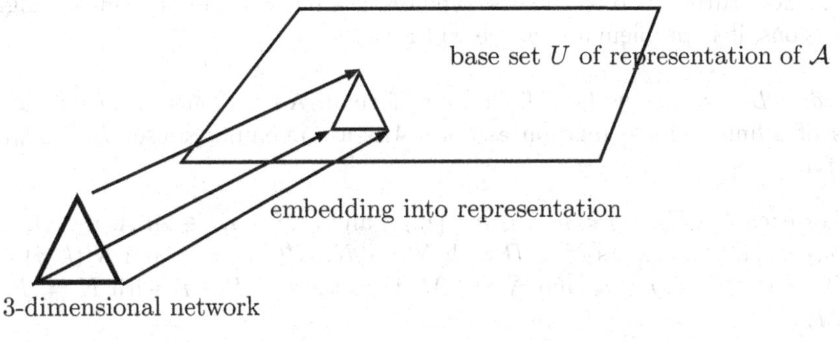

Fig. 1. Network embedding into representation

In view of this, can we make an atomic n-dimensional cylindric algebra whose atoms are networks in $\mathcal{N}_n(\mathcal{A})$, whose arbitrary elements are subsets of $\mathcal{N}_n(\mathcal{A})$, and which in the above case is representable over the base set U?

Well, we would need to know *exactly which* networks embed into h, so that they arise as some $N_{(u_0,\ldots,u_{n-1})}$. This might depend on the choice of the representation h — and when $\mathcal{A}$ is not representable, there is no such h! Remember that our reduction map f should deliver a cylindric algebra $f(\mathcal{A})$ given *any* finite simple relation algebra $\mathcal{A}$, representable or not. So we will need to 'guess' a suitable set — B, say — of networks to use.

But this is only the start of it. We also have to take account of the cylindric algebra operations.

The constants d_{ij} are easy to handle. In a representation of the cylindric algebra, we must interpret d_{ij} as $\{(u_0, \ldots, u_{n-1}) \in U^n : u_i = u_j\}$. But this is just $\{(u_0, \ldots, u_{n-1}) \in U^n : N_{(u_0,\ldots,u_{n-1})}(i,j) \leq 1'\}$. So we could let

$$\mathsf{d}_{ij} = \{N \in B : N(i,j) \leq 1'\},$$

for each $i, j < n$.

The 'cylindrifiers' c_i are a little harder. Plainly, if $(u_0, \ldots, u_{n-1}), (v_0, \ldots, v_n) \in U^n$, $i < n$, and $u_j = v_j$ for each $j \in n \setminus \{i\}$, then $N_{(u_0,\ldots,u_{n-1})} =_i N_{(v_0,\ldots,v_{n-1})}$. For any relation r in our putative cylindric algebra that holds on $(u_0, \ldots, u_{n-1})$, the relation $\mathsf{c}_i r$ must hold on $(v_0, \ldots, v_{n-1})$. So we could let

$$\mathsf{c}_i N = \{M \in B : M =_i N\},$$

for each $N \in B$. The generalisation to sets of networks is easy.

Notice that these definitions are dependent on B but independent of any representation of $\mathcal{A}$ — they make sense even if $\mathcal{A}$ is not representable.

But for a correct representation of our hoped-for cylindric algebra, *for every* sequence $(u_0, \ldots, u_{n-1}) \in U^n$ with $N_{(u_0,\ldots,u_{n-1})} = N$, and *every* $M \in B$ with $M =_i N$, there must be a sequence $(v_0, \ldots, v_n) \in U^n$ with $u_j = v_j$ for each $j \in n \setminus \{i\}$, and $N_{(v_0,\ldots,v_{n-1})} = M$. This boils down to saying that *each* $N \in B$ *embeds homogeneously into* h.

This does turn out to be the case when $n = 3$ and $B = N_3(\mathcal{A})$. But for higher dimensions, it is problematic, as we will now see.

Cylindric Bases. Let us be a little more formal. An *n-dimensional cylindrical basis* of a finite simple relation algebra $\mathcal{A}$ is a non-empty subset $B \subseteq \mathcal{N}_n(\mathcal{A})$ satisfying[3]

1. For each $N \in B$, $i, j < n$, $k \in n \setminus \{i, j\}$, and atoms $a, b \in At\,\mathcal{A}$, if $N(i,j) \leq a\,;b$, then there exists $N' \in B$ with $N' =_k N$, $N'(i,k) = a$, and $N'(k,j) = b$.
2. If $N, M \in B$, $i, j < n$, and $N =_{ij} M$, then there is $P \in B$ with $N =_i P =_j M$.

We will not use the details of this definition, but we do point the reader's attention to the similarity of clause 1 to our earlier saturation condition for $\mathcal{A}$-networks. The salient facts about cylindric bases are as follows.

1. We can view an n-dimensional cylindric basis B as an n-dimensional cylindric-type atom structure

$$\mathcal{B} = (B, R_{\mathsf{d}_{ij}}, R_{\mathsf{c}_i} : i, j < n),$$

[3] The definition we give here is not the same as Maddux's definition in (e.g.,) [6, definition 4], but it is equivalent to it for finite simple relation algebras. See, e.g., [2, lemma 12.36].

where $R_{\mathsf{d}_{ij}} = \{N \in B : N(i,j) \leq 1'\}$ and $R_{\mathsf{c}_i} = \{(N,M) \in B^2 : N =_i M\}$ — that is, R_{c_i} is just $=_i$. This is as suggested above. We write B for the basis and $\mathcal{B}$ for the corresponding atom structure. The definition of cylindric basis ensures that $\mathfrak{Cm}\,\mathcal{B}$ is always an n-dimensional cylindric algebra, with atom structure isomorphic to $\mathcal{B}$. The map $B \mapsto \mathfrak{Cm}\,\mathcal{B}$ is recursive.

2. In dimension 3:
   - Every finite simple relation algebra $\mathcal{A}$ has a 3-dimensional cylindric basis. The set $\mathcal{N}_3(\mathcal{A})$ is one such, and it's the only one, actually.
   - For $B = \mathcal{N}_3(\mathcal{A})$, the complex algebra $\mathfrak{Cm}\,\mathcal{B}$ is a 3-dimensional cylindric algebra isomorphic to what Monk's construction gives.
   - $\mathfrak{Cm}\,\mathcal{B}$ is representable iff $\mathcal{A}$ is representable as a relation algebra.

   Here is the gist of the proof. For $\Leftarrow$, we can read off a representation of $\mathfrak{Cm}\,\mathcal{B}$ from any representation of $\mathcal{A}$, because the relation algebra operations are strong enough to ensure that all networks in $\mathcal{N}_3(\mathcal{A})$ embed homogeneously into any representation of $\mathcal{A}$.

   More formally, if $h : A \to \wp(U^2)$ is a representation of $\mathcal{A}$, then define a representation $h^* : \wp(B) \to \wp(U^3)$ of $\mathfrak{Cm}\,\mathcal{B}$ by

   $$h^*(X) = \{(u_0, u_1, u_2) \in U^3 : N_{(u_0,u_1,u_2)} \in X)\}$$

   for each $X \subseteq B$, using the notation $N_{(u_0,\ldots,u_{n-1})}$ introduced above.

   For $\Rightarrow$, $\mathcal{A}$ is a subalgebra of the *(neat) relation algebra reduct* of $\mathfrak{Cm}\,\mathcal{B}$ obtained by restricting to its '2-dimensional elements'. Taking relation algebra reducts preserves representability. For more details, see [7, §4].
3. However, for dimensions $n > 3$:
   - $\mathcal{N}_n(\mathcal{A})$ may *not* be an n-dimensional cylindric basis.
   - For $n \geq 5$, not every atomic relation algebra $\mathcal{A}$ has any n-dimensional cylindric basis at all.
   - Even when $\mathcal{A}$ does have an n-dimensional cylindric basis, say B, it may be that $\mathcal{A}$ is representable but $\mathfrak{Cm}\,\mathcal{B}$ is not (though it will be an n-dimensional cylindric algebra). The 'reason' is that *not every network in B need embed (at all, or homogeneously) into a representation of $\mathcal{A}$.* Examples can be found in [6, pp. 960–961] and [7, p. 389].
   - It is true that if $\mathcal{A}$ has a cylindric basis B and $\mathfrak{Cm}\,\mathcal{B}$ is representable, then $\mathcal{A}$ is representable, as again it is a subalgebra of the relation algebra reduct of $\mathfrak{Cm}\,\mathcal{B}$. But for a reduction, this is not enough.

So, while excellent in dimension 3 and very important in general, cylindric bases do not suit our purposes in higher dimensions than 3.

4 Reduction in Arbitrary Dimensions

Recall that we wish to find a recursive construction of an n-dimensional cylindric algebra from an arbitrary finite simple relation algebra, and the construction should preserve representability both ways.

The constructions of Monk and Maddux do not achieve this aim in higher dimensions, but they do in dimension 3. We can learn from this.

An n-dimensional cylindric basis uses 'n-dimensional' networks with base set n. *All* pairs (i, j) of nodes $i, j < n$ are 'labelled' with atoms. In dimensions higher than 3, this can cause a mismatch between the kind of networks that exist in the basis and those that embed (homogeneously) into a representation of $\mathcal{A}$. But in dimension 3, there is no mismatch at all, because the relation algebra operations 'control' exactly which 3-dimensional networks embed in a representation (namely, all of them do), and moreover they ensure that every network embeds homogeneously as well.

So, let us try to devise a new kind of 'n-dimensional network' — one based on the 3-dimensional networks that work so well, without adding any extra higher-dimensional structure from the relation algebra point of view. One can have all sorts of ideas about how to do this (believe me), but they often fail, because potentially fatal higher-dimensional information is smuggled in.

4.1 Motivation from Representations

It can help to think in terms of representations. Given a representation $h : A \to \wp(U^2)$ of $\mathcal{A}$, an n-tuple $(u_0, \ldots, u_{n-1})$ of elements of U 'sees' what the representation says about its points — that is, the collection of atoms $\lambda(u_i, u_j)$ associated with pairs of points from the tuple. This gives it information on the mutual relationships of up to n points of the representation. This is dangerous, for the above-mentioned reasons. The challenge we face is to limit this information to groups of at most three points, while still having all of $u_0, \ldots, u_{n-1}$ around.

So we consider a new kind of structure. Let V be a set. Suppose that for each subset $S \subseteq V$ of cardinality exactly $n - 3$, we have a representation $h_S : A \to \wp((V \setminus S)^2)$ of $\mathcal{A}$ on the base $V \setminus S$. Here, S is a sort of 'black hole', carrying no information inside it. There need be no correlation whatever between the h_S, as S varies.

Extending our earlier notation, for $u, v \in V \setminus S$ we write $\lambda_S(u, v)$ for the unique atom $a \in At\,\mathcal{A}$ such that $(u, v) \in h_S(a)$.

Now, given an n-tuple $(u_0, \ldots, u_{n-1})$ of elements of V, the only information from $\mathcal{A}$ that $(u_0, \ldots, u_{n-1})$ can 'see' is the collection of atoms

$$\langle \lambda_S(u_i, u_j) : S \subseteq \{u_0, \ldots, u_{n-1}\},\ |S| = n - 3,\ i, j < n,\ u_i, u_j \notin S \rangle. \quad (2)$$

Crucially, only 3-dimensional information (from at most three points) about any one representation h_S is now visible to $(u_0, \ldots, u_{n-1})$. This is because at most three points of $u_0, \ldots, u_{n-1}$ can lie outside each S in (2). Moreover, the cylindric algebra operations cannot be used to garner higher-dimensional information about h_S. For that would involve 'moving' a point $u_i \in S$ to a point outside S, using a c_i. But then, S is no longer a subset of the points in the resulting n-tuple, so (see (2)) no information about h_S is available to it at all.

4.2 Holograms

So we wish to devise a new kind of n-dimensional network embodying the information in (2) above. The network will become an atom of our final cylindric algebra that will 'hold' on $(u_0, \ldots, u_{n-1})$.

But the definition of the new network cannot use any representation, since $\mathcal{A}$ may not have a representation!

So we simply throw in any 3-dimensional networks, subject only to identity constraints. We call our new-style network a *hologram,* since it incorporates many different 3-dimensional 'views'.

Definition 2. *Let $\sim$ be an equivalence relation on n. Write $H(\sim)$ for the set of all subsets $X \subseteq n$ such that $n \setminus X$ is the union of exactly $(n-3)$ $\sim$-classes. Quite possibly, $H(\sim) = \emptyset$.*

For a finite simple relation algebra $\mathcal{A}$, an (n-dimensional) hologram (over $\mathcal{A}$) *is a family*

$$\eta = (\sim, N_X : X \in H(\sim)),$$

where $\sim$ is an equivalence relation on n, each N_X is an $\mathcal{A}$-network whose set of nodes is X, and for each $X \in H(\sim)$ and $i, j \in X$, if $i \sim j$ then $N_X(i,j) \leq 1$'.

Example 1. In terms of the sketch in section 4.1 with V and the h_S and λ_S, a hologram $\eta = (\sim, N_X : X \in H(\sim))$ will 'hold' on $(u_0, \ldots, u_{n-1}) \in V^n$ as per (2) iff:

H1 For each $i, j < n$ we have $u_i = u_j$ iff $i \sim j$.
H2 For each $X \in H(\sim)$, if $S = \{u_k : k \in n \setminus X\}$, then $N_X(i,j) = \lambda_S(u_i, u_j)$ for each $i, j \in X$. That is, the map $(i \mapsto u_i : i \in X)$ is an embedding of N_X into h_S.

Note that if $X \in H(\sim)$, the set $n \setminus X$ is the union of exactly $n-3$ $\sim$-classes, and so no element of X is $\sim$-equivalent to any element of $n \setminus X$. So by H1, the set S in H2 has size $n-3$ and $u_i \notin S$ for each $i \in X$. Hence, H2 makes sense. Also note that for $i, j \in X$ we have $i \sim j \Rightarrow u_i = u_j \Rightarrow N_X(i,j) = \lambda_S(u_i, u_j) \leq 1$', which is consistent with definition 2.

4.3 Atom Structure from Holograms

Let M be the set of all (n-dimensional) holograms. We wish to form an n-dimensional cylindric-type atom structure $\mathcal{M} = (M, R_{\mathsf{d}_{ij}}, R_{\mathsf{c}_i} : i, j < n)$. From H1 above, it is clear that we should define

$$R_{\mathsf{d}_{ij}} = \{(\sim, N_X : X \in H(\sim)) \in M : i \sim j\}.$$

But what about R_{c_i}? For inspiration, we consider again the picture in section 4.1. Suppose we have two tuples $(u_0, \ldots, u_{n-1})$ and $(v_0, \ldots, v_{n-1})$ in V^n, with $u_j = v_j$ for all $j \in n \setminus \{i\}$. What is the connection between the holograms $(\sim, N_X : X \in H(\sim))$ and $(\sim', N'_X : X \in H(\sim'))$ that 'hold' in the sense of example 1 on $(u_0, \ldots, u_{n-1})$ and $(v_0, \ldots, v_{n-1})$, respectively?

Well, we certainly have $j \sim k$ iff $j \sim' k$ for all $j, k \in n \setminus \{i\}$. To see this, note that $u_j = v_j$ and $u_k = v_k$, so by H1, $j \sim k$ iff $u_j = u_k$ iff $v_j = v_k$ iff $j \sim' k$. We say for short that $\sim$ *and* $\sim'$ *agree off of* i. We cannot say any more about $\sim, \sim'$ than that, since we do not know whether $u_i = v_i$.

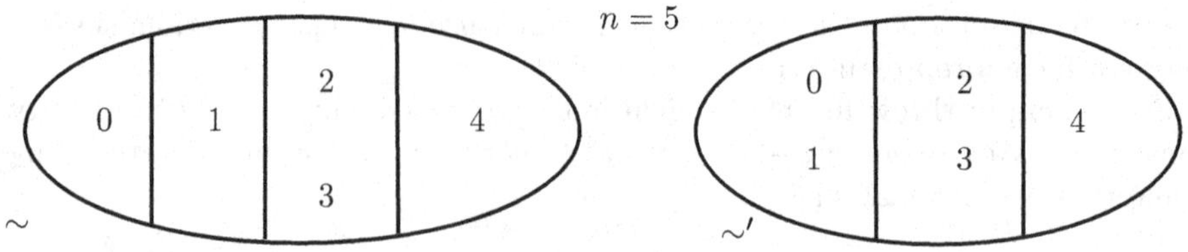

Fig. 2. Equivalence relations $\sim, \sim'$ on 5

What about the N_X and $N'_{X'}$? The only sets S of $n-3$ points (the 'black holes') whose representations h_S carry information common to both $(u_0, \ldots, u_{n-1})$ and $(v_0, \ldots, v_{n-1})$ are those S that remain unchanged by moving u_i. These 'stable sets' are the subsets of $\{u_j : j \in n \setminus \{i\}\}$ of size $n-3$. (They may or may not contain u_i, v_i.)

Now a set $X \in H(\sim)$ corresponds to the black hole $S = \{u_j : j \in n \setminus X\}$. This is a stable set just when it is equal to $\{u_j : j \in n \setminus (X \cup \{i\})\}$. And since $|S| = n-3$, this is exactly when $n \setminus (X \cup \{i\})$ already contains $n-3$ pairwise $\sim$-inequivalent elements. A set $X' \in H(\sim')$ corresponds to this same black hole S just when $X \cup \{i\} = X' \cup \{i\}$. For all such X, X', we will require $N_X(j,k) = N'_{X'}(j,k)$ for every $j, k \in (X \cap X') \setminus \{i\}$. Since plainly $X \setminus \{i\} = X' \setminus \{i\}$, this is exactly when $N_X =_i N'_{X'}$. Note here that we do not require that $i \in X \cup X'$ — see the definition of $=_{i_1,\ldots,i_k}$ in section 2.1.

We should therefore demand that $N_X =_i N'_{X'}$ whenever the above conditions are met.

Well, this all looks very messy, but we are forced into it by our idea, and the notions involved are elementary. Perhaps an example will help.

Example 2. In the notation above, suppose $n = 5$, the $\sim$-classes are $\{0\}, \{1\}, \{2,3\}, \{4\}$, and the $\sim'$-classes are $\{0,1\}, \{2,3\}, \{4\}$. See figure 2. Intuitively, u_0 is different from all other u_j, while $v_0 = v_1$. We can see that $\sim$ and $\sim'$ agree off of 0. We have $n - 3 = 2$, and

- $H(\sim) = \{\{0,1\}, \{0,2,3\}, \{0,4\}, \{1,2,3\}, \{1,4\}, \{2,3,4\}\}$,
- $H(\sim') = \{\{0,1\}, \{2,3\}, \{4\}\}$.

Let $i = 0$.

- The sets $X \in H(\sim)$ such that $n \setminus (X \cup \{i\})$ contains $(n-3) = 2$ pairwise $\sim$-inequivalent elements are $\{0,1\}, \{0,2,3\}, \{0,4\}$.
- All sets $X' \in H(\sim')$ are such that $n \setminus (X' \cup \{i\})$ contains $(n-3)$ $\sim'$-inequivalent elements. For example, for $X' = \{2,3\}$, we have $n \setminus (X' \cup \{i\}) = n \setminus \{0,2,3\} = \{1,4\}$, and plainly $1 \not\sim' 4$.

Of these, the sets $X \in H(\sim)$ and $X' \in H(\sim')$ such that $X \cup \{0\} = X' \cup \{0\}$, so they have the same complement in n and correspond to the same 'black hole', are

- $X = X' = \{0,1\}$,
- $X = \{0,2,3\}$ and $X' = \{2,3\}$,
- $X = \{0,4\}$ and $X' = \{4\}$.

So for any holograms $\eta = (\sim, N_X : X \in H(\sim))$ and $\eta' = (\sim', N'_{X'} : X' \in H(\sim'))$ with $\sim, \sim'$ as above, we have $R_{\mathsf{c}_0}(\eta, \eta')$ iff $N_{\{0,1\}} =_0 N'_{\{0,1\}}$, $N_{\{0,2,3\}} =_0 N'_{\{2,3\}}$, and $N_{\{0,4\}} =_0 N'_{\{4\}}$.

The conclusion of this discussion is the following definition.

Definition 3. *Let $\mathcal{A}$ be a finite simple relation algebra.*

1. *Let $\eta = (\sim, N_X : X \in H(\sim))$ and $\eta' = (\sim', N'_X : X \in H(\sim'))$ be n-dimensional holograms over $\mathcal{A}$. For each $i, j < n$, define*
   - *$R_{\mathsf{d}_{ij}}(\eta)$ iff $i \sim j$,*
   - *$R_{\mathsf{c}_i}(\eta, \eta')$ iff*
     (a) $\sim$ and $\sim'$ agree off of i,
     (b) for each $X \in H(\sim)$ and $X' \in H(\sim')$, if $X \cup \{i\} = X' \cup \{i\} = I$, say, and $n \setminus I$ contains $n-3$ pairwise $\sim$-inequivalent elements, then $N_X =_i N'_{X'}$.

   This defines a unary relation $R_{\mathsf{d}_{ij}}$ and a binary relation R_{c_i} on the set M of holograms.
2. *Let $\mathcal{M}(\mathcal{A})$ be the n-dimensional cylindric-type atom structure $(M, R_{\mathsf{d}_{ij}}, R_{\mathsf{c}_i} : i, j < n)$.*
3. *Define $\mathcal{C}_n(\mathcal{A}) = \mathfrak{Cm}\,\mathcal{M}(\mathcal{A})$.*

4.4 Reduction Function; Undecidability of Representability

Definition 4. *Fix a finite non-representable n-dimensional cylindric algebra $\mathcal{C}^\times$. Define a function f from finite simple relation algebras to n-dimensional cylindric algebras, by*

$$f(\mathcal{A}) = \begin{cases} \mathcal{C}_n(\mathcal{A}), & \text{if this is an } n\text{-dimensional cylindric algebra,} \\ \mathcal{C}^\times, & \text{otherwise.} \end{cases}$$

Is the function f a reduction as desired? It is recursive, since $\mathcal{C}_n(\mathcal{A})$ is finite and recursively constructible from $\mathcal{A}$, and there is an algorithm to decide whether a finite algebra is an n-dimensional cylindric algebra or not. Plainly, $f(\mathcal{A})$ is always an n-dimensional cylindric algebra — the use of $\mathcal{C}^\times$ avoids having to verify that $\mathcal{C}_n(\mathcal{A})$ is always a cylindric algebra. Now it can be shown that

$$\mathcal{A} \text{ is representable iff } \mathcal{C}_n(\mathcal{A}) \text{ is representable.} \tag{3}$$

We will discuss the proof below. So consider the cases.

1. Suppose that $\mathcal{A}$ is representable. Then $\mathcal{C}_n(\mathcal{A})$ is representable, and it follows that $\mathcal{C}_n(\mathcal{A})$ is an n-dimensional cylindric algebra. The definition of f yields $f(\mathcal{A}) = \mathcal{C}_n(\mathcal{A})$, and this is representable.

2. Suppose that $\mathcal{A}$ is not representable. There are two possibilities. If $f(\mathcal{A}) = \mathcal{C}_n(\mathcal{A})$, then by (3), $f(\mathcal{A})$ is not representable. If $f(\mathcal{A}) = \mathcal{C}^\times$ then by choice of $\mathcal{C}^\times$ it is not representable.

So indeed, modulo (3), f is our desired reduction. Since it is undecidable whether a finite simple relation algebra is representable [2, theorem 18.13], we deduce by Turing reduction that

Theorem 1. *For each finite $n \geq 3$, there is no algorithm to decide whether a finite n-dimensional cylindric algebra is representable.*

For further applications, see [3].

4.5 Co-representability of $\mathcal{A}$ and $\mathcal{C}_n(\mathcal{A})$

How can we prove (3)? For a full proof, see [3]. We will sketch some of the ideas.

Suppose that $\mathcal{C}_n(\mathcal{A})$ is representable. It can be shown that a representation of it over the base set V must be of the form described in section 4.1, with the caveat that the representations h_S may be *loose* (see section 2.1). But if $\mathcal{A}$ has loose representations, it is representable.

Conversely, assume that $\mathcal{A}$ is representable. We need to construct a representation of $\mathcal{C}_n(\mathcal{A})$. The key is to construct a V as in section 4.1 in which the h_S are 'random' (and loose).

To see why we need randomness, suppose that $(u_0, \ldots, u_{n-1})$ is an n-tuple in V on which the atom $\eta = (\sim, N_X : X \in H(\sim))$ of $\mathcal{C}_n(\mathcal{A})$ 'holds' in the sense of example 1. Suppose that $R_{\mathsf{c}_i}(\eta, \eta')$, where $\eta' = (\sim', N'_X : X \in H(\sim'))$. Then, to be a good representation, there must be some tuple $(v_0, \ldots, v_{n-1}) \in V^n$ on which η' holds, and with $v_j = u_j$ for each $j \in n \setminus \{i\}$. So there must be a suitable point $v_i \in V$.

Why should there be such a point? What are the constraints?

Well, for η' to hold on $(v_0, \ldots, v_{n-1})$ given that η holds on $(u_0, \ldots, u_{n-1})$, we require firstly that $v_i = v_j$ iff $i \sim' j$, for each $j \in n \setminus \{i\}$. The case where $i \sim' j$ for some such j is easily handled, as it can be shown that η' already holds on $(u_0, \ldots, u_{i-1}, u_j, u_{i+1}, \ldots, u_{n-1})$. So assume that $i \not\sim' j$ for every $j \in n \setminus \{i\}$. This means that v_i has to be a 'new' element of V not in the set

$$O = \{u_j : j \in n \setminus \{i\}\} = \{v_j : j \in n \setminus \{i\}\}$$

of 'old' elements.

Plainly, $|O| < n$. The worst case is when $|O| = n - 1$, so let us examine that case. Consider a 'black hole' $S \subseteq O$ of size $n - 3$. Bear in mind that there are $\binom{n-1}{n-3}$ of these — $\mathcal{O}(n^2)$. Choose $j, k \in n \setminus \{i\}$ such that

$$O \setminus S = \{u_j, u_k\}.$$

Note that $u_j = v_j$ and $u_k = v_k$. Let $X' = \{i, j, k\}$. Then $X' \in H(\sim')$. For the hologram η' to hold on $(v_0, \ldots, v_{n-1})$, the map $(i \mapsto v_i, j \mapsto v_j, k \mapsto v_k)$ must be an embedding of $N'_{X'}$ into h_S; and we require the analogous property for every S.

Can we find such a v_i? Well, let $X = \{l < n : u_l \notin S\}$. Then $j, k \in X \in H(\sim)$. We have $X \cup \{i\} = X' \cup \{i\}$, and $n \setminus (X \cup \{i\})$ contains $n - 3$ pairwise $\sim$-inequivalent elements, because $\{u_l : l \in n \setminus (X \cup \{i\})\} = S$ and $|S| = n - 3$. Since $R_{c_i}(\eta, \eta')$, it follows that $N_X =_i N'_{X'}$.

Since η holds on $(u_0, \ldots, u_{n-1})$, the map

$$(j \mapsto u_j, k \mapsto u_k) \tag{4}$$

is a partial embedding of N_X into h_S. But $N_X =_i N'_{X'}$. So the map

$$(j \mapsto v_j, k \mapsto v_k), \tag{5}$$

being the exact same map as (4) since $u_j = v_j$ and $u_k = v_k$, is also a partial embedding of $N'_{X'}$ into h_S.

Now h_S is a (loose) representation of $\mathcal{A}$ over $V \setminus S$. By basic properties of relation algebra representations, every $\mathcal{A}$-network with at most 3 nodes embeds homogeneously into every loose representation. So the partial embedding (5) of $N'_{X'}$ into h_S extends to i, and we can indeed find a point $v_i \in V \setminus S$ such that the map $(i \mapsto v_i, j \mapsto v_j, k \mapsto v_k)$ is an embedding of $N'_{X'}$ into h_S.

It all looks so rosy. But remember: we have found a point v_i for this particular S. Sure, for *each* $S \subseteq O$ of size $n-3$, we can find a suitable $v_i \in V \setminus S$ in this way: that is, $\forall S \exists v_i$. But of course we have to find a *single* point v_i that works *for every* $S \subseteq O$ *of size* $n - 3$. We need $\exists v_i \forall S$.

And that's not all. We have not yet considered the $S \subseteq \{v_0, \ldots, v_{n-1}\}$ with $|S| = n - 3$ and $v_i \in S$. We must choose v_i additionally so that for each of these S, if $X' = \{l \in n \setminus \{i\} : v_l \notin S\} \in H(\sim')$, then the map $(j \mapsto v_j : j \in X')$ embeds $N'_{X'}$ into h_S.

This seems a tall order. But if the h_S are in a sense 'randomly chosen', it is possible to find such a v_i. Similar arguments can be found in random graph theory and 0–1 laws for logics.

Actually, the mention of probability is just to give the flavour. We do not really use probability. What we actually do is to *build* the h_S in a kind of forcing construction using an infinite game. This ensures that we get the points v_i that we need. To do it, it is important that the h_S are *loose* representations of $\mathcal{A}$. For full details, see [3, proposition 4.7].

References

1. Hirsch, R., Hodkinson, I.: Representability is not decidable for finite relation algebras. Trans. Amer. Math. Soc. 353, 1403–1425 (2001)
2. Hirsch, R., Hodkinson, I.: Relation algebras by games. Studies in Logic and the Foundations of Mathematics, vol. 147. North-Holland, Amsterdam (2002)
3. Hodkinson, I.: A construction of cylindric and polyadic algebras from atomic relation algebras. Algebra Universalis 68, 257–285 (2012)

4. Lyndon, R.: The representation of relational algebras. Annals of Mathematics 51(3), 707–729 (1950)
5. Maddux, R.D.: Topics in relation algebra. Ph.D. thesis, University of California, Berkeley (1978)
6. Maddux, R.D.: Non-finite axiomatizability results for cylindric and relation algebras. J. Symbolic Logic 54(3), 951–974 (1989)
7. Maddux, R.D.: Introductory course on relation algebras, finite-dimensional cylindric algebras, and their interconnections. In: Andréka, H., Monk, J.D., Németi, I. (eds.) Algebraic Logic. Colloq. Math. Soc. J. Bolyai, vol. 54, pp. 361–392. North-Holland, Amsterdam (1991)
8. Monk, J.D.: Studies in cylindric algebra. Ph.D. thesis, University of California, Berkeley (1961)
9. Tarski, A.: On the calculus of relations. J. Symbolic Logic 6, 73–89 (1941)
10. Tarski, A., Givant, S.R.: A formalization of set theory without variables, vol. 41. Colloquium Publications, Amer. Math. Soc., Providence, Rhode Island (1987)

Towards a Probabilistic Interpretation of Game Logic

Ernst-Erich Doberkat

MATH++SOFTWARE Bochum, Germany
eed@doberkat.de

Abstract. Game logic is a modal logic the modalities of which model the interaction of two players, Angel and Demon. It is known that game logic is not adequately interpreted through relation based Kripke models. The basic mechanism behind neighborhood models, which are used instead, is given through effectivity functions. We give a brief introduction to effectivity functions based on sets, indicate some of their coalgebraic properties, and move on to a definition of stochastic effectivity functions over general measurable spaces. An interpretation of game logics in terms of these effectivity functions is sketched, and their relationship to probabilistic Kripke models and to the interpretation of the PDL fragment is indicated.

Modal Logics and Games. The formulas of a modal logics are given through the grammar

$$\varphi ::= p \mid \varphi_1 \wedge \varphi_2 \mid \neg\varphi \mid \langle a\rangle\varphi.$$

Here p is an atomic proposition, and a is a modality, which usually models actions. Thus $\langle a\rangle\varphi$ holds in a world $w \in W$ iff we can make a transition by executing action $a \in A$ into a world w' in which formula φ holds. This indicates the usual interpretation of the logic: we associate with each action a a relation $R_a \subseteq W \times W$ and define $w \models \langle a\rangle\varphi$ iff $w' \models \varphi$ for some $w' \in R_a(w) := \{w'' \mid \langle w, w''\rangle \in R_a\}$; the Boolean connectives are interpreted as usual, and each atomic proposition p is associated with a set $V(p) \subseteq W$ such that $w \models \pi$ iff $w \in V(p)$. Collect these data into a relation based *Kripke model* $\big(W, (R_a)_{a\in A}, V\big)$.

If the modalities carry a structure of their own, one would expect that this is reflected in the interpreting relations. This is the case, e.g., with Propositional Dynamic Logic (PDL) or with Game Logic (GL), which are intended to model simple programs, and two person games, respectively. We assume for the latter that we have two adversaries, Angel and Demon, playing against each other, taking turns. The grammar for games is given through

$$g ::= \gamma \mid g_1 \cap g_1 \mid g_1 \cup g_2 \mid g_1; g_2 \mid g^* \mid g^d \mid g^\times \mid \varphi?$$

with $\gamma \in \Gamma$ a primitive game [6]. Here $g_1 \cup g_2$ denotes the nondeterministic choice between games g_1 and g_2, $g_1; g_2$ is the sequential play of g_1 and g_2 in that order, and g^* is iteration of game g a finite number of times (including zero). The game $\varphi?$ tests whether or not formula φ holds, where φ is a formula from the logic. $\varphi?$ serves as a guard: $(\varphi?; g_1) \cup (\neg\varphi?; g_2)$ tests whether φ holds, if it does g_1 is

W. Kahl et al. (Eds.): RAMiCS 2015, LNCS 9348, pp. 43–47, 2015.
DOI: 10.1007/978-3-319-24704-5_3

played, otherwise, g_2 is. This describes the moves of Angel. The moves of player Demon are given by $g_1 \cap g_2$, where Demon chooses between games g_1 and g_2; this is demonic choice (in contrast to angelic choice $g_1 \cup g_2$). With $g^\times$, Demon decides to play game g a finite number of times (including not at all), and g^d indicates that Angel and Demon change places.

The informal meaning of $\langle g \rangle \varphi$ is that formula φ holds after game g is played. Let us just indicate informally by $\langle g \rangle \varphi$ that Angel has a strategy in game g which makes sure that playing g results in a state which satisfies formula φ. We assume the game to be *determined*: if one player does not have a winning strategy, then the other one has. Thus if Angle does not have a φ-strategy, then Demon has a $\neg\varphi$-strategy, and vice versa. This means that we can derive the way Demon plays the game from the way Angel does, and vice versa. Thus we may express demonic choice $g_1 \cap g_2$ through $(g_1^d \cup g_2^d)^d$, and demonic iteration $g^\times$ through angelic iteration $\left((g^d)^*\right)^d$); clearly, g^{dd} should be the same as g. In contrast to Banach-Mazur games, we do not describe formally what a strategy is.

Neighborhood Models. Game logics are usually interpreted through *neighborhood models*, which associate with each primitive game $\gamma \in \Gamma$ and each world $w \in W$ a set $N_\gamma(w)$ of subsets of W, $A \in N_\gamma(w)$ indicating that Angel has a strategy for achieving a state in A upon playing γ in state w. Thus $N_\gamma(w)$ is an upper closed subset of the power set $\mathcal{P}(W)$ of W, hence $A \in N_\gamma(w)$ and $A \subseteq B$ implies $B \in N_\gamma(w)$; the elements of $N_\gamma(w)$ are perceived as neighborhoods of w under γ. These models are more general than Kripke models: given a relation $R \subseteq W \times W$, $w \mapsto \{A \subseteq W \mid R(w) \subseteq A\}$ yields for each $w \in W$ an upper closed set. Associating with N_γ a monotone map $N_\gamma^+ : \mathcal{P}(W) \to \mathcal{P}(W)$ through $N_\gamma^+(A) := \{w \in W \mid w \in N_\gamma(w)\}$, we may perform a syntax directed translation from games to maps $\mathcal{P}(W) \to \mathcal{P}(W)$, e.g., $N_{g_1;g_2}^+ := N_{g_1}^+ \circ N_{g_2}^+$, or $N_{g^d}^+(A) := W \setminus N_g^+(W \setminus A)$. In this way, each game g gets associated with such a monotone map N_g^+. We interpret the modal formula $\langle g \rangle \varphi$ by defining $[\![\langle g \rangle \varphi]\!] := N_g^+([\![\varphi]\!])$, where, as usual, $w \in [\![\varphi]\!]$ iff $w \models \varphi$.

A coalgebraic point of view notices that the assignment $\mathbb{V} : W \mapsto \{V \subseteq \mathcal{P}(W) \mid V \text{ is upper closed}\}$ is the functorial part of a monad, and that each N_γ is a Kleisli morphism for this monad, hence a coalgebra for $\mathbb{V}$. Composition of games is interpreted through Kleisli composition in the $\mathbb{V}$-monad; the actions of Demon may be obtained through *demonization* (the demonization of $f : W \to \mathbb{V}(W)$ is given by $\partial f : w \mapsto \{A \mid W \setminus A \notin f(w)\}$). The transformation $N_\gamma \mapsto N_\gamma^+$ is given by a natural transformation of the functors $\mathcal{P} \to \mathbb{V}$.

Neighborhood models are strictly more general than Kripke models, which turn out to be not adequate for interpreting general game logics. This is the reason why: The interpretation of games through Kripke models is disjunctive, which means that $\langle g_1; (g_2 \cup g_3) \rangle \varphi$ is semantically equivalent to $\langle g_1; g_2 \cup g_1; g_3 \rangle \varphi$ for all games g_1, g_2, g_3. This, however, is evidently not desirable: Angle's decision after playing g_1 whether to play g_2 or g_3 should not be equivalent to decide whether to play $g_1; g_2$ or $g_1; g_3$. Neighborhood models in their greater generality do not display this equivalence [7].

A Stochastic Interpretation of Game Logic. We modify first the modal formulas $\langle g\rangle\varphi$ to the conditional modal formulas $\langle g\rangle_r\varphi$, indicating now that formula φ should hold after playing g with a probability not smaller than $r \in [0, 1]$. It replaces also sets of worlds by sets of probability distributions over these worlds. Playing game g in state w, $N_g(w)$ is an upper closed set, the elements of which are now probability distributions over W, $A \in N_g(w)$ indicating that Angel has a strategy for achieving a distribution of new states taken from A. So this sounds like simply replacing the set of states by the set of distributions over the states. But things are not that straightforward, unfortunately. The reason is that we need this new kind of neighborhood models be adaptable to the requirements provided by the algebraic structure of the games, in particular it should support the composition of games, and it should be closed under demonization.

This leads to the definition of stochastic effectivity functions, which model a particular kind of stochastic nondeterminism [4,2]. One first notes that the set of worlds W should carry a measurable structure, so that measures can be defined on it. The set $\mathbb{P}(W)$ of all probabilities on W then carries also a measurable structure, which is given in a fairly natural way by evaluating probabilities at events [3]. So an effectivity function P on world W should map W to the upper closed measurable subsets of $\mathbb{P}(W)$. This looks like an easy combination of two monads — the probability functor $\mathbb{P}$ is a well known monadic functor, and the upper closed functor is also monadic. Unfortunately, this does not work out well, because the composition of two monads is usually not a monad, bad luck.

The following technical construction helps to bypass this difficulty. Assume we have a measurable subset $H \subseteq \mathbb{P}(W) \times [0, 1]$, which may be thought of as a combination of measures with their numerical evaluations, e.g., $H = \{\langle\mu, q\rangle \mid \mu(A) \geq q\}$ for some measurable set A of worlds, then $H^q := \{\mu \mid \langle\mu, q\rangle \in H\}$ cuts H at q (imagine a set in the plane and look at its horizontal cuts). It can be shown that H^q is a measurable set of probabilities. We want the set $\{\langle w, q\rangle \in W \times [0, 1] \mid H^q \in P(w)\}$ be a measurable subset of $W \times [0, 1]$ for all such H; if this is the case, we call the effectivity function *t-measurable.*

Just to get the idea, assume that K is a stochastic transition kernel on W, hence $K(w)$ is a probability on W for each $w \in W$, then $w \mapsto \{A \subseteq \mathbb{P}(W) \text{ measurable} \mid K(w) \in A\}$ is such a t-measurable effectivity function (this is comparable to moving from a point to the ultrafilter generated by it). Another example comes from finite transition systems. Let the world W be finite and R a transition system on W with $R(w) \neq \emptyset$ for all $w \in W$, define the set of all weighted transitions from w through $\kappa(w) := \{\sum_{w' \in R(w)} \alpha_{w'} \cdot \delta_{w'} \mid \alpha_{w'} \geq 0 \text{ rational}, \sum_{w' \in R(w)} \alpha_{w'} = 1\}$, then $P(w) := \{A \subseteq \mathbb{P}(W) \text{ measurable} \mid \kappa(w) \subseteq A\}$ defines a t-measurable effectivity function on W. Also, if the effectivity function P is t-measurable, then $A \in \partial P(w)$ iff the complement of A is not in $P(w)$ defines a t-measurable effectivity function, the demonization of P.

As a whole, t-measurable effectivity functions have some fairly interesting algebraic properties [2], and they may be used for defining the semantics of game logics. This will be sketched now. The basic technical approach is to associate with each game g a set transformer, depending on a threshold value r, specifically, to define for $A \subseteq W$ the set $\Sigma(g \mid A, r)$ of states for which Angel has a strategy to achieve a member of A after playing g with a probability not smaller than r as the next state. For example, $\Sigma(\gamma \mid A, r)$ is defined for the primitive game $\gamma \in \Gamma$ as $\{w \in W \mid \{\mu \mid \mu(A) \geq r\} \in N_\gamma(w)\}$, so we look at all worlds for which Angel can achieve a distribution which evaluates A not smaller than r. Similarly, we define $\Sigma(g^d \mid A, r)$ as $W \setminus \Sigma(g \mid W \setminus A, r)$, thus Demon can reach a state in A with probability greater than r iff Angel cannot reach a state in $W \setminus A$ with probability greater than r. Finally — and here t-measurability kicks in — we define for the composition $\gamma; g$ with the primitive game $\gamma \in \Gamma$ and game g the transformation $\Sigma(\gamma; g \mid A, r) := \{w \in W \mid Q_g(A, r) \in N_\gamma(w)\}$, where $Q_g(A, r) := \{\mu \in \mathbb{P}(W) \mid \int_0^1 \mu(\Sigma(g \mid A, s))\, ds \geq r\}$. For an explanation, assume that $\Sigma(g \mid A, r)$ is already defined for each r as the set of states for which Angel has a strategy to achieve a state in A through playing g with probability not smaller than r. Given a distribution μ over the states, the integral $\int_0^1 \mu(\Sigma(g \mid A, s))\, ds$ is the expected value for entering a state in A through playing g for μ. The set $Q_g(A, q)$ collects all distributions, the expected value of which is not smaller than q. We collect all states such that Angel has this set in its portfolio when playing γ in this state. Selecting this set from the portfolio means that, when playing γ and subsequently g, a state in A may be reached with probability not smaller than q.

These are just some salient points in the definition of the transformation. Other cases have to be defined, depending on the games' syntax, in particular, $\Sigma(g^* \mid A, r)$ has to be determined; the details are outlined in [3, Section 4.9.4]. We have

Theorem: *If the measurable space W is complete, then $\Sigma(g \mid \cdot, r)$ transforms measurable sets into measurable sets.* $\dashv$

The reason why we need a complete measurable space here is that $\Sigma(g^* \mid A, r)$ involves some unpleasant uncountable Boolean operations, under which, however, this class of spaces is closed.

With this in mind, we can define an interpretation for modal formulas inductively through $[\![\langle g\rangle_r \varphi]\!] := \Sigma(g \mid [\![\varphi]\!], r)$, starting from some assignment of primitive propositions to measurable sets. It follows that each validity set is measurable, provided W is complete.

As in the set-valued case above, we have this property.

Proposition: *If the interpretation is Kripke generated, then it is disjunctive.* $\dashv$

Suppose that we consider only Angel's moves and forget about Demon. Then we have the PDL-fragment of game logic, which is somewhat easier to interpret. It turns out that the interpretation suggested here generalizes the known interpretations from [5,1].

Proposition: *A Kripke generated interpretation coincides on the PDL fragment with the one defined through Kleisli composition in the Giry monad.* $\dashv$

Thus the composition of programs can be described in an equivalent way through the convolution of Markov transition kernels.

Now, What? Well, it is interesting to investigate expressivity, i.e., the relationship of logical equivalence, bisimilarity and behavioral equivalence for these models. These properties have to be defined for stochastic effectivity functions (partial suggestions have been proposed in [4,2]). It would also be interesting to know whether simpler models of stochastic nondeterminism can be used for an interpretation, which would have to support the composition of games; a monad would be nice.

References

1. Doberkat, E.-E.: A stochastic interpretation of propositional dynamic logic: Expressivity. J. Symb. Logic 77(2), 687–716 (2012)
2. Doberkat, E.-E.: Algebraic properties of stochastic effectivity functions. J. Logic and Algebraic Progr. 83, 339–358 (2014)
3. Doberkat, E.-E.: Special Topics in Mathematics for Computer Science: Sets, Categories, Topologies, Measures. Springer (in print, 2015)
4. Doberkat, E.-E., Sànchez Terraf, P.: Stochastic nondeterminism and effectivity functions. J. Logic and Computation (in print, 2015) (arxiv: 1405.7141)
5. Kozen, D.: A probabilistic PDL. J. Comp. Syst. Sci. 30(2), 162–178 (1985)
6. Parikh, R.: The logic of games and its applications. In: Karpinski, M., van Leeuwen, J. (eds.) Topics in the Theory of Computation, vol. 24, pp. 111–140. Elsevier (1985)
7. Pauly, M., Parikh, R.: Game logic — an overview. Studia Logica 75, 165–182 (2003)

Theoretical Foundations

Completeness and Incompleteness in Nominal Kleene Algebra*

Dexter Kozen[1], Konstantinos Mamouras[1], and Alexandra Silva[2]

[1] Computer Science Department, Cornell University
{kozen,mamouras}@cs.cornell.edu
[2] Intelligent Systems, Radboud University Nijmegen
alexandra@cs.ru.nl

Abstract. Gabbay and Ciancia (2011) presented a nominal extension of Kleene algebra as a framework for trace semantics with statically scoped allocation of resources, along with a semantics consisting of nominal languages. They also provided an axiomatization that captures the behavior of the scoping operator and its interaction with the Kleene algebra operators and proved soundness over nominal languages. In this paper, we show that the axioms proposed by Gabbay and Ciancia are not complete over the semantic interpretation they propose. We then identify a slightly wider class of language models over which they are sound and complete.

1 Introduction

Nominal sets are a convenient framework for handling name generation and binding. They were introduced by Gabbay and Pitts [5] as a mathematical model of name binding and α-conversion.

Nominal extensions of classical automata theory have been explored quite recently [1], motivated by the increasing need for tools for languages over infinite alphabets. These play a role in various areas, including XML document processing, cryptography, and verification. An XML document can be seen as a tree with labels from the (infinite) set of all unicode strings that can appear as attribute values. In cryptography, infinite alphabets are used as *nonces*, names used only once in cryptographic communications to prevent replay attacks. In software verification, infinite alphabets are used for references, objects, pointers, and function parameters.

In this paper, we focus on axiomatizations of regular languages and how these can be lifted in the presence of a binding operator and an infinite alphabet of names. This work builds on the recent work of Gabbay and Ciancia [8], who presented a nominal extension of Kleene algebra as a framework for trace semantics with statically scoped allocation of resources, along with a semantics consisting of *nominal languages*. Gabbay and Ciancia also provided an axiomatization that captures the behavior of the scoping operator and its interaction with the usual Kleene algebra operators. They proved soundness of their axiomatization over

* This work was done while visiting Radboud University Nijmegen.

W. Kahl et al. (Eds.): RAMiCS 2015, LNCS 9348, pp. 51–66, 2015.
DOI: 10.1007/978-3-319-24704-5_4

nominal languages, but left open the question of completeness. In this paper we address this problem.

Intuitively, the challenge behind showing completeness is twofold. On one hand, one needs to find the appropriate (language) model, or in other words, the free model. On the other hand, there is a need to find an appropriate *normal form* for a given expression. Normal forms are a vehicle to completeness: two expressions are equivalent if they can be reduced to the same normal form, and the axioms are complete if they enable us to derive normal forms for all expressions.

Our approach is modular. We show that under the right definition of a language model, one can prove completeness by first transforming each expression to another expression for which only the usual Kleene algebra axioms are needed. The steps of the transformation make use of the usual axioms of Kleene algebra along with axioms proposed by Gabbay and Ciancia for the scoping operator.

We also show that the axioms are not complete for the language model proposed by Gabbay and Ciancia. We explain exactly what the problem is with their original language model, which contains what they called *non-maximal planes*. This technical difference will be clear later in the paper. We also show that the axioms are not complete for summation models in which the scoping operator is interpreted as a summation operator over a fixed set.

In devising the proof of completeness, we have developed a novel technique that might be useful in other completeness proofs. More precisely, we have made use of the well known fact that the Boolean algebra generated by finitely many regular sets consists of regular sets and is atomic. Hence, expressions can be written as sums of atoms. This is crucial in obtaining the normal form. To our knowledge this has not been used before in completeness proofs.

The paper is organized as follows. In §2 we recall basic material on Kleene algebra (KA), nominal sets, and the nominal extension of KA (NKA) of Gabbay and Ciancia. In §3 we discuss the possible language models, starting with the original one proposed in [8] and then introducing two new ones: our own alternative language model and the summation models. We give a precise description of the difference between the two language models. In §4 we present our main result on completeness. The completeness proof is given in four steps: *exposing bound variables*, *scope configuration*, *canonical choice of bound variables*, and *semilattice identities*. In §5 we present concluding remarks and directions for future work.

2 Background

In this section we review basic background material on Kleene algebra (KA), nominal sets, and the nominal extension of KA (NKA) of Gabbay and Ciancia [8]. For a more thorough introduction, the reader is referred to [7,12] for nominal sets, to [14] for Kleene (co)algebra, and to [8] for NKA.

2.1 Kleene Algebra (KA)

Kleene algebra is the algebra of regular expressions. Regular expressions are normally interpreted as regular sets of strings, but there are other useful interpretations: binary relation models used in programming language semantics, the (min, +) algebra used in shortest path algorithms, models consisting of convex sets used in computational geometry, and many others.

A *Kleene algebra* is any structure $(K, +, \cdot, ^*, 0, 1)$ where K is a set, $+$ and $\cdot$ are binary operations on K, * is a unary operation on K, and 0 and 1 are constants, satisfying the following axioms:

$$\begin{array}{lll} x + (y + z) = (x + y) + z & x(yz) = (xy)z & x + y = y + x \\ 1x = x1 = x & x + 0 = x + x = x & x0 = 0x = 0 \\ x(y + z) = xy + xz & (x + y)z = xz + yz & 1 + xx^* \leq x^* \\ y + xz \leq z \Rightarrow x^*y \leq z & y + zx \leq z \Rightarrow yx^* \leq z & 1 + x^*x \leq x^* \end{array}$$

where we define $x \leq y$ iff $x + y = y$. The axioms above not involving * are succinctly stated by saying that the structure is an idempotent semiring under $+, \cdot, 0$, and 1, the term *idempotent* referring to the axiom $x + x = x$. Due to this axiom, the ordering relation $\leq$ is a partial order. The axioms for * together say that x^*y is the $\leq$-least z such that $y + xz \leq z$ and yx^* is the $\leq$-least z such that $y + zx \leq z$.

2.2 Group Action

A *group action* of a group G on a set X is a map $G \times X \to X$, written as juxtaposition, such that $\pi(\rho x) = (\pi\rho)x$ and $1x = x$. For $x \in X$ and $A \subseteq X$, define the subgroups

$$\mathsf{fix}\, x = \{\pi \in G \mid \pi x = x\} \qquad \mathsf{Fix}\, A = \bigcap_{x \in A} \mathsf{fix}\, x = \{\pi \in G \mid \forall x \in A\ \pi x = x\}.$$

Note that $\mathsf{fix}\, A = \{\pi \in G \mid \pi A = A\}$, thus $\mathsf{Fix}\, A$ and $\mathsf{fix}\, A$ are different: they are the subgroups of G that fix A pointwise and setwise, respectively.

A *G-set* is a set X equipped with a group action $G \times X \to X$. A function $f : X \to Y$ between G-sets is called *equivariant* if $f \circ \pi = \pi \circ f$ for all $\pi \in G$.

2.3 Nominal Sets

Let $\mathbb{A}$ be a countably infinite set of *atoms* and let G be the group of all finite permutations of $\mathbb{A}$ (permutations generated by transpositions $(a\ b)$). The group G acts on $\mathbb{A}$ in the obvious way, making $\mathbb{A}$ into a G-set. If X is another G-set, we say that $A \subseteq \mathbb{A}$ *supports* $x \in X$ if $\mathsf{Fix}\, A \subseteq \mathsf{fix}\, x$. An element $x \in X$ has *finite support* if there is a finite set $A \subseteq \mathbb{A}$ that supports x. A *nominal set* is a G-set X such that every element of X has finite support.

It can be shown that if $A, B \subseteq \mathbb{A}$ and $A \cup B \neq \mathbb{A}$, then $\mathsf{Fix}(A \cap B)$ is the least subgroup of G containing both $\mathsf{Fix}\, A$ and $\mathsf{Fix}\, B$. Thus if A and B are finite and support x, then so does $A \cap B$. It follows that if x is finitely supported, there is a smallest set that supports it, which we call $\mathsf{supp}\, x$. Moreover, one can show that A supports x iff πA supports πx. In particular, $\mathsf{supp}\, \pi x = \pi\, \mathsf{supp}\, x$. Also, for $x \in X$, $\mathsf{Fix}\,\mathsf{supp}\, x \subseteq \mathsf{fix}\, x \subseteq \mathsf{fix}\,\mathsf{supp}\, x$. Both inclusions can be strict.

We write $a\#x$ and say a is *fresh for* x if $a \notin \mathsf{supp}\, x$.

2.4 Syntax of Nominal KA

NKA expressions over an alphabet Σ of primitive letters are

$$e ::= a \in \Sigma \mid e + e \mid ee \mid e^* \mid 0 \mid 1 \mid \nu a.e.$$

The scope of the binding νa in $\nu a.e$ is e. The precedence of the binding operator νa is lower than product but higher than sum; thus in products, scopes extend as far to the right as possible. For example, $\nu a.ab\ \nu b.ba$ should be read as $\nu a.(ab\ \nu b.(ba))$ and not $(\nu a.ab)(\nu b.ba)$. The set of NKA expressions over Σ is denoted Exp_Σ.

A *ν-string* is an expression with no occurrence of $+$, *, or 0, and no occurrence of 1 except to denote the null string, in which case we use ε instead:

$$x ::= a \in \Sigma \mid xx \mid \varepsilon \mid \nu a.x.$$

The set of ν-strings over Σ is denoted Σ^ν.

The *free variables* $\mathsf{FV}(e)$ of an expression or ν-string e are defined inductively as usual. We write $e[a/x]$ for the result of substituting a for variable x in e.

The nominal axioms proposed by Gabbay and Ciancia [8] are:

$$\begin{aligned} &\nu a.(d+e) = \nu a.d + \nu a.e && a\#e \Rightarrow \nu b.e = \nu a.(a\ b)e \\ &\nu a.\nu b.e = \nu b.\nu a.e && a\#e \Rightarrow (\nu a.d)e = \nu a.de \\ &a\#e \Rightarrow \nu a.e = e && a\#e \Rightarrow e(\nu a.d) = \nu a.ed. \end{aligned} \tag{1}$$

3 Models

3.1 Nominal KA

A *nominal Kleene algebra* (NKA) over atoms $\mathbb{A}$ is a structure $(K, +, \cdot, ^*, 0, 1, \nu)$ with binding operation $\nu : \mathbb{A} \times K \to K$ such that K is a nominal set over atoms $\mathbb{A}$, the KA operations and ν are equivariant in the sense that

$$\begin{aligned} &\pi(x+y) = \pi x + \pi y && \pi(xy) = (\pi x)(\pi y) && \pi 0 = 0 \\ &\pi(x^*) = (\pi x)^* && \pi(\nu a.e) = \nu(\pi a).(\pi e) && \pi 1 = 1 \end{aligned}$$

for all $\pi \in G$ (that is, the action of every $\pi \in G$ is an automorphism of K), and all the KA and nominal axioms are satisfied.

3.2 Nominal Language Model

Now we describe a nominal language interpretation $NL : \mathsf{Exp}_{\mathbb{A}} \to \mathcal{P}(\mathbb{A}^*)$ for each expression e that interprets expressions over $\mathbb{A}$ as certain subsets of $\mathbb{A}^*$. This is the language model of [8]. The definition is slightly nonstandard, as care must be taken when defining product to avoid capture.

First we give an intermediate interpretation $I : \mathsf{Exp}_{\mathbb{A}} \to \mathcal{P}(\mathbb{A}^\nu)$ of expressions as sets of ν-strings over $\mathbb{A}$. The regular operators $+$, $\cdot$, *, 0, and 1 have their usual set-theoretic interpretations, and

$$I(\nu a.e) = \{\nu a.x \mid x \in I(e)\} \qquad\qquad I(a) = \{a\}.$$

We maintain the scoping of ν-subexpressions in the ν-strings. Examples:

$$\begin{aligned}
I(\nu a.a) &= \{\nu a.a\} \\
I(\nu a.\nu b.(a+b)) &= \{\nu a.\nu b.a, \nu a.\nu b.b\} \\
I(\nu a.(\nu b.ab)(a+b)) &= \{\nu a.(\nu b.ab)a, \nu a.(\nu b.ab)b\} \\
I(\nu a.(ab)^*) &= \{\nu a.\varepsilon, \nu a.ab, \nu a.abab, \nu a.ababab, \ldots\} \\
I((\nu a.ab)^*) &= \{\varepsilon, \nu a.ab, (\nu a.ab)(\nu a.ab), (\nu a.ab)(\nu a.ab)(\nu a.ab), \ldots\}.
\end{aligned}$$

Now we describe the map $NL : \mathbb{A}^\nu \to \mathcal{P}(\mathbb{A}^*)$ on ν-strings. Given a ν-string x, first α-convert so that all bindings in x are distinct and different from all free variables in x, then delete all binding operators νa to obtain a string $x' \in \mathbb{A}^*$. For example, $(\nu a.ab)(\nu a.ab)(\nu a.ab)' = abcbdb$. Here we have α-converted to obtain $(\nu a.ab)(\nu c.cb)(\nu d.db)$, then deleted the binding operators to obtain $abcbdb$. The choice of variables in the α-conversion does not matter as long as they are distinct and different from the free variables.

Now we define for each ν-string x and expression e

$$NL(x) = \{\pi x' \mid \pi \in \mathsf{Fix}\,\mathsf{FV}(x)\} \qquad\qquad NL(e) = \bigcup_{x \in I(e)} NL(x).$$

The set $NL(x)$ is the plane $x' \circlearrowright_{\mathsf{FV}(x)}$ in the notation of [8]. Thus we let the bound variables range simultaneously over all possible values in $\mathbb{A}$ they could take on, as long as they remain distinct and different from the free variables, and we accumulate all strings obtained in this way. For example,

$$NL((\nu a.ab)(\nu a.ab)(\nu a.ab)) = \{abcbdb \mid a, c, d \in \mathbb{A} \text{ distinct and different from } b\}.$$

As mentioned, the fresh variables used in the α-conversion does not matter, thus

$$NL(x) = \{\pi y \mid \pi \in \mathsf{Fix}\,\mathsf{FV}(x)\} \tag{2}$$

for any $y \in NL(x)$.

For $x, y \in \mathbb{A}^\nu$, write $x \equiv y$ if x and y are equivalent modulo the nominal axioms (1). The following lemma says that the nominal axioms alone are sound and complete for equivalence between ν-strings in the nominal language model.

Lemma 1. *For $x, y \in \mathbb{A}^\nu$, $x \equiv y$ if and only if $NL(x) = NL(y)$.*

Proof. Soundness (the left-to-right implication) holds because each nominal axiom preserves NL, as is not difficult to check. For completeness (the right-to-left implication), suppose $NL(x) = NL(y)$. We must have $\mathsf{FV}(x) = \mathsf{FV}(y)$, because if $a \in \mathsf{FV}(x) - \mathsf{FV}(y)$, then $NL(y)$ would contain a string with no occurrence of a, whereas all strings in $NL(x)$ contain an occurrence of a. Now α-convert x and y so that all bound variables are distinct and different from the free variables, and move the bound variables to the front, so that $x = \nu A.x'$ and $y = \nu B.y'$ for some $x', y' \in \mathbb{A}^*$. By (2), $y' = \pi x'$ for some $\pi \in \mathsf{Fix}\,\mathsf{FV}(x) = \mathsf{Fix}\,\mathsf{FV}(y)$, so $x = \pi y$, and $\pi y \equiv y$ by α-conversion. □

Lemma 2. *For any $x \in \mathbb{A}^*$ and $A, B \subseteq \mathsf{FV}(x)$,*

$$A \subseteq B \Leftrightarrow NL(\nu A.x) \subseteq NL(\nu B.x)$$

(in the notation of [8], $A \subseteq B \Leftrightarrow x\circlearrowleft_{B'} \subseteq x\circlearrowleft_{A'}$, where $A' = \mathsf{FV}(x) - A$ and $B' = \mathsf{FV}(x) - B$).

Proof. If $A \subseteq B$, then $\mathsf{Fix}\,A' \subseteq \mathsf{Fix}\,B'$, therefore

$$NL(\nu A.x) = \{\pi x \mid \pi \in \mathsf{Fix}\,A'\} \subseteq \{\pi x \mid \pi \in \mathsf{Fix}\,B'\} = NL(\nu B.x).$$

Conversely, if $a \in A - B$, then $x[b/a] \in NL(\nu A.x) - NL(\nu B.x)$, where b is any element of $\mathbb{A} - \mathsf{FV}(x)$. □

Lemma 3. *Let $y \in NL(e)$ and $A \subseteq \mathsf{FV}(y)$ maximal such that $NL(\nu A.y) \subseteq NL(e)$ (in the notation of [8], this is $y\circlearrowleft_{A'} \propto NL(e)$, where $A' = \mathsf{FV}(y) - A$). Then $\nu A.y \in I(e)$, and $\nu A.y$ is the unique ν-string up to nominal equivalence for which this is true.*

Remark 1. This is the essential content of [8, Theorem 3.16]. This is important for us because it says that the set $NL(e)$ uniquely determines the maximal elements of $I(e)$ up to nominal equivalence (Lemma 4 below).

Proof. Let $x_1, \ldots, x_n \in I(e)$ be all ν-strings such that $y \in NL(x_i)$. There are only finitely many of these. Then

$$NL(\nu A.y) \subseteq NL(x_1) \cup \cdots \cup NL(x_n) \subseteq NL(e).$$

Using the nominal axioms (1), we can move the quantification in each x_i to the front of the string and α-convert so that the quantifier-free part is y. This is possible because $y \in NL(x_i)$. Thus we can assume without loss of generality that each $x_i = \nu A_i.y$ for some $A_i \subseteq \mathsf{FV}(y)$.

Let $z \in NL(\nu A.y)$ such that $(\mathsf{FV}(z) - \mathsf{FV}(\nu A.y)) \cap \mathsf{FV}(\nu A_i.y) = \emptyset$, $1 \le i \le n$. Since

$$NL(\nu A.y) \subseteq NL(x_1) \cup \cdots \cup NL(x_n) = NL(\nu A_1.y) \cup \cdots \cup NL(\nu A_n.y),$$

we must have $z \in NL(\nu A_i.y)$ for some i. But then $\mathsf{FV}(\nu A.y), \mathsf{FV}(\nu A_i.y) \subseteq \mathsf{FV}(z)$ and $\mathsf{FV}(\nu A_i.y) \subseteq \mathsf{FV}(\nu A.y)$ by choice of z, therefore $A \subseteq A_i$. Since A was maximal, $A = A_i$. □

Let $\hat{I}(e) = \{x \in I(e) \mid NL(x) \text{ is maximal in } NL(e)\}$.

Lemma 4. *$NL(e_1) = NL(e_2)$ if and only if $\hat{I}(e_1) = \hat{I}(e_2)$ modulo the nominal axioms* (1).

Proof. Suppose $NL(e_1) = NL(e_2)$. By Lemma 3, each $y \in NL(e_1)$ is contained in a unique maximal $NL(\nu A.y)$, and $\nu A.y \in \hat{I}(e_1)$. As $NL(e_1) = NL(e_2)$, these planes are also contained in $NL(e_2)$. Similarly, the maximal planes of $NL(e_2)$ are contained in $NL(e_1)$. Since the two sets contain the same set of maximal planes, they must be equal, therefore $\hat{I}(e_1) = \hat{I}(e_2)$ modulo the nominal axioms.

For the reverse implication, note that

$$NL(e) = \bigcup_{x \in I(e)} NL(x) = \bigcup_{x \in \hat{I}(e)} NL(x)$$

by the fact that every plane of e is contained in a maximal one. Then

$$NL(e_1) = \bigcup_{x \in \hat{I}(e_1)} NL(x) = \bigcup_{x \in \hat{I}(e_2)} NL(x) = NL(e_2).$$

□

3.3 Alternative Nominal Language Model

Let Σ and $\mathbb{A}$ be countably infinite disjoint sets. Letters $a, b, c, \ldots$ range over $\mathbb{A}$, $x, y, z, \ldots$ over Σ, and $u, v, w, \ldots$ over $(\Sigma \cup \mathbb{A})^*$. Quantification is only over Σ.

A *language* is a subset $A \subseteq (\Sigma \cup \mathbb{A})^*$ such that $\pi A = A$ for all $\pi \in G$. The set of languages is denoted $\mathcal{L}$.

The operations of nominal KA are defined on $\mathcal{L}$ as follows:

$$A + B = A \cup B \quad AB = \{uv \mid u \in A,\ v \in B,\ \mathsf{FV}(u) \cap \mathsf{FV}(v) \cap \mathbb{A} = \emptyset\} \quad 0 = \emptyset$$

$$A^* = \bigcup_n A^n \qquad \nu x.A = \{w[a/x] \mid w \in A,\ a \in \mathbb{A} - \mathsf{FV}(w)\},\ x \in \Sigma \qquad 1 = \{\varepsilon\}.$$

Lemma 5. *The set $\mathcal{L}$ is closed under the operations of nominal KA.*

Proof. For sum, $\pi(\bigcup_n A_n) = \bigcup_n \pi A_n = \bigcup_n A_n$. For product,

$$\begin{aligned}
\pi(AB) &= \{\pi(uv) \mid u \in A,\ v \in B,\ \mathsf{FV}(u) \cap \mathsf{FV}(v) \cap \mathbb{A} = \emptyset\} \\
&= \{(\pi u)(\pi v) \mid u \in A,\ v \in B,\ \mathsf{FV}(\pi u) \cap \mathsf{FV}(\pi v) \cap \pi\mathbb{A} = \emptyset\} \\
&= \{uv \mid u \in \pi A,\ v \in \pi B,\ \mathsf{FV}(u) \cap \mathsf{FV}(v) \cap \mathbb{A} = \emptyset\} \\
&= (\pi A)(\pi B) = AB.
\end{aligned}$$

The case of A^* follows from the previous two cases. The cases of 0 and 1 are trivial. Finally, for $\nu x.A$, we have

$$\begin{aligned}
\pi(\nu x.A) &= \{\pi(w[a/x]) \mid w \in A,\ a \in \mathbb{A} - \mathsf{FV}(w)\} \\
&= \{(\pi w)[\pi a/x] \mid w \in A,\ a \in \mathbb{A} - \mathsf{FV}(w)\} \\
&= \{w[a/x] \mid \pi^{-1} w \in A,\ \pi^{-1} a \in \mathbb{A} - \mathsf{FV}(\pi^{-1} w)\} \\
&= \{w[a/x] \mid w \in \pi A,\ a \in \pi\mathbb{A} - \pi\mathsf{FV}(\pi^{-1} w)\} \\
&= \{w[a/x] \mid w \in A,\ a \in \mathbb{A} - \mathsf{FV}(w)\} = \nu x.A.
\end{aligned}$$

□

We can interpret nominal KA expressions as languages in $\mathcal{L}$. The interpretation map $AL : \mathsf{Exp}_\Sigma \to \mathcal{L}$ is the unique homomorphism with respect to the above language operations such that $AL(x) = \{x\}$. Note that in this context, atoms $a \in \mathbb{A}$ do not appear in expressions or ν-strings.

Theorem 1. *The nominal axioms* (1) *hold in this model.*

The proof is long but not conceptually difficult.

We can also define $I : \mathsf{Exp}_\Sigma \to \Sigma^\nu$ and $\hat{I} : \mathsf{Exp}_\Sigma \to \Sigma^\nu$ exactly as in §3.2 for the nominal language model, with the modification that expressions are over Σ and not $\mathbb{A}$.

Lemma 6. $AL(e) = \bigcup_{w\in I(e)} AL(w)$.

Proof. This can be proved by a straightforward induction on the structure of e. We argue the case of products and binders explicitly.

$$\begin{aligned}
AL(e_1e_2) &= \{uv \mid u \in AL(e_1),\ v \in AL(e_2),\ \mathsf{FV}(u) \cap \mathsf{FV}(v) \cap \mathbb{A} = \emptyset\}\\
&= \{uv \mid u \in \bigcup_{p\in I(e_1)} AL(p),\ v \in \bigcup_{q\in I(e_2)} AL(q),\ \mathsf{FV}(u) \cap \mathsf{FV}(v) \cap \mathbb{A} = \emptyset\}\\
&= \bigcup_{\substack{p\in I(e_1)\\ q\in I(e_2)}} \{uv \mid u \in AL(p),\ v \in AL(q),\ \mathsf{FV}(u) \cap \mathsf{FV}(v) \cap \mathbb{A} = \emptyset\}\\
&= \bigcup_{\substack{p\in I(e_1)\\ q\in I(e_2)}} AL(pq) = \bigcup_{r\in I(e_1e_2)} AL(r).
\end{aligned}$$

$$\begin{aligned}
AL(\nu x.e) &= \nu x.AL(e)\\
&= \{w[a/x] \mid w \in AL(e),\ a \in \mathbb{A} - \mathsf{FV}(w)\}\\
&= \{w[a/x] \mid w \in \bigcup_{p\in I(e)} AL(p),\ a \in \mathbb{A} - \mathsf{FV}(w)\}\\
&= \bigcup_{p\in I(e)} \{w[a/x] \mid w \in AL(p),\ a \in \mathbb{A} - \mathsf{FV}(w)\}\\
&= \bigcup_{p\in I(e)} \nu x.AL(p) = \bigcup_{p\in I(e)} AL(\nu x.p) = \bigcup_{w\in I(\nu x.e)} AL(w).
\end{aligned}$$

□

Lemma 7. *Every plane $AL(\nu A.w)$ in $AL(e)$ is maximal; that is, $I(e) = \hat{I}(e)$.*

Proof. Replace each $x \in A$ in w with a distinct element of $\mathbb{A}$ to get w'. Then $AL(\nu A.w) = \{\pi w' \mid \pi \in G\}$. This is maximal, as all finite permutations of $\mathbb{A}$ are allowed. □

Lemma 7 characterizes the key difference between the nominal language model of [8] described in §3.2 and the alternative nominal language model of this section. It explains why the axioms are complete for the alternative model but not for the model of §3.2. In the model of §3.2, there are non-maximal planes, and these are "hidden" by the maximal planes, whereas this cannot happen in the alternative model, as all planes are maximal.

3.4 Summation Models

There are several other interesting models in which ν is interpreted as some form of summation operator: a summation model over the free KA, a summation model over languages, a summation model over an arbitrary KA, and an evaluation model. The axioms are sound over these models, but incomplete for other reasons.

4 Completeness

In this section we prove our main theorem:

Theorem 2. *The axioms of nominal Kleene algebra are sound and complete for the equational theory of nominal Kleene algebras and for the equational theory of the alternative language interpretation of §3.3.*

We thus show that if two nominal KA expressions e_1 and e_2 are equivalent in the alternative language interpretation of §3.3 in the sense that $AL(e_1) = AL(e_2)$, then e_1 and e_2 are provably equivalent in the axiomatization of Gabbay and Ciancia [8]. This says that the alternative language model of §3.3 is the free nominal KA. This is not true of Gabbay and Ciancia's language model presented in §3.2, as the inequality $a \leq \nu a.a$ holds in the language model of §3.2 but not in the summation models. Neither is it true of the summation models of §3.4, as $\nu a.aa \leq \nu a.\nu b.ab$ holds in the summation models but not in the language model. However, it is true of Gabbay and Ciancia's language model if one restricts to closed terms, as the closed terms of the language models of §3.2 and §3.3 are the same.

We show that every expression can be put into a particular canonical form that will allow us to apply the KA axioms to prove equivalence. This construction will consist of several steps: *exposing bound variables*, *scope configuration*, *canonical choice of bound variables*, and *determining semilattice identities*. Each step will involve a construction that is justified by the axioms.

For the purposes of exposition, we write $\underset{a}{(}\,e\,\underset{a}{)}$ instead of $\nu a.e$ so that it is easier to see the scope boundaries. In this notation, the nominal axioms take the following form:

$$\nu a.(d+e) = \nu a.d + \nu a.e \qquad (_a\, d+e\,)_a = (_a\, d\,)_a + (_a\, e\,)_a \tag{3}$$

$$\nu a.\nu b.e = \nu b.\nu a.e \qquad (_a (_b\, e\,)_b)_a = (_b (_a\, e\,)_a)_b \tag{4}$$

$$a\#e \Rightarrow \nu a.e = e \qquad a\#e \Rightarrow (_a\, e\,)_a = e \tag{5}$$

$$a\#e \Rightarrow \nu b.e = \nu a.(a\ b)e \qquad a\#e \Rightarrow (_b\, e\,)_b = (_a (a\ b)e\,)_a \tag{6}$$

$$a\#e \Rightarrow (\nu a.d)e = \nu a.de \qquad a\#e \Rightarrow (_a\, d\,)_a\, e = (_a\, de\,)_a \tag{7}$$

$$a\#e \Rightarrow e(\nu a.d) = \nu a.ed \qquad a\#e \Rightarrow e\, (_a\, d\,)_a = (_a\, ed\,)_a\, . \tag{8}$$

We remark that writing scope boundaries of ν-expressions as letters $(_a$ and $)_a$ is merely a notational convenience. Although it appears to allow us to violate the invariant that starred expressions and ν-expressions are mutually well-nested, in reality this is not an issue, as all our transformations are justified by the axioms, which maintain this invariant.

4.1 Exposing Bound Variables

A ν^**-string* is a string of

- letters a,
- well-nested scope delimiters $(_a$ and $)_a$, and
- starred expressions e^* whose bodies e are (inductively) sums of ν^*-strings.

We say that the bound variables of a ν^*-string are *exposed* if

(i) the first and last occurrence of each bound variable occur at the top level in the scope of their binding operator,[1] and
(ii) the bound variables of all ν^*-strings in the bodies of starred subexpressions are (inductively) exposed.

A typical ν^*-string is $(_a (_b\, abb(ab\, (_a\, ab\,)_a + b\, (_b\, ba\,)_b)^* ba\,)_b)_a$. The bound variables are exposed in this expression because the first and last occurrences of a and b occur at the top level. Inside the starred subexpression, the bound variables in the two ν^*-strings are exposed because there are no starred subexpressions.

Lemma 8. *Every expression can be written as a sum of ν^*-strings whose bound variables are exposed.*

Proof. It is straightforward to see how to use the nominal axiom (3) in the left-to-right direction and the distributivity and 0 and 1 laws of Kleene algebra to write every expression as a sum of ν^*-strings.

[1] "Top level" means not inside a starred subexpression. Inside a starred expression e^*, "top level" means not inside a starred subexpression of e.

Exposing the bound variables is a little more difficult. It may appear at first glance that one can simply unwind e^* as $1+e+ee^*e$ and then unwind the starred subexpressions of e inductively, but this is not enough. For example,

$$\begin{aligned}(a+b)^* &= 1+a+b+(a+b)(a+b)^*(a+b)\\ &= 1+a+b+a(a+b)^*a+a(a+b)^*b+b(a+b)^*a+b(a+b)^*b,\end{aligned}$$

and the subexpression $a(a+b)^*a$ does not satisfy (i). The following more complicated expression is needed:

$$\begin{aligned}(a+b)^* &= 1+a+b+aa^*a+bb^*b+ab+ba && (9)\\ &\quad + aa^*ab+aa^*ba^*a+baa^*a+abb^*b+bb^*ab^*b+bb^*ba && (10)\\ &\quad + aa^*abb^*b+aa^*b(a+b)^*ab^*b+aa^*b(a+b)^*ba^*a && (11)\\ &\quad + bb^*a(a+b)^*ab^*b+bb^*a(a+b)^*ba^*a+bb^*baa^*a && (12)\end{aligned}$$

Line (9) covers strings containing no a's or no b's or one of each. Line (10) covers strings containing one a and two or more or more b's or one b and two or more or more a's. Lines (11) and (12) cover strings containing at least two a's and at least two b's.

For the general construction, we first argue the case of $(a_1+\cdots+a_n)^*$. Write down all strings containing either zero, one, or two occurrences of each letter. For each such string, insert a starred subexpression in each gap between adjacent letters. The body of the starred expression inserted into a gap will be the sum of all letters a such that the gap falls between two occurrences of a.

For example, the second term of (11) is obtained from the string $abab$. There are three gaps, into which we insert the indicated starred expressions:

$$\begin{array}{ccccccc} a & & b & & a & & b\\ & \uparrow & & \uparrow & & \uparrow & \\ & a^* & & (a+b)^* & & b^* & \end{array}$$

In the first gap we inserted a^* because the gap falls between two occurrences of a but not between two occurrences of b. In the second gap we inserted $(a+b)^*$ because the gap falls between two occurrences of a and two occurrences of b.

This construction covers all strings whose first and last occurrences of each letter occur in the order specified by the original string before the insertion. If a letter occurs twice before the insertion, then after the insertion those two occurrences are the first and last, and they occur at the top level. If a letter occurs once before the insertion, then that is the only occurrence after the insertion, and it is at the top level. If a letter does not occur at all before the insertion, then it does not occur after.

For the general case e^*, we first perform the construction inductively on all starred subexpressions of e, writing $e^*=(e_1+\cdots+e_n)^*$ where each top-level ν^*-string e_i satisfies (i) and (ii). Now take the sum constructed above for $(a_1+\cdots+a_n)^*$ and substitute e_i for a_i in all terms. This gives an expression of the desired form. □

4.2 Scope Configuration

For this part of the construction, we first α-convert using (6) to make all bound variables distinct and different from any free variable. This is called the *Barendregt variable convention.*

Now we transform each ν^*-string to ensure that every top-level left delimiter $\underset{a}{(}$ occurs immediately to the left of a free occurrence of a that it binds:

$$\cdots \underset{a}{(}\, a \cdots \underset{b}{(}\, b \cdots \underset{c}{(}\, c \cdots \underset{c}{)} \cdots \underset{b}{)} \cdots \underset{a}{)} \cdots \tag{13}$$

That occurrence is at the top level due to the preprocessing step of §4.1. We do this without changing the order of any occurrences of variables in the string, but we may change the order of quantification.

Starting at the left end of the string, scan right, looking for top-level left delimiters. For all top-level left delimiters that we see, push them to the right as long as we do not encounter a variable bound by any of them. Stop when such a variable is encountered. For example,

$$\cdots \underset{a}{(} \cdots \underset{b}{(} \cdots \underset{c}{(} \cdots b \cdots \underset{c}{)} \cdots \underset{b}{)} \cdots \underset{a}{)} \cdots \quad\Rightarrow\quad \cdots \underset{a}{(}\underset{b}{(}\underset{c}{(}\, b \cdots \underset{c}{)} \cdots \underset{b}{)} \cdots \underset{a}{)} \cdots$$

Here we are using the nominal axiom (8) in the right-to-left direction to skip over letters and starred expressions. If such a variable is encountered, it will be at the top level because of the preprocessing step of §4.1.

In this example, we must keep the $\underset{b}{(}$ to the left of that occurrence of b, but we wish to move the $\underset{a}{(}$ and $\underset{c}{(}$ past the b. The c can be moved in using (8), but to move the a in, we must exchange the order of quantification of a and b. To do this, we push the corresponding right delimiter of b up to the right delimiter of a using the nominal axiom (7) in the left-to-right direction.

$$\cdots \underset{a}{(}\underset{b}{(}\underset{c}{(}\, b \cdots \underset{c}{)} \cdots \underset{b}{)} \cdots \underset{a}{)} \cdots \quad\Rightarrow\quad \cdots \underset{a}{(}\underset{b}{(}\underset{c}{(}\, b \cdots \underset{c}{)} \cdots \underset{b}{)}\underset{a}{)} \cdots$$

This is always possible, as there is no free occurrence of b to the right of the $\underset{b}{)}$ due to the Barendregt variable convention. Now we can exchange the order of quantification using the nominal axiom (4).

$$\cdots \underset{a}{(}\underset{b}{(}\underset{c}{(}\, b \cdots \underset{c}{)} \cdots \underset{b}{)}\underset{a}{)} \cdots \quad\Rightarrow\quad \cdots \underset{b}{(}\underset{a}{(}\underset{c}{(}\, b \cdots \underset{c}{)} \cdots \underset{a}{)}\underset{b}{)} \cdots$$

This allows us to move the a and c in past the $\underset{b}{(}$ and continue.

$$\cdots \underset{b}{(}\underset{a}{(}\underset{c}{(}\, b \cdots \underset{c}{)} \cdots \underset{a}{)}\underset{b}{)} \cdots \quad\Rightarrow\quad \cdots \underset{b}{(}\, b\, \underset{a}{(}\underset{c}{(} \cdots \underset{c}{)} \cdots \underset{a}{)}\underset{b}{)} \cdots$$

When looking for the first occurrence of a free variable bound to a left delimiter, perhaps no free occurrence is encountered before seeing a right delimiter.

In this case there is no free occurrence of the variable in the scope of the binding, so we can just forget the binding altogether.

$$\cdots\underset{a}{(}\,\underset{b}{(}\,\underset{c}{(}\,\underset{c}{)}\cdots b\cdots\underset{b}{)}\cdots\underset{a}{)}\cdots \quad\Rightarrow\quad \cdots\underset{a}{(}\,\underset{b}{(}\cdots b\cdots\underset{b}{)}\cdots\underset{a}{)}\cdots$$

This uses the nominal axiom (5).

If there exists a free occurrence of a inside a scope $\underset{a}{(}\cdots\underset{a}{)}$, then the leftmost one occurs at the top level due to the construction of §4.1. Thus, when we are done, any remaining left delimiters $\underset{a}{(}$ in the string occur immediately to the left of a free occurrence of a that is bound to that delimiter, as illustrated in (13).

Now we finish up the construction by moving the right delimiters to the left as far as possible without exchanging order of quantification. Because of the preprocessing step of §4.1, the rightmost occurrence of any variable quantified at the top level occurs at the top level. Thus every right delimiter $\underset{a}{)}$ occurs either immediately to the right of an occurrence of a bound to that delimiter or immediately to the right of another right delimiter $\underset{b}{)}$ with smaller scope.

At this point we have transformed the expression so that every ν^*-string satisfies the following properties:

(i) every ν-subformula is of the form $\nu a.ae$; that is, the leftmost symbol of every scope is a variable bound by that scope; and
(ii) the rightmost boundary of every scope is as far to the left as possible, subject to (i).

The position of the scope delimiters is canonical, because scopes are as small as possible: the left delimiters are as far to the right as they can possibly be, and the right delimiters are as far to the left as they can possibly be given the positions of the left delimiters. It follows that if two expressions are equivalent, then they generate the same ν-strings up to renaming of bound variables.

4.3 Canonical Choice of Bound Variables

Now we would like to transform the expression so that the bound variables are chosen in a canonical way. This will ensure that if two expressions are equivalent, then they generate the same ν-strings, not just up to renaming of bound variables, but absolutely. This part of the construction will thus relax the Barendregt variable convention, so that variables can be bound more than once and can occur both bound and free in a string.

Choose a set of variables disjoint from the free variables of the expression and order them in some arbitrary but fixed order $a_0, a_1, \ldots$. Moving through the expression from left to right, maintain a stack of variable names corresponding to the scopes we are currently in. When a left scope delimiter $\underset{a}{(}$ is encountered, and we are inside the scope of n ν-formulas, the variables $a_0, \ldots, a_{n-1}$ will be on the stack. We rename the bound variable a to a_n using the nominal axiom (6) for α-conversion and push a_n onto the stack. When a right scope delimiter is encountered, we pop the stack. This construction guarantees that every ν-string generated by the expression satisfies:

- For every symbol in the string, if the symbol occurs in the scope of n nested ν-expressions, then those expressions bind variables $a_0, \ldots, a_{n-1}$ in that order from outermost to innermost scope.

It follows that two semantically equivalent expressions so transformed generate exactly the same set of ν-strings.

4.4 Determining Semilattice Identities

After transforming e_1 and e_2 by the above construction, we know that if e_1 and e_2 are equivalent, then they generate the same sets of ν-strings; that is, $I(e_1) = I(e_2)$. Now we wish to show that any two such expressions can be proved equivalent using the KA and nominal axioms in conjunction with the following congruence rule for ν-formulas:

$$\frac{e_1 = e_2}{\nu a.e_1 = \nu a.e_2}. \tag{14}$$

In order to do this, there is one more issue that must be resolved. Let us first assume for simplicity that e_1 and e_2 are of ν-depth one; that is, they only contain bindings of one variable a. There may be several subexpressions in e_1 and e_2 of the form $\nu a.d$, but all with the same variable a. We will relax this restriction later.

Any substring of the form $\nu a.x$ of a ν-string generated by e_1 or e_2 must be generated by a subexpression of the form $\nu a.d$. However, there may be several different subexpressions of this form, and the string $\nu a.x$ could be generated by more than one of them. In general, the sets of ν-strings generated by the ν-subexpressions could satisfy various semilattice identities, and we may have to know these identities in order to prove equivalence.

For example, consider the two expressions $c_1 + c_2$ and $d_1 + d_2 + d_3$, where

$$\begin{array}{lll} c_1 = \nu a.a(aa)^* & c_2 = \nu a.aa(aa)^* & \\ d_1 = \nu a.a(aaa)^* & d_2 = \nu a.aa(aaa)^* & d_3 = \nu a.aaa(aaa)^* \end{array} \tag{15}$$

(c_i generates strings with i mod 2 a's and d_i generates strings with i mod 3 a's). Both c_1+c_2 and $d_1+d_2+d_3$ generate all nonempty strings of a's, but in different ways. If $c_1 + c_2$ occurs in e_1 and $d_1 + d_2 + d_3$ occurs in e_2, we would have to know that they are equivalent to prove the equivalence of e_1 and e_2.

To determine all semilattice identities such as $c_1 + c_2 = d_1 + d_2 + d_3$ that hold among the ν-subexpressions, we express every ν-subexpression in e_1 or e_2 as a sum of atoms of the Boolean algebra on sets of ν-strings generated by these ν-subexpressions. In the example above, the atoms of the generated Boolean algebra are $b_i = \nu a.a^i(a^6)^*$, $1 \leq i \leq 6$ (b_i generates strings with i mod 6 a's). Rewriting the expressions (15) as sums of atoms, we would obtain

$$c_1 = b_1 + b_3 + b_5 \quad c_2 = b_2 + b_4 + b_6 \quad d_1 = b_1 + b_4 \quad d_2 = b_2 + b_5 \quad d_3 = b_3 + b_6.$$

The equivalences are provable in pure KA plus the nominal axiom (3). Then $c_1 + c_2$ and $d_1 + d_2 + d_3$ become

$$c_1 + c_2 = (b_1 + b_3 + b_5) + (b_2 + b_4 + b_6)$$
$$d_1 + d_2 + d_3 = (b_1 + b_4) + (b_2 + b_5) + (b_3 + b_6),$$

which are clearly equivalent.

Now we observe that any ν-string $\nu a.x$ generated by e_1 or e_2 is generated by exactly one atom. Moreover, if $\nu a.f$ is an atom and $\nu a.x \in I(\nu a.f)$, and if $\nu a.x$ is generated by $\nu a.f$ in the context $u(\nu a.x)v \in I(\nu a.e_1)$, then for any other $\nu a.y \in I(\nu a.f)$, we have $u(\nu a.y)v \in I(\nu a.e_1)$ as well. This says that we may treat $\nu a.f$ as atomic. In fact, once we have determined the atoms, if we like we may replace each atom $\nu a.f$ by a single letter $a_{\nu a.f}$ in e_1 and e_2, and the resulting expressions are equivalent, therefore provable. Then a proof of the two expressions with the letters $a_{\nu a.f}$ can be transformed back to a proof with the atoms $\nu a.f$ by simply substituting $\nu a.f$ for $a_{\nu a.f}$. However, note that it is not necessary to do the actual substitution; we can carry out the same proof on the original expressions with the $\nu a.f$.

For expressions of ν-depth greater than one, we simply perform the above construction inductively, innermost scopes first. We use the KA axioms and the semilattice identities on depth-n ν-subexpressions to determine the semilattice identities on depth-$(n-1)$ ν-subexpressions, then use the nominal axiom (3) and the rule (14) to prepare these semilattice identities for use on the next level.

This completes the proof of Theorem 2.

5 Conclusion

We have presented results on completeness and incompleteness of nominal Kleene algebra as introduced by Gabbay and Ciancia [8]. There are various directions for future work.

The normalization procedure presented in this paper yields a decision procedure that, although effective, is likely to be prohibitively expensive in practice due to combinatorial explosions in the preprocessing step of §4.1 and in the intersection of regular expressions in §4.4. In a companion paper [10], we have explored the coalgebraic theory of nominal Kleene algebra with the aim of developing a more efficient coalgebraic decision procedure, which would be of particular interest for the applications mentioned in the introduction. Coalgebraic decision procedures have been devised for the related systems KAT and NetKAT [2,4,13] and have proven quite successful in applications, and we suspect that a similar approach may bear fruit here.

Another interesting direction would be to follow recent work by Joanna Ochremiak [11] involving nominal sets over atoms equipped with both relational and algebraic structure. This is an extension of the original work of Gabbay and Pitts in which atoms can only be compared for equality.

The proof we have provided is concrete and does not explore the rich categorical structure of nominal sets. It would be interesting to rephrase the proof in more abstract terms, which would also be more amenable to generalizations such as those mentioned above.

Acknowledgments. We are grateful to Jamie Gabbay for bringing the original NKA paper to our attention. We would like to thank Filippo Bonchi, Paul Brunet, Helle Hvid Hansen, Bart Jacobs, Tadeusz Litak, Daniela Petrişan, Damien Pous, Ana Sokolova, and Fabio Zanasi for many stimulating discussions, comments, and suggestions. This research was performed at Radboud University Nijmegen and supported by the Dutch Research Foundation (NWO), project numbers 639.021.334 and 612.001.113, and by the National Security Agency.

References

1. Bojanczyk, M., Klin, B., Lasota, S.: Automata theory in nominal sets. Logical Methods in Computer Science 10(3) (2014)
2. Bonchi, F., Pous, D.: Checking NFA equivalence with bisimulations up to congruence. In: POPL 2013, pp. 457–468 (January 2013)
3. Fernández, M., Gabbay, M.J.: Nominal rewriting with name generation: abstraction vs. locality. In: PPDP 2005. ACM Press (July 2005)
4. Foster, N., Kozen, D., Milano, M., Silva, A., Thompson, L.: A coalgebraic decision procedure for NetKAT. In: POPL 2015, Mumbai, India, pp. 343–355 (January 2015)
5. Gabbay, M., Pitts, A.M.: A new approach to abstract syntax involving binders. In: LICS 1999, Trento, Italy, pp. 214–224 (July 1999)
6. Gabbay, M.J.: A study of substitution, using nominal techniques and Fraenkel-Mostowski sets. Theor. Comput. Sci. 410(12-13) (March 2009)
7. Gabbay, M.J.: Foundations of nominal techniques: logic and semantics of variables in abstract syntax. Bull. Symbolic Logic 17(2), 161–229 (2011)
8. Gabbay, M.J., Ciancia, V.: Freshness and Name-Restriction in Sets of Traces with Names. In: Hofmann, M. (ed.) FOSSACS 2011. LNCS, vol. 6604, pp. 365–380. Springer, Heidelberg (2011)
9. Gabbay, M.J., Mathijssen, A.: Nominal universal algebra: equational logic with names and binding. J. Logic and Computation 19(6), 1455–1508 (2009)
10. Kozen, D., Mamouras, K., Petrişan, D., Silva, A.: Nominal Kleene coalgebra. TR, Computing and Information Science, Cornell University (February 2015), `http://hdl.handle.net/1813/39108`
11. Ochremiak, J.: Nominal sets over algebraic atoms. In: Höfner, P., Jipsen, P., Kahl, W., Müller, M.E. (eds.) RAMiCS 2014. LNCS, vol. 8428, pp. 429–445. Springer, Heidelberg (2014)
12. Pitts, A.M.: Nominal Sets: Names and Symmetry in Computer Science. Cambridge Tracts in Theoretical Computer Science, vol. 57. Cambridge University Press (2013)
13. Pous, D.: Symbolic algorithms for language equivalence and Kleene algebra with tests. In: POPL 2015, Mumbai, India, January 2015, pp. 357–368 (2015)
14. Silva, A.: Kleene Coalgebra. PhD thesis, University of Nijmegen (2010)

Closure, Properties and Closure Properties of Multirelations

Rudolf Berghammer[1] and Walter Guttmann[2]

[1] Institut für Informatik
Christian-Albrechts-Universität zu Kiel, Germany
[2] Department of Computer Science and Software Engineering
University of Canterbury, New Zealand

Abstract. Multirelations have been used for modelling games, protocols and computations. They have also been used for modelling contact, closure and topology. We bring together these two lines of research using relation algebras and more general algebras. In particular, we look at various properties of multirelations that have been used in the two lines of research, show how these properties are connected and study by which multirelational operations they are preserved. We find that many results do not require a restriction to up-closed multirelations; this includes connections between various kinds of reflexive-transitive closure.

1 Introduction

A multirelation is a relation between a set and a powerset. The powerset structure facilitates the modelling of two-player games or interaction between agents in a computation; see [5,17,19], for example. Already before these applications multirelations were used by G. Aumann to model contact and, thereby, to give beginners a more suggestive access to topology than traditional approaches do; see [1]. Properties of multirelations have been rediscovered over time, but, in our opinion, a systematic investigation is missing. The aim of the present paper is to start this research. Its methods are algebraic, in particular relation-algebraic.

The starting point is a relation-algebraic representation of multirelations and multirelational operations (Sections 2 and 3). Properties of these operations are proved using relation algebras and captured as axioms of more general structures based on lattices and semirings (Section 4). A key decision is to not specialise to up-closed multirelations at the outset, but to treat being up-closed as one among many properties a multirelation might have. This makes it possible to generalise results, for example, about closure operations (Section 5). Other properties are taken from the literature and compared systematically (Section 6). A particular question is whether they are preserved by multirelational operations (Section 7). Positive results are shown algebraically using Isabelle and automated theorem provers. Counterexamples are produced by a Haskell program. Moreover properties of topological contacts are derived from logical specifications (Section 8).

W. Kahl et al. (Eds.): RAMiCS 2015, LNCS 9348, pp. 67–83, 2015.
DOI: 10.1007/978-3-319-24704-5_5

The contributions of the paper are (1) new algebraic structures, which capture (not only up-closed) multirelations, (2) a comparison of three reflexive-transitive closure operations in these algebras, (3) a study of relationships between properties of multirelations and (4) a study of preservation of these properties by multirelational operations. Overall, this paper brings together the topological and computational lines of research on multirelations. The companion paper [7] investigates how properties from these two lines of research translate to predicate transformers. It uses relation algebras to express the correspondence of multirelations and predicate transformers, which turns out to be similar to the correspondence between contact relations and closure operations.

2 Relation-Algebraic Prerequisites

In this section we present the facts on relations and heterogeneous relation algebras that are needed in the remainder of this paper. For more details on relations and relation algebras, see [25], for example.

We write $R : A \leftrightarrow B$ if R is a (typed binary) relation with source A and target B, that is, of type $A \leftrightarrow B$. If the sets A and B are finite, we may consider R as a Boolean matrix. Since this interpretation is well suited for many purposes, we will use matrix notation and write $R_{x,y}$ instead of $(x, y) \in R$ or $x\,R\,y$.

We assume the reader to be familiar with the basic operations on relations, namely R^{c} (converse), $\overline{R}$ (complement), $R \cup S$ (union), $R \cap S$ (intersection), RS (composition), the predicates indicating $R \subseteq S$ (inclusion) and $R = S$ (equality) and the special relations O (empty relation), T (universal relation) and I (identity relation). Converse has higher precedence than composition, which has higher precedence than union and intersection. The set of all relations of type $A \leftrightarrow B$ with the operations $\overline{}$, $\cup$, $\cap$, the ordering $\subseteq$ and the constants O and T forms a complete Boolean lattice. Further well-known rules are, for example, $(R^{\mathsf{c}})^{\mathsf{c}} = R$, $\overline{R^{\mathsf{c}}} = \overline{R}^{\mathsf{c}}$, and that $R \subseteq S$ implies $R^{\mathsf{c}} \subseteq S^{\mathsf{c}}$ as well as $RP \subseteq SP$ and $QR \subseteq QS$, for all P, Q, R and S.

The theoretical framework for these rules and many others is that of a (heterogeneous) relation algebra; see [27] for details. As constants and operations of this algebraic structure we have those of concrete (that is, set-theoretic) relations. The axioms of a relation algebra are those of a complete Boolean lattice for the Boolean part, the associativity and neutrality of identity relations for composition, the equivalence of $QR \subseteq S$, $Q^{\mathsf{c}}\overline{S} \subseteq \overline{R}$ and $\overline{S}R^{\mathsf{c}} \subseteq \overline{Q}$, for all relations Q, R and S – called the *Schröder equivalences* – and that $R \neq \mathsf{O}$ implies $\mathsf{T}R\mathsf{T} = \mathsf{T}$, for all relations R.

Residuals are the greatest solutions of certain inclusions. The *left residual* of S over R, in symbols S/R, is the greatest relation X such that $XR \subseteq S$. So, we have the Galois connection $XR \subseteq S$ if and only if $X \subseteq S/R$, for all relations X. Similarly, the *right residual* of S over R, in symbols $R \setminus S$, is the greatest relation X such that $RX \subseteq S$. This implies that $RX \subseteq S$ if and only if $X \subseteq R \setminus S$, for all relations X. We will also need relations which are left and right residuals simultaneously. The *symmetric quotient* $R \div S$ of two relations R and S is defined as the greatest relation X such that $RX \subseteq S$ and $XS^c \subseteq R^c$. In terms of the basic operations we have $S/R = \overline{\overline{S}R^c}$, $R \setminus S = \overline{R^c\overline{S}}$ and $R \div S = (R \setminus S) \cap (R^c/S^c)$, for all relations R and S.

Besides empty relations, universal relations and identity relations, we need further basic relations which specify fundamental set-theoretic constructions. Assume A to be a set and let 2^A denote its powerset. Then the *membership relation* $\mathsf{E} : A \leftrightarrow 2^A$ is the relation-level equivalent to the set-theoretic predicate '$\in$'. Hence, we have $\mathsf{E}_{x,Y}$ if and only if $x \in Y$, for all $x \in A$ and $Y \in 2^A$. With the help of E we can introduce two relations on 2^A via $\mathsf{S} := \mathsf{E} \setminus \mathsf{E} : 2^A \leftrightarrow 2^A$ and $\mathsf{C} := \mathsf{E} \div \overline{\mathsf{E}} : 2^A \leftrightarrow 2^A$. A little component-wise calculation shows $\mathsf{S}_{X,Y}$ if and only if $X \subseteq Y$ and $\mathsf{C}_{X,Y}$ if and only if $Y = \overline{X}$, for all $X \in 2^A$ and $Y \in 2^A$, where $\overline{X}$ is the complement of the set X relative to its superset A. Therefore, we call S a *subset relation* and C a *set complement relation.*

3 Fundamentals of Multirelations

In this section we recall basic definitions, operations and properties of multirelations and express them in terms of relations. The presentation follows [15].

A *multirelation* (as introduced in [19,23]) is a relation of type $A \leftrightarrow 2^B$ in the sense of Section 2. It maps an element of A to a set of subsets of B. Union, intersection and complement apply to multirelations as to relations. Particular multirelations are empty relations $\mathsf{O} : A \leftrightarrow 2^B$, universal relations $\mathsf{T} : A \leftrightarrow 2^B$ and membership relations $\mathsf{E} : A \leftrightarrow 2^A$. The composition of the multirelations $Q : A \leftrightarrow 2^B$ and $R : B \leftrightarrow 2^C$ is the multirelation $Q;R : A \leftrightarrow 2^C$, given by

$$(Q;R)_{x,Z} \iff \exists Y \in 2^B : Q_{x,Y} \wedge \forall y \in Y : R_{y,Z},$$

for all $x \in A$ and $Z \in 2^C$. The *dual* of a multirelation $R : A \leftrightarrow 2^B$ is the multirelation $R^{\mathsf{d}} : A \leftrightarrow 2^B$ given by

$$R^{\mathsf{d}}{}_{x,Y} \iff \neg R_{x,\overline{Y}},$$

for all $x \in A$ and $Y \in 2^B$, where $\overline{Y}$ is the complement of Y relative to its superset B. Dual has higher precedence than composition, which has higher precedence than union and intersection. A multirelation $R : A \leftrightarrow 2^B$ is *up-closed* if

$$R_{x,Y} \wedge Y \subseteq Z \implies R_{x,Z}$$

for all $x \in A$ and $Y, Z \in 2^B$. This means that if an element of A is related to a set Y, it also has to be related to all supersets of Y. By $A \overset{\text{u}}{\leftrightarrow} 2^B$ we denote the set of all up-closed multirelations of type $A \leftrightarrow 2^B$.

The following result expresses multirelational composition, the dual and the property of being up-closed in terms of relation-algebraic operations and constants, namely right residual, membership relations, set complement relations C and subset relations S. It is proved in [15, Theorems 2, 4 and 6]; see also [16,25].

Theorem 1. *Let $Q : A \leftrightarrow 2^B$ and $R : B \leftrightarrow 2^C$ be multirelations. Then we have $Q;R = Q(\mathsf{E} \setminus R)$ and $Q^{\mathsf{d}} = \overline{\overline{Q}\mathsf{C}} = \overline{Q\mathsf{C}}$. Furthermore, Q is up-closed if and only if $Q = Q\mathsf{S}$.*

A multirelation $R : A \leftrightarrow 2^A$ models a two-player game as shown in [19]. The set A describes the possible states of the game. For each state $x \in A$ the set of subsets $Ys = \{Y \in 2^A \mid R_{x,Y}\}$ to which x is related gives the options of the first player. The first player chooses one of these subsets, a set $Y \in Ys$. This set Y gives the options of the second player, who chooses one of its elements $y \in Y$, which is the next state of the game. If the first player cannot make a choice because Ys is empty, the second player wins. If the second player cannot make a choice because Y is empty, the first player wins. Multirelations can also be used to describe the interaction of two agents in a computation (see [5,10,17]), certain kinds of contact (see [1,4]) and concurrency (see [21]).

Being relations, the multirelations of type $A \leftrightarrow 2^B$ form a bounded distributive lattice under the operations of union and intersection. The structure becomes more diversified once we take composition into account. First, familiar laws of relation algebras – that composition distributes over union and has the empty relation as a zero – no longer hold from both sides, but just from one side. Second, other laws of relation algebras – that composition is associative and has the identity relation as a neutral element – hold for up-closed multirelations, but need to be weakened in the general case as shown in [11]. On the other hand, composition remains $\subseteq$-isotone. These and related properties are summarised in the following result.

Theorem 2. *For all multirelations P, Q and R we have*

(1) $\mathsf{O};R = \mathsf{O}$ *(2)* $\mathsf{E};R = R$ *(3)* $\mathsf{T};R = \mathsf{T}$ *(4)* $R \subseteq R;\mathsf{E}$,

where in (4) equality holds if and only if R is up-closed, and also

(5) $(P \cup Q);R = P;R \cup Q;R$, *(6)* $(P \cap Q);R \subseteq P;R \cap Q;R$,

where in (6) equality holds if P and Q are up-closed, and also

(7) $(P;Q);R \subseteq P;(Q;R)$,

where in (7) equality holds if Q is up-closed, and finally

(8) $P;Q \cup P;R \subseteq P;(Q \cup R)$ *(9)* $P;(Q \cap R) \subseteq P;Q \cap P;R$.

Proof. All properties are proved in [15, Theorems 3 and 7] except (4) and (7) for general multirelations. A proof of (4) is $R \subseteq R\mathsf{S} = R(\mathsf{E} \setminus \mathsf{E}) = R;\mathsf{E}$. To prove (7) we use that $\mathsf{E}(\mathsf{E} \setminus Q)(\mathsf{E} \setminus R) \subseteq Q(\mathsf{E} \setminus R)$ implies $(\mathsf{E} \setminus Q)(\mathsf{E} \setminus R) \subseteq \mathsf{E} \setminus (Q(\mathsf{E} \setminus R))$ by the Galois connection. Hence, we get the result as follows:

$$\begin{aligned}(P;Q);R &= (P;Q)(\mathsf{E} \setminus R) = P(\mathsf{E} \setminus Q)(\mathsf{E} \setminus R) \\ &\subseteq P(\mathsf{E} \setminus (Q(\mathsf{E} \setminus R))) = P(\mathsf{E} \setminus (Q;R)) = P;(Q;R)\end{aligned}$$

□

The dual operation reverses the lattice order and distributes over composition of up-closed multirelations. Again this needs to be weakened in the general case. These and further properties are summarised in the following result.

Theorem 3. *For all multirelations Q and R we have*

$$\text{(1) } \mathsf{O}^{\mathsf{d}} = \mathsf{T} \qquad \text{(2) } \mathsf{E}^{\mathsf{d}} = \mathsf{E} \qquad \text{(3) } \mathsf{T}^{\mathsf{d}} = \mathsf{O} \qquad \text{(4) } R^{\mathsf{dd}} = R,$$

and also

$$\text{(5) } (Q \cup R)^{\mathsf{d}} = Q^{\mathsf{d}} \cap R^{\mathsf{d}} \qquad \text{(7) } (Q\,;R)^{\mathsf{d}} \subseteq Q^{\mathsf{d}}\,;R^{\mathsf{d}}$$
$$\text{(6) } (Q \cap R)^{\mathsf{d}} = Q^{\mathsf{d}} \cup R^{\mathsf{d}} \qquad \text{(8) } (Q\,;R)^{\mathsf{d}} = (Q\,;\mathsf{E})^{\mathsf{d}}\,;R^{\mathsf{d}},$$

where in (7) equality holds if Q is up-closed.

Proof. All properties are proved in [15, Theorems 5 and 7] except (7) and (8) for general multirelations. A proof of (7) and (8) is as follows:

$$\begin{aligned}
(Q\,;R)^{\mathsf{d}} &= \overline{Q\,;R}\mathsf{C} = \overline{Q(\mathsf{E}\backslash R)}\mathsf{C} = \overline{Q\mathsf{S}(\mathsf{E}\div R)}\mathsf{C} = \overline{Q\mathsf{S}}(\mathsf{E}\div R)\mathsf{C} = \overline{Q\mathsf{S}}(\overline{\mathsf{E}}\div\overline{R})\mathsf{C} \\
&= \overline{Q\mathsf{S}}(\overline{\mathsf{E}}\div\mathsf{E})(\mathsf{E}\div\overline{R})\mathsf{C} = \overline{Q\mathsf{S}}\mathsf{C}^{\mathsf{c}}(\mathsf{E}\div\overline{R})\mathsf{C} = \overline{Q\mathsf{S}}\mathsf{C}(\mathsf{E}\div\overline{R})\mathsf{C} \\
&= \overline{Q\mathsf{S}\mathsf{S}^{\mathsf{c}}}\mathsf{C}(\mathsf{E}\div\overline{R})\mathsf{C} = \overline{Q\mathsf{S}}\mathsf{C}\mathsf{S}(\mathsf{E}\div\overline{R})\mathsf{C} = \overline{Q\mathsf{S}}\mathsf{C}(\mathsf{E}\backslash\overline{R})\mathsf{C} = (Q\mathsf{S})^{\mathsf{d}}(\mathsf{E}\backslash\overline{R})\mathsf{C} \\
&= (Q\mathsf{S})^{\mathsf{d}}(\mathsf{E}\backslash(\overline{R}\mathsf{C})) = (Q(\mathsf{E}\backslash\mathsf{E}))^{\mathsf{d}}(\mathsf{E}\backslash R^{\mathsf{d}}) = (Q\,;\mathsf{E})^{\mathsf{d}}\,;R^{\mathsf{d}} \subseteq Q^{\mathsf{d}}\,;R^{\mathsf{d}}
\end{aligned}$$

This calculation uses $\overline{Q\mathsf{S}\mathsf{S}^{\mathsf{c}}} = \overline{Q\mathsf{S}}$. The inclusion '$\subseteq$' follows by applying a Schröder equivalence to $Q\mathsf{S}\mathsf{S} \subseteq Q\mathsf{S}$ and the inclusion '$\supseteq$' follows from $\mathsf{I} \subseteq \mathsf{S}$. See the proof of [15, Theorem 7.3] for an explanation of the other steps. □

4 Algebraic Structures for Investigating Multirelations

In this section we capture the properties of multirelations shown in Section 2 by five algebraic structures, which are introduced in the following.

A *bounded join-semilattice* is an algebraic structure $(S, +, 0)$ satisfying for all $x, y, z \in S$ the associativity, commutativity, idempotence and neutrality axioms:

$$x + (y + z) = (x + y) + z \qquad x + y = y + x \qquad x + x = x \qquad 0 + x = x$$

The *semilattice order*, defined by $x \leq y$ if and only if $x + y = y$, for all $x, y \in S$, has the least element 0 and the least upper bound operation '+'. The operation '+' is $\leq$-isotone.

Next, a *bounded distributive lattice* $(S, +, \curlywedge, 0, \top)$ adds to a bounded join-semilattice a dual bounded meet-semilattice $(S, \curlywedge, \top)$ as well as distribution and absorption axioms, such that for all $x, y, z \in S$ the following equations hold:

$$\begin{aligned}
x \curlywedge (y \curlywedge z) &= (x \curlywedge y) \curlywedge z & x + (y \curlywedge z) &= (x + y) \curlywedge (x + z) \\
x \curlywedge y &= y \curlywedge x & x \curlywedge (y + z) &= (x \curlywedge y) + (x \curlywedge z) \\
x \curlywedge x &= x & x + (x \curlywedge y) &= x \\
\top \curlywedge x &= x & x \curlywedge (x + y) &= x
\end{aligned}$$

The semilattice order has the alternative characterisation that $x \leq y$ if and only if $x \curlywedge y = x$, for all $x, y \in S$, the greatest element $\top$ and the greatest lower bound operation '$\curlywedge$'. The operation '$\curlywedge$' is $\leq$-isotone.

A *pre-left semiring* $(S, +, \cdot, 0, 1)$ expands a bounded join-semilattice $(S, +, 0)$ with a binary operation '$\cdot$' and a constant 1 with the following axioms for all $x, y, z \in S$:

$$\begin{aligned} x &= 1 \cdot x & (x \cdot y) + (x \cdot z) &\leq x \cdot (y + z) \\ x &\leq x \cdot 1 & (x \cdot z) + (y \cdot z) &= (x + y) \cdot z \\ (x \cdot y) \cdot z &\leq x \cdot (y \cdot z) & 0 &= 0 \cdot x \end{aligned}$$

Note the inequalities in the left column. The operation '$\cdot$' is $\leq$-isotone. We often abbreviate a product $x \cdot y$ via juxtaposition to xy.

An *idempotent left semiring* (see [18]) is a pre-left semiring $(S, +, \cdot, 0, 1)$ whose reduct $(S, \cdot, 1)$ is a monoid, which is enforced by adding the axioms

$$x = x \cdot 1 \qquad (x \cdot y) \cdot z = x \cdot (y \cdot z),$$

for all $x, y, z \in S$. Idempotent semirings are rings in which the operation '+' is idempotent instead of having an inverse. Idempotent left semirings are idempotent semirings in which the operation '$\cdot$' is $\leq$-isotone instead of distributing over the operation '+' from the left and having the right zero 0. Pre-left semirings further weaken idempotent left semirings by requiring only one inequality of the associativity and right-neutral properties. This is because multirelations do not satisfy the other inequalities in general.

Finally, combining the lattice and semiring operations, an *M0-algebra* is an algebraic structure $(S, +, \cdot, \curlywedge, 0, 1, \top)$ such that the reduct $(S, +, \curlywedge, 0, \top)$ is a bounded distributive lattice and the reduct $(S, +, \cdot, 0, 1)$ is a pre-left semiring.

The algebraic results we will derive in the following sections apply to multirelations because of the following instances. The multirelations over a set A form a bounded distributive lattice $(A \leftrightarrow 2^A, \cup, \cap, \mathsf{O}, \mathsf{T})$. By Theorem 2 these multirelations also form an M0-algebra $(A \leftrightarrow 2^A, \cup, ;, \cap, \mathsf{O}, \mathsf{E}, \mathsf{T})$ and the subset of up-closed multirelations forms an idempotent left semiring $(A \overset{u}{\leftrightarrow} 2^A, \cup, ;, \mathsf{O}, \mathsf{E})$. We refer to [22,29] for further algebraic structures underlying up-closed multirelations and to [16] for placing them in a categorical setting. See also [21], where another kind of multirelational composition '$\cdot$' is introduced that gives rise to an M0-algebra. As shown in [12], this operation is not associative for general multirelations, but satisfies $(P \cdot Q) \cdot R \subseteq P \cdot (Q \cdot R)$ and $P = P \cdot 1$ for all P, Q and R, where $1 = \mathsf{I} \div \mathsf{E}$ is the singleton multirelation.

5 Reflexive-Transitive Closures of Multirelations

As proved in [11], multirelational composition has a left residual. If we define it by $R /\!\!/ Q := R/(\mathsf{E} \setminus Q)$, for all multirelations R and Q, then we get

$$P;Q \subseteq R \iff P(\mathsf{E} \setminus Q) \subseteq R \iff P \subseteq R/(\mathsf{E} \setminus Q) \iff P \subseteq R /\!\!/ Q,$$

for all multirelations P, Q and R. In this section we use left residuals and an appropriate algebraic structure to relate three different representations of reflexive-transitive closures of multirelations.

A *residuated pre-left semiring* $(S, +, \cdot, /, 0, 1)$ expands a pre-left semiring $(S, +, \cdot, 0, 1)$ with a binary operation '/' satisfying the Galois connection

$$xy \leq z \iff x \leq z/y,$$

for all $x, y, z \in S$. It follows that the operation '/' is $\leq$-isotone in its first argument and $\leq$-antitone in its second argument. Moreover, we obtain the two properties $(x/y)y \leq x$ and $x/1 \leq x$, for all $x, y \in S$. As a consequence we get the following instance. The multirelations over a set A form a residuated pre-left semiring $(A \leftrightarrow 2^A, \cup, ;, /\!\!/, \mathsf{O}, \mathsf{E})$.

The $\leq$-isotone functions f, g and h of the following result capture left recursion, right recursion and symmetric recursion, respectively. The $\leq$-least prefixpoint μf of the function f is axiomatised using its unfold and induction properties, that is, $f(\mu f) \leq \mu f$ and that $f(x) \leq x$ implies $\mu f \leq x$, for all $x \in S$. Similar axioms are assumed for μg and μh. It is known that left and right recursion coincide for relations, but in general they do not for multirelations.

Theorem 4. *Let S be a residuated pre-left semiring and let $y \in S$. Depending on y, let f, g and h be functions on S defined by*

$$f(x) = 1 + x \cdot y \qquad g(x) = 1 + y \cdot x \qquad h(x) = 1 + y + x \cdot x,$$

for all $x \in S$. Assume that μf, μg and μh exist. Then we have $\mu f \leq \mu g = \mu h$.

Proof. We first show $\mu f \leq \mu g$. Semi-associativity of composition, the Galois property of the left residual and the prefixpoint property of μg imply

$$(y \cdot (\mu g/y)) \cdot y \leq y \cdot ((\mu g/y) \cdot y) \leq y \cdot \mu g \leq 1 + y \cdot \mu g \leq \mu g.$$

Hence, we get $y \cdot (\mu g/y) \leq \mu g/y$. Moreover, $1 \leq 1 + y \cdot \mu g \leq \mu g$ holds, whence semi-neutrality of composition gives

$$1 \cdot y = y \leq y \cdot 1 \leq 1 + y \cdot \mu g \leq \mu g.$$

So, $1 \leq \mu g/y$ and, together, we have

$$g(\mu g/y) = 1 + y \cdot (\mu g/y) \leq \mu g/y.$$

From this we obtain $\mu g \leq \mu g/y$ by the least prefixpoint property of μg. Hence

$$f(\mu g) = 1 + \mu g \cdot y \leq \mu g$$

and, therefore, $\mu f \leq \mu g$ follows by the least prefixpoint property of μf.

We next show $\mu g \leq \mu h$. This part does not use residuals. From the least prefixpoint property of μh we get $y \leq 1 + y + \mu h \cdot \mu h = h(\mu h) \leq \mu h$; hence

$$g(\mu h) = 1 + y \cdot \mu h \leq 1 + y + \mu h \cdot \mu h = h(\mu h) \leq \mu h$$

by the prefixpoint property of μh. Therefore, we arrive at $\mu g \le \mu h$ by the least prefixpoint property of μg.

We finally show $\mu h \le \mu g$ following the argument of [6, Satz 10.1.5], which is for homogeneous relations. Semi-associativity of composition, a property of the left residual and the unfold property of μg imply:

$$\begin{aligned} g(\mu g/\mu g) \cdot \mu g &= (1 + y \cdot (\mu g/\mu g)) \cdot \mu g = 1 \cdot \mu g + (y \cdot (\mu g/\mu g)) \cdot \mu g \\ &\le \mu g + y \cdot ((\mu g/\mu g) \cdot \mu g) \le \mu g + 1 + y \cdot \mu g = \mu g + g(\mu g) = \mu g \end{aligned}$$

As a consequence we obtain $g(\mu g/\mu g) \le \mu g/\mu g$ and this leads to $\mu g \le \mu g/\mu g$ by the least prefixpoint property of μg, whence $\mu g \cdot \mu g \le \mu g$. With $1 \le \mu g$ and $y \le \mu g$ shown above, it follows that

$$h(\mu g) = 1 + y + \mu g \cdot \mu g \le \mu g.$$

Therefore we have $\mu h \le \mu g$ by the least prefixpoint property of μh. □

For up-closed multirelations the equality $\mu g = \mu h$ is shown in [28]. Furthermore, for finitary up-closed multirelations $\bigcup_{n \in \mathbb{N}} g^n(\mathsf{O}) \subseteq \mu h$ is shown in [13] and $\bigcup_{n \in \mathbb{N}} g^n(\mathsf{O}) = \mu g$ is shown in [11].

We proved Theorem 4 also in Isabelle/HOL using its integrated automated theorem provers and SMT solvers, which are described in [8,20]. The same holds for the theorems we will present in the next two sections, that is, Theorem 5 to Theorem 8. We therefore omit their proofs, which are given in the Isabelle theory files available at http://www.csse.canterbury.ac.nz/walter.guttmann/algebra/.

6 Properties of Multirelations

A number of properties of multirelations were used in previous work for modelling games, protocols, computations, contact, closure and topology, see [1,5,17,19,23], for example. Algebraic definitions of these and other properties are summarised in Figure 1. Its second column states the property in terms of relations and the third column gives the corresponding definition in M0-algebras. The distributivity properties universally quantify over the multirelations P, Q and the elements y, z of the M0-algebra, respectively.

For up-closed multirelations several of the properties listed in Figure 1 are dual to each other, that is, can be obtained by applying the multirelational dual operation. This does not hold for general multirelations: for example, the conjunction of reflexive and transitive implies up-closed, but the conjunction of their duals co-reflexive and dense does not imply up-closed, which is self-dual.

In this section we investigate the connections between the properties in Figure 1 using the algebraic structure of multirelations. While many results can be derived in M0-algebras, additional axioms are needed to prove some others, leading to the following new algebraic structure. An *M1-algebra* is an M0-algebra $(S, +, \cdot, \curlywedge, 0, 1, \top)$ satisfying the axioms

$$\top = \top x \qquad x(yz) = (x(y1))z \qquad xz \curlywedge yz = (x1 \curlywedge y1)z,$$

R or x is ...	if and only if	algebraically
total	$R;\mathsf{T} = \mathsf{T}$	$x\top = \top$
co-total	$R;\mathsf{O} = \mathsf{O}$	$x0 = 0$
transitive	$R;R \subseteq R$	$xx \leq x$
dense	$R \subseteq R;R$	$x \leq xx$
reflexive	$\mathsf{E} \subseteq R$	$1 \leq x$
co-reflexive	$R \subseteq \mathsf{E}$	$x \leq 1$
idempotent	$R;R = R$	$xx = x$
up-closed	$R;\mathsf{E} = R$	$x1 = x$
$\cup$-distributive	$R;(P \cup Q) = R;P \cup R;Q$	$x(y + z) = xy + xz$
$\cap$-distributive	$R;(P \cap Q) = R;P \cap R;Q$	$x(y \curlywedge z) = xy \curlywedge xz$
a contact	$R;R \cup \mathsf{E} = R$	$xx + 1 = x$
a kernel	$R;R \cap \mathsf{E} = R;\mathsf{E}$	$xx \curlywedge 1 = x1$
a test	$R;\mathsf{T} \cap \mathsf{E} = R$	$x\top \curlywedge 1 = x$
a co-test	$R;\mathsf{O} \cup \mathsf{E} = R$	$x0 + 1 = x$
a vector	$R;\mathsf{T} = R$	$x\top = x$

Fig. 1. Fundamental properties

for all $x, y, z \in S$. An equivalent structure is obtained if just '$\leq$' is assumed instead of equality in each axiom. If all elements are up-closed, that is, $x1 = x$ holds for all $x \in S$, the last two axioms collapse to associativity of the operation '$\cdot$' and right-distributivity of '$\cdot$' over the operation '$\curlywedge$'. This shows how to obtain weaker axioms which hold for all multirelations. The following theorem summarises our results about relationships between the properties in Figure 1.

Theorem 5. *The implications shown in Figure 2 drawn as continuous (dashed) arrows hold in M0-algebras (M1-algebras). Furthermore, arrows originating in the same point indicate that the property is equivalent to the conjunction of the targets.*

Moreover, in all M1-algebras S the vector property $x\top = \top$ is equivalent to its dual $x0 = 0$ for all $x \in S$.

7 Closure Properties of Multirelational Operations

It is known that up-closed multirelations are closed under the multirelational operations we have introduced in Section 3. In this section we systematically investigate the closure properties for certain classes of multirelations, which are given by the properties presented in Figure 1. For dealing with the dual operation we need additional axioms, which lead to the expansions of M0-algebras we will introduce in this section.

First, an *M2-algebra* $(S, +, \cdot, \curlywedge, ^{\mathsf{d}}, 0, 1, \top)$ is an M0-algebra $(S, +, \cdot, \curlywedge, 0, 1, \top)$ expanded with a unary dual operation '^{d}' satisfying the axioms

$$(xy)^{\mathsf{d}} = (x1)^{\mathsf{d}} y^{\mathsf{d}} \qquad (x + y)^{\mathsf{d}} = x^{\mathsf{d}} \curlywedge y^{\mathsf{d}} \qquad x^{\mathsf{dd}} = x \qquad 1^{\mathsf{d}} = 1,$$

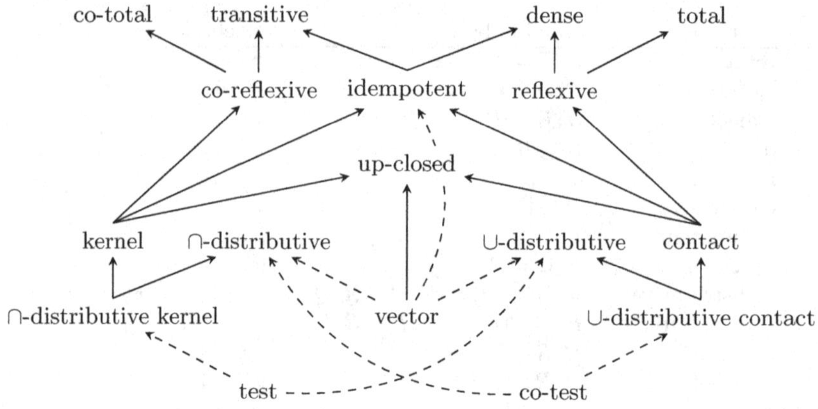

Fig. 2. Relationships between the fundamental properties

for all $x, y \in S$. Note again how distributivity of the operation '^{d}' over the operation '$\cdot$', which holds for up-closed multirelations, is weakened by replacing x with $x1$. The above axioms imply the additional axioms of M1-algebras. Thus, we obtain the following result.

Theorem 6. *All M2-algebras are M1-algebras.*

For reasoning about up-closed multirelations we use that the operation '^{d}' distributes over the operation '$\cdot$'. As a further expansion of M0-algebras, therefore, an *M3-algebra* $(S, +, \cdot, \curlywedge, ^{\mathsf{d}}, 0, 1, \top)$ is an M0-algebra $(S, +, \cdot, \curlywedge, 0, 1, \top)$ expanded with a unary dual operation '^{d}' satisfying the axioms

$$(xy)^{\mathsf{d}} = x^{\mathsf{d}} y^{\mathsf{d}} \qquad (x+y)^{\mathsf{d}} = x^{\mathsf{d}} \curlywedge y^{\mathsf{d}} \qquad x^{\mathsf{dd}} = x \qquad 1^{\mathsf{d}} = 1,$$

for all $x, y \in S$. These axioms imply the axioms of M2-algebras. Moreover, we obtain that the operation '$\cdot$' is associative with right-neutral element 1, that is, the idempotent left semiring structure.

Theorem 7. *All M3-algebras are M2-algebras and idempotent left semirings.*

The algebraic results obtained so far apply to multirelations due to the following instances. By Theorem 3, the multirelations over a set A form an M2-algebra $(A \leftrightarrow 2^A, \cup, ;, \cap, ^{\mathsf{d}}, \mathsf{O}, \mathsf{E}, \mathsf{T})$ and the up-closed multirelations over A form an M3-algebra $(A \overset{\cup}{\leftrightarrow} 2^A, \cup, ;, \cap, ^{\mathsf{d}}, \mathsf{O}, \mathsf{E}, \mathsf{T})$. The next theorem summarises the closure properties of multirelations.

Theorem 8. *Figure 3 shows which properties in Figure 1 hold for the multirelational constants and with respect to which operations these properties are closed. There an entry ■ (□) means that the property is closed under the respective operation in M2-algebras (M3-algebras). All ■ entries except those for the operation*

	O	E	T	∪	∩	;	d
total	–	■	■	■	□	□	▽
co-total	■	■	–	■	■	■	▲
transitive	■	■	■	–	■	–	▼
dense	■	■	■	■	–	–	△
reflexive	–	■	■	■	■	■	▼
co-reflexive	■	■	–	■	■	■	▲
idempotent	■	■	■	–	–	–	□
up-closed	■	■	■	■	■	■	■
∪-distributive	■	■	■	■	–	□	▽
∩-distributive	■	■	■	–	□	□	△
a contact	–	■	■	–	■	–	▼
a kernel	■	■	–	■	–	–	▲
a ∪-distributive contact	–	■	■	–	–	–	▼
a ∩-distributive kernel	■	■	–	–	–	–	▲
a test	■	■	–	■	■	■	▼
a co-test	–	■	■	■	■	■	▲
a vector	■	–	■	■	■	■	■

Fig. 3. Closure properties of multirelations

'd' *follow in M1-algebras; most of these follow already in M0-algebras. An entry* ▼/▲ (▽/△) *means that if x satisfies the property then x^{d} satisfies the property below/above in M2-algebras (M3-algebras). An entry – means that the property is not closed under the respective operation even for up-closed multirelations.*

To give an example, the dual of a co-total multirelation is total and the dual of an up-closed total multirelation is co-total. Another consequence of the closure properties are sub-algebras. For example, the set of co-total multirelations forms a pre-left semiring and so does the set of co-reflexive multirelations.

It is unknown if any of the findings □ can be strengthened to ■ in the rows for ∪-/∩-distributive in Figure 3. Moreover, it is unknown if the finding △ can be strengthened to ▲ in the row for ∩-distributive. Counterexamples for the other claims are shown in Figures 4, 5 and 6 as Boolean matrices (where a grey square denotes a 1-entry and a white square denotes a 0-entry). Most counterexamples have been found using a Haskell program which performs an exhaustive search. For ∪- and ∩-distributivity of up-closed multirelations we use the alternative characterisation provided by Aumann contacts given in Section 8.

Note that M2-algebras are not complete for multirelations. The counterexample generator Nitpick, which is described in [9], finds a counterexample showing that $x\top \curlywedge yz \leq (x\top \curlywedge y)z$ does not follow in M2-algebras. However, this property holds for multirelations since

$$P;\mathsf{T} \cap Q;R = P\mathsf{T} \cap Q(\mathsf{E} \setminus R) = (P\mathsf{T} \cap Q)(\mathsf{E} \setminus R) = (P;\mathsf{T} \cap Q);R.$$

This calculation uses that $P;\mathsf{T} = P\mathsf{T}$ as shown in [15], so intersection with this vector can be imported into the first argument of a composition.

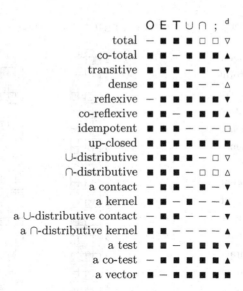

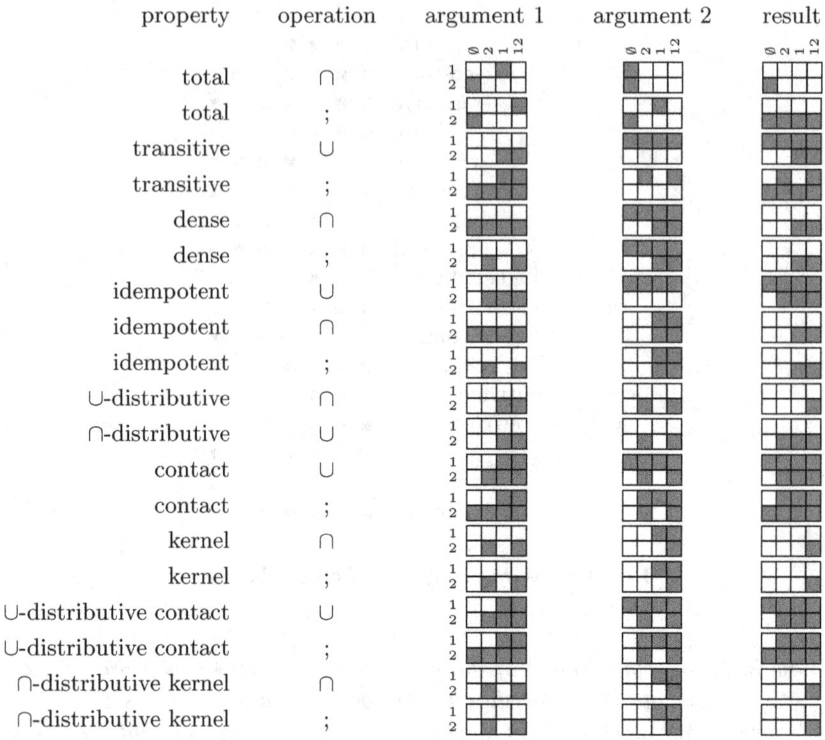
property
operation
argument 1
argument 2
result
∅ 2 1 12
1 2
total ∩
total ;
transitive ∪
transitive ;
dense ∩
dense ;
idempotent ∪
idempotent ∩
idempotent ;
∪-distributive ∩
∩-distributive ∪
contact ∪
contact ;
kernel ∩
kernel ;
∪-distributive contact ∪
∪-distributive contact ;
∩-distributive kernel ∩
∩-distributive kernel ;

Fig. 4. Counterexamples generated by a Haskell program

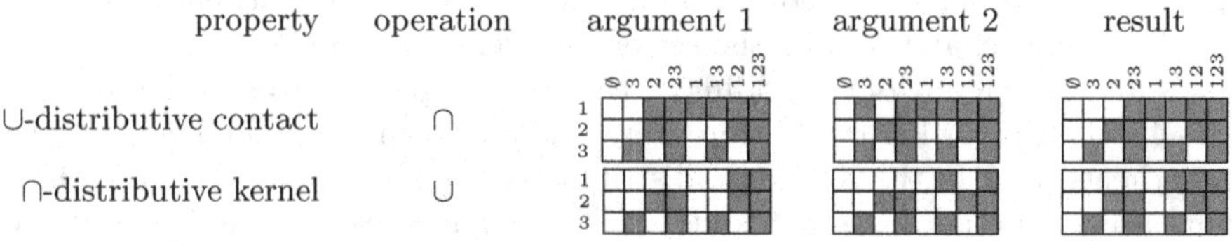

Fig. 5. Manually generated counterexamples

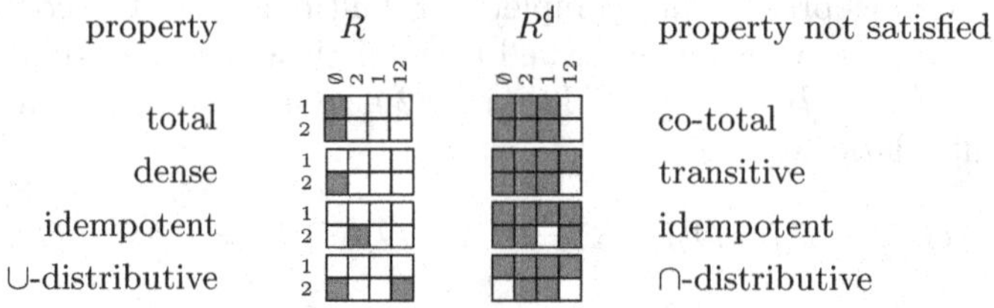
property
R
R^d
property not satisfied
∅ 2 1 12
1 2
total co-total
dense transitive
idempotent idempotent
∪-distributive ∩-distributive

Fig. 6. Counterexamples for the operation ‘^{d}’ generated by a Haskell program

Neither are M3-algebras complete for up-closed multirelations. Nitpick shows that $x\top \curlywedge x^{\mathsf{d}}0 = 0$ does not follow in M3-algebras, although it holds for up-closed multirelations. To see this, note that it is an axiom of 'algebras of monotonic Boolean transformers' of [22] or consider the following proof. Let R be an up-closed multirelation. Then we have $R(\mathsf{E}\setminus\mathsf{E}) = R;\mathsf{E} = R$. By a Schröder equivalence we get $R^{\mathsf{c}}\overline{R} \subseteq \mathsf{E}^{\mathsf{c}}\overline{\mathsf{E}} \subseteq \mathsf{T}\overline{\mathsf{E}}$. Hence, $\mathsf{T}R^{\mathsf{c}}\overline{R}\mathsf{C} \subseteq \mathsf{T}\overline{\mathsf{E}}\mathsf{C} = \mathsf{TE}$. Another Schröder equivalence gives $\overline{R}\mathsf{C}\overline{\mathsf{E}^{\mathsf{c}}\mathsf{T}} \subseteq \overline{R\mathsf{T}}$. So, the desired result is shown by

$$R;\mathsf{T} \cap R^{\mathsf{d}};\mathsf{O} = R\mathsf{T} \cap R^{\mathsf{d}}(\mathsf{E}\setminus\mathsf{O}) = R\mathsf{T} \cap \overline{R}\mathsf{C}\overline{\mathsf{E}^{\mathsf{c}}\mathsf{T}} \subseteq R\mathsf{T} \cap \overline{R\mathsf{T}} = \mathsf{O}.$$

8 Aumann Contacts and Multirelational Properties

In [1,2,3,4] G. Aumann investigated certain laws for modelling the notion of a contact in topology. Translated into the language of multirelations, he considered for a multirelation $R : A \leftrightarrow 2^A$ the following five axioms:

$$\begin{array}{ll}
(K_0) & \neg\exists x \in A : R_{x,\emptyset} \\
(K_1) & \forall x \in A : R_{x,\{x\}} \\
(K_2) & \forall x \in A : \forall Y, Z \in 2^A : R_{x,Y} \wedge Y \subseteq Z \Rightarrow R_{x,Z} \\
(K_3) & \forall x \in A : \forall Y, Z \in 2^A : R_{x,Y} \wedge (\forall y \in Y : R_{y,Z}) \Rightarrow R_{x,Z} \\
(K_4) & \forall x \in A : \forall Y, Z \in 2^A : R_{x,Y\cup Z} \Leftrightarrow R_{x,Y} \vee R_{x,Z}
\end{array}$$

Aumann called multirelations satisfying the formulas (K_1) to (K_3) 'contact relations' and multirelations satisfying the formulas (K_0) to (K_4) 'topological contact relations'. In this section we give multirelation-algebraic characterisations of these logical formulas. See [26] for the relation-algebraic treatment of a correspondence between contact relations and closure operations. Axioms (K_0), (K_2) and (K_4) generalise to multirelations of type $A \leftrightarrow 2^B$ in a straight-forward way. The following result gives the property corresponding to K_0.

Theorem 9. *A multirelation satisfies (K_0) if and only if it is co-total.*

Proof. Axiom (K_0) applied to a multirelation $R : A \leftrightarrow 2^B$ elaborates as follows:

$$\begin{aligned}
\neg\exists x \in A : R_{x,\emptyset} &\iff \forall x \in A : \neg R_{x,\emptyset} \\
&\iff \forall x \in A : \forall X \in 2^B : R_{x,X} \Rightarrow X \neq \emptyset \\
&\iff \forall x \in A : \forall X \in 2^B : R_{x,X} \Rightarrow \exists y \in B : y \in X \\
&\iff \forall x \in A : \forall X \in 2^B : R_{x,X} \Rightarrow \exists y \in B : \mathsf{T}_{x,y} \wedge \mathsf{E}_{y,X} \\
&\iff \forall x \in A : \forall X \in 2^B : R_{x,X} \Rightarrow (\mathsf{TE})_{x,X} \\
&\iff R \subseteq \mathsf{TE} \\
&\iff \mathsf{T}R \subseteq \mathsf{TE} \\
&\iff R\overline{\mathsf{TE}}^{\mathsf{c}} \subseteq \mathsf{O} \\
&\iff R(\mathsf{E}\setminus\mathsf{O}) \subseteq \mathsf{O} \\
&\iff R;\mathsf{O} \subseteq \mathsf{O}
\end{aligned}$$

Hence, the characterisation in Figure 1 shows the claim. □

The forward implication of this theorem is stated in [24], where such multirelations are called 'total'. We call the above property 'co-total' to keep the standard use of 'total' known from relations and functions. Namely,

$$R;\mathsf{T} = R(\mathsf{E} \setminus \mathsf{T}) = R\overline{\mathsf{E}^{\mathsf{c}}\overline{\mathsf{T}}} = R\overline{\mathsf{E}^{\mathsf{c}}\mathsf{O}} = R\overline{\mathsf{O}} = R\mathsf{T}$$

implies that the multirelation-algebraic property $R;\mathsf{T} = \mathsf{T}$ is equivalent to the relation-algebraic property of totality $R\mathsf{T} = \mathsf{T}$. In [23] multirelations R satisfying the property $R;\mathsf{T} = \mathsf{T}$ are called 'proper'. Next, we investigate axiom (K_1) and relate it to a property in Figure 1.

Theorem 10. *Every reflexive multirelation satisfies* (K_1)*. An up-closed multirelation satisfies* (K_1) *if and only if it is reflexive.*

Proof. Axiom (K_1) applied to a multirelation $R : A \leftrightarrow 2^A$ elaborates as follows:

$$\begin{aligned}
\forall x \in A : R_{x,\{x\}} &\iff \forall x \in A : \forall X \in 2^A : \{x\} = X \Rightarrow R_{x,X} \\
&\Longleftarrow \forall x \in A : \forall X \in 2^A : \{x\} \subseteq X \Rightarrow R_{x,X} \\
&\iff \forall x \in A : \forall X \in 2^A : x \in X \Rightarrow R_{x,X} \\
&\iff \forall x \in A : \forall X \in 2^A : \mathsf{E}_{x,X} \Rightarrow R_{x,X} \\
&\iff \mathsf{E} \subseteq R
\end{aligned}$$

Again Figure 1 shows the first claim. If R is up-closed, then the reverse implication holds since $R_{x,\{x\}}$ and $\{x\} \subseteq X$ imply $R_{x,X}$. □

Axiom (K_2) is the logical characterisation of R being an up-closed multirelation. The relation-algebraic characterisation $R = R\mathsf{S}$ is shown in [15, Theorem 6] and the multirelation-algebraic characterisation $R;\mathsf{E} = R$ in [15, Theorem 7.1]. With respect to axiom (K_3), we have the following correspondence.

Theorem 11. *A multirelation satisfies* (K_3) *if and only if it is transitive.*

Proof. Axiom (K_3) applied to a multirelation $R : A \leftrightarrow 2^A$ elaborates as follows:

$$\begin{aligned}
&\forall x \in A : \forall Y, Z \in 2^A : R_{x,Y} \wedge (\forall y \in Y : R_{y,Z}) \Rightarrow R_{x,Z} \\
\iff &\forall x \in A : \forall Y, Z \in 2^A : R_{x,Y} \wedge (\forall y \in A : y \in Y \Rightarrow R_{y,Z}) \Rightarrow R_{x,Z} \\
\iff &\forall x \in A : \forall Y, Z \in 2^A : R_{x,Y} \wedge (\forall y \in A : \mathsf{E}_{y,Y} \Rightarrow R_{y,Z}) \Rightarrow R_{x,Z} \\
\iff &\forall x \in A : \forall Y, Z \in 2^A : R_{x,Y} \wedge (\mathsf{E} \setminus R)_{Y,Z} \Rightarrow R_{x,Z} \\
\iff &\forall x \in A : \forall Z \in 2^A : (\exists Y \in 2^A : R_{x,Y} \wedge (\mathsf{E} \setminus R)_{Y,Z}) \Rightarrow R_{x,Z} \\
\iff &\forall x \in A : \forall Z \in 2^A : (R(\mathsf{E} \setminus R))_{x,Z} \Rightarrow R_{x,Z} \\
\iff &R(\mathsf{E} \setminus R) \subseteq R \\
\iff &R;R \subseteq R
\end{aligned}$$

Again Figure 1 shows the claim. □

Taken together, the axioms (K_1) to (K_3) of Aumann are equivalent to multirelations being reflexive, up-closed and transitive (or even idempotent, since reflexive implies dense). Finally, we investigate axiom (K_4). Here we obtain the following results.

Theorem 12. *Multirelations satisfying* (K_4) *are* $\cup$*-distributive. An up-closed multirelation satisfies* (K_4) *if and only if it is* $\cup$*-distributive.*

Proof. Let $R : A \leftrightarrow 2^B$ be a multirelation such that axiom (K_4) holds. Because of inclusion (8) of Theorem 2 we only have to show $R;(P \cup Q) \subseteq R;P \cup R;Q$ for all multirelations $P : B \leftrightarrow 2^C$ and $Q : B \leftrightarrow 2^C$ to verify the first claim. To this end let $x \in A$ and $X \in 2^C$ such that $(R;(P \cup Q))_{x,X}$. Then there exists $W \in 2^B$ such that $R_{x,W}$ and for all $y \in W$ also $P_{y,X}$ or $Q_{y,X}$. We define two sets $Y, Z \in 2^B$ as subsets of W as follows:

$$Y := \{y \in W \mid P_{y,X}\} \qquad Z := \{y \in W \mid Q_{y,X}\}$$

Then we get $W = Y \cup Z$. Hence, we have $R_{x,Y}$ or $R_{x,Z}$ by the assumption that (K_4) holds. In the first case this shows $(R;P)_{x,X}$, since $P_{y,X}$ for all $y \in Y$, and in the second case $(R;Q)_{x,X}$.

To prove the second claim, assume that R is up-closed and $\cup$-distributive and let $x \in A$ and $Y, Z \in 2^B$ be given. First, suppose $R_{x,Y \cup Z}$. We define the up-closed multirelations $P : A \leftrightarrow 2^B$ and $Q : A \leftrightarrow 2^B$ as follows:

$$P := \{(x, X) \in R \mid x \in X \cap Y\} \qquad Q := \{(x, X) \in R \mid x \in X \cap Z\}$$

Then we have $P_{y,Y \cup Z}$ for all $y \in Y$ and also $Q_{y,Y \cup Z}$ for all $y \in Z$. This leads to $(P \cup Q)_{y,Y \cup Z}$ for all $y \in Y \cup Z$, which gives $(R;(P \cup Q))_{x,Y \cup Z}$. By the assumption $(R;P)_{x,Y \cup Z}$ or $(R;Q)_{x,Y \cup Z}$ holds. In the first case there exists $W \in 2^B$ such that $R_{x,W}$ and $P_{y,Y \cup Z}$ for all $y \in W$. The definition of P implies that $y \in Y$ for all $y \in W$, thus $W \subseteq Y$. Since R is up-closed, this shows $R_{x,Y}$. In the second case, $R_{x,Z}$ follows analogously using the definition of Q. Altogether, $R_{x,Y \cup Z}$ implies $R_{x,Y}$ or $R_{x,Z}$. To prove the converse implication, suppose $R_{x,Y}$ or $R_{x,Z}$. In both cases we then get $R_{x,Y \cup Z}$ since R is up-closed. □

Extended to arbitrary non-empty unions, axiom (K_4) is called 'additive' in [23], which also states that additive up-closed multirelations are $\cup$-distributive.

Finally we consider the dual property of axiom (K_4), that is, the following logical formula for a given multirelation $R : A \leftrightarrow 2^B$:

$$(K_4') \quad \forall x \in A : \forall Y, Z \in 2^B : R_{x,Y} \wedge R_{x,Z} \Leftrightarrow R_{x,Y \cap Z}$$

Extended to arbitrary non-empty unions, this is called 'multiplicative' in [24], which also states that multiplicative up-closed multirelations are $\cap$-distributive. Similarly to the proof of Theorem 12 the following result can be shown.

Theorem 13. *Multirelations satisfying* (K_4') *are* $\cap$*-distributive. An up-closed multirelation satisfies* (K_4') *if and only if it is* $\cap$*-distributive.*

9 Conclusion

In this paper we investigated multirelations using relation algebras and more general algebraic structures. In particular, we considered various properties of multirelations that have been used in applications and we studied transitive closures, closure properties and Aumann contacts.

In Figure 1 we also mentioned vectors and tests and we will close with some remarks concerning these notions. Relational tests are used to represent sets. Such a test is a relation $p : A \leftrightarrow A$ with $p \subseteq \mathsf{I}$ and represents the set $\{x \in A \mid p_{x,x}\}$. A straight-forward generalisation to multirelations would take multirelations which are contained in the membership relation $\mathsf{E} : A \leftrightarrow 2^A$ as tests. But there are too many such multirelations, most of which are not up-closed. This would lead to problems, as tests are frequently used in combination with multirelational composition to restrict a computation to a set of starting states. As a solution, [14] defines multirelational tests as intersections of multirelational vectors in the sense of Figure 1 with membership relations. Hence, a multirelation $R : A \leftrightarrow 2^A$ is a *test* if $R = R;\mathsf{T} \cap \mathsf{E}$, as stated in Figure 1. Using this definition it can be shown that $P : A \leftrightarrow 2^A$ is a multirelational test if and only if there exists a relational test $p : A \leftrightarrow A$ such that $P = p\mathsf{E}$. Furthermore, as for relational tests, composition and intersection of tests coincide, that is, for multirelational tests P and Q we have $P;Q = P \cap Q$.

Acknowledgement. We thank Hitoshi Furusawa and Georg Struth for pointing out related work and the anonymous referees for their helpful comments.

References

1. Aumann, G.: Kontakt-Relationen. Sitzungsberichte der Bayerischen Akademie der Wissenschaften, Mathematisch-Naturwissenschaftliche Klasse, pp. 67–77 (1970)
2. Aumann, G.: Kontakt-Relationen (2. Mitteilung). Sitzungsberichte der Bayerischen Akademie der Wissenschaften, Mathematisch-Naturwissenschaftliche Klasse, pp. 119–122 (1971)
3. Aumann, G.: Kontaktrelationen. Mathematisch-physikalische Semesterberichte zur Pflege des Zusammenhangs von Schule und Universität 20, 182–188 (1973)
4. Aumann, G.: AD ARTEM ULTIMAM – Eine Einführung in die Gedankenwelt der Mathematik. R. Oldenbourg Verlag (1974)
5. Back, R.J., von Wright, J.: Refinement Calculus. Springer, New York (1998)
6. Berghammer, R.: Ordnungen, Verbände und Relationen mit Anwendungen, 2nd edn. Springer (2012)
7. Berghammer, R., Guttmann, W.: A relation-algebraic approach to multirelations and predicate transformers. In: Hinze, R., Voigtländer, J. (eds.) MPC 2015. LNCS, vol. 9129, pp. 50–70. Springer, Heidelberg (2015)
8. Blanchette, J.C., Böhme, S., Paulson, L.C.: Extending Sledgehammer with SMT solvers. In: Bjørner, N., Sofronie-Stokkermans, V. (eds.) CADE 2011. LNCS, vol. 6803, pp. 116–130. Springer, Heidelberg (2011)
9. Blanchette, J.C., Nipkow, T.: Nitpick: A counterexample generator for higher-order logic based on a relational model finder. In: Kaufmann, M., Paulson, L.C. (eds.) ITP 2010. LNCS, vol. 6172, pp. 131–146. Springer, Heidelberg (2010)
10. Cavalcanti, A., Woodcock, J., Dunne, S.: Angelic nondeterminism in the unifying theories of programming. Formal Aspects of Computing 18(3), 288–307 (2006)

11. Furusawa, H., Nishizawa, K., Tsumagari, N.: Multirelational models of lazy, monodic tree, and probabilistic Kleene algebras. Bulletin of Informatics and Cybernetics 41, 11–24 (2009)
12. Furusawa, H., Struth, G.: Concurrent dynamic algebra. ACM Transactions on Computational Logic 16(4:30), 1–38 (2015)
13. Furusawa, H., Tsumagari, N., Nishizawa, K.: A non-probabilistic relational model of probabilistic Kleene algebras. In: Berghammer, R., Möller, B., Struth, G. (eds.) RelMiCS/AKA 2008. LNCS, vol. 4988, pp. 110–122. Springer, Heidelberg (2008)
14. Guttmann, W.: Algebras for correctness of sequential computations. Sci. Comput. Program. 85(Part B), 224–240 (2014)
15. Guttmann, W.: Multirelations with infinite computations. Journal of Logical and Algebraic Methods in Programming 83(2), 194–211 (2014)
16. Martin, C.E., Curtis, S.A.: The algebra of multirelations. Mathematical Structures in Computer Science 23(3), 635–674 (2013)
17. Martin, C.E., Curtis, S.A., Rewitzky, I.: Modelling angelic and demonic nondeterminism with multirelations. Sci. Comput. Program. 65(2), 140–158 (2007)
18. Möller, B.: Kleene getting lazy. Sci. Comput. Program. 65(2), 195–214 (2007)
19. Parikh, R.: Propositional logics of programs: new directions. In: Karpinski, M. (ed.) FCT 1983. LNCS, vol. 158, pp. 347–359. Springer, Heidelberg (1983)
20. Paulson, L.C., Blanchette, J.C.: Three years of experience with Sledgehammer, a practical link between automatic and interactive theorem provers. In: Sutcliffe, G., Ternovska, E., Schulz, S. (eds.) Proceedings of the 8th International Workshop on the Implementation of Logics, pp. 3–13 (2010)
21. Peleg, D.: Concurrent dynamic logic. J. ACM 34(2), 450–479 (1987)
22. Preoteasa, V.: Algebra of monotonic Boolean transformers. In: Simao, A., Morgan, C. (eds.) SBMF 2011. LNCS, vol. 7021, pp. 140–155. Springer, Heidelberg (2011)
23. Rewitzky, I.: Binary multirelations. In: de Swart, H., Orłowska, E., Schmidt, G., Roubens, M. (eds.) TARSKI. LNCS, vol. 2929, pp. 256–271. Springer, Heidelberg (2003)
24. Rewitzky, I., Brink, C.: Monotone predicate transformers as up-closed multirelations. In: Schmidt, R.A. (ed.) RelMiCS/AKA 2006. LNCS, vol. 4136, pp. 311–327. Springer, Heidelberg (2006)
25. Schmidt, G.: Relational Mathematics. Cambridge University Press (2011)
26. Schmidt, G., Berghammer, R.: Contact, closure, topology, and the linking of row and column types of relations. Journal of Logic and Algebraic Programming 80(6), 339–361 (2011)
27. Schmidt, G., Hattensperger, C., Winter, M.: Heterogeneous relation algebra. In: Brink, C., Kahl, W., Schmidt, G. (eds.) Relational Methods in Computer Science, ch. 3, pp. 39–53. Springer, Wien (1997)
28. Tsumagari, N., Nishizawa, K., Furusawa, H.: Multirelational model of lazy Kleene algebra. In: Berghammer, R., Möller, B., Struth, G. (eds.) Relations and Kleene Algebra in Computer Science: PhD Programme at RelMiCS10/AKA5, pp. 73–77. Report 2008-04, Institut für Informatik, Universität Augsburg (2008)
29. von Wright, J.: Towards a refinement algebra. Sci. Comput. Program. 51(1–2), 23–45 (2004)

Relational Formalisations of Compositions and Liftings of Multirelations

Hitoshi Furusawa[1], Yasuo Kawahara[2], Georg Struth[3], and Norihiro Tsumagari[4]

[1] Department of Mathematics and Computer Science, Kagoshima University
furusawa@sci.kagoshima-u.ac.jp
[2] Professor Emeritus, Kyushu University
kawahara@i.kyushu-u.ac.jp
[3] Department of Computer Science, The University of Sheffield
g.struth@sheffield.ac.uk
[4] Center for Education and Innovation, Sojo University
tsumagari@ed.sojo-u.ac.jp

Abstract. Multirelations are studied as a semantic domain for computing systems involving two dual kinds of nondeterminism. This paper presents relational formalisations of Kleisli, Parikh and Peleg's compositions and liftings of multirelations.

1 Introduction

A multirelation is a binary relation between a set and the powerset of a set. Applications of multirelations include reasoning about games with cooperation [11,16] and reasoning about computing systems with alternation [2,7,12,13] or dual angelic and demonic nondeterminism [1].

This paper studies three kinds of compositions of multirelations. Given multirelations $R \subseteq X \times \wp(Y)$ and $S \subseteq Y \times \wp(Z)$, the compositions, respectively called Kleisli, Parikh and Peleg's composition, are defined by

$$(a, A) \in R \circ S \leftrightarrow \exists B. (a, B) \in R \wedge A = \bigcup S(B),$$
$$(a, A) \in R \diamond S \leftrightarrow \exists B. (a, B) \in R \wedge (\forall b \in B. (b, A) \in S),$$
$$(a, A) \in R * S \leftrightarrow \exists B. (a, B) \in R \wedge (\exists f. (\forall b \in B. (b, f(b)) \in S) \wedge A = \bigcup f(B)),$$

where $S(B) = \{C \in \wp(Z) \mid \exists b \in B. (b, C) \in S\}$. Kleisli's composition is inspired by the definition of the Kleisli category for a monad (triple) [8]. Parikh's composition has been proposed for the semantics of game logic [11]. Peleg's composition has been introduced in the context of concurrent dynamic logic [12,13]. It has been discussed further by Goldblatt [7] as well as Furusawa and Struth [5,6].

Although multirelations are just relations of a particular type pattern, the three notions of composition introduced are different from the usual composition of binary relations. The main contribution of this paper is the study of liftings on multirelations that translate Kleisli, Parikh and Peleg's nonstandard compositions on multirelations to a standard relational composition on lifted binary relations. This approach seems crucial for studying algebras of multirelations in the setting of enriched category theory. More precisely, for multirelations

W. Kahl et al. (Eds.): RAMiCS 2015, LNCS 9348, pp. 84–100, 2015.
DOI: 10.1007/978-3-319-24704-5_6

$R \subseteq X \times \wp(Y)$ and $S \subseteq Y \times \wp(Z)$ we wish to lift S to a relation $\lambda(S)$ of type $\wp(Y) \times \wp(Z)$ to be able to translate Kleisli, Parikh and Peleg's definition back to relational composition. We call such a relation $\lambda(S)$ a *lifting* of S and, in particular, liftings $S_\circ, S_\diamond, S_* \subseteq \wp(Y) \times \wp(Z)$ for the three kinds of compositions $\circ$, $\diamond$, $*$ are defined by

$$(B, A) \in S_\circ \leftrightarrow A = \bigcup S(B),$$
$$(B, A) \in S_\diamond \leftrightarrow \forall b \in B. (b, A) \in S,$$
$$(B, A) \in S_* \leftrightarrow \exists f. (\forall b \in B. (b, f(b)) \in S) \wedge A = \bigcup f(B).$$

We call them Kleisli, Parikh and Peleg lifting, respectively.

Martin and Curtis [9] have established categorical foundations of up-closed multirelations with Parikh's composition. They have collated essential definitions and laws of up-closed multirelations with this composition, and placed them in allegories subject to certain conditions. The relational definition of Parikh's composition has been given by them through the Parikh lifting.

Beyond their work, we add the relational definition for the other two compositions and study some of their properties by relational reasoning. In particular, Peleg's composition is studied in detail. It is known that Peleg's composition need not be associative [5]. However, when restricting our attention to the class of union-closed multirelations, this composition becomes associative and then the class forms a category with this composition. Therefore, this work is an attempt to establish categorical foundations of union-closed multirelations with Peleg's composition.

2 Preliminaries

In this article we denote by I a singleton set. A (binary) relation α from set X to set Y, written $\alpha : X \rightharpoonup Y$, is a subset $\alpha \subseteq X \times Y$. The empty relation $0_{XY} : X \rightharpoonup Y$ and the universal relation $\nabla_{XY} : X \rightharpoonup Y$ are defined by $0_{XY} = \emptyset$ and $\nabla_{XY} = X \times Y$, respectively. The converse of relation $\alpha : X \rightharpoonup Y$ is denoted by $\alpha^\sharp$. The identity relation $\{(x, x) \mid x \in X\}$ over X is denoted by id_X. For relation $\alpha : X \rightharpoonup Y$, the partial identity $\{(x, x) \mid \exists y.\ (x, y) \in \alpha\}$ is denoted by $\lfloor \alpha \rfloor$ and it is called *domain* relation of α. The standard composition of relations (which includes functions) will be denoted by juxtaposition. For example, the composite of relation $\alpha : X \rightharpoonup Y$ followed by $\beta : Y \rightharpoonup Z$ is denoted by $\alpha\beta$, and of course the composition of functions $f : X \to Y$ and $g : Y \to Z$ by fg. In addition, the traditional notation $f(x)$ is written xf as a composite of functions $x : I \to X$ and $f : X \to Y$. Note that for a relation $\alpha : X \rightharpoonup Y$, α is *univalent* iff $\alpha^\sharp\alpha \sqsubseteq \mathrm{id}_Y$, and it is *total* iff $\mathrm{id}_X \sqsubseteq \alpha\alpha^\sharp$. So, α is a *partial function* (*pfn*, for short) iff $\alpha^\sharp\alpha \sqsubseteq \mathrm{id}_Y$, and a (*total*) *function* (*tfn*, for short) iff $\alpha^\sharp\alpha \sqsubseteq \mathrm{id}_Y$ and $\mathrm{id}_X \sqsubseteq \alpha\alpha^\sharp$. Moreover, a singleton set I satisfies $0_{II} \neq \mathrm{id}_I = \nabla_{II}$ and $\nabla_{XI}\nabla_{IX} = \nabla_{XX}$ for all sets X. A tfn $x : I \to X$ is called *I-point* of X and is denoted by $x \dot{\in} X$. It is easy to see that $xx^\sharp = x\nabla_{XI} = \mathrm{id}_I$. For a relation $\rho : I \rightharpoonup X$ and an I-point $x : I \to X$, we write $x \dot{\in} \rho$ instead of $x \sqsubseteq \rho$.

Some proofs refer the axiom of subobjects (Sub) and the Dedekind formula (DF), i.e.

$$\begin{array}{ll}\text{(Sub)} & \forall \rho : I \rightharpoonup X\ \exists j : S \to X.\ (\,\rho = \nabla_{IS}j\,) \wedge (\,jj^\sharp = \mathrm{id}_S\,),\\ \text{(DF)} & \alpha\beta \sqcap \gamma \sqsubseteq \alpha(\beta \sqcap \alpha^\sharp\gamma).\end{array}$$

In fact, the subset $S \subseteq X$ and tfn $j : S \to X$ from (Sub) are $S = \{x \mid (*, x) \in \rho\}$ and $j = \{(x, x) \mid (*, x) \in \rho\}$. Note that (DF) is equivalent to

$$\text{(DF}_*\text{)}\ \alpha\beta \sqcap \gamma \sqsubseteq (\alpha \sqcap \gamma\beta^\sharp)(\beta \sqcap \alpha^\sharp\gamma).$$

Also note that the equation $\nabla_{ZY}(\nabla_{YX}\alpha \sqcap \mathrm{id}_Y) = \nabla_{ZX}\alpha$ follows from (DF). See [14] for more details on basic properties of relations.

2.1 Subidentities and Domain Relations

First, we list some basic properties of subidentities.

Proposition 1. *Let $\alpha : X \rightharpoonup Y$ be relation and $v, v' \sqsubseteq \mathrm{id}_Y$.*

(a) $\alpha \sqcap \nabla_{XY}v = \alpha v$.
(b) $v \sqsubseteq v' \leftrightarrow \nabla_{YY}v \sqsubseteq \nabla_{YY}v'$.
(c) $v = v' \leftrightarrow \nabla_{YY}v = \nabla_{YY}v'$. □

The domain relation $\lfloor\alpha\rfloor \sqsubseteq \mathrm{id}_X$ of a relation $\alpha : X \rightharpoonup Y$ can be defined explicitly as

$$\lfloor\alpha\rfloor = \alpha\alpha^\sharp \sqcap \mathrm{id}_X = \nabla_{XY}\alpha^\sharp \sqcap \mathrm{id}_X.$$

Proposition 2. *Let $\alpha, \alpha' : X \rightharpoonup Y$ and $\beta : Y \rightharpoonup Z$ be relations.*

(a) $\alpha = \lfloor\alpha\rfloor\alpha$.
(b) $\lfloor\alpha\beta\rfloor \sqsubseteq \lfloor\alpha\rfloor$ *and* $\lfloor\alpha\beta\rfloor = \lfloor\alpha\lfloor\beta\rfloor\rfloor$.
(c) $\lfloor\alpha \sqcap \alpha'\rfloor = \alpha\alpha'^\sharp \sqcap \mathrm{id}_X$.
(d) *If β is total, then $\lfloor\alpha\beta\rfloor = \lfloor\alpha\rfloor$.*
(e) *If $v \sqsubseteq \mathrm{id}_X$, then $\lfloor v\alpha\rfloor = v\lfloor\alpha\rfloor$.*
(f) $\nabla_{XX}\lfloor\alpha\rfloor = \nabla_{XY}\alpha^\sharp$. □

The following properties of partial functions are essential for this paper.

Proposition 3. *Let $\alpha, \beta : X \rightharpoonup Y$ be relations.*

(a) *If β is a pfn satisfying $\alpha \sqsubseteq \beta$ and $\lfloor\alpha\rfloor = \lfloor\beta\rfloor$, then $\alpha = \beta$.*
(b) *If β is a pfn satisfying $\alpha \sqsubseteq \beta$, then $\alpha = \lfloor\alpha\rfloor\beta$.*
(c) *If β is a pfn and $v \sqsubseteq \mathrm{id}_Y$, then $\beta v = \lfloor\beta v\rfloor\beta$.*
(d) *$f = fv$ iff $\lfloor f^\sharp\rfloor \sqsubseteq v$ for each pfn $f : X \rightharpoonup Y$ and $v \sqsubseteq \mathrm{id}_Y$.* □

2.2 Residual Composition

Let $\alpha : X \rightharpoondown Y$ and $\beta : Y \rightharpoondown Z$ be relations. The *left residual composition* $\alpha \lhd \beta$ of α followed by β is a relation such that $\delta \sqsubseteq \alpha \lhd \beta$ iff $\delta\beta^\sharp \sqsubseteq \alpha$. The *right residual composition* $\alpha \rhd \beta$ of α followed by β is a relation such that $\delta \sqsubseteq \alpha \rhd \beta$ iff $\alpha^\sharp\delta \sqsubseteq \beta$. These residual compositions satisfy $\alpha \lhd \beta = (\beta^\sharp \rhd \alpha^\sharp)^\sharp$ and

$$(x,z) \in \alpha \lhd \beta \leftrightarrow \forall y \in Y.\ (\,(x,y) \in \alpha \leftarrow (y,z) \in \beta\,),$$
$$(x,z) \in \alpha \rhd \beta \leftrightarrow \forall y \in Y.\ (\,(x,y) \in \alpha \rightarrow (y,z) \in \beta\,).$$

Proposition 4. *Let* $\alpha, \alpha' : X \rightharpoondown Y$, $\beta, \beta' : Y \rightharpoondown Z$ *and* $\gamma : Z \to W$ *be relations.*

(a) $\alpha' \sqsubseteq \alpha \wedge \beta \sqsubseteq \beta'$ *implies* $\alpha \rhd \beta \sqsubseteq \alpha' \rhd \beta'$ *and* $\alpha \sqsubseteq \alpha' \wedge \beta' \sqsubseteq \beta$ *implies* $\alpha \lhd \beta \sqsubseteq \alpha' \lhd \beta'$.
(b) $\alpha\beta \rhd \gamma = \alpha \rhd (\beta \rhd \gamma)$ *and* $\alpha \lhd \beta\gamma = (\alpha \lhd \beta) \lhd \gamma$.
(c) $(\alpha \sqcup \alpha') \rhd \beta = (\alpha \rhd \beta) \sqcap (\alpha' \rhd \beta)$ *and* $\alpha \lhd (\beta \sqcup \beta') = (\alpha \lhd \beta) \sqcap (\alpha \lhd \beta')$.
(d) $\alpha \rhd (\beta \sqcap \beta') = (\alpha \rhd \beta) \sqcap (\alpha \rhd \beta')$ *and* $(\alpha \sqcap \alpha') \lhd \beta = (\alpha \lhd \beta) \sqcap (\alpha' \lhd \beta)$.
(e) $\alpha : \mathrm{tfn}$ *implies* $\alpha \rhd \beta = \alpha\beta$ *and* $\beta^\sharp : \mathrm{tfn}$ *implies* $\alpha \lhd \beta = \alpha\beta$.
(f) $\alpha(\beta \lhd \gamma) \sqsubseteq \alpha\beta \lhd \gamma$ *and* $(\alpha \rhd \beta)\gamma \sqsubseteq \alpha \rhd \beta\gamma$.
(g) $\alpha : \mathrm{tfn}$ *implies* $\alpha(\beta \lhd \gamma) = \alpha\beta \lhd \gamma$ *and* $\gamma^\sharp : \mathrm{tfn}$ *implies* $(\alpha \rhd \beta)\gamma = \alpha \rhd \beta\gamma$.
(h) $(\alpha \rhd \beta) \lhd \gamma = \alpha \rhd (\beta \lhd \gamma)$. □

2.3 Power Functor $\wp$

The powerset $\wp(Y)$ of a set Y and the *membership relation* $\ni_Y : \wp(Y) \rightharpoondown Y$ satisfy the following laws.

(M1) $(\ni_Y \rhd \ni_Y^\sharp) \sqcap (\ni_Y \lhd \ni_Y^\sharp) \sqsubseteq \mathrm{id}_{\wp(Y)}$,
(M2) $\forall \alpha : X \rightharpoondown Y.\ (\, \lfloor \alpha^@ \rfloor = \mathrm{id}_X \,)$,

where $\alpha^@ = (\alpha \rhd \ni_Y^\sharp) \sqcap (\alpha \lhd \ni_Y^\sharp)$. Note that

$$\begin{aligned}(\alpha^@)^\sharp\alpha^@ &= ((\ni_Y \lhd \alpha^\sharp) \sqcap (\ni_Y \rhd \alpha^\sharp))((\alpha \rhd \ni_Y^\sharp) \sqcap (\alpha \lhd \ni_Y^\sharp)) \\ &\sqsubseteq (\ni_Y \rhd \alpha^\sharp)(\alpha \rhd \ni_Y^\sharp) \sqcap (\ni_Y \lhd \alpha^\sharp)(\alpha \lhd \ni_Y^\sharp) \\ &\sqsubseteq (\ni_Y \rhd \ni_Y^\sharp) \sqcap (\ni_Y \lhd \ni_Y^\sharp) \\ &\sqsubseteq \mathrm{id}_{\wp(Y)}.\end{aligned}$$

The conditions (M1) and (M2) for membership relations assert that the relation $\alpha^@$ is a tfn. The tfn $\alpha^@$ is a unique tfn such that $\alpha^@ \ni_Y = \alpha$, namely $(a,B) \in \alpha^@$ iff $B = \{b \mid (a,b) \in \alpha\}$.

The *order relation* $\Xi_Y : \wp(Y) \rightharpoondown \wp(Y)$ is defined by $\Xi_Y = \ni_Y \rhd \ni_Y^\sharp$. In fact $\Xi_Y = (\ni_Y \lhd \ni_Y^\sharp)^\sharp$ and $(A,B) \in \Xi_Y$ iff $A \subseteq B$. Define a tfn $\wp(\alpha) : \wp(X) \to \wp(Y)$ by $\wp(\alpha) = (\ni_X \alpha)^@$. Then $\wp(\alpha)$ is a unique tfn such that the following diagram commutes.

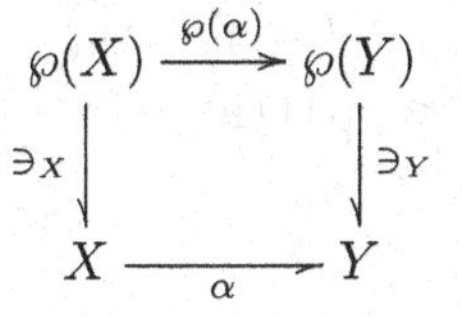

Namely $(A, B) \in \wp(\alpha)$ iff $B = \{b \mid \exists a \in A.\ (a, b) \in \alpha\}$. A tfn $1_X : X \to \wp(X)$ is defined by $1_X = \mathrm{id}_X^{@}$ and is called the *singleton map* on X. In fact, $(x, A) \in 1_X$ iff $A = \{x\}$.

For each set X, $\wp(\mathrm{id}_X)\ni_X = \ni_X \mathrm{id}_X = \mathrm{id}_{\wp(X)}\ni_X$ holds. This shows that $\wp$ preserves the identities. Also, for relations $\alpha : X \rightharpoondown Y$ and $\beta : Y \rightharpoondown Z$ $\wp(\alpha\beta)\ni_Z = \ni_X \alpha\beta = \wp(\alpha)\ni_Y \beta = \wp(\alpha)\wp(\beta)\ni_Z$. This shows that $\wp$ preserves composition. It follows that $\wp$ is a functor from the category *Rel*, which has sets as objects and relations as morphisms, to the category *Set*, which has sets as objects and (total) functions as morphisms.

The isomorphism

$$Set(X, \wp(Y)) \ni f \quad \mapsto \quad f \ni_Y \in Rel(X, Y)$$

is called the *power adjunction* together with its inverse

$$Rel(X, Y) \ni \alpha \quad \mapsto \quad \alpha^{@} \in Set(X, \wp(Y)).$$

Proposition 5. *Let $f, f' : Y \rightharpoondown \wp(Z)$ be pfns. Then $\lfloor f \rfloor = \lfloor f' \rfloor$ and $f\ni_Z = f'\ni_Z$ implies $f = f'$.*

Proof. Assume $\lfloor f \rfloor = \lfloor f' \rfloor$ and $f\ni_Z = f'\ni_Z$. By the axiom of subobjects (Sub) there exists a tfn $j : S \to Y$ such that $\lfloor f \rfloor = j^\sharp j$ and $jj^\sharp = \mathrm{id}_S$. Then both of jf and jf' are tfns. (For $\mathrm{id}_S = jj^\sharp jj^\sharp = j\lfloor f \rfloor j^\sharp \sqsubseteq jff^\sharp j^\sharp$.) As $jf\ni_Z = jf'\ni_Z$ is trivial, by the power adjunction we have $jf = jf'$ and so $f = \lfloor f \rfloor f = j^\sharp jf = j^\sharp jf' = \lfloor f' \rfloor f' = f'$. $\square$

2.4 Power Subidentities

For all subidentities $v \sqsubseteq \mathrm{id}_Y$ define a subidentity $\hat{u}_v \sqsubseteq \mathrm{id}_{\wp(Y)}$ by

$$\hat{u}_v = (\nabla_{\wp(Y)Y} v \lhd \ni_Y^\sharp) \sqcap \mathrm{id}_{\wp(Y)}.$$

The subidentity $\hat{u}_v$ is called the *power subidentity* of v. Note that $(A, A) \in \hat{u}_v$ iff $\forall a \in A.\ (a, a) \in v$.

Proposition 6. *Let v, $v' \sqsubseteq \mathrm{id}_Y$.*

(a) $\hat{u}_v \hat{u}_{v'} = \hat{u}_{vv'}$.
(b) $v \sqsubseteq v'$ *implies* $\hat{u}_v \sqsubseteq \hat{u}_{v'}$.
(c) $\hat{u}_v \wp(v) = \hat{u}_v$.
(d) $\nabla_{Z\wp(Y)} \hat{u}_v = \nabla_{ZY} v \lhd \ni_Y^\sharp$ *for all objects* Z.
(e) $\hat{u}_{id_Y} = \mathrm{id}_{\wp(Y)}$ *and* $\hat{u}_{0_{YY}} = (0_{IY}^{@})^\sharp 0_{IY}^{@}$.

Proof. (a) follows from

$$\begin{aligned}
\hat{u}_v \hat{u}_{v'} &= \hat{u}_v \sqcap \hat{u}_{v'} \\
&= (\nabla v \lhd \ni_Y^\sharp) \sqcap (\nabla v' \lhd \ni_Y^\sharp) \sqcap \mathrm{id}_{\wp(Y)} && \{\ \nabla = \nabla_{\wp(Y)Y}\ \} \\
&= ((\nabla v \sqcap \nabla v') \lhd \ni_Y^\sharp) \sqcap \mathrm{id}_{\wp(Y)} && \{\ 4\,(c)\ \} \\
&= (\nabla vv' \lhd \ni_Y^\sharp) \sqcap \mathrm{id}_{\wp(Y)} && \{\ \nabla v \sqcap \nabla v' = \nabla vv'\ \} \\
&= \hat{u}_{vv'}.
\end{aligned}$$

(b) is a corollary of (a).
(c) First, $\lfloor \hat{u}_v \wp(v) \rfloor = \lfloor \hat{u}_v \rfloor$ is trivial, since $\wp(v)$ is total. Also, by

$$\begin{aligned} \hat{u}_v \ni_Y &= \hat{u}_v \ni_Y \sqcap (\nabla_{\wp(Y)Y} v \lhd \ni_Y^\sharp) \ni_Y \\ &\sqsubseteq \hat{u}_v \ni_Y \sqcap \nabla_{\wp(Y)Y} v \\ &= \hat{u}_v \ni_Y v && \{ 1\,(a) \} \\ &\sqsubseteq \hat{u}_v \ni_Y, && \{ v \sqsubseteq \mathrm{id}_Y \} \end{aligned}$$

$\hat{u}_v \ni_Y = \hat{u}_v \ni_Y v$. So we have $\hat{u}_v \wp(v) \ni_Y = \hat{u}_v \ni_Y v = \hat{u}_v \ni_Y$. Since both of $\hat{u}_v \wp(v)$ and $\hat{u}_v$ are pfns, $\hat{u}_v \wp(v) = \hat{u}_v$ holds by 5.
(d) Since $\nabla_{ZY}(\nabla_{YX}\alpha \sqcap \mathrm{id}_Y) = \nabla_{ZX}\alpha$, we have

$$\begin{aligned} \nabla_{Z\wp(Y)} \hat{u}_v &= \nabla_{Z\wp(Y)}((\nabla_{\wp(Y)Y} v \lhd \varepsilon_Y^\sharp) \sqcap \mathrm{id}_{\wp(Y)}) \\ &= \nabla_{Z\wp(Y)}(\nabla_{\wp(Y)I}(\nabla_{IY} v \lhd \varepsilon_Y^\sharp) \sqcap \mathrm{id}_{\wp(Y)}) && \{ \nabla_{\wp(Y)I} : \mathrm{tfn} \} \\ &= \nabla_{ZI}(\nabla_{IY} v \lhd \varepsilon_Y^\sharp) \\ &= \nabla_{ZI}\nabla_{IY} v \lhd \varepsilon_Y^\sharp && \{ \nabla_{ZI} : \mathrm{tfn} \} \\ &= \nabla_{ZY} v \lhd \varepsilon_Y^\sharp. && \{ \nabla_{ZI}\nabla_{IY} = \nabla_{ZY} \} \end{aligned}$$

(e) The equation (e1) $\hat{u}_{id_Y} = \mathrm{id}_{\wp(Y)}$ follows from

$$\begin{aligned} \hat{u}_{id_Y} &= (\nabla_{\wp(Y)Y} \lhd \ni_Y^\sharp) \sqcap \mathrm{id}_{\wp(Y)} \\ &= \nabla_{\wp(Y)\wp(Y)} \sqcap \mathrm{id}_{\wp(Y)} && \{ \nabla \sqsubseteq \nabla \lhd \alpha \} \\ &= \mathrm{id}_{\wp(Y)}. \end{aligned}$$

Also, the equation (e2) $\hat{u}_{0_{YY}} = (0_{IY}^@)^\sharp 0_{IY}^@$ follows from

$$\begin{aligned} \hat{u}_{0_{YY}} &= (\nabla_{\wp(Y)Y} 0_{YY} \lhd \ni_Y^\sharp) \sqcap \mathrm{id}_{\wp(Y)} \\ &= (\nabla_{\wp(Y)I} 0_{IY} \lhd \ni_Y^\sharp) \sqcap \mathrm{id}_{\wp(Y)} \\ &= \nabla_{\wp(Y)I}(0_{IY} \lhd \ni_Y^\sharp) \sqcap \mathrm{id}_{\wp(Y)} && \{ \nabla_{\wp(Y)I} : \mathrm{tfn} \} \\ &= \nabla_{\wp(Y)I} 0_{IY}^@ \sqcap \mathrm{id}_{\wp(Y)} && \{ 0_{IY} \lhd \ni_Y^\sharp = 0_{IY}^@ \} \\ &= (0_{IY}^@)^\sharp 0_{IY}^@. && \{ 0_{IY}^@ : \mathrm{tfn},\ (\mathrm{DF}) \} \end{aligned}$$

□

3 Compositions and Liftings

The multirelational compositions can be understood as "nonstandard" compositions in the setting of categories of relations that deviate from the standard composition of relations. This section introduces suitable notions of lifting that translate multirelational compositions into the standard relational one. Consider how to define a multirelational composition for $\alpha : X \rightharpoonup \wp(Y)$ and $\beta : Y \rightharpoonup \wp(Z)$. If one can construct a relation $\lambda(\beta) : \wp(Y) \rightharpoonup \wp(Z)$ from β, then a possible composite

$$X \xrightarrow{\alpha} \wp(Y) \xrightarrow{\lambda(\beta)} \wp(Z)$$

is obtained. We call $\lambda(\beta)$ a lifting of β. Liftings enable us to use our knowledge about relations. The complexity of reasoning in particular about Peleg's second-order definition can thus be encapsulated in the lifting and standard relational composition can be used in calculations. For example, it is sufficient to show $\lambda(\beta\lambda(\gamma)) = \lambda(\beta)\lambda(\gamma)$ for associativity of a multirelational composition $\bullet$ since $\alpha \bullet (\beta \bullet \gamma) = \alpha\lambda(\beta\lambda(\gamma))$, $(\alpha \bullet \beta) \bullet \gamma = (\alpha\lambda(\beta))\lambda(\gamma)$ and the composition of relations is associative.

3.1 Kleisli Lifting

A relation $\beta_\circ = \wp(\beta\ni_Z)$ is called the *Kleisli lifting* for β. By definition, the Kleisli lifting is always a tfn. This lifting is used to give a relational definition of the Peleg lifting in Section 3.3.

Proposition 7. *Let $\beta : Y \rightharpoondown \wp(Z)$ and $\gamma : Z \rightharpoondown \wp(W)$ be relations.*

(a) $(\beta\gamma_\circ)_\circ = \beta_\circ\gamma_\circ$.
(b) $(1_Y)_\circ = \mathrm{id}_{\wp(Y)}$.
(c) $(0^@_{YZ})_\circ = 0^@_{\wp(Y)Z}$.
(d) *If β is a pfn, then $\lfloor\beta\rfloor 1_Y\beta_\circ = \beta$.*

Proof. (a) follows from

$$\begin{aligned}(\beta\gamma_\circ)_\circ &= \wp(\beta\gamma_\circ\ni_W)\\ &= \wp(\beta\wp(\gamma\ni_W)\ni_W)\\ &= \wp(\beta\ni_Z\gamma\ni_W) && \{\ \wp(\alpha)\ni_Y = \ni_X\alpha\ \}\\ &= \wp(\beta\ni_Z)\wp(\gamma\ni_W) && \{\ \wp : \text{functor}\ \}\\ &= \beta_\circ\gamma_\circ.\end{aligned}$$

(b) follows from $(1_Y)_\circ 1_Y = \wp(1_Y\ni_Y) = \wp(\mathrm{id}_Y) = \mathrm{id}_{\wp(Y)}$ since $1_Y\ni_Y = \mathrm{id}_Y$.
(c) follows from $(0^@_{YZ})_\circ = \wp(0^@_{YZ}\ni_Z) = \wp(0_{YZ}) = (\ni_Y 0_{YZ})^@ = (0_{\wp(Y)Z})^@$.
(d) Since $\lfloor\lfloor\beta\rfloor 1_Y\beta_\circ\rfloor = \lfloor\lfloor\beta\rfloor\rfloor = \lfloor\beta\rfloor$ and

$$\begin{aligned}\lfloor\beta\rfloor 1_Y\beta_\circ\ni_Z &= \lfloor\beta\rfloor 1_Y\ni_Y\beta\ni_Z\\ &= \lfloor\beta\rfloor\beta\ni_Z && \{\ 1_Y\ni_Y = \mathrm{id}_Y\ \}\\ &= \beta\ni_Z, && \{\ \lfloor\beta\rfloor\beta = \beta\ \}\end{aligned}$$

$\lfloor\beta\rfloor 1_Y\beta_\circ = \beta$ holds by Proposition 5. □

Case (a) of the last proposition ensures that Kleisli's composition $\alpha \circ \beta$ of a $\alpha : X \rightharpoondown \wp(Y)$ followed by $\beta : Y \rightharpoondown \wp(Z)$, which is defined by $\alpha \circ \beta = \alpha\beta_\circ$, is associative in general.

3.2 Parikh Lifting

A relation $\beta_\diamond = \ni_Y \rhd \beta$ is called the *Parikh lifting* for β. This lifting and the composition for this lifting have been studied by Martin and Curtis [9]. However, they have concentrated on up-closed multirelations $\alpha : X \rightharpoondown \wp(Y)$ such that $\alpha\Xi_Y = \alpha$. The following properties are satisfied by multirelations in general.

Proposition 8. *Let $\beta : Y \rightharpoonup \wp(Z)$ and $\gamma : Z \rightharpoonup \wp(W)$ be relations.*

(a) $\beta_\diamond \gamma_\diamond \sqsubseteq (\beta\gamma_\diamond)_\diamond$.
(b) $\gamma_\diamond = \Xi_Z \gamma^{\sharp @ \sharp}$.
(c) $(\beta\gamma_\diamond)_\diamond \sqsubseteq (\beta\Xi_Z)_\diamond \gamma_\diamond$.
(d) $1_Y \Xi_Y = \ni_Y^\sharp$ *and* $1_Y \sqsubseteq \ni_Y^\sharp$.
(e) $\ni_Y^\sharp \beta_\diamond = \beta$.
(f) $(\ni_Z^\sharp)_\diamond = \Xi_Z$.

Proof. (a) follows from

$$\begin{aligned}\beta_\diamond \gamma_\diamond &= (\ni_Y \rhd \beta)\gamma_\diamond \\ &\sqsubseteq \ni_Y \rhd \beta\gamma_\diamond \quad \{ (\alpha \rhd \beta)\gamma \sqsubseteq \alpha \rhd \beta\gamma \} \\ &= (\beta\gamma_\diamond)_\diamond.\end{aligned}$$

(b) follows from

$$\begin{aligned}\gamma_\diamond &= \ni_Z \rhd \gamma \\ &= \ni_Z \rhd \ni_Z^\sharp \gamma^{\sharp @ \sharp} \quad \{ \gamma^\sharp = \gamma^{\sharp @} \ni_Z \} \\ &= (\ni_Z \rhd \ni_Z^\sharp)\gamma^{\sharp @ \sharp} \quad \{ \gamma^{\sharp @} : \text{tfn} \} \\ &= \Xi_Z \gamma^{\sharp @ \sharp}. \quad \{ \ni_Z \rhd \ni_Z^\sharp = \Xi_Z \}\end{aligned}$$

(c) follows from

$$\begin{aligned}(\beta\gamma_\diamond)_\diamond &= \ni_Y \rhd \beta\gamma_\diamond \\ &= \ni_Y \rhd \beta\Xi_Z\gamma^{\sharp @ \sharp} \quad \{ \text{(b) } \gamma_\diamond = \Xi_Z \gamma^{\sharp @ \sharp} \} \\ &= (\ni_Y \rhd \beta\Xi_Z)\gamma^{\sharp @ \sharp} \quad \{ \gamma^{\sharp @} : \text{tfn} \} \\ &= (\beta\Xi_Z)_\diamond \gamma^{\sharp @ \sharp} \\ &\sqsubseteq (\beta\Xi_Z)_\diamond \Xi_Z \gamma^{\sharp @ \sharp} \quad \{ \mathrm{id}_{\wp(Z)} \sqsubseteq \Xi_Z \} \\ &= (\beta\Xi_Z)_\diamond \gamma_\diamond. \quad \{ \text{(b) } \gamma_\diamond = \Xi_Z \gamma^{\sharp @ \sharp} \}\end{aligned}$$

(d) $1_Y \Xi_Y = \ni_Y^\sharp$ follows from

$$\begin{aligned}1_Y \Xi_Y &= 1_Y(\ni_Y \rhd \ni_Y^\sharp) \\ &= 1_Y \ni_Y \rhd \ni_Y^\sharp \\ &= \mathrm{id}_Y \rhd \ni_Y^\sharp \quad \{ 1_Y \ni_Y = \mathrm{id}_Y \} \\ &= \ni_Y^\sharp. \quad \{ \mathrm{id}_Y : \text{tfn} \}\end{aligned}$$

So, $1_Y \sqsubseteq \ni_Y^\sharp$ by $\mathrm{id}_{\wp(Y)} \sqsubseteq \Xi_Y$.
(e) follows from

$$\begin{aligned}\beta &= \mathrm{id}_Y \rhd \beta \quad \{ \mathrm{id}_Y : \text{tfn} \} \\ &= 1_Y \ni_Y \rhd \beta \\ &= 1_Y(\ni_Y \rhd \beta) \quad \{ 1_Y : \text{tfn} \} \\ &\sqsubseteq \ni_Y^\sharp(\ni_Y \rhd \beta) \quad \{ \text{(d) } 1_Y \sqsubseteq \ni_Y^\sharp \} \\ &\sqsubseteq \beta.\end{aligned}$$

(f) is immediate from the definitions of the Parikh lifting and Ξ_Z. □

It is known that Parikh's composition $\alpha \diamond \beta$ of $\alpha : X \rightharpoondown \wp(Y)$ followed by $\beta : Y \rightharpoondown \wp(Z)$, which is defined by $\alpha \diamond \beta = \alpha\beta_\diamond$, need not be associative [15]. So the converse inclusion of (a) need not hold. It is associative for up-closed multirelations, and in fact, (a) and (c) imply this. Also, (e) and (f) imply that the converse of the membership relations serve as the units of Parikh's composition of up-closed multirelations. Equation (b) implies that $\alpha \diamond \beta = \alpha\beta^{\sharp @ \sharp}$ if α is up-closed.

3.3 Peleg Lifting

Before giving a relational definition of the Peleg lifting, we introduce some notation and show a property.

For a relation $\alpha : X \rightharpoondown Y$ the expressions $f \sqsubseteq_p \alpha$ and $f \sqsubseteq_c \alpha$ denote the conditions

$$(f \sqsubseteq \alpha) \wedge (f : \text{pfn}) \text{ and } (f \sqsubseteq \alpha) \wedge (f : \text{pfn}) \wedge (\lfloor f \rfloor = \lfloor \alpha \rfloor).$$

From now on, some proofs refer to the point axiom (PA) and to a variant of the (relational) axiom of choice (AC$_*$), i.e.

$$\begin{array}{ll} \text{(PA)} & \sqcup_{x \dot\in X} x = \nabla_{IX}, \\ (\text{AC}_*) & \forall \alpha : X \rightharpoondown Y.\ [(f \sqsubseteq_p \alpha) \rightarrow \exists f'.\ (f \sqsubseteq f' \sqsubseteq_c \alpha)], \end{array}$$

in addition to (Sub) and (DF). Note that (PA) is equivalent to $\text{id}_X = \sqcup_{x \in X} x^\sharp x$. Also note that (AC$_*$) implies the (relational) axiom of choice

$$\text{(AC)}\ \forall \alpha : X \rightharpoondown Y.\ [(\text{id}_X \sqsubseteq \alpha\alpha^\sharp) \rightarrow \exists f : X \rightarrow Y.\ (f \sqsubseteq \alpha)].$$

Proposition 9. *For all relations $\alpha : X \rightharpoondown Y$, the identity $\alpha = \sqcup_{f \sqsubseteq_c \alpha} f$ holds.*

Proof. The inclusion $\sqcup_{f \sqsubseteq_c \alpha} f \sqsubseteq \alpha$ is clear. It remains to show the converse inclusion. Using the point axiom (PA) we have

$$\alpha = (\sqcup_{x \in X} x^\sharp x)\alpha(\sqcup_{y \in Y} y^\sharp y) = \sqcup_{x \in X} \sqcup_{y \in Y} x^\sharp x \alpha y^\sharp y.$$

Each relation $x^\sharp x \alpha y^\sharp y$ is a pfn and $x^\sharp x \alpha y^\sharp y \sqsubseteq x^\sharp y \sqcap \alpha$. By the axiom of choice (AC$_*$), there is a pfn $f : X \rightharpoondown Y$ such that $x^\sharp x \alpha y^\sharp y \sqsubseteq f \sqsubseteq_c \alpha$. Hence we have $x^\sharp x \alpha y^\sharp y \sqsubseteq \sqcup_{f \sqsubseteq_c \alpha} f$, which proves the converse inclusion $\alpha \sqsubseteq \sqcup_{f \sqsubseteq_c \alpha} f$. □

We now give a relational definition of Peleg lifting for multirelations.

Definition 1. The *Peleg lifting* $\beta_* : \wp(Y) \rightharpoondown \wp(Z)$ of a relation $\beta : Y \rightharpoondown \wp(Z)$ is defined by $\beta_* = \sqcup_{f \sqsubseteq_c \beta} \hat{u}_{\lfloor \beta \rfloor} f_\circ$, where $f_\circ = \wp(f \ni_Z)$ (the Kleisli lifting). □

Proposition 10. *Let $\beta, \beta' : Y \rightharpoondown \wp(Z)$ be relations and $v \sqsubseteq \text{id}_Y$.*

(a) *If $\beta \sqsubseteq \beta'$, then $\beta_* \sqsubseteq \beta'_*$.*
(b) *If β is pfn, then $\beta_* = \hat{u}_{\lfloor \beta \rfloor} \beta_\circ$.*
(c) *If β is pfn, then so is β_*.*

(d) $\beta_* = \sqcup_{f \sqsubseteq_c \beta} f_*$.
(e) $\lfloor \beta_* \rfloor = \hat{u}_{\lfloor \beta \rfloor}$.
(f) $(v\beta)_* = \hat{u}_v \beta_*$.

Proof. (a) Assume $\beta \sqsubseteq \beta'$ and $f \sqsubseteq_c \beta$. By the axiom of choice (AC_*) there exists a pfn f' such that $f \sqsubseteq f' \sqsubseteq_c \beta'$. Then $f = \lfloor f \rfloor f'$ by 3 (b) and hence

$$\begin{aligned}
\hat{u}_{\lfloor \beta \rfloor} f_\circ &= \hat{u}_{\lfloor \beta \rfloor} \wp(f \ni_Z) \\
&= \hat{u}_{\lfloor \beta \rfloor} \wp(\lfloor f \rfloor f' \ni_Z) \quad \{ f = \lfloor f \rfloor f' \} \\
&= \hat{u}_{\lfloor \beta \rfloor} \wp(f' \ni_Z) \quad \{ \lfloor f' \rfloor = \lfloor \beta \rfloor, 6\,(\mathrm{g}) \} \\
&\sqsubseteq \hat{u}_{\lfloor \beta' \rfloor} \wp(f' \ni_Z) \quad \{ \beta \sqsubseteq \beta' \} \\
&= \hat{u}_{\lfloor \beta' \rfloor} f'_\circ,
\end{aligned}$$

which proves the statement.
(b) Let β be a pfn and $f \sqsubseteq_c \beta$. Then $f = \beta$ is immediate from 3 (a). Hence the statement is obvious by the definition of Peleg lifting.
(c) is a corollary of (b).
(d) follows from

$$\begin{aligned}
\beta_* &= \sqcup_{f \sqsubseteq_c \beta} \hat{u}_{\lfloor \beta \rfloor} f_\circ \\
&= \sqcup_{f \sqsubseteq_c \beta} \hat{u}_{\lfloor f \rfloor} f_\circ \quad \{ \lfloor f \rfloor = \lfloor \beta \rfloor \} \\
&= \sqcup_{f \sqsubseteq_c \beta} f_* \,. \quad \{ (\mathrm{b}) \}
\end{aligned}$$

(e) follows from

$$\begin{aligned}
\lfloor \beta_* \rfloor &= \lfloor \sqcup_{f \sqsubseteq_c \beta} \hat{u}_{\lfloor \beta \rfloor} f_\circ \rfloor \\
&= \sqcup_{f \sqsubseteq_c \beta} \lfloor \hat{u}_{\lfloor \beta \rfloor} f_\circ \rfloor \\
&= \sqcup_{f \sqsubseteq_c \beta} \hat{u}_{\lfloor \beta \rfloor} \lfloor f_\circ \rfloor \quad \{ 2\,(\mathrm{e}) \} \\
&= \sqcup_{f \sqsubseteq_c \beta} \hat{u}_{\lfloor \beta \rfloor} \quad \{ f_\circ = \wp(f \ni_Y) : \mathrm{tfn} \} \\
&= \hat{u}_{\lfloor \beta \rfloor}.
\end{aligned}$$

(f) With

$$\hat{u}_v \hat{u}_{\lfloor \beta \rfloor} f_\circ = \hat{u}_{\lfloor \beta \rfloor} \hat{u}_v f_\circ = \hat{u}_{\lfloor \beta \rfloor} \hat{u}_v \wp(f \ni_Z) = \hat{u}_{\lfloor \beta \rfloor} \hat{u}_v \wp(v f \ni_Z) = \hat{u}_{\lfloor v\beta \rfloor} (vf)_\circ,$$

we have

$$\hat{u}_v \beta_* = \sqcup_{f \sqsubseteq_c \beta} \hat{u}_v \hat{u}_{\lfloor \beta \rfloor} f_\circ = \sqcup_{f \sqsubseteq_c \beta} \hat{u}_{\lfloor v\beta \rfloor} (vf)_\circ \sqsubseteq \sqcup_{g \sqsubseteq_c v\beta} \hat{u}_{\lfloor v\beta \rfloor} g_\circ = (v\beta)_*$$

and

$$\begin{aligned}
(v\beta)_* &= \lfloor (v\beta)_* \rfloor (v\beta)_* \quad \{ \alpha = \lfloor \alpha \rfloor \alpha \} \\
&= \hat{u}_{\lfloor v\beta \rfloor} (v\beta)_* \quad \{ (\mathrm{e}) \} \\
&\sqsubseteq \hat{u}_v \beta_*. \quad \{ v\beta \sqsubseteq \beta,\ (\mathrm{a}) \}
\end{aligned}$$

□

The following proposition indicates that the singleton map serves as the unit of Peleg's composition.

Proposition 11. *Let $\beta : Y \rightharpoondown \wp(Z)$ be a relation and $v \sqsubseteq \mathrm{id}_Y$.*

(a) $1_Y \hat{u}_v = v 1_Y$.
(b) $1_Y \beta_* = \beta$.

(c) $(v1_Y)_* = \hat{u}_v$.
(d) $(1_Y)_* = \mathrm{id}_{\wp(Y)}$.

Proof. (a) follows from

$$\begin{aligned}
1_Y\hat{u}_v &= 1_Y((\nabla_{\wp(Y)Y}v \lhd \ni_Y^\sharp) \sqcap \mathrm{id}_{\wp(Y)}) \\
&= (1_Y\nabla_{\wp(Y)Y}v \lhd \ni_Y^\sharp) \sqcap 1_Y && \{\, 1_Y = \mathrm{id}_Y^@ : \mathrm{tfn} \,\} \\
&= (\nabla_{YY}v \lhd \ni_Y^\sharp) \sqcap 1_Y \\
&= ((\nabla_{YY}v \lhd \ni_Y^\sharp)1_Y^\sharp \sqcap \mathrm{id}_Y)1_Y && \{\, (\mathrm{DF}) \,\} \\
&= ((\nabla_{YY}v \lhd \ni_Y^\sharp 1_Y^\sharp) \sqcap \mathrm{id}_Y)1_Y && \{\, 1_Y : \mathrm{tfn} \,\} \\
&= ((\nabla_{YY}v \lhd \mathrm{id}_Y^\sharp) \sqcap \mathrm{id}_Y)1_Y && \{\, 1_Y\ni_Y = \mathrm{id}_Y \,\} \\
&= (\nabla_{YY}v \sqcap \mathrm{id}_Y)1_Y && \{\, \mathrm{id}_Y^\sharp = \mathrm{id}_Y \,\} \\
&= v1_Y.
\end{aligned}$$

(b) By 5, $\lfloor f \rfloor 1_Y f_\circ = f$ holds since it is clear that $\lfloor \lfloor f \rfloor 1_Y f_\circ \rfloor = \lfloor f \rfloor$ and $\lfloor f \rfloor 1_Y f_\circ \ni_Z = \lfloor f \rfloor 1_Y \ni_Y f \ni_Z = \lfloor f \rfloor f \ni_Z = f \ni_Z$. So, we have

$$\begin{aligned}
1_Y\beta_* &= \sqcup_{f \sqsubseteq_c \beta} 1_Y \hat{u}_{\lfloor f \rfloor} f_\circ \\
&= \sqcup_{f \sqsubseteq_c \beta} \lfloor f \rfloor 1_Y f_\circ && \{\, (\mathrm{a})\ 1_Y\hat{u}_v = v1_Y \,\} \\
&= \sqcup_{f \sqsubseteq_c \beta} f \\
&= \beta. && \{\, 9 \,\}
\end{aligned}$$

(c) follows from

$$\begin{aligned}
(v1_Y)_* &= \hat{u}_v \wp(v1_Y\ni_Y) && \{\, \lfloor v1_Y \rfloor = v \,\} \\
&= \hat{u}_v \wp(v) && \{\, 1_Y = \mathrm{id}_Y^@ \,\} \\
&= \hat{u}_v. && \{\, 6\,(\mathrm{b}) \,\}
\end{aligned}$$

(d) is a corollary of (c). □

It is known that Peleg's composition need not be associative [5]. In the rest of this paper, we examine associativity more closely. The following properties are used for it.

Proposition 12. *Let $f : Y \rightharpoonup \wp(Z)$ be a pfn.*

(a) $(v1_Y)_*\beta_* = (v\beta)_*$.
(b) $v \sqsubseteq \lfloor f \rfloor$ *implies* $(vf)_* = \hat{u}_v f_\circ = \hat{u}_v f_*$.

Proof. (a) follows from

$$\begin{aligned}
(v1_Y)_*\beta_* &= \hat{u}_v\beta_* && \{\, 11\,(\mathrm{c}) \,\} \\
&= \sqcup_{f \sqsubseteq_c \beta} \hat{u}_v \hat{u}_{\lfloor \beta \rfloor} f_\circ \\
&= \sqcup_{f \sqsubseteq_c \beta} \hat{u}_{\lfloor \beta \rfloor} \hat{u}_v (vf)_\circ && \{\, 6\,(\mathrm{c}) \,\} \\
&= \sqcup_{f \sqsubseteq_c \beta} \hat{u}_{v\lfloor \beta \rfloor} f_\circ \\
&= (v\beta)_*.
\end{aligned}$$

(b) Assume $v \sqsubseteq \lfloor f \rfloor$. Then

$$\begin{aligned}
(vf)_* &= \hat{u}_{\lfloor vf \rfloor}(vf)_\circ && \{\, 10\,(\mathrm{b}) \,\} \\
&= \hat{u}_v(vf)_\circ && \{\, \lfloor vf \rfloor = v\lfloor f \rfloor = v \,\} \\
&= \hat{u}_v f_\circ && \{\, 6\,(\mathrm{g})\ \hat{u}_v\wp(v) = \hat{u}_v \,\} \\
&= \hat{u}_v \hat{u}_{\lfloor f \rfloor} f_\circ && \{\, \hat{u}_v \sqsubseteq \hat{u}_{\lfloor f \rfloor} \,\} \\
&= \hat{u}_v f_*.
\end{aligned}$$

□

Proposition 13. *Let* $f : Y \rightharpoondown \wp(Z)$ *and* $g : Z \rightharpoondown \wp(W)$ *be pfns and* $\gamma : Z \rightharpoondown \wp(W)$ *a relation.*

(a) $f\gamma_* = \sqcup_{g \sqsubseteq_c \gamma} \lfloor f\hat{u}_{\lfloor\gamma\rfloor} \rfloor f g_\circ$.
(b) $f_*\gamma_* = \sqcup_{g \sqsubseteq_c \gamma} \lfloor f_*\hat{u}_{\lfloor\gamma\rfloor} \rfloor f_\circ g_\circ$.
(c) $f_* g_* = \lfloor f_*\hat{u}_{\lfloor g\rfloor} \rfloor f_\circ g_\circ$.
(d) $(fg_*)_* = \hat{u}_{\lfloor f\hat{u}_{\lfloor g\rfloor}\rfloor} f_\circ g_\circ$.

Proof. (a) follows from $f\gamma_* = \sqcup_{g \sqsubseteq_c \gamma} f\hat{u}_{\lfloor\gamma\rfloor} g_\circ = \sqcup_{g \sqsubseteq_c \gamma} \lfloor f\hat{u}_{\lfloor\gamma\rfloor} \rfloor f g_\circ$ by 3 (c).
(b) follows from

$$\begin{aligned} f_*\gamma_* &= \sqcup_{g \sqsubseteq_c \gamma} f_*\hat{u}_{\lfloor\gamma\rfloor} g_\circ \\ &= \sqcup_{g \sqsubseteq_c \gamma} \lfloor f_*\hat{u}_{\lfloor\gamma\rfloor} \rfloor f_* g_\circ && \{\ 3\,(c)\ fv = \lfloor fv \rfloor f\ \} \\ &= \sqcup_{g \sqsubseteq_c \gamma} \lfloor f_*\hat{u}_{\lfloor\gamma\rfloor} \rfloor \hat{u}_{\lfloor f\rfloor} f_\circ g_\circ \\ &= \sqcup_{g \sqsubseteq_c \gamma} \lfloor f_*\hat{u}_{\lfloor\gamma\rfloor} \rfloor f_\circ g_\circ. && \{\ \lfloor f_*\hat{u}_{\lfloor\gamma\rfloor} \rfloor \sqsubseteq \lfloor f_* \rfloor = \hat{u}_{\lfloor f\rfloor}\ \} \end{aligned}$$

(c) is a particular case of (b) when γ is a pfn.
(d) follows from

$$\begin{aligned} (fg_*)_* &= (\lfloor f\hat{u}_{\lfloor g\rfloor} \rfloor f g_\circ)_* && \{\ (a)\ \} \\ &= \hat{u}_{\lfloor f\hat{u}_{\lfloor g\rfloor}\rfloor} (fg_\circ)_\circ && \{\ 12\,(a)\ \} \\ &= \hat{u}_{\lfloor f\hat{u}_{\lfloor g\rfloor}\rfloor} f_\circ g_\circ. && \{\ 7\,(a)\ \} \end{aligned}$$

□

Proposition 14. *Let* $f : Y \rightharpoondown \wp(Z)$ *be a pfn and* $v \sqsubseteq \mathrm{id}_Z$. *Then the identity* $\lfloor f_*\hat{u}_v \rfloor = \hat{u}_{\lfloor f\hat{u}_v\rfloor}$ *holds.*

Proof. Set $\nabla = \nabla_{\wp(Y)\wp(Z)}$ for short.
(1) $\nabla\hat{u}_v\wp(f{\ni_Z})^\sharp = (\nabla\hat{u}_v \lhd f^\sharp) \lhd \ni_Y^\sharp$:

$$\begin{aligned} \nabla\hat{u}_v\wp(f{\ni_Z})^\sharp &= (\nabla v \lhd \ni_Z^\sharp)\wp(f{\ni_Z})^\sharp && \{\ 6\,(d)\ \} \\ &= \nabla v \lhd \ni_Z^\sharp\wp(f{\ni_Z})^\sharp && \{\ 4\,(e)\ \} \\ &= \nabla v \lhd \ni_Z^\sharp f^\sharp \ni_Y^\sharp && \{\ \wp\ \} \\ &= (\nabla v \lhd \ni_Z^\sharp) \lhd f^\sharp \ni_Y^\sharp && \{\ 4\,(b)\ \} \\ &= \nabla\hat{u}_v \lhd f^\sharp \ni_Y^\sharp && \{\ 6\,(d)\ \} \\ &= (\nabla\hat{u}_v \lhd f^\sharp) \lhd \ni_Y^\sharp. && \{\ 4\,(b)\ \} \end{aligned}$$

(2) $\nabla\hat{u}_v f^\sharp = \nabla f^\sharp \sqcap (\nabla\hat{u}_v \lhd f^\sharp)$:

$$\begin{aligned} \nabla\hat{u}_v f^\sharp &\sqsubseteq \nabla f^\sharp \sqcap (\nabla\hat{u}_v f^\sharp f \lhd f^\sharp) && \{\ \alpha \sqsubseteq \alpha\beta \lhd \beta^\sharp\ \} \\ &\sqsubseteq \nabla f^\sharp \sqcap (\nabla\hat{u}_v \lhd f^\sharp) && \{\ f : \text{pfn}\ \} \\ &\sqsubseteq (\nabla \sqcap (\nabla\hat{u}_v \lhd f^\sharp)f)f^\sharp && \{\ (\text{DF})\ \} \\ &\sqsubseteq \nabla\hat{u}_v f^\sharp. && \{\ (\alpha \lhd \beta)\beta^\sharp \sqsubseteq \alpha\ \} \end{aligned}$$

By (1) and (2),

$$\begin{aligned} \lfloor f_*\hat{u}_v \rfloor &= \hat{u}_{\lfloor f\rfloor} \sqcap \lfloor \wp(f{\ni_Z})\hat{u}_v \rfloor \\ &= \hat{u}_{\lfloor f\rfloor} \sqcap \nabla\hat{u}_v\wp(f{\ni_Z})^\sharp \sqcap \mathrm{id}_{\wp(Y)} \\ &= (\nabla f^\sharp \lhd \ni_Y^\sharp) \sqcap ((\nabla\hat{u}_v \lhd f^\sharp) \lhd \ni_Y^\sharp) \sqcap \mathrm{id}_{\wp(Y)} && \{\ (1)\ \} \\ &= ((\nabla f^\sharp \sqcap (\nabla\hat{u}_v \lhd f^\sharp)) \lhd \ni_Y^\sharp) \sqcap \mathrm{id}_{\wp(Y)} && \{\ 4\,(c)\ \} \\ &= (\nabla\hat{u}_v f^\sharp \lhd \ni_Y^\sharp) \sqcap \mathrm{id}_{\wp(Y)} && \{\ (2)\ \} \\ &= \hat{u}_{\lfloor f\hat{u}_v\rfloor} \end{aligned}$$

holds. This completes the proof. □

This is the first property needed to show associativity of Peleg's composition.

Proposition 15. *If $f : Y \rightharpoondown \wp(Z)$ and $g : Z \rightharpoondown \wp(W)$ are pfns, then $f_* g_* = (f g_*)_*$.*

Proof. It follows from

$$\begin{aligned} f_* g_* &= \lfloor f_* \hat{u}_{\lfloor g \rfloor} \rfloor f_\circ g_\circ && \{\ 13\,(a)\ \} \\ &= \hat{u}_{\lfloor f \hat{u}_{\lfloor g \rfloor} \rfloor} f_\circ g_\circ && \{\ 14\ \lfloor f_* \hat{v} \rfloor = \hat{u}_{\lfloor f \hat{v} \rfloor}\ \} \\ &= (f g_*)_*. && \{\ 13\,(b)\ \} \end{aligned}$$

□

Thus, for $\alpha : X \rightharpoondown \wp(Y)$, the associativity $(\alpha * f) * g = \alpha * (f * g)$ holds if $f : Y \rightharpoondown \wp(Z)$ and $g : Z \rightharpoondown \wp(W)$ are pfns.

Corollary 1. *For relations $\beta : Y \rightharpoondown \wp(Z)$ and $\gamma : Z \rightharpoondown \wp(W)$ the inclusion $\beta_* \gamma_* \sqsubseteq (\beta \gamma_*)_*$ holds.*

Proof. It follows from

$$\begin{aligned} \beta_* \gamma_* &= (\sqcup_{f \sqsubseteq_c \beta} f_*)(\sqcup_{g \sqsubseteq_c \gamma} g_*) && \{\ 10\,(d)\ \} \\ &= \sqcup_{f \sqsubseteq_c \beta} \sqcup_{g \sqsubseteq_c \gamma} f_* g_* \\ &= \sqcup_{f \sqsubseteq_c \beta} \sqcup_{g \sqsubseteq_c \gamma} (f g_*)_* && \{\ 15\ \} \\ &\sqsubseteq (\beta \gamma_*)_*. && \{\ f \sqsubseteq \beta,\ g \sqsubseteq \gamma\ \} \end{aligned}$$

□

So, we have the inclusion $(\alpha * \beta) * \gamma \sqsubseteq \alpha * (\beta * \gamma)$.

The condition for associativity may be relaxed slightly from 15.

Proposition 16. *For a relation $\beta : Y \rightharpoondown \wp(Z)$ and a pfn $g : Z \rightharpoondown \wp(W)$ the identity $\beta_* g_* = (\beta g_*)_*$ holds.*

Proof. As $\beta_* g_* \sqsubseteq (\beta g_*)_*$ by Corollary 1, we need to show the converse inclusion $(\beta g_*)_* \sqsubseteq \beta_* g_*$. Since $(\beta g_*)_* = \sqcup_{h \sqsubseteq_c \beta g_*} h_*$, it suffices to see that $h_* \sqsubseteq \beta_* g_*$ for each pfn $h \sqsubseteq_c \beta g_*$. Assume that $h \sqsubseteq_c \beta g_*$. By the axiom of choice (AC$_*$) there is a pfn $f : Y \rightharpoondown \wp(Z)$ such that $f \sqsubseteq \beta \sqcap h g_*^\sharp$ and $\lfloor f \rfloor = \lfloor \beta \sqcap h g_*^\sharp \rfloor$. Then the following holds.
(1) $\lfloor f \rfloor = \lfloor h \rfloor$:

$$\begin{aligned} \lfloor f \rfloor &= \lfloor \beta \sqcap h g_*^\sharp \rfloor \\ &= \beta g_* h^\sharp \sqcap \mathrm{id}_Y && \{\ \lfloor \alpha \sqcap \beta \rfloor = \alpha \beta^\sharp \sqcap \mathrm{id}\ \} \\ &= \lfloor \beta g_* \sqcap h \rfloor \\ &= \lfloor h \rfloor. && \{\ h \sqsubseteq \beta g_*\ \} \end{aligned}$$

(2) $h \sqsubseteq f g_*$:

$$\begin{aligned} h &= \lfloor h \rfloor h \\ &\sqsubseteq f f^\sharp h && \{\ (1)\ \lfloor h \rfloor = \lfloor f \rfloor \sqsubseteq f f^\sharp\ \} \\ &\sqsubseteq f g_* h^\sharp h && \{\ f \sqsubseteq h g_*^\sharp\ \} \\ &\sqsubseteq f g_*, && \{\ h : \text{pfn}\ \} \end{aligned}$$

(3) $h_* \sqsubseteq \beta_* g_*$:

$$\begin{aligned} h_* &\sqsubseteq (fg_*)_* \quad \{ (2)\ h \sqsubseteq fg_* \} \\ &= f_* g_* \quad \{ 15 \} \\ &\sqsubseteq \beta_* g_*. \quad \{ f \sqsubseteq \beta \} \end{aligned}$$

This completes the proof. □

Thus, the associativity $(\alpha * \beta) * g = \alpha * (\beta * g)$ holds if $g : Z \rightharpoondown \wp(W)$ is pfn.

4 Associativity of Peleg's Composition

Finally we show a more general associative law for Peleg's composition. The following notion has been suggested by Tsumagari [15].

Definition 2. A relation $\gamma : Z \rightharpoondown \wp(W)$ is called *union-closed* if $\lfloor\rho\rfloor(\rho\ni_W)^@ \sqsubseteq \gamma$ for all relations $\rho : Z \rightharpoondown \wp(W)$ such that $\rho \sqsubseteq \gamma$. □

Note that $\gamma : Z \rightharpoondown \wp(W)$ is union-closed iff for each $a \in Z$

$$\mathcal{B} \neq \emptyset \text{ and } \mathcal{B} \subseteq \{B \mid (a, B) \in \gamma\} \text{ imply } (a, \bigcup \mathcal{B}) \in \gamma.$$

For example, every pfn is union-closed, since the identity $\lfloor\rho\rfloor(\rho\ni_W)^@ = \rho$ holds for all pfns $\rho : Z \rightharpoondown \wp(W)$ by $\lfloor\lfloor\rho\rfloor(\rho\ni_W)^@\rfloor = \lfloor\rho\rfloor$ and

$$\lfloor\rho\rfloor(\rho\ni_W)^@\ni_W = \lfloor\rho\rfloor(\rho\ni_W)^@\ni_W = \lfloor\rho\rfloor\rho\ni_W = \rho\ni_W.$$

Proposition 17. *If a relation $\gamma : Z \rightharpoondown \wp(W)$ is union-closed, then for all relations $\rho : Z \rightharpoondown \wp(W)$ with $\rho \sqsubseteq \gamma$ there exists a pfn $g : Z \rightharpoondown \wp(W)$ such that $g \sqsubseteq_c \gamma$ and $\lfloor\rho\rfloor g\ni_W = \rho\ni_W$.*

Proof. As $\lfloor\rho\rfloor(\rho\ni_W)^@$ is a pfn, by the axiom of choice (AC_*) there exists a pfn g such that $\lfloor\rho\rfloor(\rho\ni_W)^@ \sqsubseteq g$ and $g \sqsubseteq_c \gamma$. Hence

$$\begin{aligned} \lfloor\rho\rfloor g\ni_W &= \lfloor\rho\rfloor(\rho\ni_W)^@\ni_W \quad \{ \lfloor\rho\rfloor g = \lfloor\rho\rfloor(\rho\ni_W)^@ \} \\ &= \lfloor\rho\rfloor\rho\ni_W \\ &= \rho\ni_W. \quad \{ \lfloor\rho\rfloor\rho = \rho \} \end{aligned}$$

□

For tfns $f : X \to Y$, $h : X \to X$, and relations $\alpha : X \rightharpoondown Y$, $\beta : Y \rightharpoondown Z$, the following *interchange law* holds:

$$[(f \sqsubseteq \alpha) \wedge (h \sqsubseteq f\beta)] \leftrightarrow [(h \sqsubseteq \alpha\beta) \wedge (f \sqsubseteq h\beta^\sharp \sqcap \alpha)].$$

This interchange law is needed for the proof of the next proposition; and so is the strict point axiom (PA_*), i.e.

$$(PA_*)\ \forall \rho : I \rightharpoondown X.\ (\rho = \sqcup_{x \,\dot{\in}\, \rho} x).$$

Note that (PA_*) implies (PA).

Proposition 18. *Let $\gamma : Z \rightharpoondown \wp(W)$ be a relation, and $f : Y \rightharpoondown \wp(Z)$ and $h : Y \rightharpoondown \wp(W)$ pfns. If γ is union-closed, $h \sqsubseteq f\gamma_*$ and $\lfloor h \rfloor = \lfloor f \rfloor$, then $h_* \sqsubseteq f_*\gamma_*$.*

Proof. For an I-point $A : I \to \wp(X)$, set $u_A = \lfloor (A \ni_X)^\sharp \rfloor$. Let $B : I \to \wp(Y)$ be an I-point (tfn) such that $u_B \sqsubseteq \lfloor h \rfloor$.
(1) $\forall y \sqsubseteq B\ni_Y \exists g_y.\ (\, g_y \sqsubseteq_c \gamma\,) \wedge (\, yh = yfg_{y\circ}\,)$:
Assume $y \sqsubseteq B\ni_Y$. Then $y^\sharp y = \lfloor y^\sharp \rfloor \sqsubseteq \lfloor (B\ni_Y)^\sharp \rfloor = u_B \sqsubseteq \lfloor h \rfloor = \lfloor f \rfloor$. This means that yh and yf are I-points (atoms). Thus

$$\begin{aligned} h \sqsubseteq f\gamma_* &\to yh \sqsubseteq yf\gamma_* \\ &\to yh \sqsubseteq \sqcup_{g \sqsubseteq_c \gamma} yfg_\circ && \{\, g_* \sqsubseteq g_\circ \,\} \\ &\to \exists g_y.\ (\, g_y \sqsubseteq_c \gamma\,) \wedge (\, yh \sqsubseteq yfg_{y\circ}\,) && \{\, yh : \text{atom} \,\} \\ &\to yh = yfg_{y\circ}. && \{\, yh, yfg_{y\circ} : \text{tfn} \,\} \end{aligned}$$

(2) $\forall z \sqsubseteq Bf_\circ \ni_Z.\ \mu_z = z(f\ni_Z)^\sharp \sqcap B\ni_Y \neq 0_{IY}$:

$$\begin{aligned} z &= z \sqcap Bf_\circ \ni_Z && \{\, z \sqsubseteq Bf_\circ \ni_Z \,\} \\ &= z \sqcap B\ni_Y f\ni_Z \\ &\sqsubseteq (z(f\ni_Z)^\sharp \sqcap B\ni_Y)f\ni_Z && \{\, \text{(DF)} \,\} \\ &= \mu_z f \ni_Z. \end{aligned}$$

So, since $z \neq 0_{IZ}$, $\mu_Z \neq 0_{IY}$.
(3) $\exists g_B.\ (\, g_B \sqsubseteq_c \gamma\,) \wedge (\forall z \sqsubseteq Bf_\circ \ni_Z.\ zg_B\ni_W = \sqcup_{y \sqsubseteq \mu_z} zg_y \ni_W\,)$:
Set $\rho_B = \sqcup_{y \sqsubseteq B\ni_Y} u_{yf} g_y$. It is trivial that $\rho_B \sqsubseteq \gamma$ and $\lfloor \rho_B \rfloor = \sqcup_{y \sqsubseteq B\ni_Y} u_{yf} \lfloor \gamma \rfloor$.

$$\begin{aligned} \rho_B &= \sqcup_{y \sqsubseteq B\ni_Y} \sqcup_{z \sqsubseteq yf\ni_Z} z^\sharp z g_y && \{\, u_{uf} = \sqcup_{z \sqsubseteq yf\ni_Z} z^\sharp z \,\} \\ &= \sqcup_{z \sqsubseteq Bf_\circ \ni_Z} \sqcup_{y \sqsubseteq \mu_z} z^\sharp z g_y. && \{\, \text{interchange law} \,\} \end{aligned}$$

Hence $z\rho_B = \sqcup_{y \sqsubseteq \mu_z} zg_y$ for all $z \sqsubseteq Bf_\circ \ni_Z$. On the other hand, by 17 we have

$$\exists g_B.\ g_B \sqsubseteq_c \gamma \wedge \rho_B \ni_W = \lfloor \rho_B \rfloor g_B \ni_W.$$

Hence for all $z \sqsubseteq Bf_\circ \ni_Z$

$$\begin{aligned} zg_B\ni_W &= z\lfloor \rho_B \rfloor g_B \ni_W && \{\, z^\sharp z \sqsubseteq \lfloor \rho_B \rfloor \,\} \\ &= z\rho_B \ni_W && \{\, \rho_B \ni_W = \lfloor \rho_B \rfloor g_B \ni_W \,\} \\ &= \sqcup_{y \sqsubseteq \mu_z} zg_y \ni_W. && \{\, z\rho_B = \sqcup_{y \sqsubseteq \mu_z} zg_y \,\} \end{aligned}$$

(4) $Bh_\circ = Bf_\circ g_{B\circ}$:

$$\begin{aligned} Bh_\circ \ni_W &= B\ni_Y h \ni_W && \{\, h_\circ = \wp(h\ni_W) \,\} \\ &= \sqcup_{y \sqsubseteq B\ni_Y} yh\ni_W && \{\, (\text{PA}_*) \,\} \\ &= \sqcup_{y \sqsubseteq B\ni_Y} yfg_{y\circ}\ni_W && \{\, (1) \,\} \\ &= \sqcup_{y \sqsubseteq B\ni_Y} yf\ni_Z g_y \ni_W \\ &= \sqcup_{y \sqsubseteq B\ni_Y} \sqcup_{z \sqsubseteq yf\ni_Z} zg_y\ni_W && \{\, (\text{PA}_*) \,\} \\ &= \sqcup_{z \sqsubseteq Bf_\circ\ni_Z} \sqcup_{y \sqsubseteq \mu_z} zg_y\ni_W && \{\, \text{interchange law} \,\} \\ &= \sqcup_{z \sqsubseteq Bf_\circ\ni_Z} zg_B\ni_W && \{\, (3) \,\} \\ &= Bf_\circ\ni_Z g_B \ni_W && \{\, (\text{PA}_*) \,\} \\ &= Bf_\circ g_{B\circ}\ni_W. \end{aligned}$$

Hence $Bh_\circ = Bf_\circ g_{B\circ}$, since both sides of the last identity are tfns.
(5) $h_* \sqsubseteq f_*\gamma_*$:

$$\begin{aligned}
h_* &= \lfloor h_* \rfloor h_\circ \\
&= \textstyle\bigsqcup_{u_B \sqsubseteq \lfloor h \rfloor} B^\sharp B h_\circ && \{\ \lfloor h_* \rfloor = \textstyle\bigsqcup_{u_B \sqsubseteq \lfloor h \rfloor} B^\sharp B\ \} \\
&= \textstyle\bigsqcup_{u_B \sqsubseteq \lfloor h \rfloor} B^\sharp B f_\circ g_{B\circ} && \{\ (4)\ \} \\
&\sqsubseteq \textstyle\bigsqcup_{g \sqsubseteq_c \gamma} \bigsqcup_{u_B \sqsubseteq \lfloor h \rfloor} B^\sharp B f_\circ g_\circ \\
&= \textstyle\bigsqcup_{g \sqsubseteq_c \gamma} \lfloor h_* \rfloor f_\circ g_\circ && \{\ \lfloor h_* \rfloor = \textstyle\bigsqcup_{u_B \sqsubseteq \lfloor h \rfloor} B^\sharp B\ \} \\
&= \textstyle\bigsqcup_{g \sqsubseteq_c \gamma} \lfloor f_* \lfloor \gamma_* \rfloor \rfloor f_\circ g_\circ && \{\ \lfloor h_* \rfloor = \lfloor f_* \lfloor \gamma_* \rfloor \rfloor\ \} \\
&= f_*\gamma_*. && \{\ 13\,(\mathrm{b})\ \}
\end{aligned}$$

□

Assume that $h \sqsubseteq_c \beta\gamma_*$ for relations $\beta : Y \rightharpoondown \wp(Z)$ and $\gamma : Z \rightharpoondown \wp(W)$. By (AC$_*$), there is a pfn $f : Y \rightharpoondown \wp(Z)$ such that $f \sqsubseteq \beta \sqcap h\gamma_*^\sharp$ and $\lfloor f \rfloor = \lfloor \beta \sqcap h\gamma_*^\sharp \rfloor$. Then, by similar calculation as for (1) and (2) in the proof of 16, we have $\lfloor h \rfloor = \lfloor f \rfloor$ and $h \sqsubseteq f\gamma_*$. Thus, by 18, $h_* \sqsubseteq \beta_*\gamma_*$ whenever γ is union-closed. Moreover, this implies $(\beta\gamma_*)_* = \sqcup_{h \sqsubseteq_c \beta\gamma_*} h_* \sqsubseteq \sqcup_{f \sqsubseteq_c \beta} f_*\gamma_* = (\sqcup_{f \sqsubseteq_c \beta} f_*)\gamma_* = \beta_*\gamma_*$. Therefore, together with Corollary 1, we have $\beta_*\gamma_* = (\beta\gamma_*)_*$ if γ is union-closed.

5 Conclusion

We have studied three kinds of composition through suitable liftings using relational calculi. We have introduced relational definitions of the Kleisli and Peleg lifting. Then, we have shown that Kleisli's composition is associative, and that the singleton map serves the unit of Peleg's composition. We have also shown some basic properties of Parikh's composition without restriction to up-closed multirelations, in contrast to Martin and Curtis [9]. It is known that Peleg's composition need not be associative [5]. Introducing the notion of union-closed multirelations, we have shown that Peleg's composition becomes associative if the third argument is union-closed. It is obvious that the singleton map is union-closed. Thus, the set of union-closed multirelations forms a category together with Peleg's composition.

The main contribution of this work is the translation from complex non-standard reasoning to well known tools, namely

- reasoning with a complex higher-order set-theoretic definition or a non-associative operation of sequential composition can be replaced by standard relational reasoning, and
- categories of multirelations can be defined and standard category-theoretic tools apply.

This paper has provided all notions and discussions in relational style. However, we mentioned neither allegories [3] nor Dedekind categories [10], which are categorical frameworks suitable for relations, because of the use of the strict point axiom (PA$_*$) which makes a Dedekind category (equivalently, a locally complete division allegory) isomorphic to some full subcategory of the category Rel of sets and relations [4].

Acknowledgement. The presentation of this article has benefitted from the comments of reviewers. The authors acknowledge support by the Royal Society and JSPS KAKENHI grant number 25330016 for this research. They are grateful to Koki Nishizawa and Toshinori Takai for enlightening discussions. The fourth author would like to thank Ichiro Hasuo and members of his group at the University of Tokyo for their generous support.

References

1. Back, R.-J., von Wright, J.: Refinement Calculus: A Systematic Introduction. Springer (1998)
2. Chandra, A.K., Kozen, D., Stockmeyer, L.J.: Alternation. J. ACM 28(1), 114–133 (1981)
3. Freyd, P., Scedrov, A.: Categories, allegories. North-Holland, Amsterdam (1990)
4. Furusawa, H., Kawahara, Y.: Point axioms and related conditions in Dedekind categories. J. Log. Algebr. Meth. Program 84(3), 359–376 (2015)
5. Furusawa, H., Struth, G.: Concurrent Dynamic Algebra. ACM Transactions on Computational Logic (in Press)
6. Furusawa, H., Struth, G.: Taming Multirelations. CoRR abs/1501.05147 (2015)
7. Goldblatt, R.: Parallel Action: Concurrent Dynamic Logic with Independent Modalities. Studia Logica 51(3/4), 551–578 (1992)
8. Mac Lane, S.: Categories for the working mathematician. Springer (1971)
9. Martin, C.E., Curtis, S.A.: The algebra of multirelations. Mathematical Structures in Computer Science 23(3), 635–674 (2013)
10. Olivier, J.-P., Serrato, D.: Catégories de Dedekind. Morphismes dans les Catégories de Schröder. C. R. Acad. Sci. Paris 260, 939–941 (1980)
11. Parikh, R.: Propositional Game Logic. In: FOCS 1983, pp. 195–200. IEEE Computer Society (1983)
12. Peleg, D.: Communication in Concurrent Dynamic Logic. J. Comput. Syst. Sci. 35(1), 23–58 (1987)
13. Peleg, D.: Concurrent dynamic logic. J. ACM 34(2), 450–479 (1987)
14. Schmidt, G.: Relational Mathematics. Encyclopedia of Mathematics and its Applications, vol. 132. Cambridge University Press (2011)
15. Tsumagari, N.: Probability meets Non-Probability via Complete IL-Semi-rings. Ph.D. Thesis, Graduate School of Science and Engineering, Kagoshima University, Japan (2012)
16. van Benthem, J., Ghosh, S., Liu, F.: Modelling simultaneous games in dynamic logic. Synthese 165(2), 247–268 (2008)

Relations among Matrices over a Semiring

Dylan Killingbeck[1], Milene Santos Teixeira[2,*], and Michael Winter[1,**]

[1] Department of Computer Science,
Brock University,
St. Catharines, Ontario, Canada, L2S 3A1
{dk10qt,mwinter}@brocku.ca
[2] Department of Computer Science,
Federal University of Santa Maria,
Santa Maria, Brazil
milene.tsi@gmail.com

Abstract. If $I^{\times}(S)$ denotes the set of (multiplicative) idempotent elements of a commutative semiring S, then a matrix over S is idempotent with respect to the Hadamard product iff all its coefficients are in $I^{\times}(S)$. Since the collection of idempotent matrices can be seen as an embedded structure of binary relations inside the category of matrices over S, we are interested in the relationship between the two structures. In particular, we are interested under which properties the idempotent matrices form a (distributive) allegory.

1 Introduction

Matrices or arrays have been used for centuries in order to solve simultaneous equations. In 1858 Arthur Cayley [4] started to see matrices themselves as mathematical objects by defining operations such as addition and multiplication for matrices and investigating their basic properties. On the other hand, matrices have also been used for representing certain algebraic structures. For example, it is well known that linear maps between (finite dimensional) vector spaces can be represented by matrices with coefficients from the underlying field. A generalization of this is to replace the field by a ring or even a commutative semiring. Another example with a long tradition is the matrix representation of (finite) binary relations [2,18,19]. In fact, it was shown in [21,22] that any suitable category of relations can be represented by matrices. In particular, set-theoretic relations are Boolean matrices, and its generalization to fuzzy resp. L-fuzzy relations leads to matrices with coefficients from the unit interval $[0\ldots1]$ of the real numbers or the lattice L, respectively. It is worth noting that the truth values as well as the unit interval and distributive lattices in general are (commutative and idempotent) semirings. This indicates that matrices over semirings generalize both linear algebra and categories of relations. Because of this common generalization the two theories share a number of similar properties. Furthermore, a matrix over a semiring is idempotent with

* The author gratefully acknowledges support from the CAPES Foundation, Ministry of Education of Brazil, Brasilia - DF. Zip Code 70.040-020.

** The author gratefully acknowledges support from the Natural Sciences and Engineering Research Council of Canada.

W. Kahl et al. (Eds.): RAMiCS 2015, LNCS 9348, pp. 101–118, 2015.
DOI: 10.1007/978-3-319-24704-5_7

respect to the Hadamard product iff all its coefficients are idempotent. This indicates that relations can be identified as the idempotent matrices among all matrices over the semiring.

Relations can be used to model and verify qualitative properties of a problem at hand. For example, relations are widely used to reason about graphs and their properties such as bipartiteness or the existence of Hamiltonian cycles or kernels. On the other hand, linear algebra can be used to represent quantitative properties of the problem. For example, a matrix may describe the probability of a failure of each connection in an interconnected network. A theory that is capable of handling both aspects, i.e., both kinds of matrices, would be very useful for reasoning and software development in this context. However, the operations of a semiring do not necessarily allow to define the basic relation-algebraic operations such as relational composition and join on idempotent matrices. In this paper we are interested in semirings so that the collection of idempotent matrices actually forms a distributive allegory. In addition, we will study some basic properties of the operations involved.

The remainder of the paper is organized as follows. In Section 2 we recall some basic definitions and properties from semirings, lattices and allegories. In order to define the join and composition of relations we introduce sup-semirings in Section 3. We provide two different but equivalent approaches to these structures. Furthermore, we investigate some basic properties of sup-semirings and matrices over sup-semirings. In particular, we study the relationship between relational sums and biproducts with respect to the linear operations.

2 Mathematical Preliminaries

In this section we want to recall some basic definitions and properties of semirings, lattices and allegories. For more details we refer to [3,6,7,9].

Definition 1. *A structure* $\langle S, +, *, 0, 1\rangle$ *is called a semiring iff*

1. $\langle S, +, 0\rangle$ *is a commutative monoid, i.e., we have*
   (a) $x + (y + z) = (x + y) + z$ *for all* $x, y, z \in S$, *(Associativity)*
   (b) $x + 0 = 0 + x = x$ *for all* $x \in S$, *(Identity Law)*
   (c) $x + y = y + x$ *for all* $x, y \in S$. *(Commutativity)*
2. $\langle S, *, 1\rangle$ *is a monoid, i.e., we have*
   (a) $x * (y * z) = (x * y) * z$ *for all* $x, y, z \in S$, *(Associativity)*
   (b) $x * 1 = 1 * x = x$ *for all* $x \in S$. *(Identity Law)*
3. *Multiplication left- and right-distributes over addition, i.e., we have*
   (a) $x * (y + z) = (x * y) + (x * z)$ *for all* $x, y, z \in S$, *(Left Distributivity)*
   (b) $(x + y) * z = (x * z) + (y * z)$ *for all* $x, y, z \in S$. *(Right Distributivity)*
4. *Zero is an annihilator for multiplication, i.e., we have*
   (a) $x * 0 = 0 * x = 0$ *for all* $x \in S$. *(Annihilator Law)*

A semiring is called commutative if $*$ *is commutative, i.e., id we have* $x * y = y * x$ *for all* $x, y \in S$.

We assume that $*$ binds tighter than $+$, and we will use associativity and commutativity of the operations without mentioning. Furthermore, we will use the abbreviation x^2 for $x * x$. An element $x \in D$ is called (multiplicative) idempotent iff $x^2 = x$. We will denote the set of all (multiplicative) idempotent elements of S by $I^\times(S)$ (or $I(S)$ for short).

Lattices and semilattices are defined as usual. Such a structure is called bounded if it has a least and a greatest element. Notice that a lattice is a commutative semiring in which addition and multiplication are both idempotent. This can be generalized for commutative semirings as follows.

Lemma 1. *Let $\langle S, +, *, 0, 1\rangle$ be a commutative semiring. Then $\langle I(S), *, 0, 1\rangle$ is a semilattice with least element* 0 *and greatest element* 1.

Proof. First of all, $I(S)$ is closed under $*$ because the commutativity of $*$ immediately implies $(x * y) * (x * y) = x^2 * y^2 = x * y$ for all $x, y \in I(S)$. The order in $I(S)$ is given by $x \leqslant y$ iff $x * y = x$ so that $x * 1 = x$ and $0 * x = 0$ for all x shows that 0 and 1 are the smallest resp. greatest element in $I(S)$. □

A common approach to relations is based on allegories. These categories generalize the category of binary relations between sets. We will write $R : A \to B$ to indicate that a morphism R of a category $\mathcal{R}$ has source A and target B and we will use $\mathcal{R}[A, B]$ for the collection of all such morphisms. Composition is denoted by ;, which has to be read from left to right. The identity morphism on A is written as $\mathbb{I}_A$.

Definition 2. *An allegory $\mathcal{R}$ is a category satisfying the following:*

1. *For all objects A and B the class $\mathcal{R}[A, B]$ is a semilattice. Meet and the induced ordering are denoted by $\sqcap, \sqsubseteq$, respectively. The elements in $\mathcal{R}[A, B]$ are called relations.*
2. *There is a monotone operation $\breve{}$ (called converse) such that for all relations $Q : A \to B$ and $S : B \to C$ the following holds:*

$$(Q; S)\breve{} = S\breve{}; Q\breve{} \quad \text{and} \quad (Q\breve{})\breve{} = Q.$$

3. *For all relations $Q : A \to B$ and $R, S : B \to C$ we have $Q; (R \sqcap S) \sqsubseteq Q; R \sqcap Q; S$.*
4. *For all relations $Q : A \to B$ and $R : B \to C$ and $S : A \to C$ the modular law $Q; R \sqcap S \sqsubseteq Q; (R \sqcap Q\breve{}; S)$ holds.*

If $\mathcal{R}[A, B]$ are distributive lattices with join $\sqcup$ and least element $\perp\!\!\!\perp_{AB}$ and we have

5. *$Q; \perp\!\!\!\perp_{BC} = \perp\!\!\!\perp_{AC}$ for all relations $Q : A \to B$,*
6. *$Q; (R \sqcup S) = Q; R \sqcup Q; S$ for all relations $Q : A \to B$ and $R, S : B \to C$,*

then $\mathcal{R}$ is called a distributive allegory.

In linear algebra as well as in the theory of relations biproducts are essential in order to combine matrices in an abstract manner [12,15,19,20]. Different (but isomorphic) versions of biproducts can even lead to different algorithms computing certain aspects of matrices [12]. They are also essential in representing abstract categories by categories of matrices [21,22].

In a category with a zero object, i.e., an object that is initial and terminal, a biproduct of two objects A and B is an object that is simultaneously a product and a coproduct of A and B. If every hom-set of the category is a commutative monoid, then biproducts can be defined equationally as follows.

Definition 3. *Let C be a category with a zero object so that every hom-set $C[A, B]$ has $+$ and $⫫_{AB}$ forming a commutative monoid. Furthermore, assume that composition of morphisms is bilinear, i.e., we have $f;(g+h) = f;g + f;h$ and $(g+h);k = g;k + h;k$ for all $f : A \to B$ and $g, h : B \to C$ and $k : C \to D$. Then an object $A \oplus B$ together with morphisms $\pi : A \oplus B \to A$ and $\rho : A \oplus B \to B$ and $\iota : A \to A \oplus B$ and $\kappa : B \to A \oplus B$ is called a biproduct of A and B iff*

$$\iota;\pi = \mathbb{I}_A, \quad \kappa;\rho = \mathbb{I}_B, \quad \iota;\rho = ⫫_{AB}, \quad \kappa;\pi = ⫫_{BA}, \quad \pi;\iota + \rho;\kappa = \mathbb{I}_{A\oplus B}.$$

Notice that a relational sum [19,20] is a biproduct with respect to $\sqcup$ where $\pi = \iota^{\smile}$ and $\rho = \kappa^{\smile}$.

2.1 Matrices over Semirings

In linear algebra matrices with coefficients from a field correspond to linear maps between the corresponding vector spaces. In a more general approach the field is replaced by a semiring. This leads to semiring modules and linear maps. In either case we obtain a commutative monoid structure on the set of matrices with equal size induced by the addition of the semiring. We denote this operation also by $+$, i.e., if $M = [a_{ij}]_{mn}$ denotes a matrix of size $m \times n$ with coefficients a_{ij} from S, then we define

$$[a_{ij}]_{mn} + [b_{ij}]_{mn} = [a_{ij} + b_{ij}]_{mn}$$

and we have

$$(M+N)+P = M+(N+P), \quad M+N = N+M, \quad M+⫫ = ⫫+M = M,$$

where $⫫ = [0]_{mn}$ is the matrix with 0's everywhere. Furthermore, if two finite matrices are of appropriate size, i.e., $M = [a_{ij}]_{mn}$ and $N = [b_{jk}]_{np}$, then matrix multiplication can be defined as usual by

$$[a_{ij}]_{mn}[b_{jk}]_{np} = [\sum_{j=1}^{n} a_{ij} * b_{jk}]_{mp}.$$

Matrix multiplication together with the identity matrix forms a category. Furthermore, we have $M⫫ = ⫫ = ⫫M$ and matrix multiplication is bilinear, i.e., we have

$$M(N+P) = MN + MP, \quad (N+P)Q = NQ + PQ.$$

Last but not least, we may also define the converse (or transpose) of a matrix and the Hadamard product of matrices of equal size by

$$[a_{ij}]^{\smile}_{mn} = [a_{ji}]_{nm}, \qquad [a_{ij}]_{mn} \cdot [b_{ij}]_{mn} = [a_{ij} * b_{ij}]_{mn}.$$

Converse distributes over $+$ and we have $(M \cdot N)^{\smile} = N^{\smile} \cdot M^{\smile}$. Similar to the sum of matrices, the Hadamard product inherits its properties directly from the multiplication of S. If S is a commutative semiring, then so are the matrices of size $m \times n$ with respect to the matrix sum, the Hadamard product, $⫫$ and $\top\!\!\!\top = [1]_{mn}$.

The idempotent matrices (with respect to $\cdot$) are the matrices where every coefficient is from $I(S)$. Therefore, these matrices form a semilattice with least element $\mathbb{L}$ and greatest element $\mathbb{T}$. The following example provides a justification for calling idempotent matrices relations.

Example 1. Consider the matrices over the field of real numbers $\mathbb{R}$. The idempotent matrices are exactly those matrices that only contain 0's and 1's. If we interpret 0 as false and 1 as true, these matrices can be seen as binary relations. For example, consider the following labeled graph and the following two 4×4 matrices:

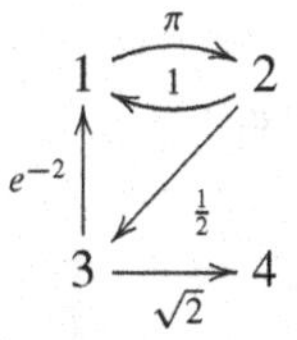

$$\begin{bmatrix} 0 & \pi & 0 & 0 \\ 1 & 0 & \frac{1}{2} & 0 \\ e^{-2} & 0 & 0 & \sqrt{2} \\ 0 & 0 & 0 & 0 \end{bmatrix} \qquad \begin{bmatrix} 0&1&0&0 \\ 1&0&1&0 \\ 1&0&0&1 \\ 0&0&0&0 \end{bmatrix}$$

Both matrices can be seen as matrices on the set $\{1, 2, 3, 4\}$. Each row (column) refers to the element given by the row (column) index. An entry in the matrix represents the connection between the elements in the graph. For example, the π in Row 1 and Column 2 of the first matrix indicates that there is an edge labeled π from 1 to 2 in the graph. A 0 indicates that there is no edge between the corresponding nodes, e.g., the 0 in Row 3 and Column 2 indicates that there is no edge from 3 to 2. The second matrix is idempotent, and, hence, represents a relation on the set $\{1, 2, 3, 4\}$. This relation represents the corresponding unlabeled graph, i.e., the 1 in Row 1 and Column 2 indicates that there is an edge between 1 and 2 and the 0 in Row 3 and Column 2 indicates that there is no edge from 3 to 2. The Hadamard product for relations computes the meet of relations. In this example, relations actually form a Boolean algebra. However, notice that the join of matrices is not induced by any semiring operation or property. If we denote by $\sqcup$ the maximum operation on the set $\{0, 1\}$, we do have $(x \sqcup y) + x * y = x + y$ for $x, y \in \{0, 1\}$ so that we can define $x \sqcup y = (x + y) - x * y$ because of the additive group structure of $\mathbb{R}$. This is not available for arbitrary semirings.

The first matrix is a matrix representation of the labeled graph. The second matrix represents the graph without labels as a relation, i.e., we may obtain the second matrix from the first by only considering connections and ignoring labels. With other words, the qualitative information given by the labels is replaced by the simple (quantitative) connectivity information.

3 Sup-Semirings

In this section we want to investigate the relationship between arbitrary and idempotent elements of a semiring. In addition, we are interested under which circumstances the idempotent elements form a distributive lattice.

3.1 Flattening

We are interested in relating arbitrary matrices to their corresponding relation similarly to the example in the previous section. Our approach first uses a flattening operation on the semiring mapping arbitrary elements to idempotent elements.

Definition 4. *Let $\langle S, +, *, 0, 1\rangle$ be a commutative semiring. An operation $(.)'$ is called a flattening operation iff*

1. $x' * x' = x'$ *for all* $x \in S$,
2. $x * z = x$ *iff* $x' * z = x'$ *for all* $z \in I(S)$.

The right-hand side of (2) in the definition above is equivalent to $x' \leqslant z$ since $x', z \in I(S)$ and $I(S)$ is a semilattice. Therefore, by definition the operation $(.)'$ assigns to an $x \in S$ the smallest idempotent element z so that $x * z = x$, i.e., the smallest idempotent element that keeps x multiplicatively invariant.

Lemma 2. *Let $(.)'$ be a flattening operation. Then we have:*

1. $x' = x$ *iff* $x \in I(S)$.
2. $x'' = x'$.

Proof. 1. If $x' = x$, then we have $x * x = x' * x' = x' = x$, i.e., $x \in I(S)$. Conversely, from $x * x = x$ we get $x' * x = x'$. Similarly, $x' * x' = x'$ implies $x * x' = x$ since $x' \in I(S)$. We conclude $x' = x' * x = x * x' = x$.
2. This follows immediately from (1) since $x' \in I(S)$. □

In a lot of examples the set of idempotent elements only consists of 0 and 1 (see also Lemma 4(3)). In this case we can use the canonical flattening operation.

Lemma 3. *If $I(S) = \{0, 1\}$, then the canonical flattening operation*

$$x' := \begin{cases} 1 \textit{ iff } x \neq 0, \\ 0 \textit{ iff } x = 0, \end{cases}$$

is the only flattening operation.

Proof. First we show that the canonical flattening operation is a flattening operation. The first property is obviously satisfied. Now suppose $z \in I(S) = \{0, 1\}$ and compute

$$\begin{aligned} x * z = x &\iff z = 1 \text{ or } (z = 0 \text{ and } x = 0) \\ &\iff x' * z = x', \end{aligned}$$

verifying the second property. From (2) of the definition of a flattening operation we obtain by using $z = 0$ that $x = 0$ iff $x' = 0$ for every flattening operation. Since $I(S) = \{0, 1\}$ this shows that there is only one flattening operation. □

Example 2. In this example we want to consider the Bayesian, possibilistic or Viterbi semiring $S = \langle [0, 1], \max, *, 0, 1\rangle$ [7,9]. This semiring can be used to model probabilities in networks. Notice that this semiring is isomorphic to the tropical semiring $\langle \mathbb{R}^+ \cup \{\infty\}, \min, +, \infty, 0\rangle$ via the negative logarithm function. Furthermore, we have $I(S) = \{0, 1\}$ so that we can use the canonical flattening operation on S. Let us assume we want to investigate the hypercube network in which every connection has a non-failure rate of 90%, i.e., with a probability of 90% a communication between two adjacent nodes in the network is successful. This situation can be modeled using matrices over the possibilistic semiring. For example, the 3-dimensional hypercube is represented by the following matrices:

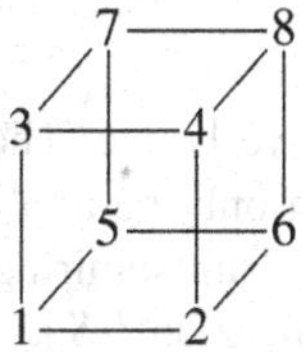

$$\begin{bmatrix} 0 & \frac{9}{10} & \frac{9}{10} & 0 & \frac{9}{10} & 0 & 0 & 0 \\ \frac{9}{10} & 0 & 0 & \frac{9}{10} & 0 & \frac{9}{10} & 0 & 0 \\ \frac{9}{10} & 0 & 0 & \frac{9}{10} & 0 & 0 & \frac{9}{10} & 0 \\ 0 & \frac{9}{10} & \frac{9}{10} & 0 & 0 & 0 & 0 & \frac{9}{10} \\ \frac{9}{10} & 0 & 0 & 0 & 0 & \frac{9}{10} & \frac{9}{10} & 0 \\ 0 & \frac{9}{10} & 0 & 0 & \frac{9}{10} & 0 & 0 & \frac{9}{10} \\ 0 & 0 & \frac{9}{10} & 0 & \frac{9}{10} & 0 & 0 & \frac{9}{10} \\ 0 & 0 & 0 & \frac{9}{10} & 0 & \frac{9}{10} & \frac{9}{10} & 0 \end{bmatrix} \qquad \begin{bmatrix} 0&1&1&0&1&0&0&0 \\ 1&0&0&1&0&1&0&0 \\ 1&0&0&1&0&0&1&0 \\ 0&1&1&0&0&0&0&1 \\ 1&0&0&0&0&1&1&0 \\ 0&1&0&0&1&0&0&1 \\ 0&0&1&0&1&0&0&1 \\ 0&0&0&1&0&1&1&0 \end{bmatrix}$$

As in the previous example the idempotent matrix represents the basic structure of the network ignoring all probabilities, i.e., the second matrix can be obtained from the first by applying the flattening operation $(.)'$ to each of its elements. In Example 4 we will investigate this example even further.

It is worth mentioning that there is a relationship between flattening operations and (dual) discriminator algebras. Recall that a term t resp. d in an algebra is discriminator resp. dual discriminator term if the following is satisfied for all x, y, u:

$$t(x,y,u) = \begin{cases} u \text{ iff } x = y \\ x \text{ otherwise} \end{cases} \qquad d(x,y,u) = \begin{cases} x \text{ iff } x = y \\ u \text{ otherwise} \end{cases}$$

(discriminator) (dual discriminator)

Note that a dual discriminator can always be obtained from a discriminator but not necessarily vice versa [8]. The canonical flattening operation can easily be defined using the dual discriminator, i.e., $x' = d(x, 0, 1)$. Using a discriminator one can define a so-called switching term $s(x, y, u, v) = t(t(x, y, u), t(x, y, v), v)$ that satisfies

$$s(x,y,u,v) = \begin{cases} u \text{ iff } x = y \\ v \text{ otherwise} \end{cases}$$

If $I(S)$ is a finite linear order, then the switching term can be used to define a flattening operation. We want to illustrate this by an example. Suppose $I(D)$ is the linear ordering $0 < a < b < c < 1$ and s is a switching term for D. Then we define

$$x' = s(x * 0, x, 0, s(x * a, x, a, s(x * b, x, b, s(x * c, x, c, 1)))).$$

Suppose $x \in D$ so that the smallest idempotent element z with $x * z = x$ is b, i.e., we should obtain $x' = b$. Notice that for every idempotent element $b \leqslant y$ we have $x = x * b = x * b * y = x * y$. This implies

$$\begin{aligned} x' &= s(x * 0, x, 0, s(x * a, x, a, s(x * b, x, b, s(x * c, x, c, 1)))) \\ &= s(x * 0, x, 0, s(x * a, x, a, s(x * b, x, b, c))) && b \leqslant c \\ &= s(x * 0, x, 0, s(x * a, x, a, b)) && b \leqslant b \\ &= s(x * 0, x, 0, b) && x * a \neq x \\ &= b. && x \neq 0 \end{aligned}$$

A similar definition can be used for semirings with if-then-else [10]. However, a general definition without requiring any additional properties seems not obvious. Further investigation into this relationship is left for future work.

3.2 Distributivity

We are now interested under which conditions $I(S)$ is a distributive lattice. The join operation on $I(S)$, if it exists, is denoted by $\vee$. Notice that $\vee$ can only be applied to elements from $I(S)$ in contrast to the additional operation $\sqcup$ of a sup-semiring (see Section 3.3) which can be applied to all elements and coincides with $\vee$ on $I(S)$.

Lemma 4. *Suppose $\langle S, +, *, 0, 1\rangle$ is a commutative semiring.*

1. *If S is a ring, i.e., a semiring with additive inverses, then $I(S)$ is a distributive lattice with $x \vee y = x + y - x * y$.*
2. *If $+$ satisfies the absorption law $x + x * y = x$ for all $x, y \in S$, then $I(S)$ is a distributive lattice with $x \vee y = x + y$.*
3. *If S is multiplicative cancelative, i.e., $x * y = x * z$ implies $y = z$ for every $x \neq 0$, then $I(S) = \{0, 1\}$ is the Boolean algebra with two elements.*

Proof. 1. Suppose $x, y \in I(S)$. Then we have

$$\begin{aligned}
(x \vee y)^2 &= (x + y - x * y)^2 \\
&= x^2 + x * y - x^2 * y + x * y + y^2 - x * y^2 \\
&\quad - x^2 * y - x * y^2 + x^2 * y^2 && \text{distributivity} \\
&= x + x * y - x * y + x * y + y - x * y \\
&\quad - x * y - x * y + x * y && x, y \text{ idempotent} \\
&= x + y - x * y \\
&= x \vee y,
\end{aligned}$$

i.e., $x \vee y \in I(S)$. The operation $\vee$ is commutative because $+$ and $*$ are. Furthermore, associativity of $\vee$ follows from

$$\begin{aligned}
(x \vee y) \vee z &= (x + y - x * y) + z - (x + y - x * y) * z \\
&= x + y + z - x * y - x * z - y * z + x * y * z && \text{distributivity} \\
&= x + y + z - y * z - x * y - x * z + x * y * z \\
&= x + (y + z - y * z) - x * (y + z - y * z) && \text{distributivity} \\
&= x \vee (y \vee z).
\end{aligned}$$

The two absorption laws are shown by the computation

$$\begin{aligned}
x \vee x * y &= x + x * y - x^2 * y \\
&= x + x * y - x * y && x \text{ idempotent} \\
&= x, \\
x * (x \vee y) &= x * (x + y - x * y) \\
&= x^2 + x * y - x^2 * y && \text{distributivity}
\end{aligned}$$

$$= x + x * y - x * y \qquad x \text{ idempotent}$$
$$= x.$$

Finally, distributivity follows from

$$\begin{aligned} x * (y \vee z) &= x * (y + z - y * z) \\ &= x * y + x * z - x * y * z && \text{distributivity} \\ &= x * y + x * z - x * y * x * z && x \text{ idempotent} \\ &= x * y \vee x * z. \end{aligned}$$

2. If $+$ satisfies the first absorption law, then only the second absorption law remains to be shown. We have

$$\begin{aligned} x * (x + y) &= x^2 + x * y && \text{distributivity} \\ &= x + x * y && x \text{ idempotent} \\ &= x. && \text{assumption} \end{aligned}$$

3. Assume that $0 \neq x \in I(S)$. Then $x * x = x = x * 1$ so that $x = 1$ follows since x is cancelable. □

The second case of the previous lemma is not an unusual situation. For example, the possibilitic semiring, and, hence, the tropical as well as the arctic semiring [7,9], satisfy the absorption law.

3.3 Sup-Semirings

In this section we want to investigate an alternative approach to the combination of a join operation on idempotent elements and a flattening operation. Instead, we are assuming another operation $\sqcup$ that is defined on the semiring.

Definition 5. *A structure $\langle D, +, *, \sqcup, 0, 1\rangle$ is called a sup-semiring iff*

1. *$\langle D, +, *, 0, 1\rangle$ is a commutative semiring.*
2. *$\langle D, \sqcup\rangle$ is a commutative semigroup, i.e., we have*
   (a) *$x \sqcup (y \sqcup z) = (x \sqcup y) \sqcup z$ for all $x, y, z \in D$,* *(Associativity)*
   (b) *$x \sqcup y = y \sqcup x$ for all $x, y \in D$,* *(Commutativity)*
3. *$(x \sqcup y) * (x \sqcup y) = x \sqcup y$ for all $x, y \in D$,* *(Relative Idempotency)*
4. *$x * (x \sqcup y) = x$ for all $x, y \in D$,* *(Absorption)*
5. *if $x * x = x$, then $x \sqcup (x * y) = x$ for all $x, y \in D$,* *(Relative Absorption)*
6. *if $x * x = x$ and $y * y = y$ and $z * z = z$, then $x * (y \sqcup z) = x * y \sqcup x * z$ for all $x, y, z \in D$.* *(Relative Distributivity)*

In the next lemma we have summarized some basic properties of sup-semirings.

Lemma 5. *1. The following two conditions are equivalent:*
   (a) *$x * y = x$ and y idempotent,*
   (b) *$x \sqcup y = y$.*

2. *x is idempotent iff $x \sqcup x = x$.*
3. *$x \sqcup z = z$ and $y \sqcup z = z$ implies $z * (x \sqcup y) = x \sqcup y$.*
4. *$x \sqcup 0 = x \sqcup x$.*
5. *$0 \sqcup 0 = 0$ and $x \sqcup 1 = 1$.*
6. *$x \sqcup y = (x \sqcup 0) \sqcup (y \sqcup 0)$.*

Proof. 1. Assume $x * y = x$ and $y^2 = y$. Then we have

$$\begin{aligned} y &= y \sqcup y * x && \text{rel. absorption} \\ &= x \sqcup y. \end{aligned}$$

Conversely, assume $x \sqcup y = y$. Then we compute

$$\begin{aligned} x &= x * (x \sqcup y) && \text{absorption} \\ &= x * y, \\ y &= y * (x \sqcup y) && \text{absorption} \\ &= y * y. \end{aligned}$$

2. This is a special case of (1) for $x = y$.
3. The assumptions immediately imply $z \sqcup (x \sqcup y) = z$ so that we obtain

$$\begin{aligned} x \sqcup y &= (x \sqcup y) * (z \sqcup (x \sqcup y)) && \text{absorption} \\ &= (x \sqcup y) * z. \end{aligned}$$

4. From $(x \sqcup 0)^2 = x \sqcup 0$ by relative idempotency and $x * (x \sqcup 0) = x$ by absorption we obtain $x \sqcup (x \sqcup 0) = x \sqcup 0$ using (1). From (3) we conclude $(x \sqcup 0) * (x \sqcup x) = x \sqcup x$. On the other hand, we have $(x \sqcup x)^2 = x \sqcup x$ by relative idempotency. From absorption we obtain $x * (x \sqcup x) = x$ and, hence, $x \sqcup (x \sqcup x) = x \sqcup x$ by (1). In addition, $0 * (x \sqcup x) = 0$ implies $0 \sqcup (x \sqcup x) = x \sqcup x$ by using (1) again. From (3) we get $(x \sqcup x) * (x \sqcup 0) = x \sqcup 0$, i.e., together we have $x \sqcup x = (x \sqcup 0) * (x \sqcup x) = (x \sqcup x) * (x \sqcup 0) = x \sqcup 0$.
5. From $0 * 0 = 0$ we obtain $0 \sqcup 0 = 0$ by (2). Since $1 * 1 = 1$ we get $1 = 1 \sqcup (1 * x) = x \sqcup 1$ using relative absorption.
6. $x \sqcup y$ is idempotent by rel. idempotency. We obtain

$$\begin{aligned} x \sqcup y &= (x \sqcup y) \sqcup 0 && \text{by (2) and (4)} \\ &= (x \sqcup y) \sqcup (0 \sqcup 0) && \text{by (5)} \\ &= (x \sqcup 0) \sqcup (y \sqcup 0). && \text{associativity and commutativity} \end{aligned}$$

□

A sup-semiring induces a flattening operation if we define $x' = x \sqcup 0$.

Theorem 1. *Let $\langle D, +, *, \sqcup, 0, 1\rangle$ be a sup-semiring. Then the idempotent elements, i.e., the structure $\langle I(D), *, \sqcup, 0, 1\rangle$, form a distributive lattice. Furthermore, $x' = x \sqcup 0$ is a flattening operation for $\langle D, +, *, 0, 1\rangle$.*

Proof. First we want to show that $I(D)$ is a distributive lattice. This follows immediately from axioms absorption, rel. absorption and rel. distributivity. It remains to show that x' is aflattening operation. First of all, by rel. idempotency $x' \in I(D)$. Furthermore, assume $z \in I(D)$ with $x * z = x$. Then Lemma 5(1) implies $x \sqcup z = z$. We conclude $(x \sqcup x) * z = x \sqcup x$ from Lemma 5(3), and, hence, $x' * z = x'$ because $x' = x \sqcup x$ by Lemma 5(4). Conversely, assume $z \in I(D)$ with $x' * z = x'$. Then Lemma 5(1) implies $x' \sqcup z = z$. Since $z \in I(S)$ we have $z' = z$ by Lemma 5(2&4). We obtain $x \sqcup z = x' \sqcup z' = x' \sqcup z = z$ by applying Lemma 5(6). Using Lemma 5(1) again verifies $x * z = x$. □

In the rest of the paper we will use the abbreviation $x' = x \sqcup 0$ for every sup-semring. On the other hand, any sup-semiring is generated by a flattening operation if the idempotent elements form a distributive lattice.

Theorem 2. *Let $\langle S, +, *, 0, 1\rangle$ be a commutative semiring with a flattening operation. Furthermore, assume that $\langle I(S), \vee, *, 0, 1\rangle$ is a distributive lattice. Then $\langle S, +, *, \sqcup, 0, 1\rangle$ with $x \sqcup y = x' \vee y'$ is a sup-semiring.*

Proof. First of all, we have $x \sqcup y \in I(S)$ by definition. This immediately implies rel. idempotency. Consider the computation

$$\begin{aligned} x' * (x \sqcup y) &= x' * (x' \vee y') \\ &= x'. && I(S) \text{ lattice} \end{aligned}$$

Since $x \sqcup y \in I(S)$ and $(.)'$ is a flattening operation we conclude $x * (x \sqcup y) = x$. Now suppose $x^2 = x$, i.e., $x \in I(S)$. Then from $(x * y) * x = x^2 * y = x * y$ and the fact that $(.)'$ is a flattening operation we obtain $(x * y)' * x = (x * y)'$. This implies

$$\begin{aligned} x &= x * (x \sqcup x * y) && \text{absorption} \\ &= x * (x' \vee (x * y)') \\ &= x * x' \vee x * (x * y)' && I(S) \text{ distributive lattice} \\ &= x' \vee (x * y)' && \text{see above} \\ &= x \sqcup x * y. \end{aligned}$$

Finally, rel. distributivity follows from the distributivity of $I(S)$. □

Two of the axioms of a sup-semiring are implications. One of them, the rel. distributivity, can be replaced by an equation.

Lemma 6. *In the context of the other axioms in Def. 5 rel. distributivity is equivalent to*

$$(*)\quad x' * (y \sqcup z) = x' * y' \sqcup x' * z'.$$

Proof. First we show that (*) is valid. Since x' is idempotent by rel. idempotency we get

$$\begin{aligned} x' * (y \sqcup z) &= x' * (y' \sqcup z') && \text{Lemma 5(6)} \\ &= x' * y' \sqcup x' * z'. && \text{rel. distributivity} \end{aligned}$$

Now, assume that all other axioms and (*) are valid. If x, y and z are idempotent, then $x' = x, y' = y$ and $z' = z$ by Lemma 5(2&4). This immediately implies rel. distributivity. □

The axiom of rel. absorption cannot be replaced by an equation as the following lemma shows.

Lemma 7. *The theory of sup-semirings is not equational, i.e., the class of sup-semirings does not form a variety.*

Proof. We show that the class of sup-semirings is not a variety, and, hence, not definable by equations, by showing that the class is not closed under forming quotients. Consider the semiring $\langle \mathbb{N}, +, *, 0, 1\rangle$. This semiring is multiplicative cancellative so that $I(\mathbb{N}) = \{0, 1\}$ follows from Lemma 4(3). The operations

$$x \sqcup y := \begin{cases} 0, \text{ iff } x = y = 0, \\ 1, \text{ otherwise.} \end{cases}$$

makes $\mathbb{N}$ into a sup-semiring. Let $\equiv$ be the equivalence relation that has the three equivalence classes $[0] = \{0\}, [1] = \{1\}$ and $[n] = \{n \in \mathbb{N} \mid n > 1\}$. It is easy to see that $\equiv$ is a congruence and that the induced operations on the equivalence classes are

+	[0]	[1]	[n]
[0]	[0]	[1]	[n]
[1]	[1]	[n]	[n]
[n]	[n]	[n]	[n]

*	[0]	[1]	[n]
[0]	[0]	[0]	[0]
[1]	[0]	[1]	[n]
[n]	[0]	[n]	[n]

⊔	[0]	[1]	[n]
[0]	[0]	[1]	[1]
[1]	[1]	[1]	[1]
[n]	[1]	[1]	[1]

Now, $[n]$ is idempotent but $[n] \sqcup [n] * [0] = [n] \sqcup [0] = [1] \neq [n]$, i.e., rel. absorption is not true. □

In Lemma 4(1) we have shown that the join operation on idempotent elements is given by $x \vee y = x + y - x * y$ if the semiring has additive inverses, i.e., is a ring. The next lemma (for $n = 2$) shows that there is a similar relationship between $\sqcup$ and $+$ in every sup-semiring if we require $x * y = 0$ in addition.

Lemma 8. *Let $\langle D, +, *, \sqcup, 0, 1\rangle$ be a sup-semiring and $n \geqslant 2$. If x_i is idempotent for $i \in \{1, \dots, n\}$ and $x_i * x_j = 0$ for all $i \neq j$, then $\sum_{i=1}^{n} x_i = \bigsqcup_{i=1}^{n} x_i$.*

Proof. We show this by induction on n. For $n = 2$ we have to show that $x + y = x \sqcup y$ for all idempotent elements x, y with $x * y = 0$. First verify that $x + y$ is idempotent by computing

$$\begin{aligned} (x + y) * (x + y) &= (x + y) * x + (x + y) * y && \text{distributivity} \\ &= x^2 + y * x + x * y + y^2 && \text{distributivity} \\ &= x + y. && \text{assumptions} \end{aligned}$$

This implies

$$\begin{aligned} x + y &= x * (x \sqcup y) + y * (x \sqcup y) && \text{absorption} \\ &= (x + y) * (x \sqcup y) && \text{distributivity} \\ &= (x + y) * x \sqcup (x + y) * y && \text{rel. distributivity} \\ &= (x^2 + y * x) \sqcup (x * y + y^2) && \text{distributivity} \\ &= x \sqcup y. && \text{assumptions} \end{aligned}$$

For the induction step we first compute

$$\left(\bigsqcup_{i=1}^{n} x_i\right) * x_{n+1} = \bigsqcup_{i=1}^{n} x_i * x_{n+1} \qquad \text{rel. distributivity and } x_i \text{ idempotent}$$
$$= 0. \qquad x_i * x_{n+1} = 0$$

This implies

$$\begin{aligned}
\sum_{i=1}^{n+1} x_i &= \left(\sum_{i=1}^{n} x_i\right) + x_{n+1} \\
&= \left(\bigsqcup_{i=1}^{n} x_i\right) + x_{n+1} && \text{induction hypothesis} \\
&= \left(\bigsqcup_{i=1}^{n} x_i\right) \sqcup x_{n+1} && \text{case } n = 2 \text{ and property above} \\
&= \bigsqcup_{i=1}^{n+1} x_i. && \square
\end{aligned}$$

3.4 Matrices over Sup-Semirings

As discussed in Section 2.1 the matrices over a semiring form a category. Furthermore, the collection of idempotent matrices with respect to the Hadamard product forms a semilattice. The additional operation $\sqcup$ of a sup-semiring gives rise to two new operations on matrices. In order to define these operations we first have to define $\sqcup$ for an arbitrary finite number of parameters. Since $\sqcup$ does not have a neutral element this needs clarification if the number of arguments of $\bigsqcup$ is zero or one. Notice that this clarification was not necessary for Lemma 8 since $\bigsqcup$ has at least two arguments in the property mentioned there. We define

$$\bigsqcup_{i=1}^{0} x_i = 0, \text{ and } \bigsqcup_{i=1}^{n+1} x_i = \left(\bigsqcup_{i=1}^{n} x_i\right) \sqcup x_{n+1} \text{ for } n \geqslant 0.$$

Notice that the definition above implies $\bigsqcup_{i=1}^{1} x_i = x_1 \sqcup 0 = x_1'$. The main reason for this definition is that it generalizes Lemma 5(6) to $\bigsqcup_{i=1}^{n} x_i = \bigsqcup_{i=1}^{n} x_i'$ for arbitrary $n \geqslant 0$. Now, we define

$$[a_{ij}]_{mn} \sqcup [b_{ij}]_{mn} = [a_{ij} \sqcup b_{ij}]_{mn}, \qquad [a_{ij}]_{mn}; [b_{jk}]_{np} = [\bigsqcup_{j=1}^{n} a_{ij} * b_{jk}]_{mp}.$$

Notice that $[a];[b] = [a * b \sqcup 0] = [(a * b)'] = [a] \cdot [b] \sqcup [0] = ([a] \cdot [b])'$ for 1×1 matrices due to our definition regarding $\bigsqcup$. With these operations the collection of relations forms a distributive allegory. Within this substructure all regular definitions for and properties of relations can be used. For example, if Q is a relation, then Q is said to be univalent iff $Q^{\smile};Q \sqsubseteq \mathbb{I}$. However, notice that the composition ; is not even associative if we consider all matrices. Similarly, the identity matrix is the identity for ; and relations but not if we consider all matrices. The following example demonstrates these properties.

Example 3. Consider the set $A = \{0, 1, 2, 3\}$ and the operations:

$$\begin{array}{c|cccc} + & 0 & 1 & 2 & 3 \\ \hline 0 & 0 & 1 & 2 & 3 \\ 1 & 1 & 1 & 1 & 1 \\ 2 & 2 & 1 & 2 & 2 \\ 3 & 3 & 1 & 2 & 3 \end{array} \qquad \begin{array}{c|cccc} * & 0 & 1 & 2 & 3 \\ \hline 0 & 0 & 0 & 0 & 0 \\ 1 & 0 & 1 & 2 & 3 \\ 2 & 0 & 2 & 2 & 0 \\ 3 & 0 & 3 & 0 & 0 \end{array} \qquad \begin{array}{c|cccc} \sqcup & 0 & 1 & 2 & 3 \\ \hline 0 & 0 & 1 & 2 & 1 \\ 1 & 1 & 1 & 1 & 1 \\ 2 & 2 & 1 & 2 & 1 \\ 3 & 1 & 1 & 1 & 1 \end{array}$$

This structure is a sup-semiring and we have

$$(2 * 3)' = 0' = 0 \neq 2 = 2 * 1 = 2' * 3'.$$

This immediately implies $([2];[3]);[1] = [(2*3)'];[1] = [0];[1] = [0*1] = [0]$ and $[2];([3];[1]) = [2];[(3*1)'] = [2];[3'] = [2*1] = [2]$, i.e., composition of matrices is not associative. Furthermore, we have

$$\begin{bmatrix} 3 & 0 \\ 0 & 0 \end{bmatrix} ; \begin{bmatrix} 1 & 0 \\ 0 & 1 \end{bmatrix} = \begin{bmatrix} 3*1 \sqcup 0*0 & 3*0 \sqcup 0*1 \\ 0*1 \sqcup 0*0 & 0*0 \sqcup 0*1 \end{bmatrix} = \begin{bmatrix} 1 & 0 \\ 0 & 0 \end{bmatrix}.$$

The previous example indicates that composing a matrix with the identity performs a flattening to each entry of the matrix. The following lemma verifies this property in general.

Lemma 9. *Suppose Q is a matrix over a sup-semiring. Then we have $Q;\mathbb{I} = \mathbb{I};Q = Q'$ where $Q' = Q \sqcup [0]$.*

Proof. We only show $Q;\mathbb{I} = Q'$. The second equation follows analogously. From the computation

$$\begin{aligned} (Q;\mathbb{I})_{ik} &= \bigsqcup_{j=1}^{n} Q_{ij} * \mathbb{I}_{jk} = Q_{ik} * 1 \sqcup 0 && \text{definition } \mathbb{I} \text{ matrix} \\ &= Q_{ik} \sqcup 0 = Q'_{ik} \end{aligned}$$

we immediately conclude $Q;\mathbb{I} = Q'$. □

Lemma 8 shows that $+$ and $\sqcup$ coincide for idempotent and disjoint elements, i.e., if x, y are idempotent and $x*y = 0$. A similar result can be shown for matrix multiplication and the composition operation ;.

Theorem 3. *Suppose Q, R are relations. If Q is univalent, then $QR = Q;R$.*

Proof. Since all coefficients from Q and R are idempotent, and, hence, elements of the distributive lattice $I(D)$, we obtain for $j_1 \neq j_2$

$$\begin{aligned}(Q_{ij_1} * R_{j_1k}) * (Q_{ij_2} * R_{j_2k}) &= Q^{\smile}_{j_1i} * Q_{ij_2} * R_{j_1k} * R_{j_2k} \\ &\leqslant \left(\bigsqcup_i Q^{\smile}_{j_1i} * Q_{ij_2}\right) * R_{j_1k} * R_{j_2k} \\ &= (Q^{\smile};Q)_{j_1j_2} * R_{j_1k} * R_{j_2k} \\ &= 0 \qquad\qquad\qquad\qquad j_1 \neq j_2.\end{aligned}$$

From Lemma 8 we immediately get $(QR)_{ik} = \sum\limits_j Q_{ij} * R_{jk} = \bigsqcup\limits_j Q_{ij} * R_{jk} = (Q;R)_{ik}$. □

In [5] the notion of the "shape" of matrices over the complex numbers using relations was introduced. For an arbitrary complex matrix A and a relation R, we say that A has shape R iff $A \cdot R = A$. Since the component-wise flattening operation is a flattening operation on the semiring of matrices this is equivalent to $A' \cdot R = A'$. The latter equation only uses relations so that it is equivalent to $A' \sqsubseteq R$. Following [5] we call a matrix A diagonal if it has shape $\mathbb{I}$, i.e., $A' \sqsubseteq \mathbb{I}$. This notion led to define a univalent relation Q as a relation so that $Q^\dagger Q$ is diagonal where $Q^\dagger$ is the conjugate transposed matrix. Notice $I(\mathbb{C}) = \{0,1\}$ so that $Q^\dagger = Q^{\smile}$ for (complex) relations. However, this generalization is based on the fact that adding non-zero idempotent elements does not result in 0, which might not be the case in an arbitrary sup-semiring. In fact, this might not even be the case in an arbitrary field. Suppose $\mathbb{F}_2$ is the field with two elements together with the obvious definition of $\sqcup$. Then we have

$$\mathbb{T}^{\smile}\mathbb{T} = \begin{bmatrix} 1\,1 \\ 1\,1 \end{bmatrix}\begin{bmatrix} 1\,1 \\ 1\,1 \end{bmatrix} = \begin{bmatrix} 1*1+1*1 & 1*1+1*1 \\ 1*1+1*1 & 1*1+1*1 \end{bmatrix} = \begin{bmatrix} 0\,0 \\ 0\,0 \end{bmatrix}$$

which is of shape $\mathbb{I}$ but not a univalent relation in the original sense since $\mathbb{T}^{\smile};\mathbb{T} = \mathbb{T} \not\sqsubseteq \mathbb{I}$. On the other hand, also the notion of a scalar relation was generalized. A scalar relation α is a relation so that $\alpha \sqsubseteq \mathbb{I}$ and $\mathbb{T};\alpha = \alpha;\mathbb{T}$. The generalization to arbitrary matrices defines a matrix A to be a scalar iff A is diagonal and $\mathbb{T}A = A\mathbb{T}$. If α is a scalar relation, then we have $\mathbb{T}\alpha = \mathbb{T};\alpha = \alpha;\mathbb{T} = \alpha\mathbb{T}$ because of Theorem 3 and the fact that α and $\alpha^{\smile}$ are univalent. Most of Proposition 2 of [5] now carries over in the more general setting of sup-semirings. Instead of providing an explicit proof we refer to [5] and leave the obvious generalization of the proofs provided there to the reader.

Lemma 10. *Suppose D, D_1, D_2 are diagonal, Q, R are relations and A, B, C, E are arbitrary matrices. Then we have*

1. $D \cdot A^{\smile} = DA \cdot \mathbb{I}$,
2. $D = D^{\smile}$,
3. $D = D\mathbb{T} \cdot \mathbb{I}$,
4. $A \cdot D = A^{\smile} \cdot D$,
5. $(A\mathbb{T} \cdot B)C = A\mathbb{T} \cdot BC$ *and* $E(\mathbb{T}A \cdot B) = \mathbb{T}A \cdot EB$,
6. $A(B\mathbb{T} \cdot C) = (A \cdot \mathbb{T}B^{\smile})C$,
7. D_1D_2 *is diagonal,*

8. $D_1 \cdot D_2 = D_1 D_2$,
9. *if* D *is a scalar, then* $DA = AD$ *for all* A,
10. $\mathbb{T}A\mathbb{T} \cdot \mathbb{I}$ *is a scalar,*
11. $(A \cdot B)\mathbb{T} = (AB^{\smile} \cdot \mathbb{I})\mathbb{T}$ *and* $\mathbb{T}(A \cdot B) = \mathbb{T}(A^{\smile}B \cdot \mathbb{I})$.

Property (p) of Proposition 2 in [5] is a generalization of the well-known Dedekind rule. Unfortunately, this property cannot be shown in the context of arbitrary sup-semirings because it refers to the order structure of the real numbers.

3.5 Biproducts

In this section we want to investigate the relationship between biproducts with respect to $+$ and linear composition on the one hand and relational sums, i.e., biproducts with respect to $\sqcup$ and ;, on the other hand. This relationship seems important if we consider an abstract theory for matrices over sup-semirings and aim for a pseudo-representation theorem similar to the one shown for relational categories in [21,22]. Since an abstract theory for these matrices will be based on axioms for the linear operations as well as the relational operations it seems important under which circumstances biproduct with respect to both structures coincide. Only in such a case a simultaneous representation of both structures seems possible.

Notice that we prove the following theorem without referring to matrices. We only use the abstract properties that have been shown before so that the theorem remains true if we move from matrices to an axiomatic theory providing the properties used in the theorem follow from the axioms.

Theorem 4. *Suppose* $A + B$ *together with* $\iota : A \to A + B$ *and* $\kappa : B \to A + B$ *is a relational sum of* A *and* B*. Then* $A + B$ *together with* $\iota^{\smile}, \kappa^{\smile}, \iota, \kappa$ *is a biproduct with respect to* $+$ *and linear composition.*

Proof. By definition both injections ι and κ are injective and univalent. From Theorem 3 we immediately obtain

$$\iota\iota^{\smile} = \iota;\iota^{\smile} = \mathbb{I}_A, \quad \kappa\kappa^{\smile} = \kappa;\kappa^{\smile} = \mathbb{I}_B, \quad \iota\kappa^{\smile} = \iota;\kappa^{\smile} = \mathbb{⊥⊥}_{AB}.$$

Furthermore, from $\iota^{\smile};\iota \cdot \kappa^{\smile};\kappa \sqsubseteq \iota^{\smile};(\iota \cdot \iota;\kappa^{\smile};\kappa) = \mathbb{⊥⊥}_{A+B\,A+B}$ we conclude

$$\begin{aligned} \iota^{\smile}\iota + \kappa^{\smile}\kappa &= \iota^{\smile};\iota + \kappa^{\smile};\kappa && \iota,\kappa \text{ injective} \\ &= \iota^{\smile};\iota \sqcup \kappa^{\smile};\kappa && \text{Lemma 8} \\ &= \mathbb{I}_{A+B}, \end{aligned}$$

i.e., that $\iota, \kappa, \iota^{\smile}, \kappa^{\smile}$ forms a biproduct. □

Since ι, κ give rise to a biproduct with respect to the linear operations the constructions $\iota^{\smile}A + \kappa^{\smile}B$ and $C\iota + D\kappa$ resp. $\iota^{\smile}E\iota + \kappa^{\smile}F\kappa$ allows us to abstractly combine matrices into larger ones. In particular, if the morphisms A to F are relations, then we can show analogously to the proof of Theorem 4 that $\iota^{\smile}A + \kappa^{\smile}B = \iota^{\smile};A \sqcup \kappa^{\smile};B$ and $C\iota + D\kappa = C;\iota \sqcup D;\kappa$ and $\iota^{\smile}E\iota + \kappa^{\smile}F\kappa = \iota^{\smile};E;\iota \sqcup \kappa^{\smile};F;\kappa$, i.e., linear and relational "stacking" of matrices coincide [11,12]. We want to illustrate this in the following example.

Example 4. We want to investigate the hypercube example (Example 2) in more detail. In particular, we want to define the n-dimensional hypercube with success rate $\frac{9}{10}$ recursively. If $[\frac{9}{10}]$ denotes the scalar induced by the fraction $\frac{9}{10}$, i.e., the diagonal matrix with $\frac{9}{10}$ on the diagonal, then we define $H_0 := \perp\!\!\!\perp_{11}$ and

$$H_{n+1} := \iota^{\smile}(H_n\iota + [\frac{9}{10}]\kappa) + \kappa^{\smile}([\frac{9}{10}]\iota + H_n\kappa),$$

i.e., the 1-dimensional hypercube is just a single node (no edges) and the $n+1$-dimensional hypercube is obtained by two copies of the n-dimensional hypercube and connecting corresponding nodes by a $\frac{9}{10}$-edge. This matrix can be used for a quantitative analysis of the problem at hand. For example, by applying the operation $A \mapsto A + AA$ multiple times we obtain the matrix connecting two nodes with a rate α if there is a way of sending information from the start to the end node with success rate α. Furthermore, by applying the flattening operation to H_n we obtain the underlying unlabeled graph. Notice that we immediately get

$$H'_{n+1} = \iota^{\smile}; (H'_n; \iota \sqcup \kappa) \sqcup \kappa^{\smile}; (\iota + H'_n; \kappa),$$

i.e., a recursive and relational definition of the hypercube. This can now be used to investigate qualitative properties such as Hamiltonian cycles. In fact, it is not hard to show that the definition $C_2 = H'_2$ provides a Hamiltonian cycle for the 2-dimensional hypercube. Using the recursive definition

$$C_{n+1} = \iota^{\smile}; ((C_n \cdot \overline{e}); \iota \sqcup e; e^{\smile}; \kappa) \sqcup \kappa^{\smile}; (e^{\smile}; e; \iota \sqcup (C_n \cdot \overline{e})^{\smile}; \kappa)$$

for $n > 2$ with e an atom included in C_n, i.e., one edge, we obtain Hamiltonian cycles for hypercube of higher dimension. We omit the corresponding proofs because of lack of space.

4 Conclusion and Outlook

In this paper we started the investigation of matrices over semirings and the embedded structure of relations based on multiplicative idempotent elements. Several basic properties of such matrices over sup-semirings were investigated. A natural next step is to propose axioms for such a category of matrices. One goal of this endeavor could be a pseudo-representation theorem similar to that of [21,22]. A first step towards such a theorem has already been done by relating relational sums and biproducts with respect to the linear operations abstractly.

Last but not least, we would like to mention that the research for this paper and probably future work highly benefitted from a computer system for handling matrices with arbitrary coefficients specified by the user. This system was started by the third author, further developed by the first and the second author. It allows to specify categories of coefficients and operations on them that are then lifted to operations on matrices.

References

1. Anderson, M., Feil, T.: Turning Lights Out with Linear Algebra. Mathematics Magazine 71(4), 300–303 (1998)
2. Berghammer, R., Neumann, F.: RelView – an OBDD-based computer algebra system for relations. In: Ganzha, V.G., Mayr, E.W., Vorozhtsov, E.V. (eds.) CASC 2005. LNCS, vol. 3718, pp. 40–51. Springer, Heidelberg (2005)
3. Birkhoff, G.: Lattice Theory, 3rd edn., vol. XXV. American Mathematical Society Colloquium Publications (1940)
4. Cayley, A.: A Memoir on the Theory of Matrices. Phil. Trans. of the Royal Soc. of London 148(1), 17–37 (1858)
5. Desharnais, J., Grinenko, A., Möller, B.: Relational style laws and constructs of linear algebra. JLAMP 83(2), 154–168 (2014)
6. Freyd, P., Scedrov, A.: Categories, Allegories. North-Holland (1990)
7. Golan, J.S.: Semirings and their Application. Kluwer (1999)
8. Grätzer, G.: Universal Algebra, 2nd edn. Springer (2008)
9. Hebish, U., Weinert, H.J.: Semirings - Algebraic Theory and Application in Computer Science. Series in Algebra, vol. 5. World Scientific (1993)
10. Jackson, M., Stokes, T.: Semigroups with if-then-else and halting programs. IJAC 19(7), 937–961 (2009)
11. Macedo, H.D., Oliveira, J.N.: Matrices As Arrows! A Biproduct Approach to Typed Linear Algebra. In: Bolduc, C., Desharnais, J., Ktari, B. (eds.) MPC 2010. LNCS, vol. 6120, pp. 271–287. Springer, Heidelberg (2010)
12. Macedo, H.D., Oliveira, J.N.: Typing Linear Algebra: a Biproduct-oriented Approach. Science of Comp. Programming 78, 2160–2191 (2012)
13. Macedo, H.D., Oliveira, J.N.: A linear algebra approach to OLAP. Formal Asp. of Comput. 27(2), 283–307 (2015)
14. Mac Lane, S.: Categories for the Working Mathematician, 2nd edn. Graduate Texts in Mathematics, vol. 5. Springer (1998)
15. Oliveira, J.N.: Towards a Linear Algebra of Programming. Formal Asp. of Comput. 24(4-6), 433–458 (2012)
16. Oliveira, J.N.: Typed linear algebra for weighted (probabilistic) automata. In: Moreira, N., Reis, R. (eds.) CIAA 2012. LNCS, vol. 7381, pp. 52–65. Springer, Heidelberg (2012)
17. Schmidt, G., Ströhlein, T.: On kernels of graphs and solutions of games: a synopsis based on relations and fixpoints. SIAM J. Alg. Discrete Meth. (6), 54–65 (1985)
18. Schmidt, G., Ströhlein, T.: Relationen und Graphen. Springer (1989); English version: Relations and Graphs. Discrete Mathematics for Computer Scientists, EATCS Monographs on Theoret. Comput. Sci. Springer (1993)
19. Schmidt, G.: Relational Mathematics. Encyplopedia of Mathematics and Its Applications, vol. 132 (2011)
20. Winter, M.: Strukturtheorie heterogener Relationenalgebren mit Anwendung auf Nichtdetermismus in Programmiersprachen. Dissertationsverlag NG Kopierladen GmbH, München (1998)
21. Winter, M.: Relation Algebras are Matrix Algebras over a suitable Basis. University of the Federal Armed Forces Munich, Report Nr. 1998-05 (1998)
22. Winter, M.: A Pseudo Representation Theorem for various Categories of Relations. TAC Theory and Applications of Categories 7(2), 23–37 (2000)

Completeness via Canonicity for Distributive Substructural Logics: A Coalgebraic Perspective

Fredrik Dahlqvist and David Pym

University College London

Abstract. We prove strong completeness of a range of substructural logics with respect to their relational semantics by completeness-via-canonicity. Specifically, we use the topological theory of canonical (in) equations in distributive lattice expansions to show that distributive substructural logics are strongly complete with respect to their relational semantics. By formalizing the problem in the language of coalgebraic logics, we develop a modular theory which covers a wide variety of different logics under a single framework, and lends itself to further extensions.

1 Introduction

This work lies at the intersection of resource semantics/modelling, substructural logics, and the theory of canonical extensions and canonicity. These three areas respectively correspond to the semantic, proof-theoretic, and algebraic sides of the problem we tackle: to give a systematic, modular account of the relation between resource semantics and logical structure. We do not delve into the proof theory of substructural logics, but rather deal with the algebraic formulations of many such substructural proof systems ([29] summarizes the correspondence between classes of residuated lattices and substructural logics). A version of this work that includes detailed proofs can be found as a UCL Research Note [12].

Resource Semantics and Modelling. Resource interpretations of substructural logics — see, for example, [18,30,31,15,7] — are well-known and exemplified in the context of program verification and semantics by Ishtiaq and O'Hearn's pointer logic [23] and Reynolds' separation logic [32], each of which amounts to a model of a specific theory in Boolean BI. Resource semantics and modelling with resources has become an active field of investigation in itself (see, for example, [8]). Certain requirements, discussed below, seem natural (and useful in practice) in order to model naturally arising examples of resource.

1. We need to be able to compare at least some resources. Indeed, in a completely discrete model of resource (i.e., where no two resources are comparable) it is impossible to model key concepts such as 'having enough resources'. On the other hand, there is no reason to assume that *any two* resources be comparable (e.g., heaps). This suggests at least a preorder structure on models. In fact, we take the view that comparing two resources is fundamental, and in particular, if two resources cannot be distinguished in this way then they can be identified. We thus add antisymmetry and work with posets.

W. Kahl et al. (Eds.): RAMiCS 2015, LNCS 9348, pp. 119–135, 2015.
DOI: 10.1007/978-3-319-24704-5_8

2. We need to be able to combine (some) resources to form new resources (e.g., union of heaps with disjoint domains [23]). We denote the combination operation by $*$. An equivalent, but often more useful, point of view is to be able to specify how resources can be 'split up' into pairs of constituent resources. Moreover, since comparing resources is more important than establishing their equality, it makes sense to be able to list for a given resource r, the pairs (s_1, s_2) of resources which combine to form a resource $s_1 * s_2 \leq r$.
3. All reasonable examples of resources possess 'unit' resources with respect to the combination operation $*$; that is, special resources that leave other resources unchanged under the combination operation.
4. The last requirement is crucial, but slightly less intuitive. In the most well-behaved examples of resource models (e.g., $\mathbb{N}$), if we are given a resource r and a 'part' s of r, there exists a resource s' that 'completes' s to make r; that is, we can find a resource s' such that $s * s' = r$. More generally, given two resources r, s, we want to be able to find the the best s' such that $s*s' \leq r$. In a model of resource without this feature, it is impossible to provide an answer to legitimate questions such as 'how much additional resource is needed to make statement ϕ hold?'. Mathematically, this requirement says that the resource composition is a residuated mapping in both its arguments.

The literature on resource modelling, and on separation logic in particular, is vast, but two publications ([6] and [4]) are strongly related to this work. Both show completeness of 'resource logics' by using Sahlqvist formulas, which amounts to using completeness-via-canonicity ([3,24]).

Completeness-via-canonicity and Substructural Logics. The logical side of resource modelling is the world of substructural logics, such as BI, and of their algebraic formulations; that is, residuated lattices, residuated monoids, and related structures. The past decade has seen a fair amount of research into proving the completeness of relational semantics for these logics (for BI, for example, [31,15]), using, among other approaches, techniques from the duality theory of lattices. In [13], Dunn *et al.* prove completeness of the full Lambek calculus and several other well-known substructural logics with respect to a special type of Kripke semantics by using duality theory. This type of Kripke semantics, which is two-sorted in the non-distributive case, was studied in detail by Gerhke in [16]. The same techniques have been applied to prove Kripke completeness of fragments of linear logic in [5]. Finally, the work of Suzuki [33] explores in much detail completeness-via-canonicity for substructural logics. Our work follows in the same vein but with with some important differences. Firstly, we use a dual adjunction rather than a dual equivalence to connect syntax and semantics. This is akin to working with Kripke frames rather than descriptive general frames in modal logics: the models are simpler and more intuitive, but the tightness of the fit between syntax and semantics is not as strong. Secondly, we use the topological approach to canonicity of [17,21,34] because we feel it is the most flexible and modular approach to building canonical (in)equations. Thirdly, we only consider distributive structures. This is to some extent a matter a taste. Our choice is driven by the desire to keep the theory relatively simple (the non-distributive case is more involved), by the fact that from a resource modelling perspective the non-distributive case does not seem to occur 'in the wild', and finally because we place ourselves in the framework of coalgebraic logic, where the category of distributive lattices forms a particularly nice 'base category'.

The Coalgebraic Perspective. Coalgebraic methods bring many advantages to the study of completeness-via-canonicity. First, it greatly clarifies the connection between canonicity as an algebraic method and the existence of 'canonical models'; that is, strong completeness. Second, it provides a generic framework in which to prove completeness-via-canonicity for a vast range of logics ([11]). Third, it is intrinsically modular; that is, it provides theorems about complicated logics by combining results for simpler ones ([9,10]).

2 Substructural Logics: A Coalgebraic Perspective

We use the 'abstract' version of coalgebraic logic developed in, for example, [27], [28] and [25]; that is, we require the following basic situation:

$$L \circlearrowright \mathscr{C} \underset{G}{\overset{F}{\rightleftarrows}} \mathscr{D}^{op} \circlearrowleft T^{op} \quad (F \dashv G) \tag{1}$$

The left hand-side of the diagram is the syntactic side, and the right-hand side the semantic one. The category $\mathscr{C}$ represents a choice of 'reasoning kernel'; that is, of logical operations which we consider to be fundamental, whilst L is a syntax constructing functor which builds terms over the reasoning kernel. Objects in $\mathscr{D}$ are the carriers of models and T specifies the coalgebras on these carriers in which the operations defined by L are interpreted. The functors F and G relate the syntax and the semantics, and F is left adjoint to G. We will denote such an adjunction by $F \dashv G : \mathscr{C} \to \mathscr{D}$. Note, as mentioned in the introduction, that we only need a dual adjunction, not a full duality.

2.1 Syntax

Reasoning Kernels. There are three choices for the category $\mathscr{C}$ which are particularly suited to our purpose, the category **DL** of distributive lattices, the category **BDL** of bounded distributive lattices, and the category **BA** of boolean algebras. The choice of **DL** as our most basic category was justified in the introduction, but we should also mention an important technical advantage of **DL**, **BDL** and **BA** from the perspective of coalgebraic logic: each category is locally finite; that is, finitely generated objects are finite. This is a very desirable technical property for the presentation of endofunctors on this category and for coalgebraic strong completeness theorems. We denote by $\mathsf{F} \dashv \mathsf{U}$ the usual free-forgetful adjunction between **DL** (resp. **BDL**, resp. **BA**) and **Set**.

True and False. The choice of including (or not) $\top$ and $\bot$ to the logic is clearly provided by the choice of reasoning kernel.

Algebras. Recall that an algebra for an endofunctor $L : \mathscr{C} \to \mathscr{C}$ is an object A of $\mathscr{C}$ together with a morphism $\alpha : LA \to A$. We refer to endofunctors $L : \mathscr{C} \to \mathscr{C}$ as *syntax constructors*.

Intuitionistic Implication. We do not consider the intuitionistic implication as a fundamental operation; in particular, the category of Heyting algebras does not form a reasoning kernel. This choice is motivated by the fact that the semantics of intuitionistic logic can be given in terms of Kripke frames, that the intuitionistic implication is not usually part of the basic language of substructural logics, and that the category **HA** of Heyting algebras is not as well-behaved as our choices of reasoning kernels. We therefore add the implication as an additional (modal) operation on (bounded) distributive lattices via the syntax constructor:

$$L_{\text{Hey}} : \mathbf{DL} \to \mathbf{DL}, \begin{cases} A \mapsto \mathsf{F}\{a \to b \mid a, b \in \mathsf{U}A\}/\equiv \\ L_{\text{RL}} f : L_{\text{Hey}} A \to L_{\text{Hey}} B, [a]_\equiv \mapsto [f(a)]_\equiv, \end{cases}$$

where $\equiv$ is the fully invariant equivalence relation in **DL** generated by the following Heyting Distribution Laws for *finite* subsets X of A:

HDL1. $a \to \bigwedge X = \bigwedge [a \to X]$
HDL2. $\bigvee X \to a = \bigwedge [X \to a]$.

where we use the notation $\bigwedge [a \to X] := \bigwedge_{x \in X} a \to x$ and the convention that $\bigwedge \emptyset = \top$ and $\bigvee \emptyset = \bot$ when the objects of the reasoning kernel are bounded. The language defined by L_{Hey} for a set V of propositional variables is the free L_{Hey}-algebra over $\mathsf{F}V$; that is, the language of intuitionistic propositional logic quotiented by the axioms of distributive lattices and HDL1-2. Note that an L_{Hey}-algebra is *not* a Heyting algebra, the axioms HDL1-2 only capture some of the Heyting algebra structure. Instead, an L_{Hey}-algebra is simply a distributive lattice with a binary map satisfying the distribution laws above (which happen to be valid in HAs). The remaining features of HAs will be captured in a second stage via *canonical frame conditions*. The reason for proceeding in this step-by-step way will become clear in the sequel and is similar in spirit to the approach of [1]. The main difference is that in [1], the axioms of Heyting algebras are separated into rank 1 and non-rank 1 axioms, leading to the notion of *weak Heyting algebras* which obey the axioms HDL1-2 and also $a \to a = \top$. In this work, we want to build a minimal 'pre-Heyting' logic with a strongly complete semantics and well-behaved (viz. smooth, see Section) operations, and L_{Hey}-algebras perform this role.

Resource Operations. The operations on resources specified in the introduction; that is, a combination operation and its left and right residuals, are introduced via the following syntax constructor:

$$L_{\text{RL}} : \mathscr{C} \to \mathscr{C}, \begin{cases} L_{\text{RL}} A = \mathsf{F}\{I, a * b, a \backslash b, a / b \mid a, b \in \mathsf{U}A\}/\equiv \\ L_{\text{RL}} f : L_{\text{RL}} A \to L_{\text{RL}} B, [a]_\equiv \mapsto [f(a)]_\equiv, \end{cases}$$

where $\equiv$ is the fully invariant equivalence relation in $\mathscr{C}$ generated by following the Distribution Laws for finite subsets X of A:

DL1. $\bigvee X * a = \bigvee[X * a]$
DL2. $a * \bigvee X = \bigvee[a * X]$
DL3. $a\backslash \bigwedge X = \bigwedge[a\backslash X]$
DL4. $\bigvee X\backslash a = \bigwedge[X\backslash a]$
DL5. $\bigwedge X/a = \bigwedge[X/a]$
DL6. $a/\bigvee X = \bigwedge[a/X]$.

The language defined by L_{RL} is the free L_{RL}-algebra over $\mathsf{F}V$, which is the language of the distributive full Lambek calculus (or residuated lattices) quotiented under the axioms of $\mathscr{C}$ and DL1-6. An L_{RL}-algebra is simply an object of $\mathscr{C}$ endowed with a nullary operation I and binary operations $*, \backslash$ and $/$ satisfying the distribution laws above. Again, note that an L_{RL}-algebra is *not* a distributive residuated lattice. Only some features of this structure have been captured by the axioms above. But several are still missing, and will be added subsequently as *canonical frame conditions.* Both L_{Hey}-algebras and L_{RL}-algebras are examples of *Distributive Lattice Expansions*, or DLEs; that is, distributive lattices endowed with a collection of maps of finite arities. When $\mathscr{C} = \mathbf{BA}$, L_{RL}-algebras are an example *Boolean Algebra Expansions*, or BAEs.

Modularity. The syntax developed above is completely modular. For example, if we wish to study boolean BI, it is natural to consider $L_{\mathrm{RL}} : \mathbf{BA} \to \mathbf{BA}$ as our syntax constructor. If we wish to study intuitionistic BI, then we should consider

$$L_{\mathrm{Hey}} + L_{\mathrm{RL}} : \mathbf{BDL} \to \mathbf{BDL}, \text{ where } (L_{\mathrm{Hey}} + L_{\mathrm{RL}})A = L_{\mathrm{Hey}}A + L_{\mathrm{RL}}A,$$

where the coproduct is taken in **BDL**, and is thus a ‘free product’ generating precisely the expected language. Finally, we may wish to add modal operators to the language (see the ‘relevant modal logic’ in [33]), for example $\Diamond$. In this case, we can in the same way add the syntax constructor for modal logic, namely,

$$L_{\Diamond} : \mathscr{C} \to \mathscr{C}, A \mapsto \mathsf{F}\{\Diamond a \mid a \in \mathsf{U}A\}/\{\Diamond(\bigvee X) = \bigvee[\Diamond X]\}$$

2.2 Coalgebraic Semantics

Semantic Domain. As we mentioned in the introduction, it is reasonable to assume that a model of resources should be a poset, and thus taking $\mathscr{D} = \mathbf{Pos}$ is intuitively justified. This is a particularly attractive choice of ‘semantic domain’ given that the category **Pos** is related to **DL** by the dual adjunction $\mathsf{Pf} \dashv \mathcal{U} : \mathbf{DL} \to \mathbf{Pos}^{\mathrm{op}}$, where Pf is the functor sending a distributive lattice to its poset of prime filters, and **DL**-morphisms to their inverse images, and $\mathcal{U}$ is the functor sending a poset to the distributive lattice of its up-sets and monotone maps to their inverse images. In the case in which a distributive lattice is a boolean algebra, it is well-known that prime filters are maximal (i.e., ultrafilters) and the partial order on the set of ultrafilter is thus discrete; that is, ultrafilters are only related to themselves. Thus the dual adjunction $\mathsf{Pf} \dashv \mathcal{U}$ becomes the well-known adjunction $\mathsf{Uf} \dashv \mathsf{P}_c : \mathbf{BA} \to \mathbf{Set}^{\mathrm{op}}$.

Coalgebras. Recall that a coalgebra for an endofunctor $T : \mathscr{D} \to \mathscr{D}$, is an object W of $\mathscr{D}$ together with a morphism $\gamma : W \to TW$. The endofunctors that we will consider are built from products and ‘powersets’ and will be referred to as *model constructors.* Note that **Pos** has products, which are simply the **Set**

products with the obvious partial order on pairs of elements. The 'powerset' functor which we will consider is the *convex powerset* functor: $\mathsf{P}_c : \mathbf{Pos} \to \mathbf{Pos}$, sending a poset to its set of convex subsets, where a subset U of a poset $(X, \leq)$ is convex if $x, z \in U$ and $x \leq y \leq z$ implies $y \in U$. The set $\mathsf{P}_c X$ is given a poset structure via the *Egli-Milner* order (see [2]).

Coalgebras for the Intuitionistic Implication. We define the following mod-el constructor, which will interpret $\to$:

$$T_{\mathrm{Hey}} : \mathbf{Pos} \to \mathbf{Pos}, \begin{cases} T_{\mathrm{Hey}} W = \mathsf{P}_c(W^{\mathrm{op}} \times W) \\ T_{\mathrm{Hey}} f : T_{\mathrm{Hey}} W \to T_{\mathrm{Hey}} W', U \mapsto (f \times f)[U]. \end{cases}$$

where W^{op} is the poset whose carrier is W and whose order is dual to that of W.

Coalgebras for the Resource Operations. We define the following model constructor, which is used to interpret $I, *, \backslash$ and $/$:

$$T_{\mathrm{RL}} : \mathscr{D} \to \mathscr{D}, \begin{cases} T_{\mathrm{RL}} W = \mathbb{2} \times \mathsf{P}_c(W \times W) \times \mathsf{P}_c(W^{\mathrm{op}} \times W) \times \mathsf{P}_c(W \times W^{\mathrm{op}}) \\ T_{\mathrm{RL}} f : T_{\mathrm{RL}} W \to T_{\mathrm{RL}} W', U \mapsto (\mathsf{Id}_{\mathbb{2}} \times (f \times f)^3)[U]. \end{cases}$$

The intuition is that the first component of the structure map of a T_{RL}-coalgebra (to the (po)set $\mathbb{2}$) separates states into units and non-units. The second component sends each 'state' $w \in W$ to the pairs of states which it 'contains', the next two components are used to interpret $\backslash$ and $/$, respectively, and turn out to be very closely related to the second component. Note that if $\mathscr{D} = \mathbf{Pos}$, the structure map of coalgebras are monotone, intuitively this means bigger resources can be split up in more ways.

The Semantic Transformations. In the abstract flavour of coalgebraic logic, the semantics is provided by a natural transformation $\delta : LG \to GT^{\mathrm{op}}$ called the *semantic transformation.* We show below how this defines an interpretation map, but we first define our two semantic transformations. As already noted above, a $\mathscr{C}$-morphism $\delta_W^{\mathrm{Hey}} : L_{\mathrm{Hey}} GW \to GT_{\mathrm{Hey}} W$ is equivalent to a function over the set of generators $\{U \to V \mid U, V \in \mathsf{U}GW\}$ satisfying the distributivity laws HDL1-2, and similarly for $\delta_W^{\mathrm{RL}} : L_{\mathrm{RL}} GW \to GT_{\mathrm{RL}} W$ and the distributivity laws DL1-6. We now define

$$\delta_W^{\mathrm{Hey}}(U \to V) = \{(x, y) \in T_{\mathrm{Hey}} W \mid x \in U \Rightarrow y \in V\}$$

and similarly (by using the usual projections maps $\pi_i, 1 \leq i \leq 4$)

$$\begin{aligned} \delta_W^{\mathrm{RL}}(I) &= \{t \in T_{\mathrm{RL}} W \mid \pi_1(t) = 0 \in \mathbb{2}\} \\ \delta_W^{\mathrm{RL}}(u * v) &= \{t \in T_{\mathrm{RL}} W \mid \exists (x, y) \in \pi_2(t), x \in u, y \in v\} \\ \delta_W^{\mathrm{RL}}(u \backslash w) &= \{t \in T_{\mathrm{RL}} W \mid \forall (x, y) \in \pi_3(t), x \in u \Rightarrow y \in w\} \\ \delta_W^{\mathrm{RL}}(w / v) &= \{t \in T_{\mathrm{RL}} W \mid \forall (x, y) \in \pi_4(t), x \in v \Rightarrow y \in w\}. \end{aligned}$$

Proposition 1. *The natural transformations δ^{Hey} and δ^{RL} are well-defined, in particular each map δ_W^{Hey} satisfies the distributivity laws HDL1-2, and each map δ_W^{RL} satisfies the distributivity laws the distributivity laws DL1-6.*

The semantic transformations are thus well-defined. We now show how the interpretation map arises from the semantic transformation. Recall that, for a given syntax constructor $L : \mathscr{C} \to \mathscr{C}$, the language of L is the free L-algebra over $\mathsf{F}V$. This is equivalent to saying that it is the initial $L(-) + \mathsf{F}V$-algebra. We use initiality to define the interpretation map by putting an $L(-) + \mathsf{F}V$-algebra structure on the 'predicates' of a T-coalgebra $\gamma : W \to TW$; that is, on the carrier set GW. By definition of the coproduct, this means defining a morphism $LGW \to GW$ and a morphism $\mathsf{F}V \to GW$. By adjointness it is easy to see that the latter is simply a valuation $v : V \to \mathsf{U}GW$. For the former we simply use the semantic transformation and G applied to the coalgebra. The interpretation map $[\![-]\!]_W$ is thus given by the catamorphism:

$$
\begin{array}{ccc}
L\mu(L(-) + \mathsf{F}V) + \mathsf{F}V & \xrightarrow{L[\![-]\!]_W + \mathsf{Id}_{\mathsf{F}V}} & LGW + \mathsf{F}V \\
 & & \downarrow{\scriptstyle \delta_W + \mathsf{Id}_{\mathsf{F}V}} \\
\big\downarrow & & GTW + \mathsf{F}V \\
 & & \downarrow{\scriptstyle G\gamma + v} \\
\mu(L(-) + \mathsf{F}V) & \xrightarrow[{[\![-]\!]_W}]{} & GW
\end{array}
$$

Modularity. Model constructors and semantic transformations can be assembled in a way that is dual to the the syntax constructors. For example, if we wish to interpret both the intuitionistic implication and the resource operations, we use a coalgebra of type $\gamma_1 \times \gamma_2 : W \to T_{\mathrm{Hey}}W \times T_{\mathrm{RL}}W$. The overall semantics is then inherited from that of the constituents via the following diagram:

$$
\begin{array}{ccc}
(L_{\mathrm{Hey}}+L_{\mathrm{RL}})\mu(L_{\mathrm{Hey}}+L_{\mathrm{RL}}(-)+\mathsf{F}V)+\mathsf{F}V & \xrightarrow{L_{\mathrm{Hey}} + L_{\mathrm{RL}}[\![-]\!]_W + \mathsf{Id}_{\mathsf{F}V}} & L_{\mathrm{Hey}}GW+L_{\mathrm{RL}}GW+\mathsf{F}V \\
 & & \downarrow{\scriptstyle \delta_W^{\mathrm{Hey}}+\delta_W^{\mathrm{RL}}+\mathsf{Id}_{\mathsf{F}V}} \\
\big\downarrow & & GT_{\mathrm{Hey}}W+GT_{\mathrm{RL}}W+\mathsf{F}V \\
 & & \downarrow{\scriptstyle G(\gamma_1\times\gamma_2)\circ(G\pi_1+G\pi_2)+v} \\
\mu(L_{\mathrm{Hey}}+L_{\mathrm{RL}}(-)+\mathsf{F}V) & \xrightarrow[{[\![-]\!]_W}]{} & GW
\end{array}
$$

3 Canonicity

3.1 Canonical Extension of Distributive Lattices

We now briefly present the salient facts about canonical extensions. For more details the reader is referred to [19] for BAs, [26,24] for BAOs, and [20,21] for

DLEs. The main rationale for studying canonical extensions is to embed a lattice-based structure, typically a language quotiented by some axioms, into a similar structure which is more 'set-like'; that is, whose elements can be viewed as parts of a set, or of a set with some additional structure. In this way, we can establish a connection between the syntax and the semantics; that is, build models from formulas. But what does being 'set-like' mean? Two criteria emerge as being fundamental: completeness and being generated from below (i.e., by joins) by something akin to 'elements'. Canonical extensions satisfy these two conditions. For a distributive lattice A, the idea behind the construction of its canonical extension A^σ is to build a completion of A which is not 'too big' and not 'too different' from A. Technically, we want A to be *dense* and *compact* in A^σ.

Density. To build a completion of A it is natural to formally add to A all meets, all joins, all meets of all joins, all joins of all meets, etc.. In the case of the canonical extension we require that this procedure stops after two iterations (i.e., we want a Δ_1-completion; see [22]). Intuitively, this prevents the completion from becoming 'too big'. Based on this intuition we introduce the following terminology: given a sub-lattice A of a complete distributive lattice C, we define the meets in C of elements of A as the *closed elements* of C and denote this set by $K(A)$ (or simply K when there is no ambiguity); dually, we define the joins in C of elements of A as the *open elements* of C and denote this set by $O(A)$. Finally, we say that A is *dense* in C if $C = O(K(A)) = K(O(A))$.

Compactness. The canonical extension A^σ is also required not to be too different from A in the sense that facts about arbitrary meets and joins of elements of A in A^σ must already be 'witnessed' by finite meets and joins in A. Formally, if A is a sub-lattice of C, A is *compact* in C if, for every $X, Y \subseteq A$ such that $\bigwedge X \leq \bigvee Y$, there exist *finite* subsets $X_0 \subseteq X, Y_0 \subseteq Y$ such that $\bigwedge X_0 \leq \bigvee Y_0$. An equivalent definition is that A is compact in C if for every closed element $p \in K(A)$ and open element $u \in O(A)$ such that $p \leq u$, there exists an element $a \in A$ such that $p \leq a \leq u$. The *canonical extension* A^σ of a distributive lattice A is the complete distributive lattice such that A is dense and compact in A^σ. We can summarize what we need to know about A^σ in the following theorem:

Theorem 1 ([20,17,21]). *The canonical extension A^σ of a distributive lattice A can be concretely represented as the lattice $A^\sigma \simeq \mathcal{U}\mathsf{Pf}A$; in particular, it is completely distributive.*

Note that this theorem requires the Prime Ideal Theorem for distributive lattices which is a non-constructive principle, albeit one that is strictly weaker than the axiom of choice. Note also that since the canonical extension of a BA is complete and completely distributive, it is also *atomic* (see [19] Ch. 14); that is, it is a complete atomic boolean algebra. It is concretely represented by $A^\sigma = \mathsf{P}_c\mathsf{Uf}A$, in which case it is not simply 'set-like', but an actual algebra of subsets.

3.2 Canonical Extension of Distributive Lattice Expansions

We now sketch the theory canonical extensions for Distributive Lattice Expansions (DLE) — for the details, see [20,21]. Each map $f : \mathsf{U}A^n \to \mathsf{U}A$ can be extended to a map $(\mathsf{U}A^\sigma)^n \to \mathsf{U}A^\sigma$ in two canonical ways:

$$f^\sigma(x) = \bigvee\{\bigwedge f[d,u] \mid K^n \ni d \leq x \leq u \in O^n\}$$
$$f^\pi(x) = \bigwedge\{\bigvee f[d,u] \mid K^n \ni d \leq x \leq u \in O^n\},$$

where $f[d,u] = \{f(a) \mid a \in A^n, d \leq a \leq u\}$. Note that since A is compact in A^σ the intervals $[d,u]$ are never empty, which justifies these definitions. For a signature Σ, the *canonical extension* of a Σ-DLE $(A, (f_s : \mathsf{U}A^{\mathrm{ar}(n)} \to \mathsf{U}A)_{s\in\Sigma})$ is defined to be the Σ-DLE $(A^\sigma, (f_s^\sigma : \mathsf{U}(A^\sigma)^{\mathrm{ar}(n)} \to \mathsf{U}A^\sigma)_{s\in\Sigma})$, and similarly for BAEs. We summarize some important facts about canonical extensions of maps in the following proposition, proofs can be found in, for example, [17,21,34]:

Proposition 2. *Let A be a distributive lattice, and $f : \mathsf{U}A^n \to \mathsf{U}A$.*

1. $f^\sigma \restriction A^n = f^\pi \restriction A^n = f$.
2. $f^\sigma \leq f^\pi$ *under pointwise ordering.*
3. *If f is monotone in each argument, then* $f^\sigma \restriction (K \cup O)^n = f^\pi \restriction (K \cup O)^n$.

We call a monotone map $f : \mathsf{U}A^n \to \mathsf{U}A$ *smooth in its ith argument* $(1 \leq i \leq n)$ if, for every $x_1, \ldots, x_{i-1}, x_{i+1}, \ldots, x_n \in K \cup O$,

$$f^\sigma(x_1, \ldots, x_{i-1}, x_i, x_{i+1}, \ldots, x_n) = f^\pi(x_1, \ldots, x_{i-1}, x_i, x_{i+1}, \ldots, x_n),$$

for every $x_i \in A^\sigma$. A map $f : \mathsf{U}A^n \to \mathsf{U}A$ is called *smooth* if it is smooth in each of its arguments.

In order to study effectively the canonical extension of maps, we need to define six topologies on A^σ. First, we define $\sigma^\uparrow = \{\uparrow p \mid p \in K\}$, $\sigma^\downarrow = \{\downarrow u \mid u \in O\}$ and $\sigma = \sigma^\uparrow \cup \sigma^\downarrow$; that is, the join of $\sigma^\uparrow$ and $\sigma^\downarrow$ in the lattice of topologies on A^σ. It is easy to check that the sets above do define topologies and that $\sigma = \{\uparrow p \cap \downarrow u \mid K \ni p \leq u \in O\}$. The next set of topologies is well-known to domain theorists: a *Scott open* in A^σ is a subset $U \subseteq A^\sigma$ such that (1) U is an upset and (2) for any up-directed set D such that $\bigvee D \in U$, $D \cap U \neq \emptyset$. The collection of Scott opens forms a topology called the *Scott topology*, which we denote $\gamma^\uparrow$. The dual topology will be denoted by $\gamma^\downarrow$, and their join by γ. It is not too hard to show (see [17,34]) that $\gamma^\uparrow \subseteq \sigma^\uparrow$, $\gamma^\downarrow \subseteq \sigma^\downarrow$, and $\gamma \subseteq \sigma$. We denote the product of topologies by $\times$, and the n-fold product of a topology τ by τ^n. The following result shows why these topologies are important: they essentially characterize the canonical extensions of maps:

Proposition 3 ([17]). *For any DL A and any map $f : \mathsf{U}A^n \to \mathsf{U}A$,*

1. f^σ *is the largest $(\sigma^n, \gamma^\uparrow)$-continuous extension of f,*
2. f^π *is the smallest $(\sigma^n, \gamma^\downarrow)$-continuous extension of f*
3. f *is smooth iff it has a unique (σ^n, γ)-continuous extension.*

From this important result, it is not hard to get the following key theorem, sometimes known as *Principle of Matching Topologies*, which underlies the basic 'algorithm' for canonicity:

Theorem 2 (Principle of Matching Topologies,[17,34]). *Let A be a distributive lattice, and $f : \mathsf{U}A^n \to \mathsf{U}A$ and $g_i : \mathsf{U}A^{m_i} \to \mathsf{U}A, 1 \leq i \leq n$ be arbitrary maps. Assume that there exist topologies τ_i on A, $1 \leq i \leq n$ such that each g_i^σ is (σ^{m_i}, τ_i)-continuous, then*

1. *if f^σ is $(\tau_1 \times \ldots \times \tau_n, \gamma^\uparrow)$-continuous, then $f^\sigma(g_1^\sigma, \ldots, g_n^\sigma) \leq (f(g_1, \ldots, g_n))^\sigma$,*
2. *if f^σ is $(\tau_1 \times \ldots \times \tau_n, \gamma^\downarrow)$-continuous, then $f^\sigma(g_1^\sigma, \ldots, g_n^\sigma) \geq (f(g_1, \ldots, g_n))^\sigma$*
3. *if f^σ is $(\tau_1 \times \ldots \times \tau_n, \gamma)$-continuous, then $f^\sigma(g_1^\sigma, \ldots, g_n^\sigma) = (f(g_1, \ldots, g_n))^\sigma$.*

The last piece of information we need to effectively use the Principle of Matching Topologies is to determine when maps are continuous for a certain topology, based on the distributivity laws they satisfy. For our purpose the following results will be sufficient:

Proposition 4 ([20,17,21,34]). *Let A be a distributive lattice, and let $f : \mathsf{U}A^n \to \mathsf{U}A$ be a map. For every $(n-1)$-tuple $(a_i)_{1 \leq i \leq n-1}$, we denote by $f_{\mathrm{a}}^k : A \to A$ the map defined by $x \mapsto f(a_1, \ldots, a_{k-1}, x, a_k, \ldots, a_{n-1})$.*

1. *If f_{a}^k preserves binary joins, then $(f^\sigma)_{\mathrm{a}}^k$ preserve all non-empty joins and is $(\sigma^\downarrow, \sigma^\downarrow)$-continuous.*
2. *If f_{a}^k preserves binary meets, then $(f^\sigma)_{\mathrm{a}}^k$ preserve all non-empty meets and is $(\sigma^\uparrow, \sigma^\uparrow)$-continuous.*
3. *If f_{a}^k anti-preserves binary joins (i.e., turns them into meets), then $(f^\sigma)_{\mathrm{a}}^k$ anti-preserve all non-empty joins and is $(\sigma^\downarrow, \sigma^\uparrow)$-continuous.*
4. *If f_{a}^k anti-preserves binary meets (i.e., turns them into joins), then $(f^\sigma)_{\mathrm{a}}^k$ anti-preserve all non-empty meets and is $(\sigma^\uparrow, \sigma^\downarrow)$-continuous.*
5. *In each case f is is smooth in its kth argument.*

3.3 Canonical (in)equations

To say anything about the canonicity of equations, we need to compare interpretations in A with interpretations in A^σ. It is natural to try to use the extension $(\cdot)^\sigma$ to mediate between these interpretations, but $(\cdot)^\sigma$ is defined on maps, not on terms. Moreover, not every valuation on A^σ originates from valuation on A. We would therefore like to recast the problem in such a way that (1) terms are viewed as maps, and (2) we do not need to worry about valuations.

Term Functions. The solution is to adopt the language of *term functions* (as first suggested in [24]). Given a signature Σ, let $\mathsf{T}(V)$ denote the language of Σ-DLEs (or Σ-BAEs) over a set V of propositional variables. We view each term $t \in \mathsf{T}(V)$ as defining, for each Σ-DLE A, a map $t^A : A^n \to A$. This allows us to consider its canonical extension $(t^A)^\sigma$, and also allows us to reason without having to worry about specifying valuations. Formally, given a signature Σ and a set V a propositional variables, we inductively define the term function associated with an element t built from variables $x_1, \ldots, x_n \in V$ as follows:

- $x_i^A = \pi_i^n : A^n \to A, 1 \le i \le n$;
- $(f(t_1, \ldots, t_m))^A = f^A \circ \langle t_1^A, \ldots, t_m^A \rangle$.

where π_i is the usual projection on the ith component, f^A is the interpretation of the symbol f in A and $\langle t_1^A, \ldots, t_m^A \rangle$ is usual the product of m maps. Note that in this definition we work in **Set**, and the building blocks of term functions are thus the variables in V (interpreted as projections) and all operation symbols, including $\vee, \wedge$ and possibly $\neg$.

Proposition 5. *Let s, t be terms in the language defined by a signature Σ and A be a Σ-DLE,*

$$A \models s = t \text{ iff } s^A = t^A .$$

Canonical (in)equations. An equation $s = t$ where $s, t \in \mathsf{T}(V)$ is called *canonical* if $A \models s = t$ implies $A^\sigma \models s = t$, and similarly for inequations. Following [24], we say that $t \in \mathsf{T}(V)$ is *stable* if $(t^A)^\sigma = t^{A^\sigma}$, that t is *expanding* if $(t^A)^\sigma \le t^{A^\sigma}$, and that t is *contracting* if $(t^A)^\sigma \ge t^{A^\sigma}$, for any A. The inequality between maps is taken pointwise. The following proposition illustrates the usefulness of these notions:

Proposition 6 ([24]). *If $s, t \in \mathsf{T}(V)$ are stable then the equation $s = t$ is canonical. Similarly, let $s, t \in \mathsf{T}(V)$ such that s is contracting and t is expanding, then the inequality $s \le t$ is canonical.*

4 Completeness via-canonicity

4.1 Axiomatizing HAs and Distributive Residuated Lattices

So far we have only captured part of the structure of Heyting algebras and distributive residuated lattices, namely we have enforced the distribution properties of $\to, *, \backslash$ and $/$ by our definition of the syntax constructors L_{Hey} and L_{RL}. In order to capture the rest of the structures we now add *frame conditions* to the coalgebraic models. To do this we need to find axioms which, when added to HDL1-2 and DL1-6 axiomatize HAs and distributive residuated lattices respectively. Due to the constraints that these axioms must be canonical, we choose the following Heyting Frame Conditions:

HFC1. $a \to a = \top$,
HFC2. $a \wedge (a \to b) = a \wedge b$
HFC3. $(a \to b) \wedge b = b$

and, for distributive lattices, the Frame Conditions:

FC1. $a * I = a$, $I * a = a$,
FC2. $I \le a \backslash a$, $I \le a/a$,
FC3. $a * (b \backslash c) \le (a * b) \backslash c$,
FC4. $(c/b) * a \le c/(a * b)$,
FC5. $(a/b) * b \le a$, and
FC6. $b * (b \backslash a) \le a$,

Proposition 7. *The axioms HDL1-2 and HFC1-3 axiomatize Heyting algebras, and similarly, the axioms DL1-6 and FC1-6 axiomatize distributive residuated lattices.*

We now show one of the crucial steps.

Proposition 8. *The axioms HFC1-3 and FC1-6 are canonical.*

Proof. The proof is an application of Theorem 2 and Proposition 6.

FC1: Since $*$ preserves binary joins in each argument, it is smooth by Prop. 4, and it follows that it is (σ^2, γ)-continuous. Since π_1^σ and I^σ are trivially (σ, σ)-continuous, it follows from Theorem 2 that $(* \circ \langle \pi_1, I \rangle)^\sigma = *^\sigma \circ \langle \pi_1, 1 \rangle^\sigma$. Each side of the equation is thus stable and the result follows from Prop. 6.

FC2: I is stable and thus contracting, and $(\backslash \circ \langle \pi_1, \pi_1 \rangle)^\sigma = \backslash^\sigma \circ \langle \pi_1, \pi_1 \rangle^\sigma$, since π_1^σ is (σ, σ)−continuous and $\backslash^\sigma$ is smooth. The RHS of the inequality is thus stable, and a fortiori expanding, and the inequality is thus canonical.

FC3-4: Since $*^\sigma$ preserve joins in each argument, it preserves up-directed ones, and is thus $((\gamma^\uparrow)^2, \gamma^\uparrow)$-continuous. Since $\backslash^\sigma$ is smooth it is in particular $(\sigma^2, \gamma^\uparrow)$-continuous. Since π_1^σ is $(\sigma, \gamma^\uparrow)$-continuous, we get that $*^\sigma \circ \langle \pi_1^\sigma, \backslash^\sigma \circ \langle \pi_2^\sigma, \pi_3^\sigma \rangle \rangle$ is $(\sigma^3, \gamma^\uparrow)$-continuous and thus contracting. For the RHS, note that since $\backslash^\sigma$ preserves meets in its first argument, it must in particular preserve down-directed ones, thus $\backslash^\sigma$ is $(\gamma^\downarrow, \gamma^\downarrow)$-continuous in its first argument. Similarly, since $\backslash^\sigma$ anti-preserve joins in its second argument, it must in particular anti-preserve up-directed ones, and is thus $(\gamma^\uparrow, \gamma^\downarrow)$-continuous in its second argument. This means that $\backslash^\sigma$ is $(\gamma^2, \gamma^\downarrow)$-continuous. We thus have that the full term is $(\sigma^3, \gamma^\downarrow)$ continuous, and thus expanding. The inequation is therefore canonical.

FC5-6: The LHS is contracting by the same reasoning as above, and the RHS is stable and thus expanding.

4.2 Strong Completeness Results

The Jónsson-Tarski Theorem. We first establish the strong completeness of the logics defined by our syntax constructors L_{Hey} and L_{RL} with respect to their T_{Hey}- and T_{RL}-coalgebraic models. The proof is an application of the coalgebraic Jónsson-Tarksi theorem. Recall from Theorem 1 and Diagram (1), that the canonical extension of an object A in any of our reasoning kernels $\mathscr{C}$ is given by GFA. This justifies the following:

Theorem 3 (Coalgebraic Jónsson-Tarksi theorem, [28]). *Assuming the basic situation of Diagram (1) and a semantic transformation $\delta : LG \to GT$, if its adjoint transpose $\hat{\delta} : TF \to FL$ has a right-inverse $\hat{\delta}^{-1} : FL \to TF$, then for every L-algebra $\alpha : LA \to A$, the embedding $\eta_A : A \to GFA$ of A into its canonical extension can be lifted to the following L-algebra embedding:*

$$\begin{array}{ccccccc} LA & & & \xrightarrow{\quad\alpha\quad} & & & A \\ \downarrow{\scriptstyle L\eta_A} & & & & & & \downarrow{\scriptstyle \eta_A} \\ LGFA & \xrightarrow[\delta_{FA}]{} & GTFA & \xrightarrow[G\hat{\delta}_A^{-1}]{} & GFLA & \xrightarrow[GF\alpha]{} & GFA \end{array} \tag{2}$$

We call the coalgebra $\hat{\delta}^{-1} \circ F\alpha : FA \to TFA$ a *canonical model* of (the L-algebra) A. If A is the free L-algebra over $\mathsf{F}V$ we recover the usual notion of canonical model. The 'truth lemma' follows from the definition of η.

We now prove the existence of canonical models for the logics defined by L_{Hey} and L_{RL}. The result generalizes lemma 5.1 of [14], which builds canonical models for countable DLs with a unary operator, and lemma 4.26 of [3], which builds canonical models for countable BAs with n-ary operators. We essentially show how to build canonical models for arbitrary DLs with n-ary expansions all of whose arguments either (1) preserve joins or anti-preserve meets, or (2) preserve meets or anti-preserve joins.

Theorem 4. *The logic defined by L_{Hey} (resp. L_{RL}) is sound and strongly complete with respect to the class of all T_{Hey}- (resp. T_{RL}-) coalgebras.*

Proof (Sketch). The proof follows a Prime Ideal Theorem argument. To interpret $*$ on $\mathsf{Pf}A$ for some A in **DL** we define $\gamma_A^* : \mathsf{Pf}A \to \mathsf{P}_c(\mathsf{Pf}A \times \mathsf{Pf}A), F \mapsto \{(F_1, F_2) \mid a \in F_1, b \in F_2 \Rightarrow a * b \in F\}$. It is easy to check that if $\exists F_1, F_2$ s.th. $(F_1, F_2) \in \gamma_A(F)$ and $a \in F_1, b \in F_2$, then $a * b \in F$ and $F \models a * b$. The converse is harder: given $a * b \in F$, we must build prime filters F_1, F_2 s.th. $a \in F_1, b \in F_2$ and $c * d \notin F \Rightarrow c \notin F_1$ or $d \notin F_2$. We consider the set $\mathscr{P}(a, b)$ of pairs of filter-ideal pairs $((F_1, I_1), (F_2, I_2))$ s.th.

1. $\uparrow a \subseteq F_1 \subseteq \{c \mid \forall d \in F_2, c * d \in F\}$
2. $\uparrow b \subseteq F_2 \subseteq \{d \mid \forall c \in F_1, c * d \in F\}$
3. $I_1 = \{c \mid \exists d \in F_2 \text{ s.th. } c * d \notin F\}$
4. $I_2 = \{d \mid \exists c \in F_1 \text{ s.th. } c * d \notin F\}$

It can be shown that $\mathscr{P}(a, b)$ is not-empty, forms a poset, has the property that I_1, I_2 are ideals such that $F_1 \cap I_1 = F_2 \cap I_2 = \emptyset$, and is closed under union of chains. Zorn's lemma then yields a maximum element which provides the desired prime filters. The same technique can be applied to define $\gamma_A^{\backslash}, \gamma_A^{/}$ interpreting $\backslash, /$, and it is easy to check that $\langle 0, \gamma_A^*, \gamma_A^{\backslash}, \gamma_A^{/} \rangle$ is a right inverse of $\hat{\delta}_A^{\text{DL}}$.

The Jónsson-Tarski Embedding and Canonical Extensions. We now apply the theory of canonicity to show that HAs and distributive residuated lattices are strongly complete with respect to the (proper) classes of T_{Hey}- and T_{RL}-coalgebras validating HFC1-3 and FC1-6 respectively. We need one important technical result, which shows that the Jónsson-Tarski embedding of Theorem 3 is the canonical extension defined in Section 3.2.

Proposition 9. *The structure map of the Jónsson-Tarski extension of an L_{Hey}- or L_{RL}-algebra is equal to the canonical extension of its structure map (in the sense of Section 3.2).*

Proof (Sketch). Recall Diagram (2) and that a DL-morphism $\alpha : L_{\text{RL}}A \to A$ is equivalent to being given a constant and binary operations $\alpha_*, \alpha_{\backslash}, \alpha_{/}$ on $\mathsf{U}A$ satisfying DL1-DL6. Similarly, $\mathcal{U}\mathsf{Pf}\alpha \circ \mathcal{U}\gamma_A \circ \delta_{\mathsf{Pf}A}^{\text{RL}}$ is equivalent to a constant and three binary operations $\mathcal{U}\mathsf{Pf}\alpha \circ \mathcal{U}\gamma_A \circ \delta_*, \mathcal{U}\mathsf{Pf}\alpha \circ \mathcal{U}\gamma_A \circ \delta_{\backslash}, \mathcal{U}\mathsf{Pf}\alpha \circ \mathcal{U}\gamma_A \circ \delta_{/}$ on A^σ. By commutativity of (2), the latter are extensions of the former. It is not hard to show that if an extension of a map on $\mathsf{U}A$ preserves or anti-preserves all non-empty meets or joins, then it is smooth and thus unique by Proposition 6. Direct calculation shows that $\delta_*, \delta_{\backslash}$ and $\delta_{/}$ all have such preservation properties in each argument. Moreover, $\mathcal{U}\mathsf{Pf}\alpha$ and $\mathcal{U}\gamma_A$ being inverse images preserve any meet or join. We thus get that $\mathcal{U}\mathsf{Pf}\alpha \circ \mathcal{U}\gamma_A \circ \delta_*$ is smooth and thus equal to α^σ as desired, and similarly for the other operations.

Strong Completeness. We are now ready to state our main result.

Theorem 5 (Strong completeness theorem). *Intuitionistic logic is strongly complete with respect to the class of T_{Hey}-coalgebras validating HFC1-3. The Distributive Full Lambek Calculus is strongly complete with respect to the class of T_{RL}-coalgebras validating FC1-6.*

Proof (Sketch). We treat the case of the distributive full Lambek calculus; intuitionistic logic is treated similarly. Let Φ, Ψ be sets of L_{RL}-formulas such that FC1-6+$\Phi \nvdash \Psi$. We need to find a model in which FC1-6 are valid, and which satisfies all formulas of Φ and no formula of Ψ at a certain point. Consider the Lindenbaum-Tarski L_{RL}-algebra $\mathcal{L}$ defined by FC1-6. These axioms are clearly valid in $\mathcal{L}$, and since they are canonical by Prop. 8, they are also valid in $\mathcal{L}^\sigma$, which by Prop. 9 is just its coalgebraic Jónsson-Tarski extension. It follows that FC1-6 are valid on the *model* $\mathsf{Pf}\mathcal{L} \to T_{\mathrm{RL}}\mathsf{Pf}\mathcal{L}$. To find the desired point, note that the filter generated by Φ in $\mathcal{L}$ is proper and does not intersect the ideal $\langle \Psi \rangle$ generated by Ψ, or else our staring assumption would be contradicted. We can thus find $\mathsf{Pf}\mathcal{L} \ni p_\Phi \supseteq \Phi$ s.th. $p_\Phi \cap \langle \Psi \rangle = \emptyset$, and $p_\Phi \models \Phi$, $p_\Phi \not\models \Psi$ follows.

Describing T_{Hey}-coalgebras Validating the Heyting Frame Conditions. Let us examine what T_{Hey}-coalgebras validating HFC1-3 look like. For every $\gamma : W \to T_{\mathrm{Hey}}W$ in this class, every $w \in W$ and every valuation, $w \models a \to a$. By considering a formula satisfied at a single point in the model is easy to see that $(x, y) \in \gamma(w) \Rightarrow x = y$; that is, the structure map of the coalgebra only really defines a binary relation to interpret $\to$. Thus T_{Hey}-coalgebras validating HFC1 are equivalent to P_c-coalgebras where $w \models a \to b$ iff $\forall x \in \gamma(w), x \models a \Rightarrow x \models b$. The distributivity laws of $\to$ together with HFC2-3 encode the well-known residuation property of $\to$ with respect to $\wedge$. Combined with HFC1 and the associated reformulation in terms of P_c-coalgebra, the residuation property states that:

$$w \models a \wedge b \;\Rightarrow\; w \models c \quad \text{iff} \quad w \models b \;\Rightarrow\; (\forall x \in \gamma(w)\ (x \models a \;\Rightarrow\; x \models c)).$$

Assuming the left-hand side, for the right-hand side to hold it is necessary that if $w \models b$, then $\forall x \in \gamma(w), x \models b$; that is, successor states satisfy the so-called 'persistency' condition. It also follows that $x \in \gamma(x)$; that is, the relation is reflexive. Finally, from HFC3 we get that $a \wedge b \leq c$ iff $b \leq a \to c$ iff $b \leq a \to (c \wedge (a \to c))$. By unravelling the interpretation of this last inequality, we get that the relation interpreting $\to$ must also be transitive. Thus we have recovered the traditional Kripke semantics of intuitionistic logic via a pre-order and persistent valuations by using the theory of canonicity for distributive lattices.

Describing T_{RL}-coalgebras Validating FC1-6. Axiom FC1 means that at every w in a T_{RL}-coalgebra, amongst all the pairs of states into which w can be 'separated' there must exist a *unit* state i, viz. $\pi_1(\gamma(i)) = 0$, such that $(w, i) \in \pi_2(\gamma(w))$. Similarly, there must exist a unit state i' such that $(i', w) \in \pi_2(\gamma(w))$. This condition can be found in this form in, for example, [6]. The other axioms are simply designed to capture the residuation condition in such a way that canonicity can be used, so a model in which FC2-6 are valid is simply a model in which the residuation conditions hold. By considering models with only three points it is easy to see that these conditions imply that

$$(y, z) \in \pi_1(\gamma(x)) \text{ iff } (x, z) \in \pi_2(\gamma(y)) \text{ iff } (y, x) \in \pi_3(\gamma(z)),$$

that is, the last three components of a T_{RL}-coalgebra's structure map are determined by any one of them. If we choose the second as defining the last two, a T_{RL}-coalgebra validating FC1-6, really is a coalgebra for the functor $T'_{\mathrm{RL}} : \mathscr{D} \to \mathscr{D}, T'_{\mathrm{RL}}W = 2 \times \mathsf{P}_c(W \times W)$ in which the interpretation of the operators is given by:

1. $w \models a * b$ iff $\exists (x, y) \in \gamma(w)$ s.t. $x \models a$ and $y \models b$
2. $w \models a/b$ iff $\forall (x, y)$ s.th $(w, y) \in \gamma(x)$ if $y \models b$ then $x \models a$
3. $w \models b\backslash a$ iff $\forall (x, y)$ s.th $(y, w) \in \gamma(x)$ if $y \models b$ then $x \models a$.

Modularity. The coalgebraic setting allows us to combine completeness-via-cano-nicity results from simple logics to get results for more complicated logics. It can be shown that the coalgebraic Jónsson-Tarski theorem is modular in the sense that if logics defined by syntax constructors L_1 and L_2 and interpreted in T_1- and T_2-coalgebras respectively via semantic transformations δ_1 and δ_2 whose adjoint transposes have right-inverses, then the logic defined by $(L_1 + L_2)$ is strongly complete w.r.t. $(T_1 \times T_2)$-coalgebras.

Theorem 6 (Strong completeness of intuitionistic BI). *Intuitionistic BI is strongly complete w.r.t. the class of* $T_{\mathrm{Hey}} \times T_{\mathrm{RL}}$*-coalgebra satisfying HFC1-3 and FC1-6.*

Additional Frame Conditions. We can consider more axioms to restrict further the classes of models we might be interested in. The following (in)equations can all easily be verified to be canonical and each corresponds to admitting a structural rule to the full distributive Lambek calculus: (1) Commutativity: $a * b = b * a$; (2) Increasing idempotence: $a \leq a * a$ (defines relevant logic); and (3) Integrality: $a \leq I$ (defines affine logic). More generally, we have presented a general methodology to get completeness results for axioms that could capture the behaviour of certain sub-classes of resources (e.g., heaps in separation logic).

5 Conclusion and Future Work

We have shown how distributive substructural logics can be formalized and given a semantics in the framework of coalgebraic logic, and highlighted the modularity of this approach. By choosing a syntax whose operators explicitly follow distribution rules, we can use the elegant topological theory of canonicity for DLs, and in particular the notion of smoothness and of topology matching, to build a set of canonical (in)equation capturing intuitionistic logic and the distributive full Lambek calculus. The coalgebraic approach makes the connection between algebraic canonicity and canonical models explicit, categorical and generalizable.

The modularity provided by our approach is twofold. Firstly, we have a generic method for building canonical (in)equations by using the Principle of Matching Topologies. Getting completeness results with respect to simple Kripke models for variations of the distributive full Lambek calculus (e.g., distributive affine logic) becomes very straightforward. Secondly, adding more operators to the fundamental language simply amounts to taking a *coproduct* of syntax constructors (e.g., $L_{\mathrm{RL}} + L_{\mathrm{Hey}}$ to define intuitionistic BI) and interpreting it with a *product* of model constructors (e.g., $T_{\mathrm{RL}} \times T_{\mathrm{Hey}}$). This seems particularly suited to logics which build on BI such as the bi-intuitionistic boolean BI of [4].

The operators $*, \backslash, /$, and $\to$ all satisfy simple distribution laws, but our approach can also accommodate operators with more complicated distribution laws and non-relational semantics. For example, the theory presented in this work could be extended to cover a graded version of $*$, say $*_k$, whose interpretation would be 'there are at least k ways to separate a resource such that...', the semantics would be given by coalgebras of the type $\mathbb{2} \times \mathcal{B}(- \times -)$ where $\mathcal{B}$ is the 'bag' or multiset functor. Similarly, a graded version $\to_k$ of the intuitionistic implication whose meaning would be '... implies ... apart from at most k exceptions' and interpreted by $\mathcal{B}(- \times -)$-coalgebras could also be covered by our approach. Crucially, such operators do satisfy (more complicated) distribution laws which lead to generalizations of the results in Section 3.2, and the possibility of building canonical (in)equations. The coalgebraic infrastructure then allows the rest of the theory to stay essentially unchanged.

References

1. Bezhanishvili, N., Gehrke, M.: Free Heyting Algebras: Revisited. In: Kurz, A., Lenisa, M., Tarlecki, A. (eds.) CALCO 2009. LNCS, vol. 5728, pp. 251–266. Springer, Heidelberg (2009)
2. Bílková, M., Kurz, A., Petrişan, D., Velebil, J.: Relation Liftings on Preorders and Posets. In: Corradini, A., Klin, B., Cîrstea, C. (eds.) CALCO 2011. LNCS, vol. 6859, pp. 115–129. Springer, Heidelberg (2011)
3. Blackburn, P., de Rijke, M., Venema, Y.: Modal Logic. Cambridge Tracts in Theoretical Computer Science, vol. 53. CUP (2001)
4. Brotherston, J., Villard, J.: Bi-intuitionistic boolean bunched logic. Research Note RN/14/06, University College London (2014)
5. Coumans, D., Gehrke, M., van Rooijen, L.: Relational semantics for a fragment of linear logic. In: Proceedings of PhDs in Logic III (2011)
6. Calcagno, C., Gardner, P., Zarfaty, U.: Context logic as modal logic: completeness and parametric inexpressivity. ACM SIGPLAN Not. 42(1), 123–134 (2007)
7. Collinson, M., Pym, D.: Algebra and logic for resource-based systems modelling. Mathematical Structures in Computer Science 19(5), 959–1027 (2009)
8. Collinson, M., Monahan, B., Pym, D.: A Discipline of Mathematical Systems Modelling. College Publications (2012)
9. Cîrstea, C., Pattinson, D.: Modular construction of complete coalgebraic logics. Theoret. Comp. Sci. 388(1-3), 83–108 (2007)
10. Dahlqvist, F., Pattinson, D.: On the fusion of coalgebraic logics. In: Corradini, A., Klin, B., Cîrstea, C. (eds.) CALCO 2011. LNCS, vol. 6859, pp. 161–175. Springer, Heidelberg (2011)

11. Dahlqvist, F., Pattinson, D.: Some sahlqvist completeness results for coalgebraic logics. In: Pfenning, F. (ed.) FOSSACS 2013 (ETAPS 2013). LNCS, vol. 7794, pp. 193–208. Springer, Heidelberg (2013)
12. Dahlqvist, F., Pym, D.: Completeness via canonicity for distributive substructural logics: a coalgebraic perspective. Research Note RN/15/04, Department of Computer Science, UCL (2015), `http://www.cs.ucl.ac.uk/fileadmin/UCL-CS/research/Research_Notes/rn-15-04.pdf`
13. Dunn, J.M., Gehrke, M., Palmigiano, A.: Canonical Extensions and Relational Completeness of Some Substructural Logics. J. Symb. Logic 70(3), 713–740 (2005)
14. Dunn, J.M.: Positive modal logic. Studia Logica 55(2), 301–317 (1995)
15. Galmiche, D., Méry, D., Pym, D.: The Semantics of BI and Resource Tableaux. Mathematical Structures in Computer Science 15, 1033–1088 (2005)
16. Gehrke, M.: Generalized Kripke frames. Studia Logica 84(2), 241–275 (2006)
17. Gehrke, M., Harding, J.: Bounded lattice expansions. Journal of Algebra 238(1), 345–371 (2001)
18. Girard, J.-Y.: Linear logic. Theoret. Comp. Sci. 50, 1–102 (1987)
19. Givant, S., Halmos, P.: Introduction to Boolean Algebras. Springer (2009)
20. Gehrke, M., Jónsson, B.: Bounded distributive lattices with operators. Mathematica Japonica 40(2), 207–215 (1994)
21. Gehrke, M., Jónsson, B.: Bounded distributive lattice expansions. Mathematica Scandinavica 94, 13–45 (2004)
22. Gehrke, M., Jansana, R., Palmigiano, A.: Δ_1 -completions of a poset. Order 30(1), 39–64 (2013)
23. Ishtiaq, S.S., O'Hearn, P.W.: BI as an assertion language for mutable data structures. ACM SIGPLAN Not. 36(3), 14–26 (2001)
24. Jónsson, B.: On the canonicity of Sahlqvist identities. Stud. Log. 53(4), 473–492 (1994)
25. Jacobs, B., Sokolova, A.: Exemplaric expressivity of modal logics. J. Log. Computat. 20, 1041–1068 (2010)
26. Jónsson, B., Tarski, A.: Boolean algebras with operators. part 1. Amer. J. Math. 33, 891–937 (1951)
27. Kupke, C., Kurz, A., Pattinson, D.: Algebraic semantics for coalgebraic logics. In: ENTCS, vol. 106, pp. 219–241 (2004)
28. Kupke, C., Kurz, A., Pattinson, D.: Ultrafilter Extensions for Coalgebras. In: Fiadeiro, J.L., Harman, N.A., Roggenbach, M., Rutten, J. (eds.) CALCO 2005. LNCS, vol. 3629, pp. 263–277. Springer, Heidelberg (2005)
29. Ono, H.: Substructural logics and residuated lattices — an introduction. Trends in Logic 20, 177–212 (2003)
30. O'Hearn, P.W., Pym, D.J.: The logic of bunched implications. Bulletin of Symbolic Logic 5(2), 215–244 (1999)
31. Pym, D., O'Hearn, P., Yang, H.: Possible Worlds and Resources: The Semantics of BI. Theoret. Comp. Sci. 315(1), 257–305 (2002); Erratum: p. 285, l. -12: ', for some $P', Q = P; P'$' should be '$P \vdash Q$'
32. J.: C Reynolds. Separation logic: A logic for shared mutable data structures. In: Proc. 17th LICS, pp. 55–74. IEEE (2002)
33. Suzuki, T.: Canonicity results of substructural and lattice-based logics. The Review of Symbolic Logic 4, 1–42 (2011)
34. Venema, Y.: Algebras and coalgebras. In: van Benthem, J., Blackburn, P., Wolter, F. (eds.) Handbook of Modal Logic. Elsevier (2006)

Generalised N-ary Relations and Allegories

Bartosz Zieliński

Department of Computer Science
Faculty of Physics and Applied Informatics
University of Łódź
ul. Pomorska nr 149/153, 90-236 Łódź, Poland
bzielinski@uni.lodz.pl

Abstract. Allegories abstract useful features of the enriched category of sets and binary relations. N-ary relations can be easily defined in any allegory with relational products as binary relations between appropriate product objects. Unfortunately, in many applications (especially those related to databases) such an indirect way of thinking about N-ary relations is somewhat awkward. In this paper we develop a formalism for allegorical generalisations of N-ary relations particularly well suited for database applications.

Keywords: Allegories, Relations, Databases.

1 Introduction

Allegories [9] can be thought of as a categorical generalisation of relation algebras ([13],[10]). Thus arrows in an allegory are like binary relations, but using allegories in various applications instead of actual binary relations allows the same formalism to be much more widely applicable — for instance, when the relations are locale-valued or fuzzy.

Recently a new database modeling formalism based on allegories was proposed [16], [17]. For database applications it is important to have the possibility of representing n-ary relations for arbitrary finite n. This requires the existence of relational products ([9], [8], cf. [3], [11]) in an allegory. Relational products generalise Cartesian products.

If required relational products exist in a given allegory then one can represent an n-ary relation, for $n > 2$, as a binary relation between product objects. The main problem is that this representation is non-canonical — it is necessary to divide the "legs" of an n-ary relation between source and target of the representing binary relation and there are many ways to do it. In particular, different partitions of legs may be required to make some operations in allegory legal, like joins and intersections with particular arrows.

In this paper we develop a formalism for allegorical generalisations of N-ary relations which is particularly well suited for database applications, and which seems not to have some of the disadvantages mentioned above. The formalism was partially inspired by [5], [4].

W. Kahl et al. (Eds.): RAMiCS 2015, LNCS 9348, pp. 136–150, 2015.
DOI: 10.1007/978-3-319-24704-5_9

2 Preliminaries

The reader should be familiar with basic category theory, (see e.g., [2] for an introduction). In the preliminaries (partially taken from [15]) we recall some basic definitions to fix the clean but idiosyncratic notation we use. We also provide introductory material on allegories which are not widely known and for which [9] is the basic textbook.

For simplicity, in what follows we do not distinguish between sets and classes.

2.1 Categories and Graphs

A *graph* $\mathscr{G}$ consists of a set of vertices $\mathrm{Obj}[\mathscr{G}]$, a set of arrows $\mathrm{Arr}[\mathscr{G}]$, and a pair of maps $\overleftarrow{(\cdot)}, \overrightarrow{(\cdot)} : \mathrm{Arr}[\mathscr{G}] \longrightarrow \mathrm{Obj}[\mathscr{G}]$, called source and target, respectively. We denote by $\mathrm{Arr}_{\mathscr{G}}(A,B)$ the set of arrows with source A and target B, where $A,B \in \mathrm{Obj}[\mathscr{G}]$.

A *category* $\mathscr{C}$ is a graph with associative arrow composition $f;g \in \mathrm{Arr}_{\mathscr{C}}(\overleftarrow{f}, \overrightarrow{g})$ defined whenever $\overrightarrow{f} = \overleftarrow{g}$ (note the diagrammatic order), and identity map $\mathrm{id} : \mathrm{Obj}[\mathscr{C}] \to \mathrm{Arr}[\mathscr{C}]$ such that $\mathrm{id}(\overleftarrow{f});f = f;\mathrm{id}(\overrightarrow{f}) = f$ for all $f \in \mathrm{Arr}[\mathscr{C}]$. We write $\mathrm{id}_A := \mathrm{id}(A)$. We will often omit the semicolon composition operator abbreviating $fg := f;g$.

Diagrams will be frequently used to declare composability of arrows, e.g.,

$$\bullet \xrightarrow{f} \bullet \overset{g}{\underset{h}{\rightrightarrows}} \bullet \quad :\equiv \quad (\overrightarrow{f} = \overleftarrow{g}) \wedge (\overrightarrow{f} = \overleftarrow{h}) \wedge (\overrightarrow{g} = \overrightarrow{h}),$$

but unlike in [9], the diagrams are not considered commutative by default.

A *categorical n-ary product* is a family of arrows $\{\pi_i\}_{i \in \{1,\ldots,n\}}$ with a common source such that for any other family of arrows $\{f_i\}_{i \in \{1,\ldots,n\}}$ with a common source and such that $\overrightarrow{f_i} = \overrightarrow{\pi_i}$ for all $i \in \{1,\ldots,n\}$ there exists a unique arrow h such that $f_i = h\pi_i$ for all $i \in \{1,\ldots,n\}$.

2.2 Allegories

An *allegory* [9] $\mathscr{A}$ is a category enriched with intersection and reciprocation operators:

$$\cdot \sqcap \cdot : \mathrm{Arr}_{\mathscr{A}}(A,B) \times \mathrm{Arr}_{\mathscr{A}}(A,B) \to \mathrm{Arr}_{\mathscr{A}}(A,B),$$
$$(\cdot)^\circ : \mathrm{Arr}_{\mathscr{A}}(A,B) \to \mathrm{Arr}_{\mathscr{A}}(B,A),$$

for all $A,B \in \mathrm{Obj}[\mathscr{A}]$, which are required to satisfy the following conditions: Intersections make each homset a meet semi-lattice (see e.g. [7]), where we denote the associated partial order by $\sqsubseteq$, i.e., $R \sqsubseteq S :\equiv R \sqcap S = R$, for all $R,S \in \mathrm{Arr}_{\mathscr{A}}(\overleftarrow{R}, \overrightarrow{R})$. In addition, $\cdot \sqcap \cdot$ and $(\cdot)^\circ$ are to satisfy

$$R^{\circ\circ} = R, \qquad (1a) \qquad\qquad R(S \sqcap T) \sqsubseteq RS \sqcap RT \qquad (1d)$$
$$(RS)^\circ = S^\circ R^\circ, \qquad (1b) \qquad\qquad RS \sqcap T \sqsubseteq (R \sqcap TS^\circ)S \qquad (1e)$$
$$(R \sqcap S)^\circ = R^\circ \sqcap S^\circ, \qquad (1c)$$

for all $R,S,T \in \mathrm{Arr}[\mathscr{A}]$ such that the above formulas are well defined. Applying the reciprocation to both sides of the Equations (1d) (right semi-distributivity) and (1e) (right modular identity), using the identities (1a)-(1c) and redefining symbols yields easily the following right versions:

$$(S \sqcap T)R \sqsubseteq SR \sqcap TR, \qquad (1f) \qquad RS \sqcap T \sqsubseteq R(S \sqcap R^\circ T). \qquad (1g)$$

Allegories generalise the allegory $\mathscr{R}$ of sets (objects) and binary relations (arrows). Because of it we may refer to arrows in any allegory as "relations". In $\mathscr{R}$ we write aRb iff $(a,b) \in R$. The identity in $\mathscr{R}$ is $\mathrm{id} : A \mapsto \{(a,a) | a \in A\}$, intersection is the set intersection, i.e., $R \sqcap S := R \cap S$, reciprocation is defined by $aR^\circ b :\equiv bRa$ and composition of relations $R,S \in \mathrm{Arr}[\mathscr{R}]$ such that $\overrightarrow{R} = \overleftarrow{S}$ is defined by:

$$a(RS)c \quad :\equiv \quad \exists b \in \overrightarrow{R} \,.\, aRb \wedge bSc. \qquad (2)$$

Another example which will feature in this paper is the allegory $\mathscr{R}[\Lambda]$ of Λ-valued relations, where Λ is an arbitrary locale (i.e., a complete, distributive lattice in which infima distribute over arbitrary — that is also infinite — suprema). In $\mathscr{R}[\Lambda]$ objects are sets and $\mathrm{Arr}_{\mathscr{R}[\Lambda]}(A,B)$ is the set of functions $R : A \times B \to \Lambda$. The allegorical operations are defined as follows:

$$\mathrm{id}_A(a,a') := \begin{cases} \top & \text{if } a = a' \\ \bot & \text{if } a \neq a' \end{cases}, \qquad R^\circ(b,a) := R(a,b),$$

$$(R \sqcap S)(a,b) := R(a,b) \wedge S(a,b), \quad (R;T)(a,c) := \bigvee_{b \in B} R(a,b) \wedge T(b,c). \qquad (3)$$

Note that we denote the top and bottom elements of a locale Λ by $\top$ and $\bot$, respectively.

We distinguish the following classes of arrows in an allegory:

- If $\mathrm{id}_{\overleftarrow{R}} \sqsubseteq RR^\circ$ then R is called *total.*
- If $R^\circ R \sqsubseteq \mathrm{id}_{\overrightarrow{R}}$ then R is called *functional.*
- If R is functional and total it is called a *map*. The set of all maps in an allegory $\mathscr{A}$ is denoted by $\mathrm{Map}[\mathscr{A}]$.
- If $RR^\circ \sqsubseteq \mathrm{id}_{\overleftarrow{R}}$ then R is called *injective.*
- If $\mathrm{id}_{\overrightarrow{R}} \sqsubseteq R^\circ R$ then R is called *surjective.*

If $R \in \mathrm{Arr}[\mathscr{A}]$ is an isomorphism in an allegory $\mathscr{A}$ then both R and R° are maps and $R^{-1} = R^\circ$. Note that id_A for all $A \in \mathrm{Obj}[\mathscr{A}]$ is a map and the composition of maps is a map. Thus maps in $\mathscr{A}$ form a subcategory of $\mathscr{A}$. For any $A,B \in \mathrm{Obj}[\mathscr{A}]$ we denote by $\top_{A,B}$ the top element of $\mathrm{Arr}_{\mathscr{A}}(A,B)$, if it exists. In $\mathscr{R}$ we have $\top_{AB} := A \times B$. We will make use of the following results (cf. [9]):

Lemma 1. *$R(S \sqcap T) = RS \sqcap RT$ for all* $\bullet \xrightarrow{R} \bullet \overset{S}{\underset{T}{\rightrightarrows}} \bullet$ *such that R is functional.*

Similarly, $(S \sqcap T)R = SR \sqcap TR$ for all $\bullet \overset{S}{\underset{T}{\rightrightarrows}} \bullet \xrightarrow{R} \bullet$ *such that R° is functional.*

Lemma 2. *([14]) Suppose that* 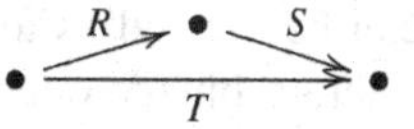 *and that relation S is functional. Then $RS \sqcap T = (R \sqcap TS^\circ)S$.*

Lemma 3. *Suppose that* 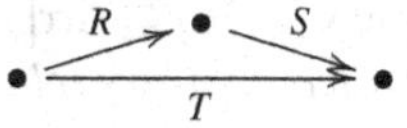 *and that S is total. Then $RS \sqsubseteq T$ implies that $R \sqsubseteq TS^\circ$.*

Definition 1. *([9]) An object* 1 *in allegory $\mathscr{A}$ is called a unit whenever* $\mathrm{id}_1 = \top_{11}$ *and for any $A \in \mathrm{Obj}[\mathscr{A}]$ there exists some $u_A \in \mathrm{Arr}_{\mathscr{A}}(A, 1)$ which is total. An allegory where a unit exists is called unitary.*

Lemma 4. *For any $A \in \mathrm{Obj}[\mathscr{A}]$ an arrow u_A is a map. Moreover $u_A = \top_{A1}$.*

Lemma 5. *For any $A, B \in \mathrm{Obj}[\mathscr{A}]$ we have $u_A; (u_B)^\circ = \top_{AB}$.*

3 Relational Products

3.1 Basic Definitions

The following is a standard definition of a relational binary product in an allegory. It is well known [9] that the relational binary product is the categorical product in the subcategory of maps. Note that the definition assumes the existence of the top relation between components of the product.

Definition 2. *([9]) Let $\mathscr{A}$ be an allegory. A pair of arrows $\bullet \xleftarrow{\pi_1} C \xrightarrow{\pi_2} \bullet$ (called projection arrows) is called a relational product if and only if it satisfies the following conditions:*

$$\pi_1^\circ \pi_1 = \mathrm{id}_{\overrightarrow{\pi_1}}, \quad \pi_2^\circ \pi_2 = \mathrm{id}_{\overrightarrow{\pi_2}}, \tag{4a}$$

$$\pi_1 \pi_1^\circ \sqcap \pi_2 \pi_2^\circ = \mathrm{id}_C, \tag{4b}$$

$$\pi_1^\circ \pi_2 = \top_{\overrightarrow{\pi_1}\overrightarrow{\pi_2}} \tag{4c}$$

We will often name the common source of π_i's as $\overrightarrow{\pi_1} \times \overrightarrow{\pi_2}$ and abuse the language by refering to it as "the product", even though it is determined only up to an isomorphism, and the projection arrows might not be determined uniquely by the common source. In order to distinguish projections of different relational products we will use $\pi_1^{A \times B}$ and $\pi_2^{A \times B}$ to denote the projections with $A \times B$ as the common source. Also, whenever $A \neq B$, we will assume that $A \times B = B \times A$, $\pi_1^{A \times B} = \pi_2^{B \times A}$ and $\pi_2^{A \times B} = \pi_1^{B \times A}$.

Note that in $\mathscr{R}$ (as well as in $\mathscr{R}[\Lambda]$) the relational product is isomorphic with the cartesian product. Another special case is given by the following observation.

Lemma 6. *Suppose that $\mathscr{A}$ is a unitary allegory. Then for any $A \in \mathrm{Obj}[\mathscr{A}]$ the pair*

$$A \xleftarrow{\mathrm{id}_A} A \xrightarrow{u_A} 1$$

is a relational product.

The unit in a unitary allegory can be viewed as a 0-ary relational product. To get n-ary products for $n > 2$ one can iterate binary ones. Sometimes, a direct algebraic characterization of n-ary relational products becomes handy.

Definition 3. *(cf. [8]) A finite family of arrows $\{\pi_i\}_{i\in I} \subseteq \mathrm{Arr}[\mathscr{A}]$ with a common source C, that is, such that $\overleftarrow{\pi_i} = C$ for all $i \in I$, is called an n-ary relational product if and only if it satisfies the following conditions:*

$$\forall i \in I \,.\, \pi_i^\circ \pi_i = \mathrm{id}_{\overrightarrow{\pi_i}}, \tag{5a}$$

$$\bigsqcap_i \pi_i \pi_i^\circ = \mathrm{id}_C, \tag{5b}$$

$$\forall k \in I \,.\, \Big(\bigsqcap_{i\in I\setminus\{k\}} \pi_i \pi_i^\circ\Big) \pi_k = \top_{\overleftarrow{\pi_k}\,\overrightarrow{\pi_k}} \tag{5c}$$

It is also easy to see that binary relational products are 2-ary relational products and vice versa. It was proven in [15] (cf. [8]) in a slightly more general setting that n-ary relational products are categorical products in the subcategory of maps. In particular, we have

Lemma 7. *([15, Theorem 18]) if $\{\pi_i\}_{i\in I}$ is an n-ary relational product with a common source C and $\{\, A \xrightarrow{f_i} \overrightarrow{\pi_i} \,\}_{i\in I}$ is a family of maps then the unique map $A \xrightarrow{f} C$ such that $f\pi_i = f_i$ for all $i \in I$ is given by the formula*

$$f = \bigsqcap_{i\in I} f_i \pi_i^\circ . \tag{6}$$

Another result which we will use in what follows is the following:

Lemma 8. *(cf. [15, Lemma 17]) Let $\{\, C \xrightarrow{\pi_i} \bullet \,\}_{i\in I}$ be an n-ary relational product. For any $\varnothing \neq J \subsetneq I$ and a family of total arrows $\{\, A \xrightarrow{R_i} \overrightarrow{\pi_i} \,\}_{i\in J}$ the arrow $R_J := \bigsqcap_{i\in J} R_i \pi_i^\circ$ is also total.*

Note that [15, Lemma 17] actually assumes that R_i's are maps and π_i's form a weak n-ary relational product. It is, however, easy to check that if π_i's form a (strong) n-ary relational product then the proof works when R_i's are merely assumed to be total.

4 Generalisation of Sharpness

First, let us recall the notion of sharpness of relational product. Suppose that $\{\pi_i\}_{i\in I}$ is an n-ary relational product, where $I = \{1, \ldots, n\}$. Let $\{R_i\}_{i\in I}$ and $\{S_i\}_{i\in I}$ be two families of arrows such that R_i's have a common source, S_i's have a common source, and $\overrightarrow{S_i} = \overrightarrow{R_i} = \overrightarrow{\pi_i}$, for all $i \in I$, i.e.,

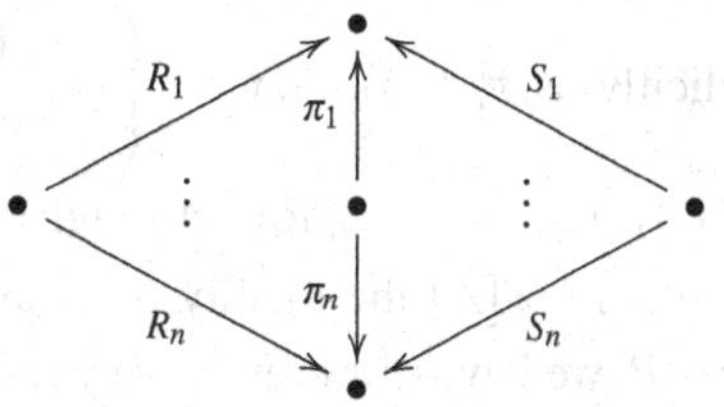

(7)

Ideally, we would like the factorisation through the relational product to commute with the composition of R_i's with S_i°'s, that is, we would like the product to satisfy the *sharpness condition*:

$$\left(\bigsqcap_{i\in I} R_i\pi_i^\circ\right)\left(\bigsqcap_{j\in I} \pi_j S_j^\circ\right) = \bigsqcap_{i\in I} R_i S_i^\circ. \tag{8}$$

Unfortunately sharpness condition is not satisfied in general allegories for general R_i's and S_i's (cf. [8]).

The sharpness condition does not seem sufficient to prove the results in the next subsection. Therefore here we will introduce a more general condition:

Definition 4. *Let $\{\pi_i\}_{i\in I}$ be a n-ary relational product and let $\{R_i\}_{i\in I}$ and $\{S_i\}_{i\in I}$ be families of arrows as in Diagram (7). We say that $\{\pi_i\}_{i\in I}$ satisfies the generalised sharpness condition for families $\{R_i\}_{i\in I}$ and $\{S_i\}_{i\in I}$ iff, for all non-empty $I_1, I_2 \subseteq I$ we have*

$$\left(\bigsqcap_{i\in I_1} R_i\pi_i^\circ\right)\left(\bigsqcap_{j\in I_2} \pi_j S_j^\circ\right) = \left(\bigsqcap_{i\in I_1\cap I_2} R_i S_i^\circ\right).$$

Here (as elsewhere) an intersection of an empty family of arrows is a top arrow (as it is the infimum of an empty family).

Note that in general

$$\left(\bigsqcap_{i\in I_1} R_i\pi_i^\circ\right)\left(\bigsqcap_{j\in I_2} \pi_j S_j^\circ\right) \sqsubseteq \left(\bigsqcap_{i\in I_1\cap I_2} R_i S_i^\circ\right) \tag{9}$$

Indeed, the inequality obviously holds when $I_1 \cap I_2 = \varnothing$. If $I_1 \cap I_2 \neq \varnothing$ then for any $k \in I_1 \cap I_2$ we have $\left(\bigsqcap_{i\in I_1} R_i\pi_i^\circ\right)\left(\bigsqcap_{j\in I_2} \pi_j S_j^\circ\right) \sqsubseteq R_k S_k^\circ$, and thus $\left(\bigsqcap_{i\in I_1} R_i\pi_i^\circ\right)\left(\bigsqcap_{j\in I_2} \pi_j S_j^\circ\right) \sqsubseteq \bigsqcap_{k\in I_1\cap I_2} R_k S_k^\circ$. Hence, only the inequality in the other direction is non-trivial.

It is obvious that the generalised sharpness condition cannot be satisfied for arbitrary families $\{R_i\}_{i\in I}$ and $\{S_i\}_{i\in I}$ of arrows (e.g., take disjoint I_1 and I_2 and consider R_i's and S_i's to be bottom arrows in a distributive allegory). However, as the following result shows, it is not unreasonable to assume the generalised sharpness for total arrows:

Proposition 1. *In $\mathscr{R}[\Lambda]$, for any locale Λ, the generalised sharpness condition is satisfied for arbitrary families of total arrows.*

Proof. First recall that in $\mathscr{R}[\Lambda]$ the common source of π_i's is isomorphic with $\times_{i\in I}\overrightarrow{\pi_i}$ and π_i's can be given explicitly as $\pi_i((x_m)_{m\in I},x) = \begin{cases} \top & \text{if } x_i = x \\ \bot & \text{if } x_i \neq x \end{cases}$. Then one easily checks that $\left(\sqcap_{i\in I_1} R_i\pi_i^\circ\right)(x,(y_n)_{n\in I}) = \bigwedge_{i\in I_1} R_i(x,y_i)$ and $\left(\sqcap_{i\in I_2} S_i\pi_i^\circ\right)(z,(y_n)_{n\in I}) = \bigwedge_{i\in I_2} S_i(z,y_i)$. Noting now that in $\mathscr{R}[\Lambda]$ the totality of an arrow R is equivalent with $\bigvee_{y\in\overrightarrow{R}} R(x,y) = \top$ for all $x \in \overleftarrow{R}$ we have (for brevity we write (y_n) instead of $(y_n)_{n\in I}$):

$$
\begin{aligned}
&\left(\left(\bigsqcap_{i\in I_1} R_i\pi_i^\circ\right)\left(\bigsqcap_{j\in I_2}\pi_j S_j^\circ\right)\right)(x,z)\\
&= \bigvee_{(y_n)\in\times_{i\in I}\overrightarrow{\pi_i}} \left(\left(\bigsqcap_{j\in I_1} R_j\pi_j^\circ\right)(x,(y_n)) \wedge \left(\bigsqcap_{k\in I_2} S_k\pi_k^\circ\right)(z,(y_n))\right)\\
&= \bigvee_{(y_n)\in\times_{i\in I}\overrightarrow{\pi_i}} \left(\bigwedge_{j\in I_1} R_j(x,y_j) \wedge \bigwedge_{k\in I_2} S_k(z,y_k)\right)\\
&= \bigwedge_{i\in I_1\setminus I_2}\left(\bigvee_{y_i\in\overrightarrow{\pi_i}} R_i(x,y_i)\right) \wedge \bigwedge_{j\in I_2\setminus I_1}\left(\bigvee_{y_j\in\overrightarrow{\pi_j}} S_j(z,y_j)\right)\\
&\qquad \wedge \bigwedge_{k\in I_1\cap I_2}\left(\bigvee_{y_k\in\overrightarrow{\pi_k}} (R_k(x,y_k)\wedge S_k(z,y_k))\right)\\
&= \top\wedge\top\wedge \bigwedge_{k\in I_1\cap I_2}\left(\bigvee_{y_k\in\overrightarrow{\pi_k}} (R_k(x,y_k)\wedge S_k(z,y_k))\right)\\
&= \bigwedge_{k\in I_1\cap I_2} (R_k S_k^\circ)(x,z).
\end{aligned}
$$

Unfortunately, the author does not know if the sharpness condition for total arrows implies generalised sharpness for total arrows.

4.1 Iterating and De-Iterating Relational Products

The notions of iterating and de-iterating relational products are best explained by Figures 1 and 2. It is well known that iterating n-ary relational products gives relational products (see e.g., [8] or, in slightly more general context, [15]). Here we prove that iteration preserves the generalised sharpness of total arrows:

Lemma 9. *Suppose that* $\{\, C \xrightarrow{\pi_i} \bullet \,\}_{i\in I}$ *is an* $|I|$*-ary relational product, let* $k\in I$ *and let* $\{\, \overrightarrow{\pi_k} \xrightarrow{\rho_j} \bullet \,\}_{j\in J}$ *be a* $|J|$*-ary relational product. Suppose that* π_i*'s and* ρ_i*'s satisfy the generalised sharpness condition for total arrows. Then* $\{\pi_k\rho_j\}_{j\in J}\cup\{\pi_i\}_{i\in I\setminus\{k\}}$ *is an* $(|I|+|J|-1)$*-ary relational product also satisfying the generalised sharpness condition for total arrows.*

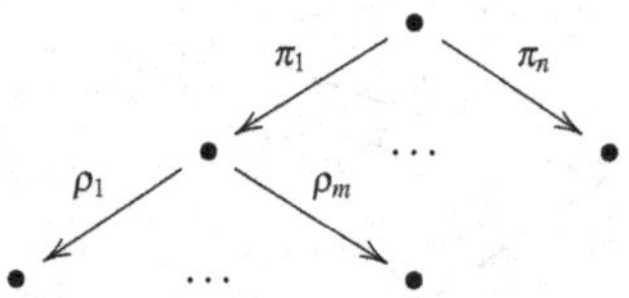

Fig. 1. Iterating relational products. Is $\{\pi_1\rho_i\}_{1\leqslant i\leqslant m} \cup \{\pi_j\}_{2\leqslant j\leqslant n}$ an $(m+n-1)$-ary relational product if π_i's and ρ_i's are, respectively, n-ary and m-ary relational products?

Proof. Without loss of generality we may assume that $I \cap J = \varnothing$. Let $I_1, I_2 \subseteq I\backslash\{k\}$ and $J_1, J_2 \subseteq J$ be such that $I_1 \cup J_1$ and $I_2 \cup J_2$ are both non-empty. Let

$$\{\, A \xrightarrow{R_i} \vec{\pi_i} \,\}_{i\in I\backslash\{k\}} \cup \{\, A \xrightarrow{R_j} \vec{\rho_j} \,\}_{j\in J}, \quad \{\, B \xrightarrow{S_m} \vec{\pi_m} \,\}_{m\in I\backslash\{k\}} \cup \{\, B \xrightarrow{S_n} \vec{\rho_n} \,\}_{n\in J}$$

be two families of total arrows. Then

$$\begin{aligned}
&\Big(\bigsqcap_{i\in I_1} R_i\pi_i^\circ \sqcap \bigsqcap_{j\in J_1} R_j\rho_j^\circ\pi_k^\circ\Big)\Big(\bigsqcap_{m\in I_2}\pi_m S_m^\circ \sqcap \bigsqcap_{n\in J_2}\pi_k\rho_n S_n^\circ\Big)\\
&\qquad\{\text{By Lemma 1}\}\\
=\;&\Big(\bigsqcap_{i\in I_1} R_i\pi_i^\circ \sqcap \big(\bigsqcap_{j\in J_1} R_j\rho_j^\circ\big)\pi_k^\circ\Big)\Big(\bigsqcap_{m\in I_2}\pi_m S_m^\circ \sqcap \pi_k\bigsqcap_{n\in J_2}\rho_n S_n^\circ\Big)\\
&\qquad\{\text{By the generalised sharpness of } \pi_i\text{'s for total arrows}\\
&\qquad\quad \text{as by Lemma 8 } \bigsqcap_{j\in J_1} R_j\rho_j^\circ \text{ and } \bigsqcap_{j\in J_2} S_j\rho_j^\circ \text{ are total}\}\\
=\;&\bigsqcap_{i\in I_1\cap I_2} R_iS_i^\circ \sqcap \Big(\bigsqcap_{j\in J_1} R_j\rho_j^\circ\Big)\Big(\bigsqcap_{n\in J_2}\rho_n S_n^\circ\Big)\\
&\qquad\{\text{By the generalised sharpness of } \rho_i\text{'s for total arrows}\}\\
=\;&\bigsqcap_{i\in I_1\cap I_2} R_iS_i^\circ \sqcap \bigsqcap_{j\in J_1\cap J_2} R_jS_j^\circ\\
=\;&\bigsqcap_{i\in(I_1\cup J_1)\cap(I_2\cup J_2)} R_iS_i^\circ .
\end{aligned}$$

Lemma 10. *Let* $\{\, A \xrightarrow{\pi_i} \bullet \,\}_{i\in I}$ *be an* $|I|$*-ary relational product. Furthermore, let* $J \subsetneq I$ *and let* $\{\, B \xrightarrow{\rho_j} \vec{\pi_j} \,\}_{j\in J}$ *be a* $|J|$*-ary relational product. Define* $\pi_J := \bigsqcap_{j\in J}\pi_j\rho_j^\circ$*. Then if* π_i*'s and* ρ_i*'s satisfy the generalised sharpness condition for maps, it follows that* $\{\pi_J\} \cup \{\pi_i\}_{i\in I\backslash J}$ *is an* $(|I|-|J|+1)$*-ary relational product which also satisfies the generalised sharpness condition for maps.*

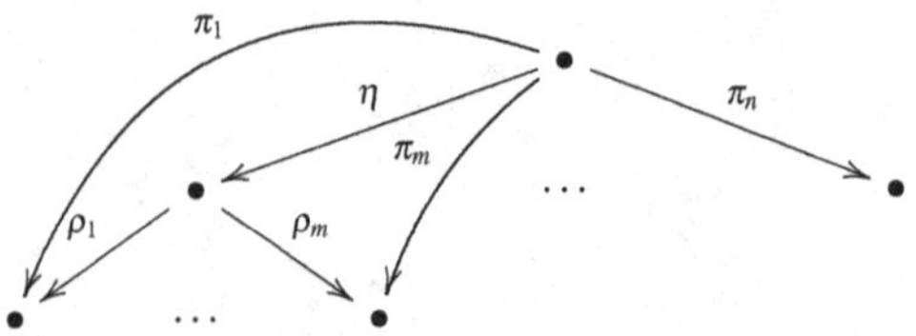

Fig. 2. De-iterating relational products: Assuming that π_i's and ρ_j's are relational products and η is the unique map such that $\pi_i = \eta\rho_i$, for all $1 \leqslant i \leqslant m$, is $\{\eta, \pi_{m+1}, \ldots, \pi_n\}$ a relational product?

Proof. First we check the conditions of the Definition 3. It suffices to check Equation (5a) for π_J as $\pi_i^\circ \pi_i = \mathrm{id}_{\overrightarrow{\pi_i}}$ for all $i \in I\backslash J$ by assumption. By the generalised sharpness of π_i's for maps we have that (as ρ_i's are maps)

$$\pi_J^\circ \pi_J = \Big(\bigsqcap_{i\in J}\rho_i\pi_i^\circ\Big)\Big(\bigsqcap_{j\in J}\pi_j\rho_j^\circ\Big) = \bigsqcap_{i\in J}\rho_i\rho_i^\circ = \mathrm{id}_B,$$

where the last equality follows from Equation (5b) for ρ_i's.

From the generalised sharpness of ρ_i's for maps it follows that

$$\pi_J\pi_J^\circ = \Big(\bigsqcap_{i\in J}\pi_i\rho_i^\circ\Big)\Big(\bigsqcap_{j\in J}\rho_j\pi_j^\circ\Big) = \bigsqcap_{i\in J}\pi_i\pi_i^\circ. \tag{10}$$

Thus, Equation (5b) follows:

$$\bigsqcap_{i\in I\backslash J}\pi_i\pi_i^\circ \sqcap \pi_J\pi_J^\circ = \bigsqcap_{i\in I\backslash J}\pi_i\pi_i^\circ \sqcap \bigsqcap_{j\in J}\pi_j\pi_j^\circ = \bigsqcap_{i\in I}\pi_i\pi_i^\circ = \mathrm{id}_A.$$

Finally, to verify Equation (5c), we need to consider two cases. First, let $k \in I\backslash J$. Then

$$\Big(\bigsqcap_{i\in I\backslash J\backslash\{k\}}\pi_i\pi_i^\circ \sqcap \pi_J\pi_J^\circ\Big)\pi_k = \Big(\bigsqcap_{i\in I\backslash J\backslash\{k\}}\pi_i\pi_i^\circ \sqcap \bigsqcap_{j\in J}\pi_j\pi_j^\circ\Big)\pi_k = \Big(\bigsqcap_{i\in I\backslash\{k\}}\pi_i\pi_i^\circ\Big)\pi_k = \top_{A\overrightarrow{\pi_k}}$$

by the virtue of Equation (10). On the other hand,

$$\Big(\bigsqcap_{i\in I\backslash J}\pi_i\pi_i^\circ\Big)\pi_J = \Big(\bigsqcap_{i\in I\backslash J}\pi_i\pi_i^\circ\Big)\Big(\bigsqcap_{j\in J}\pi_j\rho_j^\circ\Big) = \top_{AB}$$

using the generalised sharpness of π_i's for maps.

Now we will prove that $\{\pi_J\} \cup \{\pi_i\}_{i\in I\backslash J}$ satisfies the sharpness condition for maps. Let $I_1, I_2 \subseteq (I\backslash J) \cup \{J\}$ be non-empty, and let

$$\{\, C \xrightarrow{f_i} \overrightarrow{\pi_i}\,\}_{i\in I\backslash J} \cup \{\, C \xrightarrow{f_J} B\,\},\quad \{\, D \xrightarrow{g_i} \overrightarrow{\pi_i}\,\}_{i\in I\backslash J} \cup \{\, D \xrightarrow{g_J} B\,\}$$

be two families of maps. There are four cases to consider:

1. $J \notin I_1$ and $J \notin I_2$. Then $\left(\bigsqcap_{i\in I_1} f_i\pi_i^\circ\right)\left(\bigsqcap_{j\in I_2}\pi_j g_j^\circ\right) = \bigsqcap_{i\in I_1\cap I_2} f_i g_i^\circ$ follows easily from $\{\pi_i\}_{i\in I}$ satisfying the generalised sharpness condition for maps.

2. $J \in I_1$ and $J \notin I_2$. Then

$$\left(\prod_{i \in I_1} f_i \pi_i^\circ\right)\left(\prod_{j \in I_2} \pi_j g_j^\circ\right) = \left(\prod_{i \in I_1 \setminus \{J\}} f_i \pi_i^\circ \sqcap f_J \prod_{j \in J} \rho_j \pi_j^\circ\right)\left(\prod_{k \in I_2} \pi_k g_k^\circ\right)$$

$$= \left(\prod_{i \in I_1 \setminus \{J\}} f_i \pi_i^\circ \sqcap \prod_{j \in J} f_J \rho_j \pi_j^\circ\right)\left(\prod_{k \in I_2} \pi_k g_k^\circ\right) = \prod_{i \in I_1 \cap I_2} f_i g_i^\circ .$$

Here we used Lemma 1 in the second equality, and we invoked the generalised sharpness for maps ($f_J \rho_i$'s are maps by Lemma 7) satisfied by π_i's, as well as the fact that J is disjoint from I_1 and I_2.
3. $J \notin I_1$ and $J \in I_2$. Proven similarly as the previous one.
4. $J \in I_1$ and $J \in I_2$. Then

$$\left(\prod_{i \in I_1} f_i \pi_i^\circ\right)\left(\prod_{j \in I_2} \pi_j g_j^\circ\right)$$

$$= \left(\prod_{i \in I_1 \setminus \{J\}} f_i \pi_i^\circ \sqcap f_J \prod_{j \in J} \rho_j \pi_j^\circ\right)\left(\prod_{k \in I_2 \setminus \{J\}} \pi_k g_k^\circ \sqcap \left(\prod_{n \in J} \pi_n \rho_n^\circ\right) g_J^\circ\right)$$

{By Lemma 1}

$$= \left(\prod_{i \in I_1 \setminus \{J\}} f_i \pi_i^\circ \sqcap \prod_{j \in J} f_J \rho_j \pi_j^\circ\right)\left(\prod_{k \in I_2 \setminus \{J\}} \pi_k g_k^\circ \sqcap \left(\prod_{n \in J} \pi_n \rho_n^\circ g_J^\circ\right)\right)$$

{By the generalised sharpness of π_i's for maps}

$$= \prod_{i \in (I_1 \setminus J) \cap (I_2 \setminus J)} f_i g_i^\circ \sqcap \prod_{j \in J} f_J \rho_j \rho_j^\circ g_J^\circ$$

{By Lemma 1 and Equation 5b}

$$= \prod_{i \in I_1 \cap I_2} f_i g_i^\circ .$$

5 Relational Schemas and n-ary Relations in Allegories

5.1 Moving Legs Around

If an allegory has relational products then n-ary relations, for $n > 2$, can be represented as arrows (binary relations) between products. The main problem here is that this representation is non-canonical — we need to divide the n "legs" of the relation into two groups. Moreover, different representations of the same, in some sense, n-ary relation are required for some relational operations, such as joins or intersections with particular arrows, to be applicable. For instance, let R be a ternary relation with legs typed as A, B and C and let $S : A \to D$ and $T : E \to B$ be binary relations. In order for S to be composable (joinable) with R the latter should be represented as $R : B \times C \to A$. On the other hand, joinability of T with R seems to require representation $R : B \to A \times C$. Note that one of the representation change operations is built into any allegory — it is the

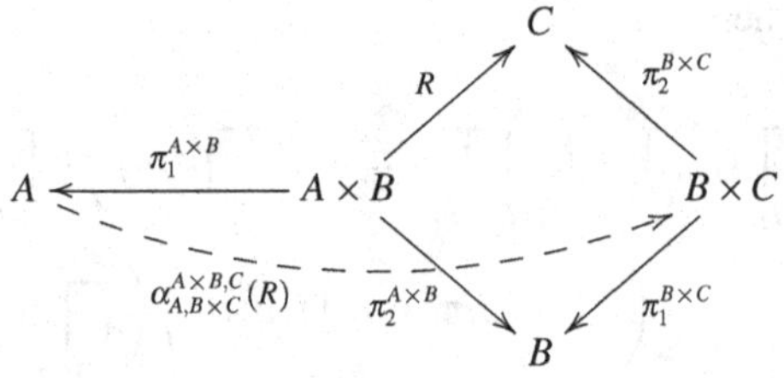

Fig. 3. Diagram of construction of $\alpha^{A\times B,C}_{A,B\times C}(R)$

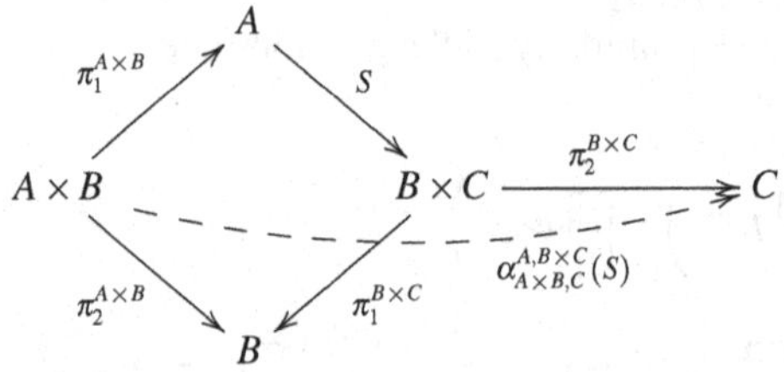

Fig. 4. Diagram of construction of $\alpha^{A,B\times C}_{A\times B,C}(S)$

reciprocation, which allows to flip the legs of a binary relation. We still need, however, an operation which allows to move a leg of the n-ary relation from a source of representation to the target and vice-versa. In this subsection we will write those operations explicitly, and we will prove that they are one-another's inverses.

First we define the leg moving operations:

Definition 5. *Let $\mathscr{A}$ be an allegory, let $A,B,C \in \mathrm{Obj}[\mathscr{A}]$ and suppose that binary relational products $A \times B$ and $B \times C$ exist in $\mathscr{A}$. Then we define the pair of maps*

$$\mathrm{Arr}_{\mathscr{A}}(A \times B, C) \underset{\alpha^{A,B\times C}_{A\times B,C}}{\overset{\alpha^{A\times B,C}_{A,B\times C}}{\rightleftarrows}} \mathrm{Arr}_{\mathscr{A}}(A, B \times C) \quad \textit{with the formulas}$$

$$\alpha^{A\times B,C}_{A,B\times C}(R) := (\pi_1^{A\times B})^{\circ}; (R; (\pi_2^{B\times C})^{\circ} \sqcap \pi_2^{A\times B}; (\pi_1^{B\times C})^{\circ}), \tag{11a}$$

$$\alpha^{A,B\times C}_{A\times B,C}(S) := (\pi_1^{A\times B}; S \sqcap \pi_2^{A\times B}; (\pi_1^{B\times C})^{\circ}); \pi_2^{B\times C}. \tag{11b}$$

The definition is best understood by looking at the Diagrams 3 and 4. Note that $\alpha^{A,B\times C}_{A\times B,C}(S) = \left(\alpha^{C\times B,A}_{C,B\times A}(S^{\circ})\right)^{\circ}$. The meaning of the α maps is explained by noting that in $\mathscr{R}$, for all $a \in A$, $b \in B$ and $c \in C$ we have $a\alpha^{A\times B,C}_{A,B\times C}(R)(b,c) \equiv (a,b)Rc$.

Lemma 11. *(cf. [12, Proposition 6.1]) Assume that A, B, C are objects in some allegory $\mathscr{A}$ such that $\alpha^{A\times B,C}_{A,B\times C}$ and $\alpha^{A,B\times C}_{A\times B,C}$ are well defined. Then*

$$\alpha^{A\times B,C}_{A,B\times C} = \left(\alpha^{A,B\times C}_{A\times B,C}\right)^{-1}.$$

Proof. We will prove that $\alpha^{A\times B,C}_{A,B\times C} \circ \alpha^{A,B\times C}_{A\times B,C} = \mathrm{id}$. *We leave the similar checking of the other identity to the reader. Let* $S \in \mathrm{Arr}_{\mathscr{A}}(A, B\times C)$. *Then*

$$
\begin{aligned}
& \alpha^{A\times B,C}_{A,B\times C}\left(\alpha^{A,B\times C}_{A\times B,C}(S)\right) \\
& \qquad \{\textit{By Equation (11a)}\} \\
= \;& (\pi_1^{A\times B})^\circ\left(\alpha^{A,B\times C}_{A\times B,C}(S)(\pi_2^{B\times C})^\circ \sqcap \pi_2^{A\times B}(\pi_1^{B\times C})^\circ\right) \\
& \qquad \{\textit{By Equation (11b)}\} \\
= \;& (\pi_1^{A\times B})^\circ\left((\pi_1^{A\times B}S \sqcap \pi_2^{A\times B}(\pi_1^{B\times C})^\circ)\pi_2^{B\times C}(\pi_2^{B\times C})^\circ \sqcap \pi_2^{A\times B}(\pi_1^{B\times C})^\circ\right) \\
& \qquad \{\textit{Because } \pi_2^{B\times C} \textit{ is total}\} \\
\sqsupseteq \;& (\pi_1^{A\times B})^\circ\left((\pi_1^{A\times B}S \sqcap \pi_2^{A\times B}(\pi_1^{B\times C})^\circ) \sqcap \pi_2^{A\times B}(\pi_1^{B\times C})^\circ\right) \\
& \qquad \{\textit{By idempotency of } \sqcap\} \\
= \;& (\pi_1^{A\times B})^\circ\left(\pi_1^{A\times B}S \sqcap \pi_2^{A\times B}(\pi_1^{B\times C})^\circ\right) \\
& \qquad \{\textit{By Lemma 2}\} \\
= \;& S \sqcap (\pi_1^{A\times B})^\circ\pi_2^{A\times B}(\pi_1^{B\times C})^\circ \\
& \qquad \{\textit{By Lemma 3 and Equation 4c as } \pi_1^{B\times C} \textit{ is total}\} \\
= \;& S.
\end{aligned}
$$

Let us denote for brevity $\lambda := \pi_1^{A\times B}S \sqcap \pi_2^{A\times B}(\pi_1^{B\times C})^\circ$. *Then we have*

$$
\begin{aligned}
& \alpha^{A\times B,C}_{A,B\times C}\left(\alpha^{A,B\times C}_{A\times B,C}(S)\right) \\
& \qquad \{\textit{By (11a), (11b) and the definition of } \lambda\} \\
= \;& (\pi_1^{A\times B})^\circ\left(\lambda\pi_2^{B\times C}(\pi_2^{B\times C})^\circ \sqcap \pi_2^{A\times B}(\pi_1^{B\times C})^\circ\right) \\
& \qquad \{\textit{By Equation (1g)}\} \\
\sqsubseteq \;& (\pi_1^{A\times B})^\circ\lambda\left(\pi_2^{B\times C}(\pi_2^{B\times C})^\circ \sqcap \lambda^\circ\pi_2^{A\times B}(\pi_1^{B\times C})^\circ\right) \\
& \qquad \{\textit{By monotonicity of all operations as } \lambda^\circ \sqsubseteq \pi_1^{B\times C}(\pi_2^{A\times B})^\circ\} \\
\sqsubseteq \;& (\pi_1^{A\times B})^\circ\lambda\left(\pi_2^{B\times C}(\pi_2^{B\times C})^\circ \sqcap \pi_1^{B\times C}(\pi_2^{A\times B})^\circ\pi_2^{A\times B}(\pi_1^{B\times C})^\circ\right) \\
& \qquad \{\textit{By Equation (4a)}\} \\
= \;& (\pi_1^{A\times B})^\circ\lambda\left(\pi_2^{B\times C}(\pi_2^{B\times C})^\circ \sqcap \pi_1^{B\times C}(\pi_1^{B\times C})^\circ\right) \\
& \qquad \{\textit{By Equation 4b and the definition of } \lambda\} \\
= \;& (\pi_1^{A\times B})^\circ\left(\pi_1^{A\times B}S \sqcap \pi_2^{A\times B}(\pi_1^{B\times C})^\circ\right) \\
& \qquad \{\textit{By Lemma 2}\} \\
= \;& S \sqcap (\pi_1^{A\times B})^\circ\pi_2^{A\times B}(\pi_1^{B\times C})^\circ
\end{aligned}
$$

$$\{\text{By Lemma 3 and Equation 4c as } \pi_1^{B\times C} \text{ is total}\}$$
$$= \quad S.$$

5.2 Relational Schemas and Canonical Presentation of n-ary Relations

In relational algebra *á la* Codd [6] "legs" (columns) of an n-ary relation are not really ordered (although they are in most practical implementations), i.e., they are not identified by position. Instead, they are identified by name. A set of column names of a given relation together with the assignement of a type to a column is called a schema of this relation. A given relation schema may have many instances, i.e., relations with a given column names and types. Here we will mimic those ideas.

Definition 6. *Let $\mathbb{T}$ be a fixed set of basic types (e.g.,* `integer`*,* `varchar`*, etc.). A relation schema (X,α) over $\mathbb{T}$ consists of a finite set X of column names together with a mapping $\alpha : X \to \mathbb{T}$ assigning types to column names.*

Definition 7. *A category $\mathscr{S}[\mathbb{T}]$ of relation schemas over $\mathbb{T}$ has as objects relation schemas over $\mathbb{T}$. Morphisms $f : (X,\alpha) \to (Y,\beta)$ between relation schemas are injective maps $f : X \to Y$ between sets such that $\beta \circ f = \alpha$.*

Note that $\mathscr{S}[\mathbb{T}]$ is a subcategory of a slice category Set/$\mathbb{T}$.

The following definition attempts to give an almost canonical, natural representation of an n-ary relation as an arrow in an allegory with some additional structure (cf. vectorization of binary relations, see e.g., [12, Section 6]).

Definition 8. *Let (X,α) be a relation schema over $\mathbb{T}$. Let $\mathscr{A}$ be a unitary allegory such that there exists a mapping $[\![\cdot]\!] : \mathbb{T} \to \mathrm{Obj}[\mathscr{A}]$ interpreting basic types as objects in $\mathscr{A}$. Let 1 be a unit in $\mathscr{A}$. A pair $(R, \{\ \overrightarrow{R} \xrightarrow{\pi_i} [\![\alpha(i)]\!]\ \}_{i\in X})$ is called an instance of (X,α) iff $\overleftarrow{R} = 1$ and $\{\pi_i\}_{i\in X}$ is an $|X|$-ary relational product.*

Note that this representation is not completely canonical as it depends on the choice of relational product. This is not a great problem as in many allegories (e.g., $\mathscr{R}$ or $\mathscr{R}[\Lambda]$) there exists a canonical choice of products. Also note that now the set of column names is a part of the definition of an instance. In particular, any bijection $f : (X,\alpha) \to (Y,\beta)$ gives rise to the renaming transformation $\hat{f}$ of instances (which corresponds to the renaming operation in the Codd's relational algebra):

$$\hat{f}((R, \{\ \overrightarrow{R} \xrightarrow{\pi_i} [\![\alpha(i)]\!]\ \}_{i\in X})) := (R, \{\ \overrightarrow{R} \xrightarrow{\rho_j} [\![\beta(j)]\!]\ \}_{j\in Y}), \tag{12}$$

where $\rho_j := \pi_{f^{-1}(j)}$, $j \in Y$.

Assume for simplicity that $\mathbb{T} = \mathrm{Obj}[\mathscr{A}]$ and that $[\![\cdot]\!] : \mathbb{T} \to \mathrm{Obj}[\mathscr{A}]$ is given by identity. Suppose that $A \xleftarrow{\pi_1} A \times B \xrightarrow{\pi_2} B$ is a relational product of A and B. Observe that any arrow $R \in \mathrm{Arr}_{\mathscr{A}}(A,B)$ gives rise to an instance $(\alpha^{A,B}_{1,A\times B}(R), \{\pi_i\}_{i\in\{1,2\}})$ of a schema $(\{1,2\}, \{1 \mapsto A, 2 \mapsto B\})$, where we utilize Lemma 6. Moreover, by Lemma 11, this assignement is invertible (see also [12, Proposition 6.1]), and thus we loose no information when we change into this representation.

5.3 Joins and Intersections of n-ary Relations in Allegories

One can intersect only the instances of the same schemas with common product components. Let $(R,\{\pi_i\}_{i\in X})$ and $(S,\{\pi_i\}_{i\in X})$ be instances of the same relation schema (X,α). Then

$$(R,\{\pi_i\}_{i\in X}) \sqcap (S,\{\pi_i\}_{i\in X}) := (R\sqcap S,\{\pi_i\}_{i\in X}). \tag{13}$$

This explicit dependence on products might be troublesome. Note however, that relational products of a given collection of objects in an allegory are categorical products in the subcategory of maps, and thus are unique up to an isomorphism (c.f. [8]).

Suppose now that (X,α) and (Y,β) are relation schemas over $\mathbb{T}$ such that $X\cap Y$ is non-empty and $\alpha|_{X\cap Y}=\beta|_{X\cap Y}$. Moreover, let

$$\{\,A \xrightarrow{\rho_i^A} [\![\alpha(i)]\!]\,\}_{i\in X\setminus Y},\quad \{\,B \xrightarrow{\rho_j^B} [\![\alpha(j)]\!]\,\}_{j\in X\cap Y},\quad \{\,C \xrightarrow{\rho_k^C} [\![\beta(k)]\!]\,\}_{k\in Y\setminus X},$$

be relational products that satisfy the generalised sharpness property for maps. Consider instances $(R,\{\pi_i\}_{i\in X})$ of (X,α) and $(S,\{\sigma_i\}_{i\in Y})$ of (Y,β) such that π_i's and σ_i's satisfy the generalised sharpness condition for maps. Define

$$\pi_1^{A\times B} := \bigsqcap_{i\in X\setminus Y} \pi_i(\rho_i^A)^\circ,\quad \pi_2^{A\times B} := \bigsqcap_{i\in X\cap Y} \pi_i(\rho_i^B)^\circ,$$
$$\pi_1^{B\times C} := \bigsqcap_{i\in X\cap Y} \sigma_i(\rho_i^B)^\circ,\quad \pi_2^{B\times C} := \bigsqcap_{i\in Y\setminus X} \sigma_i(\rho_i^C)^\circ,$$

Then it follows from Lemma 10 that $\{\pi_1^{A\times B},\pi_2^{A\times B}\}$ and $\{\pi_1^{B\times C},\pi_2^{B\times C}\}$ are relational products. Finally, assume that $A \xleftarrow{\mu_1} D \xrightarrow{\mu_2} \vec{S}$ is also a relational product. Then we define a natural join of $(R,\{\pi_i\}_{i\in X})$ and $(S,\{\sigma_i\}_{i\in Y})$ as an instance $(R\bowtie S,\{\nu_i\}_{i\in X\cup Y})$ of a schema $(X\cup Y,\gamma)$ where

$$\gamma(i) := \begin{cases}\alpha(i) & \text{if } i\in X\\ \beta(i) & \text{if } i\in Y\end{cases},\quad \nu_i := \begin{cases}\mu_1\rho_i^A & \text{if } i\in X\setminus Y\\ \mu_2\sigma_i & \text{if } i\in Y\end{cases},$$
$$R\bowtie S := \alpha_{1,A\times(B\times C)}^{A,B\times C}(\alpha_{A,B}^{1,A\times B}(R)(u_B S\sqcap(\pi_1^{B\times C})^\circ)) : 1\to D.$$

It is easy to verify that in $\mathscr{R}$ this definition corresponds to the usual definition of a natural join of two relations in Codd's relational algebra [6].

6 Conclusion

In the paper we have presented a new approach, partially inspired by [5], [4] to n-ary relations in allegories, which can be useful for database modeling, e.g., extending the approach of [16] and [17]. Considering n-ary relations in the (almost) general allegorical framework allows us to transparently use Codd's relational algebra operations with various generalised relation-like constructs, particularly with fuzzy (locale-valued) relations. This can be important for assigning semantics to some fuzzy extensions of relational query languages, for instance PREDICTION JOIN construct of DMX [1].

The most important results of the paper are Lemma 10 and Lemma 11 as well as the introduction of the condition of generalised sharpness for total maps. These are crucial for the new approach to n-ary relations to work as expected.

In the future work we would like to examine the properties of relational operations on n-ary relations in allegories defined in the previous section. In particular, we would like to check how many of Codd's axioms can be transported to general allegorical setting.

Finally, we would like to check if it is possible to develop results similar to the ones in this paper while working with weakened definitions of relational products, such as those in [14], [15].

Acknowledgements. I would like to thank prof. Paweł Maślanka for helpful discussions. I would also like to thank the anonymous reviewers for their suggestions.

References

1. Data Mining Extensions (DMX) Reference, `https://msdn.microsoft.com/en-us/library/ms132058.aspx`
2. Barr, M., Wells, C.: Category theory for computing science. Prentice-Hall International Series in Computer Science. Prentice Hall (1995)
3. Berghammer, R., Haeberer, A., Schmidt, G., Veloso, P.: Comparing two different approaches to products in abstract relation algebra. In: Algebraic Methodology and Software Technology (AMAST 1993), pp. 167–176. Springer (1994)
4. Brown, C., Hutton, G.: Categories, allegories and circuit design. In: Proceedings of the Symposium on Logic in Computer Science, LICS 1994, pp. 372–381. IEEE (1994)
5. Brown, C., Jeffrey, A.: Allegories of circuits. In: Matiyasevich, Y.V., Nerode, A. (eds.) LFCS 1994. LNCS, vol. 813, pp. 56–68. Springer, Heidelberg (1994)
6. Codd, E.F.: A relational model of data for large shared data banks. Commun. ACM 13(6), 377–387 (1970), `http://doi.acm.org/10.1145/362384.362685`
7. Davey, B., Priestley, H.: Introduction to Lattices and Order. Cambridge mathematical text books. Cambridge University Press (2002)
8. Desharnais, J.: Monomorphic characterization of n-ary direct products. Inf. Sci. 119(3-4), 275–288 (1999), `http://dx.doi.org/10.1016/S0020-02559900020-1`
9. Freyd, P., Scedrov, A.: Categories, Allegories. North-Holland Mathematical Library. Elsevier Science (1990)
10. Givant, S.: The calculus of relations as a foundation for mathematics. Journal of Automated Reasoning 37(4), 277–322 (2006), `http://dx.doi.org/10.1007/s10817-006-9062-x`
11. Schmidt, G., Ströhlein, T.: Relations and graphs. Springer, Heidelberg (1993)
12. Schmidt, G., Winter, M.: Relational mathematics continued. arXiv:1403.6957 (2014)
13. Tarski, A., Givant, S.R.: A formalization of set theory without variables, vol. 41. American Mathematical Soc. (1987)
14. Winter, M.: Products in categories of relations. The Journal of Logic and Algebraic Programming 76(1), 145–159 (2008)
15. Zieliński, B., Maślanka, P.: Weak n-ary relational products in allegories. Axioms 3(4), 342–357 (2014)
16. Zieliński, B., Maślanka, P., Sobieski, Ś.: Allegories for database modeling. In: Cuzzocrea, A., Maabout, S. (eds.) MEDI 2013. LNCS, vol. 8216, pp. 278–289. Springer, Heidelberg (2013), `http://dx.doi.org/10.1007/978-3-642-41366-7_24`
17. Zieliński, B., Maślanka, P., Sobieski, Ś.: Modalities for an allegorical conceptual data model. Axioms 3(2), 260–279 (2014), `http://www.mdpi.com/2075-1680/3/2/260`

Mechanised Relation-Algebraic Order Theory in Ordered Categories without Meets

Musa Al-hassy and Wolfram Kahl*

McMaster University, Hamilton, Ontario, Canada
alhassy@gmail.com, kahl@mcmaster.ca

Abstract. In formal concept analysis, complete lattices of "concepts" are represented by entity-attribute relations called "contexts". Using the dependently-typed programming language Agda, we build on a previous formalisation of the category of contexts to obtain a fully verified abstract implementation of the duality between contexts and complete lattices in the abstract setting of locally ordered categories with converse, residuals, symmetric quotients, and direct powers.

1 Introduction

Locally-ordered categories with converse (OCCs) were identified in [Kah04] as a common substrate between the allegories of Freyd and Scedrov [FS90] and typed Kleene algebras [Koz98] and variants. This common substrate is important since in OCCs, a large variety of relation-algebraic specification and reasoning patterns is already possible. The distinguishing feature of OCC-based formalisation is:

"No binary meets (intersections, $\sqcap$), no binary joins (unions, $\sqcup$)."

For the "contexts" of Wille's "formal concept analysis" [Wil05], Moshier proposes a relational homomorphism concept [Mos13, Jip12] in conventional mathematical style. We showed in [Kah14b] that enriching OCCs with power operators and power orders following [BdM97] implies that also left- and right-residuals become available, and that this extended setting is sufficient to formalise the category of contexts with their relational homomorphisms.

Moshier goes on to prove that this category is dual to that of complete semilattices with meet-preserving homomorphisms [Mos13, Jip12]; the current paper shows that this can still be formalised without binary joins and meets, even though it may not be immediately obvious how to deal in particular with antisymmetry in this setting. The conventional relation-algebraic characterisation of a relation E as a partial order has:

Reflexivity: $\mathsf{Id} \sqsubseteq \mathsf{E}$ Transitivity: $\mathsf{E} \mathbin{;} \mathsf{E} \sqsubseteq \mathsf{E}$ Antisymmetry: $\mathsf{E} \sqcap \mathsf{E}^{\smile} \sqsubseteq \mathsf{Id}$

In addition to left- and right-residuals, we do require also symmetric quotients. These have originally been introduced by Berghammer, Schmidt, and Zierer

* This research is supported by the National Science and Engineering Research Council of Canada, NSERC.

W. Kahl et al. (Eds.): RAMiCS 2015, LNCS 9348, pp. 151–168, 2015.
DOI: 10.1007/978-3-319-24704-5_10

[BSZ86, BSZ89] as the intersection of two residuals in the context of heterogeneous relation algebras (see also [SS93, Sect. 4.4]). We first proposed a meet-free definition for symmetric quotients that also does not rely on separate residuals in [FK98]; this has then been set into the context of ordered semigroupoids with converse in the Agda formalisation "RATH-Agda" of relation-algebraic theories [Kah11, Kah14c], but has so far not yet been applied in such a meet-free context.

As first demonstrated by Schmidt *et al.* [BSZ86, BSZ89, SS93], symmetric quotients can be used for specifying set membership. The resulting concept of *direct power* is slightly stronger than that of the power operators based on power transpose, where antisymmetry of the power order requires tabular allegories according to Bird and de Moor [BdM97, p. 106].

We use the meet-free definition of symmetric quotients to achieve a meet-free formulation of antisymmetry and of other order-theoretic constructions that are normally defined using meets. Using this in the construction of the categorical duality between contexts and complete semilattices requires some non-trivial proofs, but overall results in a satisfying formalisation.

Developments of "familiar theory" in "familiar, but reduced" axiom systems easily fall into the trap of inadvertantly using derived laws that are not derivable anymore from the reduced axiom system. Since our development is quite large, only a mechanically checkable formalisation can plausibly convince the reader that we did not "cheat", or overlook anything. We choose Agda [Nor07] for our formalisation, which accepts as input an easily recognisable variant of the calculational proofs that would otherwise be written in LaTeX. By also presenting our mathematical development in this Agda notation, we strive to demonstrate that mechanised developments (in Agda) can be readable and writable, making the "cost" of switching to mechanically checked proofs tolerable and well-spent, even for the development of new mathematics, since the use of a proof-checking environment significantly increases the confidence of both the developer and the reader. The source files of our full development [KAh15], are available on-line at: http://relmics.mcmaster.ca/RATH-Agda/#AContext

Overview: We provide a quick introduction to most of the Agda notation we use in this paper in Sect. 2, via a presentation of the definition of OCCs in Agda notation. In the context of the meet-free definition of symmetric quotients in Sect. 3, we use our proofs of example properties of symmetric quotients to explain how calculational proofs are presented to Agda. These tools are used for meet-free definitions and theorems about orders (Sect. 4), set membership (Sect. 5), and complete semilattices (Sect. 6). After a short review of our previous development of abstract contexts in Sect. 7, we briefly present the main ingredients of the proof of the categoric duality between contexts and complete semilattices in Sect. 8.

2 OCCs in Agda Notation

Agda [Nor07] is a dependently-typed programming language that is also a proof checker in a variant of Martin-Löf type theory. A number of its design choices (on top of the dependent type theory) make it a convenient vehicle not only for verified functional programming, but also for mechanisation of mathematics in a rather natural way:

- Identifiers can be almost arbitrary white-space-free strings in Unicode encoding. (As a consequence, most lexemes need to be separated by white space.)
- Mixfix syntax: Operator names include underscores for argument positions.
- Implicit arguments allow information that can be inferred from the context to be omitted, an important ingredient of "mathematically natural syntax".
- The module system allows nesting, parameters, qualified and unqualified import and re-export with or without renaming and/or instantiation, allowing natural theory structuring and modularisation.
- Data constructors and record field labels can be overloaded.
- The interactive front-end supports type-directed editing and also automates some aspects of proof construction.

An ordered category with converse (OCC) consists of the following, using Agda notation throughout:

- A type Obj of *objects*, which should be considered as abstracting from sets,
- For any two objects A B : Obj, a type Mor A B of *morphisms* from A to B, which should be considered as abstracting from relations between A and B, together with an equivalence relation _≈_ that serves as morphism equality (since as usual in type-theoretic formalisations of category theory, we use the setoid appproach for homsets), and an inclusion relation _⊑_ on Mor A B forming a partial order with respect to _≈_.
- A binary composition operator _⨾_ : Mor A B → Mor B C → Mor A C for any A B C : Obj, where application of the infix operator "_⨾_" to two arguments R : Mor A B and S : Mor B C is written "R ⨾ S"; note that we use *forward composition* $A \xrightarrow{R} B \xrightarrow{S} C$. Composition is associative, ⨾-assoc : (Q ⨾ R) ⨾ S ≈ Q ⨾ R ⨾ S, and associates to the right, which is why we did not need to add parentheses on the right-hand side. Composition also preserves equality and inclusion, as witnessed by the following proof term constructors:

  ```
  ⨾-cong₁ : R₁ ≈ R₂ → R₁ ⨾ S ≈ R₂ ⨾ S      ⨾-monotone₁ : R₁ ⊑ R₂ → R₁ ⨾ S ⊑ R₂ ⨾ S
  ⨾-cong₂ : S₁ ≈ S₂ → R ⨾ S₁ ≈ R ⨾ S₂      ⨾-monotone₂ : S₁ ⊑ S₂ → R ⨾ S₁ ⊑ R ⨾ S₂
  ```

  (We will also use variants like ⨾-cong₂₂₁ where the subscript digit sequence indicates the term position of the respective rule application.) Note that the function type constructor → also serves as logical implication between types that are considered as formulae (where the elements are considered as proofs).
- For each object A : Obj a morphism Id {A} : Mor A A satisfying leftId : Id ⨾ R ≈ R and rightId : Q ⨾ Id ≈ Q. If R : Mor A B, then we could have made the implicit argument to Id, indicated by the braces {...}, explicit by writing Id {A} ⨾ R ≈ R, but this is normally omitted, just like in mathematics.
- For any A B : Obj, a *converse* operator _˘ : Mor A B → Mor B A satisfying:

  ```
  ˘-cong  : R ≈ S → R ˘ ≈ S ˘      -- preservation of equality
  ˘˘      :     (R ˘) ˘ ≈ R        -- involution
  ```

```
⨾-˘          :  (R ⨾ S) ˘ ≈ S ˘ ⨾ R ˘    -- contravariance
˘-monotone : R ⊑ S → R ˘ ⊑ S ˘           -- preservation of inclusion
```

For the ⨾-contravariance ⨾-˘ of converse we also use the following abbreviations:

```
⨾˘-˘   : (S  ⨾ R ˘) ˘  ≈  R  ⨾ S ˘
˘⨾-˘   : (S ˘ ⨾ R  ) ˘  ≈  R ˘ ⨾ S
˘⨾˘-˘ : (S ˘ ⨾ R ˘) ˘  ≈  R  ⨾ S
```

The names of most of these properties are intended to evoke the left-hand sides of their (conclusion) equality; the hyphen "-" is used in this context for "pauses of breath" similar to the way that pauses of breath in natural language indirectly indicate syntactic structure. For the time being, we feel that this is more readable that using parentheses-like Unicode codepoints in identifiers (parentheses themselves cannot be used).

Frequently we will in addition assume existence of *residuals* of composition; for derived laws for residuals in OCCs, see [Kah04, Kah11, Kah14c]. For any A B C : Obj and morphisms S : Mor A C and Q : Mor A B and R : Mor B C:

– The *left residual* S / R : Mor A B is defined by:
```
/-cancel-outer : (S / R) ⨾ R ⊑ S
/-universal : ∀ {X : Mor A B} → X ⨾ R ⊑ S → X ⊑ S / R
```
– The *right residual* Q \ S : Mor B C is defined by:
```
\-cancel-outer : Q ⨾ (Q \ S) ⊑ S
\-universal : ∀ {Y : Mor B C} → Q ⨾ Y ⊑ S → Y ⊑ Q \ S
```

3 Symmetric Quotients in OCCs without Meets

For the symmetric quotient of Q and S, we use the notation Q ╳ S of [Kah14c], instead of the notation "syq(Q, S)" previously used by Schmidt *et al.*.

In an OCC (which does not need to have residual operators), the **symmetric quotient** Q ╳ S : $B \leftrightarrow C$ of two relations Q : $A \leftrightarrow B$ and S : $A \leftrightarrow C$ is defined by

Y ⊑ Q ╳ S iff Q ⨾ Y ⊑ S and Y ⨾ S ˘ ⊑ Q ˘ for all Y : $B \leftrightarrow C$.

The Agda formulation requires names for all axioms; for pragmatic reasons we split the "iff" into two directions, and split the "implies ... and ..." into two conjuncts; we give the resulting pieces three easily recognisable names, and for now omit the introduction of the bound variables and their types for the three laws:

```
_╳_  : {A B C : Obj} → Mor A B → Mor A C → Mor B C
╳-cancel-left  : Q ⨾ (Q ╳ S)  ⊑ S
╳-cancel-right : (Q ╳ S) ⨾ S ˘ ⊑ Q ˘
╳-universal    : Q ⨾ R ⊑ S →  R ⨾ S ˘ ⊑ Q ˘ → R ⊑ Q ╳ S
```

Agda directly supports literate programming in that Agda source files may be LaTeX source files with Agda code embedded in {code} environments. The typesetting in this paper was produced via pre-processing using *lhs2TeX* [Löh12]. The following is the full body of the Agda **record** definition for symmetric quotient operators — like the remaining "code blocks" in this paper, it is an only slightly typographically-enhanced rendering of the Agda source code. Since *lhs2TeX* handles vertical alignment automatically, Agda source code turns out easier to write than LaTeX source of similar mathematics; the fact that writing Agda requires more type information than customary in much mathematical writing can be regarded as a weakness, but we prefer to see it as a strength, since it relieves the reader to a certain degree from performing type inference and guessing the scopes of bound variables which are all too often left unclear.

This full version of the symmetric quotient operator definition includes also a congruence law — symmetric quotient operators are neither monotone nor antitone in any argument, so ⨳-cong appears to not be consequence of the other laws. (For the one-sided residuals __ and _/_, their definition implies monotonicity in the "upper" argument and antitonicity in the "lower" argument, and these in turn imply congruence via antisymmetry of _⊑_.)

```
infix 9 _⨳_                    -- operator precedence level
field
  _⨳_ : {A B C : Obj} → Mor A B → Mor A C → Mor B C
  ⨳-cong          : {A B C : Obj} {Q₁ Q₂ : Mor A B} {S₁ S₂ : Mor A C}
                  → Q₁ ≈ Q₂ → S₁ ≈ S₂ → Q₁ ⨳ S₁ ≈ Q₂ ⨳ S₂
  ⨳-cancel-left   : {A B C : Obj} {Q : Mor A B} {S : Mor A C}
                  → Q ⨾ (Q ⨳ S)  ⊑ S
  ⨳-cancel-right  : {A B C : Obj} {Q : Mor A B} {S : Mor A C}
                  → (Q ⨳ S) ⨾ S ˘ ⊑ Q ˘
  ⨳-universal     : {A B C : Obj} {Q : Mor A B} {S : Mor A C} {R : Mor B C}
                  → Q ⨾ R ⊑ S → R ⨾ S ˘ ⊑ Q ˘ → R ⊑ Q ⨳ S
```

When there is no loss of clarity, we shall elide such variable introductions, as in the display at the beginning of this section.

In a division allegory, where right-residual __ and left-residual _/_ operators are both available, as well as binary meet _⊓_, we have the theorem that a symmetric quotient operator as defined above satisfies the conventional symmetric quotient definition:

```
⨳≈\⊓/ : ∀ {Q S}    →    Q ⨳ S  ≈  Q \ S ⊓ Q ˘ / S ˘
```

Even though we do not assume the existence of all binary meets in OCCs, symmetric quotients still are meets. But since all symmetric quotients are difunctional, and in most OCCs, most morphisms are not difunctional, demanding existence of all symmetric quotients is still quite remote from demanding existence of all meets. Basic reasoning about symmetric quotients typically bifurcates into branches through the two-premise rule ⨳-universal. As a simple example, we show one side of the proof that converse of symmetric quotients just swaps the arguments. The following Agda source block contains first the statement of this theorem, named "⨳-˘-⊑", with typed universally quantified variables A, B, C, Q,

S, and then the proof term proving this theorem, constructed using the two-premise rule ╳-universal applied to proofs of the inclusions S ⨾ (Q ╳ S) ˘ ⊑ Q and (Q ╳ S) ˘ ⨾ Q ˘ ⊑ S presented as calculational proofs where each "reason" (enclosed in ⟨...⟩) is a full proof term for the respective equality or inclusion. The first equality steps in both calculations are "backwards", as signalled by the "˘" preceding the ⟨...⟩. We chose to not clutter this paper with explicit introductions of all proof term elements used in the calculational proofs presented; most of the few unintroduced ones should be reasonably clear from the context, and all are defined in the sources [KAh15].

```
╳-˘-⊑ : {A B C : Obj} {Q : Mor A B} {S : Mor A C} → (Q ╳ S) ˘ ⊑ S ╳ Q
╳-˘-⊑ {A} {B} {C} {Q} {S} = ╳-universal
  (⊑-begin
      S ⨾ (Q ╳ S) ˘
    ≈˘⟨ ⨾-˘-˘ ⟩
      ((Q ╳ S) ⨾ S ˘) ˘
    ⊑⟨ ˘-monotone ╳-cancel-right ⟩
      Q ˘ ˘
    ≈⟨ ˘˘ ⟩
      Q
  □)
  (⊑-begin
      (Q ╳ S) ˘ ⨾ Q ˘
    ≈˘⟨ ⨾-˘ ⟩
      (Q ⨾ (Q ╳ S)) ˘
    ⊑⟨ ˘-monotone ╳-cancel-left ⟩
      S ˘
  □)
```

The following inclusion will occasionally be useful:

```
╳-cancel-inner : ∀ {A B C Z} {Q : Mor A B} {S : Mor A C} {P : Mor Z A}
                 → Q ╳ S ⊑ (P ⨾ Q) ╳ (P ⨾ S)
╳-cancel-inner {_} {_} {_} {_} {Q} {S} {P} = ╳-universal
  (⨾-assoc ⟨≈⊑⟩ ⨾-monotone₂ ╳-cancel-left)
  (⊑-begin
      (Q ╳ S) ⨾ (P ⨾ S) ˘
    ≈⟨ ⨾-cong₂ ⨾-˘ ⟩
      (Q ╳ S) ⨾ S ˘ ⨾ P ˘
    ⊑⟨ ⨾-assocL ⟨≈⊑⟩ ⨾-monotone₁ ╳-cancel-right ⟩
      Q ˘ ⨾ P ˘
    ≈˘⟨ ⨾-˘ ⟩
      (P ⨾ Q) ˘
  □)
```

The converse inclusion would frequently be nice to have, but unfortunately it holds in general only for univalent and surjective P. Similarly stringent are the laws for "multiplication from the outside", shown without proofs:

```
╳-in-left  : {A B C D : Obj} {Q : Mor C B} {S : Mor C D} {F : Mor A B}
             → isMapping F → F ⨾ (Q ╳ S) ≈ (Q ⨾ F ˘) ╳ S
╳-in-right : {A B C D : Obj} {Q : Mor A B} {S : Mor A C} {F : Mor C D}
             → isBijective F → (Q ╳ S) ⨾ F ≈ Q ╳ (S ⨾ F)
```

We have been able to re-prove in our formalisation [KAh15] all common laws about symmetric quotients where the statement does not involve meet (or join), in particular those collected in [FK98].

A new general theorem that finds uses where the assumptions of the rules above do not hold will be used for showing "antisymmetry" of suborders later:

```
retract╳ : {A B C₁ C₂ : Obj}
           {F₁ G₁ : Mor B C₁} {F₂ G₂ : Mor B C₂} {H₁ H₂ : Mor A B}
         → F₁ ⊑ G₁                              -- premise F₁⊑G₁
         → F₂ ⊑ G₂                              -- premise F₂⊑G₂
         → H₁ ⨾ G₂ ⨾ F₂ ˘        ⊑ H₂           -- premise H₁⨾G₂⨾F₂˘⊑H₂
         → F₁ ⨾ G₁ ˘ ⨾ H₂ ˘      ⊑ H₁ ˘         -- premise F₁⨾G₁˘⨾H₂˘⊑H₁˘
         → F₁ ⨾ (G₁ ╳ G₂) ⨾ F₂ ˘ ⊑ H₁ ╳ H₂      -- conclusion
retract╳ {A} {B} {C₁} {C₂} {F₁} {G₁} {F₂} {G₂} {H₁} {H₂}
      F₁⊑G₁  F₂⊑G₂  H₁⨾G₂⨾F₂˘⊑H₂  F₁⨾G₁˘⨾H₂˘⊑H₁˘  =  ╳-universal
  (⊑-begin
    H₁ ⨾ F₁ ⨾ (G₁ ╳ G₂) ⨾ F₂ ˘
  ⊑⟨ ⨾-monotone₂₁ F₁⊑G₁ ⟩
    H₁ ⨾ G₁ ⨾ (G₁ ╳ G₂) ⨾ F₂ ˘
  ⊑⟨ ⨾-monotone₂ (⨾-assocL ⟨≈⊑⟩ ⨾-monotone₁ ╳-cancel-left) ⟩
    H₁ ⨾ G₂ ⨾ F₂ ˘
  ⊑⟨ H₁⨾G₂⨾F₂˘⊑H₂ ⟩
    H₂
  □) (⊑-begin
    (F₁ ⨾ (G₁ ╳ G₂) ⨾ F₂ ˘) ⨾ H₂ ˘
  ⊑⟨ ⨾-monotone₁ (⨾-monotone₂₂ (˘-monotone F₂⊑G₂)) ⟩
    (F₁ ⨾ (G₁ ╳ G₂) ⨾ G₂ ˘) ⨾ H₂ ˘
  ⊑⟨ ⨾-monotone₁₂ ╳-cancel-right ⟨⊑≈⟩ ⨾-assoc ⟩
    F₁ ⨾ G₁ ˘ ⨾ H₂ ˘
  ⊑⟨ F₁⨾G₁˘⨾H₂˘⊑H₁˘ ⟩
    H₁ ˘
  □)
```

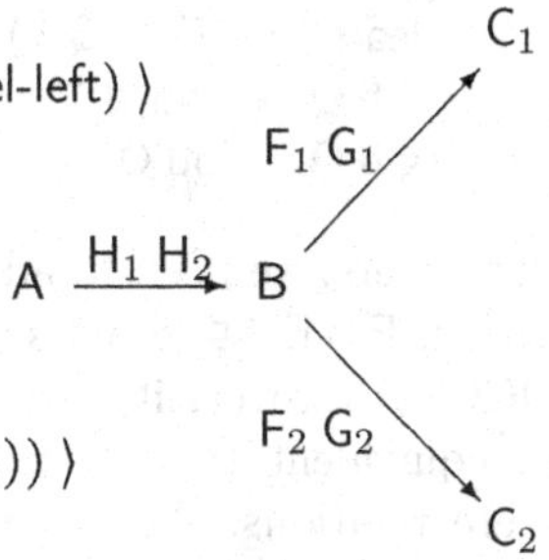

4 Orders in OCCs with Symmetric Quotients and Residuals

Preorders pose no problem in OCCs, and the "minorants" (lower bounds, lbd) and "majorants" (upper bounds, ubd) operators of [SS93, Sect. 3.3] are formulated using residual operators and have their properties shown as usual — following

Gunther Schmidt, we use the letter E for order relations, due to the fact that its shape is close to "⊑":

```
record IsPreorder {A : Obj} (E : Mor A A) : Set k₂ where
  field refl  : Id ⊑ E       -- reflexivity
        trans : E ⨾ E ⊑ E   -- transitivity
  ubd lbd : {I : Obj} → Mor I A → Mor I A
  ubd Q = Q ˘ \ E
  lbd Q = Q ˘ \ E ˘
```

In allegories, and in the relation algebra setting of [SS93], the operators for greatest and least elements are usually defined using meets (expansions for lub and glb are shown in comments):

```
gre Q = Q ⊓ ubd Q
lea Q = Q ⊓ lbd Q
lub Q = lea (ubd Q)         -- ≈ ubd R ⊓ lbd (ubd R)
glb Q = gre (lbd Q)         -- ≈ lbd R ⊓ ubd (lbd R)
```

From this, a presentation using symmetric quotients is then normally proven — we use that presentation as our definition instead:

```
gre lea lub glb : {I : Obj} → Mor I A → Mor I A
gre Q = (E ⨾ Q ˘)   ╳ E
lea Q = (E ˘ ⨾ Q ˘) ╳ E ˘
lub Q = ubd Q ˘     ╳ E ˘       -- ≈ (E ˘ / Q) ╳ E ˘
glb Q = lbd Q ˘     ╳ E         -- ≈ (E / Q) ╳ E
```

For orders, we need to add a condition for antisymmetry. The conventional condition, E ⊓ E ˘ ⊑ Id, uses the binary meet operator _⊓_ that is not available in OCCs. However, it turns out that the condition E ╳ E ⊑ Id is, for preorders E, equivalent to E ⊓ E ˘ ⊑ Id in division allegories, and therefore for concrete relations. We use this condition as our third order axiom; the equality antisym≈ : E ╳ E ≈ Id can then be derived.

```
record IsOrder {A : Obj} (E : Mor A A) : Set k₂ where
  field refl    : Id ⊑ E       -- reflexivity
        trans   : E ⨾ E ⊑ E   -- transitivity
        antisym : E ╳ E ⊑ Id   -- antisymmetry
```

Interestingly, some properties become easier to prove using symmetric quotients than using meets, for example:

```
lub-mapping : {I : Obj} {R : Mor I A} → isMapping R → lub R ≈ R
lub-mapping {I} {R} R-map = ≈-begin
    lub R
  ≈⟨⟩
    ubd R ˘ ╳ E ˘
```

```
  ≈⟨ ╳-cong₁ (˘-cong (ubd-mapping R-map) ⟨≈≈⟩ ⨾-˘) ⟩
    (E ˘ ⨾ R ˘) ╳ E ˘
  ≈˘⟨ ╳-in-left R-map ⟩
    R ⨾ (E ˘ ╳ E ˘)
  ≈⟨ ⨾-cong₂ ˘-antisym≈ ⟨≈≈⟩ rightId ⟩
    R
  □
```

One example for a property that becomes harder with symmetric quotients is the fact that least upper bounds are the greatest lower bounds of all upper bounds:

```
lub-≈-glb-ubd : {I : Obj} {Q : Mor I A} → lub Q ≈ glb (ubd Q)
lub-≈-glb-ubd {I} {Q} = ≈-begin
    lub Q
  ≈⟨⟩
    ubd Q ˘ ╳ E ˘
  ≈⟨ ⊑-antisym
    (╳-universal
      (⊑-begin
          (E / ubd Q) ⨾ (ubd Q ˘ ╳ E ˘)
        ⊑⟨ ⨾-monotone₂ ˘╳˘-⊑-/ ⟩
          (E / ubd Q) ⨾ (ubd Q / E)
        ⊑⟨ /-cancel-middle ⟨⊑≈⟩ order-/ ⟩
          E
        □)
      (⊑-begin
          (ubd Q ˘ ╳ E ˘) ⨾ E ˘
        ⊑⟨ ⨾-monotone₁ ╳-⊑-\ ⟩
          (ubd Q ˘ \ E ˘) ⨾ E ˘
        ≈⟨ lbd-downclosed ⟩
          ubd Q ˘ \ E ˘
        ≈˘⟨ /-˘ ⟩
          (E / ubd Q) ˘
        □))
    (╳-universal
      (⊑-begin
          ubd Q ˘ ⨾ ((E / ubd Q) ╳ E)
        ⊑⟨ ⨾-monotone₂ (╳-⊑-/ ⟨⊑≈˘⟩ \-˘ ⟨⊑≈⟩ ˘-cong \/-≈) ⟩
          ubd Q ˘ ⨾ ((E \ E) / ubd Q) ˘
        ⊑⟨ ⨾-˘ ⟨≈˘⊑⟩ ˘-monotone (/-cancel-outer ⟨⊑≈⟩ order-\) ⟩
          E ˘
        □)
      ((⊑-begin
          ((E / ubd Q) ╳ E) ⨾ (E ˘) ˘
        ⊑⟨ ⨾-monotone ╳-⊑-\ (⊑-reflexive ˘˘) ⟩
          ((E / ubd Q) \ E) ⨾ E
        ≈⟨ ⨾-cong₁ \S∘S/∘\S ⟨≈≈⟩ ubd-upclosed ⟩
          ubd Q
        □) ⟨⊑≈˘⟩ ˘˘))
  ⟩
```

```
    (E / ubd Q) ⨉ E
  ≈˘⟨ ⨉-cong₁ lbd-ubd-˘ ⟩
    lbd (ubd Q) ˘ ⨉ E
  ≈⟨⟩
    glb (ubd Q)
  □
```

We will need a suborder construction "theorem": If F_0 is an injective mapping into A, where E : Mor A A is an order, then $\mathsf{F}_0 ⨾ \mathsf{E} ⨾ \mathsf{F}_0\ \breve{}$ is an order again.

Proof. The preorder preservation follows easily; preservation of antisymmetry is surprisingly hard, and requires use of retract⨉ from Sect. 3, which is prepared by the last two calculation steps.

```
antisym = ⊑-begin
    (F₀ ⨾ E ⨾ F₀ ˘) ⨉ (F₀ ⨾ E ⨾ F₀ ˘)
  ≈⟨ ⨉-cong ⨾-assocL ⨾-assocL ⟩
    ((F₀ ⨾ E) ⨾ F₀ ˘) ⨉ ((F₀ ⨾ E) ⨾ F₀ ˘)
  ≈⟨ ⨉-in-left F-isM ⟨≈˘≈˘⟩ ⨾-cong₂ (⨉-M-in-right F-isM) ⟩
    F₀ ⨾ ((F₀ ⨾ E) ⨉ (F₀ ⨾ E)) ⨾ F₀ ˘
  ⊑⟨ retract⨉ rightSupId rightSupId
    (⊑-begin
        (E ⨾ F₀ ˘) ⨾ (F₀ ⨾ E) ⨾ F₀ ˘
      ⊑⟨ ⨾-assoc ⟨≈⊑⟩ ⨾-monotone₂ (⨾-₁₂₁assoc₂₂ ⟨≈⊑⟩ proj₁ F-unival) ⟩
        E ⨾ E ⨾ F₀ ˘
      ⊑⟨ ⨾-assocL ⟨≈⊑⟩ ⨾-monotone₁ trans ⟩
        E ⨾ F₀ ˘
      □)
    (⊑-begin
        F₀ ⨾ (F₀ ⨾ E) ˘ ⨾ (E ⨾ F₀ ˘) ˘
      ≈⟨ ⨾-cong₂ (⨾-cong ⨾-˘ ⨾˘-˘ ⟨≈≈⟩ ⨾-assoc) ⟩
        F₀ ⨾ E ˘ ⨾ F₀ ˘ ⨾ F₀ ⨾ E ˘
      ⊑⟨ ⨾-monotone₂₂ (⨾-assocL ⟨≈⊑⟩ proj₁ F-unival) ⟩
        F₀ ⨾ E ˘ ⨾ E ˘
      ⊑⟨ ⨾-monotone₂ ˘-trans ⟩
        F₀ ⨾ E ˘
      ≈˘⟨ ⨾˘-˘ ⟩
        (E ⨾ F₀ ˘) ˘
      □)
    ⟩
    (E ⨾ F₀ ˘) ⨉ (E ⨾ F₀ ˘)
  ≈⟨ ⨉-in-left F-isM ⟨≈˘≈˘⟩ ⨾-cong₂ (⨉-M-in-right F-isM) ⟩
    F₀ ⨾ (E ⨉ E) ⨾ F₀ ˘
  ⊑⟨ ⨾-cong₂ (⨾-cong₁ antisym≈ ⟨≈≈⟩ leftId) ⟨≈⊑⟩ isInjective-to-I F-inj ⟩
    Id
  □
```

5 Direct Powers and Polarities

The approach of Schmidt *et al.* to formalise set membership using symmetric quotients [BSZ86, BSZ89, SS93] carries over to our setting with only one modification: In OCCs, totality of R is defined as Id ⊑ R ⨾ R ˘.

```
record DirectPower : Set (i ⊍ j ⊍ k₂) where
  field ℙ : Obj → Obj                                 -- power object constructor
        ∈ : {A : Obj} → Mor A (ℙ A)                    -- membership "relation"
        ∈-extensional : ∈ ╳ ∈ ⊑ Id                     -- sets defined by extension
        ∈-comprehensive : ∀ {Q} → isTotal (Q ╳ ∈)      -- all possible sets
  Ω : {A : Obj} → Mor (ℙ A) (ℙ A)                      -- the set inclusion "relation"
  Ω = ∈ \ ∈
  Λ₀ : {I A : Obj} → Mor I A → Mor I (ℙ A)             -- "power transpose"
  Λ₀ R = R ˘ ╳ ∈
  Λ : {I A : Obj} → Mor I A → Mapping I (ℙ A)          -- power transpose mapping
  Λ R = record {mor = Λ₀ R; prf = ...}
```

The membership ∈ together with the "power transpose" Λ produce a power operator PowerOp of the type the development in [Kah14b] is based on. A key ingredient of the mathematical treatment of concept lattices are the *polarities* _↓ and _↑, set-theoretically

$$S \uparrow (A) = \text{"the } S\text{-successors of all of } A\text{"} = \{s \mid \forall e : e \in A : e\,S\,s\} = \Lambda(\in \backslash S)(A)$$

and likewise $S \downarrow (B)$ = "the S-predecessors of all of B". These two operations constitute an antitone Galois connection, as already proved in [Kah14b].

```
module _ {A B : Obj} where
  _↑ :                  Mor A B  → Mapping (ℙ A) (ℙ B)
  _↓ :                  Mor A B  → Mapping (ℙ B) (ℙ A)
  Galois-↓-↑ : {R : Mor A B} → Ω ⨾ (R ↓₀) ˘ ≈ R ↑₀ ⨾ Ω ˘
```

In [Kah14b], the polarities are defined using Λ and right-residuals; they now satisfy useful properties involving symmetric quotients, for example (where, as before, the subscript "$_0$" indicates the underlying morphism of a mapping):

```
S ↓₀     ≈   Λ₀ (∈ \ S ˘)        ≈   (S / ∈ ˘) ╳ ∈
S ↑₀     ≈   Λ₀ (∈ \ S)          ≈   (S ˘ / ∈ ˘) ╳ ∈
S ↓↑₀    ≈   S ↓₀ ⨾ S ↑₀         ≈   (S ˘ / (∈ \ S ˘)) ╳ ∈
S ↑↓₀    ≈   S ↑₀ ⨾ S ↓₀         ≈   (S / (∈ \ S)) ╳ ∈
             S ↑₀ ⨾ (S ↓₀) ˘     ≈   (S ˘ / ∈ ˘) ╳ (S / ∈ ˘)
```

6 Complete Semilattices

A (lower) complete semilattice is a partial order E where each subset of the carrier has a lower bound, that is, glb R is total for every R. We call such a structure "*abstract* complete semilattice" (ACSL) to emphasise that this is not set-based, but rather set in an abstract OCC setting:

```
record ACSL : Set (i ⊍ j ⊍ k₂) where
  field Carrier    : Obj
        ≼          : Mor Carrier Carrier
        ≼-isOrder  : IsOrder ≼
  open IsOrder ≼-isOrder          -- to make in particular glb available
  field glb-total  : {I : Obj} (R : Mor I Carrier) → isTotal (glb R)
```

A homomorphism of such semilattices consisting of an order relation ≼ and an operation glb providing arbitrary meets is a mapping preserving this structure; order preservation is "monotone", preservation of arbitrary meets is "continuous". Although we will show below how to derive monotonicity from continuity, we still keep monotonicity as a constituent since applications depending on the computational content of the proofs may want to supply more efficient implementations of the monotonicity proof than the one we derive from continuity.

```
record ACSLHom (A B : ACSL) : Set (i ⊍ j ⊍ k₁ ⊍ k₂) where
  field map : Mapping A.Carrier B.Carrier
  map₀ : Mor A.Carrier B.Carrier
  map₀ = Mapping.mor map
  field monotone   : A.≼ ⨾ map₀    ⊑ map₀ ⨾ B.≼
        continuous : {I : Obj} {S : Mor I A.Carrier}
                   → A.glb S ⨾ map₀ ≈ B.glb (S ⨾ map₀)
```

For the purpose of proving monotonicity from continuity, we first show a lemma that corresponds to the fact that for set-based orders, the greatest lower bound of the image of the "up-set" of any element exists and is the image of that element.

```
glb-≼⨾continuous : B.glb (A.≼ ⨾ map₀) ≈ map₀
glb-≼⨾continuous = ≈-begin
    B.glb (A.≼ ⨾ map₀)
  ≈˘⟨ continuous ⟩
    A.glb A.≼ ⨾ map₀
  ≈⟨ ⨾-cong₁ A.glb-order ⟨≈≈⟩ leftId ⟩
    map₀
  □
```

The proof of monotonicity from continuity only needs totality of map; it does not even need completeness (totality of glb). The proof below essentially proves monotonicity in the shape map₀ ˘ ⨾ A.≼ ⨾ map₀ ⊑ B.≼ by replacing the first map₀ with B.glb (A.≼ ⨾ map₀) using the lemma above, and then using the glb definition in B. The step using B.order-\ at the end of the calculation corresponds to using an "indirect inclusion" argument.

```
monotone′ = ⊑-begin
    A.≼ ⨾ map₀
  ⊑⟨ proj₁ (mappingTotal map) ⟨⊑≈⟩ ⨾-assoc ⟩
    map₀ ⨾ map₀ ˘ ⨾ A.≼ ⨾ map₀
```

```
≈⟨ ⨾-cong21 (˘-cong glb-≼⨾continuous ⟨≈˘≈⟩ (╳-˘ ⟨≈≈⟩ ╳-cong2 ˘\˘-˘)) ⟩
  map0 ⨾ (B.≼ ╳ (B.≼ / (A.≼ ⨾ map0))) ⨾ A.≼ ⨾ map0
⊑⟨ ⨾-monotone2 (\-universal (⊑-begin
     B.≼ ⨾ (B.≼ ╳ (B.≼ / (A.≼ ⨾ map0))) ⨾ A.≼ ⨾ map0
   ⊑⟨ ⨾-assocL ⟨≈⊑⟩ ⨾-monotone1 ╳-cancel-left ⟩
     (B.≼ / (A.≼ ⨾ map0)) ⨾ A.≼ ⨾ map0
   ⊑⟨ /-cancel-outer ⟩
     B.≼
   □)) ⟩
  map0 ⨾ (B.≼ \ B.≼)
≈⟨ ⨾-cong2 B.order-\ ⟩
  map0 ⨾ B.≼
□
```

Producing a category INF of complete semilattices of type ACSL with meet-preserving ACSLHom homomorphisms is straight-forward: Morphism equality is equality of the underlying maps, and monotonicity and meet preservation hold by simple standard proofs for identities and composition.

7 Review: Contexts and Context Homomorphisms

We now briefly review the essentials of the formalisation previously presented in [Kah14b] of Moshier's category of formal contexts with relational homomorphisms. That formalisation was set in the context of OCCs with power operators and left- and right-residuals, where polarities are defined using residuals and ∧. An "abstract" *context* is merely a typed "relation":

```
record AContext : Set (i ⊍ j) where
  field ent : Obj            -- "entities"
        att : Obj            -- "attributes"
        inc : Mor ent att    -- "incidence"
```

Instantiating this for the OCC of sets and relations, a concrete context A consists of two sets A.ent and A.att and a relation A.inc : A.ent ↔ A.att. In this set-based variant of such contexts, a "concept" in the sense of formal concept analysis is either a subset of A.ent that is closed under A.inc ↑ ⨾ A.inc ↓, or a subset of A.att that is closed under A.inc ↓ ⨾ A.inc ↑, where these two views are isomorphic to each other. For each context, the induced concept lattice is complete.

Given two contexts A and B, a "context homomorphism" between them is expected to be a homomorphism of the induced concept lattices. Moshier proposes [Mos13, Jip12] to represent such a homomorphism by a relation R : A.ent ↔ B.att between the source of the first and the target of the second. For composition of these to be compatible with the concept lattice view and well-defined, Moshier identifies a pair of *compatibility conditions* that each morphism needs to satisfy, formalised in the srcCompat and trgCompat conditions:

```
record AContextHom (A B : AContext) : Set (i ⊍ j ⊍ k1 ⊍ k2) where
  field mor : Mor A.ent B.att
```

$$\mathsf{srcCompat} : \mathsf{mor}\downarrow \mathbin{;_1} \mathsf{A.inc}\uparrow \mathbin{;_1} \mathsf{A.inc}\downarrow \approx_1 \mathsf{mor}\downarrow$$
$$\mathsf{trgCompat} : \mathsf{B.inc}\downarrow \mathbin{;_1} \mathsf{B.inc}\uparrow \mathbin{;_1} \mathsf{mor}\downarrow \approx_1 \mathsf{mor}\downarrow$$

These can be taken to express that there is no distinction between the three natural paths from the bottom center $\mathbb{P}\ \mathsf{B.att}$ to the top center $\mathbb{P}\ \mathsf{A.ent}$ in the following diagram:

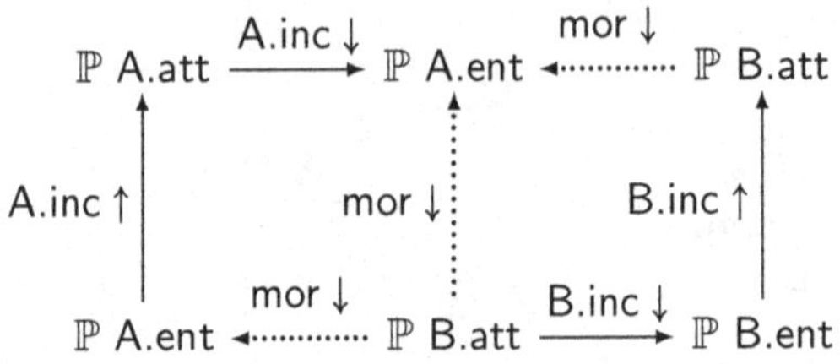

The mechanised proof that these homomorphisms together with $\mathsf{AContexts}$ as objects give rise to a category CXT was presented in [Kah14b] and is available in [Kah14a, KAh15].

8 Duality between Contexts and Complete Semilattices

For showing the categorical duality between abstract contexts and abstract complete semilattices, we follow the general approach outlined by Moshier [Mos13], but return to assume the setting of OCCs with left- and right-residuals, symmetric quotients, and direct powers. This provides a derived power operator for instantiating the development of [Kah14b] summarised in the previous section.

From CXT to INF^{op}: Every context A determines a complete lattice $\mathcal{C}_{\mathsf{ent}}\,\mathsf{A}$, with carrier denoted $\mathsf{A.}\uparrow\downarrow\mathsf{\text{-}image}$, which is defined as the sub-lattice of $\mathbb{P}\ \mathsf{A.ent}$ restricted to the $\mathsf{A.inc}\uparrow\downarrow$-closed subsets of $\mathsf{A.ent}$. (We use $\mathsf{A.}\Gamma : \mathsf{A.}\uparrow\downarrow\mathsf{\text{-}image} \to \mathbb{P}\ \mathsf{A.ent}$ as the subobject injection.) Since the meet in this lattice is just the intersection, while the join is derived, it is more natural to view $\mathcal{C}_{\mathsf{ent}}\,\mathsf{A}$ as just a complete meet semilattice.

Due to the co-continuity of polarities, the mapping $\mathsf{R} \mapsto \mathsf{B.inc}\uparrow \mathbin{;} \mathsf{R.mor}\downarrow$, for a context homomorphism R from context A to B, is infimum preserving, and so an INF-arrow, that is, a meet-preserving homomorphism of complete lower semilattices. However, $\mathsf{B.inc}\uparrow \mathbin{;} \mathsf{R.mor}\downarrow$ is a lattice homomorphism from $\mathbb{P}\ \mathsf{B.ent}$ to $\mathbb{P}\ \mathsf{A.ent}$, so the resulting functor is contravariant, $\mathcal{C}_{\mathsf{ent}} : CXT \to INF^{\mathsf{op}}$.

For the formalisation, $\mathcal{C}_{\mathsf{ent}}$ presented two main challenges: Restriction of the power order Ω to the suborder defined by the range of $\mathsf{A.inc}\uparrow\downarrow$ required the non-trivial $\mathsf{subOrder}$ construction shown at the end of Sect. 4. For well-definedness, it was then necessary to show that "arbitrary intersections of closed sets are closed again". We proved this in the more general setting of Sect. 4, by assuming an arbitrary closure operator C, that is, a mapping C that is a monotone ($\mathsf{E} \mathbin{;} \mathsf{C} \sqsubseteq \mathsf{C} \mathbin{;} \mathsf{E}$, which implies $\mathsf{C\text{-}monotone}^{\smile} : \mathsf{C}^{\smile} \mathbin{;} \mathsf{E}^{\smile} \sqsubseteq \mathsf{E}^{\smile} \mathbin{;} \mathsf{C}^{\smile}$), idempotent, and contained in the order E. Closure of arbitrary meets of closed elements then can be formalised as the statement $\mathsf{Q} \mathbin{;} \mathsf{C} \approx \mathsf{Q} \to \mathsf{glb}\ \mathsf{Q} \mathbin{;} \mathsf{C} \sqsubseteq \mathsf{glb}\ \mathsf{Q}$, and proved using the fact that $\mathsf{glb}\ \mathsf{Q}$ is defined as the symmetric quotient $\mathsf{lbd}\ \mathsf{Q}^{\smile} \mathbin{\between} \mathsf{E}$:

```
glb-closed-⊑ : {I : Obj} {Q : Mor I A} → Q ⨾ C ≈ Q → glb Q ⨾ C ⊑ glb Q
glb-closed-⊑ {I} {Q} Q⨾C≈Q = ˘╳-universal (⊑-begin
        lbd Q ˘ ⨾ (lbd Q ˘ ╳ E) ⨾ C
      ⊑⟨ ⨾-assocL ⟨≈⊑⟩ ⨾-monotone₁ ╳-cancel-left ⟩
        E ⨾ C
      ⊑⟨ ⨾-monotone₂ C⊑E ⟨⊑⊑⟩ trans ⟩
        E
    □) (⨾-assoc ⟨≈⊑⟩ \-universal ((⊑-begin
        Q ˘ ⨾ (lbd Q ˘ ╳ E) ⨾ (C ⨾ E ˘)
      ⊑⟨ ⨾-cong₁ (˘-cong Q⨾C≈Q ⟨≈˘≈⟩ ⨾-˘) ⟨≈⊑⟩ ⨾-monotone₂₁ ˘╳-⊑-/ ⟩
        (C ˘ ⨾ Q ˘) ⨾ ((Q ˘ \ E ˘) / E ˘) ⨾ (C ⨾ E ˘)
      ⊑⟨ ⨾-₂₂assoc₁₂₁ ⟨≈⊑⟩ ⨾-monotone₂₁ /-outer-⨾ ⟩
        C ˘ ⨾ ((Q ˘ ⨾ (Q ˘ \ E ˘)) / E ˘) ⨾ (C ⨾ E ˘)
      ⊑⟨ ⨾-monotone₂₁ (/-monotone \-cancel-outer ⟨⊑≈⟩ order˘-/) ⟩
        C ˘ ⨾ E ˘ ⨾ (C ⨾ E ˘)
      ⊑⟨ ⨾-monotone₁&₂₁ C-monotone˘ ⟩
        E ˘ ⨾ C ˘ ⨾ (C ⨾ E ˘)
      ⊑⟨ ⨾-monotone₂ (⨾-assocL ⟨≈⊑⟩ proj₁ C.unival) ⟩
        E ˘ ⨾ E ˘
      ⊑⟨ ˘-trans ⟩
        E ˘
    □)))
```

The opposite inclusion, glb-closed-⊒, follows from the additional assumption that glb Q is total due to the fact that then the left-hand side of glb-closed-⊑ is total and the right-hand side univalent. We conjecture that without that additional assumption, glb-closed-⊒ can still be shown in allegories, but not in OCCs.

The full definition of the mapping underlying the image ACSL homomorphism of R : AContextHom A B also involves the subobject injections $\mathsf{A}.\Gamma$ and $\mathsf{B}.\Gamma$ (where due to the partiality of $\mathsf{A}.\Gamma$ ˘, the totality proof invokes the fact that the images of R.mor $\downarrow_0$ are A.inc↑↓-closed due to source compatibility):

```
Φ₀ : Mor B.↑↓-image A.↑↓-image
Φ₀ = B.Γ ⨾ B.inc ↑₀ ⨾ R.mor ↓₀ ⨾ A.Γ ˘
```

From INF^{op} to CXT: For the opposite direction, every INF object is a lattice, and as such endowed with an ordering relation. Following Moshier, who calls this the "standard polarity", we use this to define the object mapping:

```
fromACSL : ACSL → AContext
fromACSL A = record {ent = A.Carrier; att = A.Carrier; inc = A.≼}
```

Since we are constructing a contravariant functor from INF^{op} to CXT, given an INF map Φ : A → B, the image in CXT needs to be a relation from B.Carrier to A.Carrier, which is the type of Φ ˘. For turning this into a context homomorphism satisfying the compatibility conditions, the natural choice is ≼B ⨾ Φ_0 ˘. For showing the compatibility conditions, we use the fact that for every B : ACSL with carrier B_0, there is a Galois connection between arbitrary join B.lub (∈ ˘) : ℙ B_0 → B_0 and the downset operator Λ (B.≼ ˘) : B_0 → ℙ B_0. We show one proof of source compatibility, where we extensively use the residual- and symmetric-quotient-based definitions for the polarities $\downarrow_0$ and $\uparrow\downarrow_0$.

```
fromACSLHom : {A B : ACSL}
  → ACSLHom A B → AContextHom (fromACSL B) (fromACSL A)
fromACSLHom {A} {B} Φ = record
  {mor = ≼B ⨾ Φ₀ ˘   -- Φ₀ is the underlying mapping of Φ
  ;srcCompat = ≈-begin
      (≼B ⨾ Φ₀ ˘) ↓₀ ⨾ ≼B ↑↓₀
    ≈⟨ ⨾-cong₂ ↑↓≈Ⅹ ⟩
      (≼B ⨾ Φ₀ ˘) ↓₀ ⨾ ((≼B / (∈ \ ≼B)) Ⅹ ∈)
    ≈⟨ Ⅹ-in-left (Mapping.prf ((≼B ⨾ Φ₀ ˘) ↓)) ⟩
      ((≼B / (∈ \ ≼B)) ⨾ (≼B ⨾ Φ₀ ˘) ↓₀ ˘) Ⅹ ∈
    ≈⟨ Ⅹ-cong₁ (/-inner-⨾ (Mapping.prf ((≼B ⨾ Φ₀ ˘) ↓))) ⟩
      (≼B / ((≼B ⨾ Φ₀ ˘) ↓₀ ⨾ (∈ \ ≼B))) Ⅹ ∈
    ≈⟨ Ⅹ-cong₁ (/-cong₂ ↓⨾∈\) ⟩
      (≼B / (((≼B ⨾ Φ₀ ˘) / ∈ ˘) \ ≼B)) Ⅹ ∈
    ≈˘⟨ Ⅹ-cong₁ (/-cong₂ (\-cong₁ (/-flip (Mapping.prf Φ)))) ⟩
      (≼B / ((≼B / (∈ ˘ ⨾ Φ₀)) \ ≼B)) Ⅹ ∈
    ≈⟨ Ⅹ-cong₁ S/∘\S∘S/ ⟩
      (≼B / (∈ ˘ ⨾ Φ₀)) Ⅹ ∈
    ≈⟨ Ⅹ-cong₁ (/-flip (Mapping.prf Φ)) ⟩
      ((≼B ⨾ Φ₀ ˘) / ∈ ˘) Ⅹ ∈
    ≈˘⟨ ↓≈Ⅹ ⟩
      (≼B ⨾ Φ₀ ˘) ↓₀
    □
  ;trgCompat = ...
  }
```

The resulting "mountains of residuals" can be less readable than translating these polarities into Λ and the order operators lbd, lub, etc., but in our experience, such residual-based proofs are far easier to find using a systematic approach, whereas the multitude of order operators re-awakens the problem frequently cited about relation-algebraic proofs that there are "too many choices".

Equivalence: With functors in opposite directions, proving equivalence of the categories requires exhibiting natural isomorphisms between the functor compositions and the respective identity functors.

On the context side, this natural isomorphism needs to provide, for each A : AContext, a pair of context isomorphisms between fromACSL ($\mathcal{C}_{\mathsf{ent}}$ A) and A. The carrier of the former is a sub-lattice of ℙ A.ent, so an adapted element relation is a natural choice for the context incidence of the "backwards" direction (S below). The "forward" direction (R below) needs to relate subsets of A.ent with A.att; although one might be tempted to move to the A.att side via A.inc, using A.Γ ⨾ ∈ ˘ ⨾ A.inc, but it is important to do this at the lattice level via A.inc $\uparrow_0$:

```
R : Mor A.↑↓-image A.att          S : Mor A.ent A.↑↓-image
R = A.Γ ⨾ A.inc ↑₀ ⨾ ∈ ˘           S = ∈ ⨾ A.Γ ˘
```

On the ACSL side, for $\mathsf{A} : \mathsf{ACSL}$, with $\mathsf{A}' = \mathsf{fromACSL}\ \mathsf{A}$, the carrier of $\mathcal{C}_{\mathsf{ent}}\,\mathsf{A}'$ is the sub-lattice of $\mathsf{A}'.\uparrow\downarrow$-closed subsets of the carrier of A. The mappings underlying the ACSL homomorphisms between $\mathcal{C}_{\mathsf{ent}}\,\mathsf{A}'$ and A in INF^{op} are:

$$\mathsf{L}_0 : \mathsf{Mor}\ \ \mathsf{A.Carrier}\ \ \mathsf{A}'.\uparrow\downarrow\text{-}\mathsf{image} \qquad \mathsf{R}_0 : \mathsf{Mor}\ \ \mathsf{A}'.\uparrow\downarrow\text{-}\mathsf{image}\ \ \mathsf{A.Carrier}$$
$$\mathsf{L}_0 = \mathsf{A.downset}_0 \mathbin{;} \mathsf{A}'.\Gamma\,\breve{} \qquad \mathsf{R}_0 = \mathsf{A.lub}\,(\mathsf{A}'.\Gamma \mathbin{;} \in\breve{})$$

We have shown that these satisfy all the properties required of natural isomorphisms [KAh15], and therewith provide a mechanised confirmation of the categoric duality of CXT and INF.

9 Conclusion

At the time of writing, although all components of the natural isomorphisms for the duality between CXT and INF have been checked by Agda [KAh15], we have not yet been able to typecheck the **record** that assembled all these components into a single duality proof within the RAM available on our machines. Many of the component proofs probably still can be made shorter and more readable, and we hope that this will also allow us to also check the full duality proof **record**.

We also do not yet have an actual implementation of residuals and direct powers for an OCC where morphisms are data structures — once this is completed, we can actually "run" the functors we defined, and use our mathematical definitions to perform conversions.

Another interesting question to explore is whether there are (interesting) models of OCCs with residuals, symmetric quotients, and direct powers, but without all binary meets.

Nevertheless, the formalisation we presented is a nontrivial development in calculational mathematics, even without the aspect of mechanisation. Although the version of Moshier's draft [Mos13] we had access to provided some guidance for the essential definitions, it frequently contained only little indication of possible routes to verify its results. In addition, our self-imposed restriction to OCCs with only residuals, symmetric quotients and direct powers meant that we had to re-prove quite a few basic properties from scratch. Manually producing proofs in such a setting that feels almost completely natural to relation-algebraists always bears the danger that laws that "suddenly are not available" are used without noticing. Especially for this kind of setting, a proof checker is invaluable, and although calculational Agda proofs in such theories as supported by the RATH-Agda libraries [Kah14c] are perhaps not yet as readable as proofs produced directly in LaTeX; we feel that they are already "more writable": While in LaTeX, one would spend a lot of time getting all occurrences of braces and & right just for obtaining the desired layout, this time is spent with Agda on getting the *mathematically relevant* syntax, typing, and logical correctness of the proofs right.

References

[BSZ86] Berghammer, R., Schmidt, G., Zierer, H.: Symmetric Quotients. Technical Report TUM-INFO 8620, TU München, Fak. Informatik, 18 p. (1986)

[BSZ89] Berghammer, R., Schmidt, G., Zierer, H.: Symmetric Quotients and Domain Constructions. Inform. Process. Lett. 33, 163–168 (1989)

[BdM97] Bird, R.S., de Moor, O.: Algebra of Programming, International Series in Computer Science, vol. 100. Prentice Hall (1997)

[FS90] Freyd, P.J., Scedrov, A.: Categories, Allegories, North-Holland Mathematical Library, vol. 39. North-Holland, Amsterdam (1990)

[FK98] Furusawa, H., Kahl, W.: A Study on Symmetric Quotients. Technical Report 1998-06, Fakultät für Informatik, Univ. der Bundeswehr München (1998)

[Jip12] Jipsen, P.: Categories of Algebraic Contexts Equivalent to Idempotent Semirings and Domain Semirings. In: Kahl, W., Griffin, T.G. (eds.) RAMICS 2012. LNCS, vol. 7560, pp. 195–206. Springer, Heidelberg (2012)

[Kah04] Kahl, W.: Refactoring Heterogeneous Relation Algebras around Ordered Categories and Converse. J. Rel. Methods in Comp. Sci. 1, 277–313 (2004)

[Kah11] Kahl, W.: Dependently-Typed Formalisation of Relation-Algebraic Abstractions. In: de Swart, H. (ed.) RAMICS 2011. LNCS, vol. 6663, pp. 230–247. Springer, Heidelberg (2011)

[Kah14a] Kahl, W.: Abstract Context Lattices Formalised for Concrete Applications — AContext-1.0. Mechanically checked Agda theories, 25 pp. literate document output (2014). `http://relmics.mcmaster.ca/RATH-Agda/#AContext`

[Kah14b] Kahl, W.: A Mechanised Abstract Formalisation of Concept Lattices. In: Höfner, P., Jipsen, P., Kahl, W., Müller, M.E. (eds.) RAMiCS 2014. LNCS, vol. 8428, pp. 242–260. Springer, Heidelberg (2014)

[Kah14c] Kahl, W.: Relation-Algebraic Theories in Agda — RATH-Agda-2.0.1. Mechanically checked Agda theories, with 456 pages literate document output (2014). `http://relmics.mcmaster.ca/RATH-Agda/`

[KAh15] Kahl, W., Al-hassy, M.: Order Theory and Concept Lattices in Ordered Categories Without Meets, Formalised in Agda — AContext-2.1. Mechanically checked Agda theories, 181 pages literate document output (2015). `http://relmics.mcmaster.ca/RATH-Agda/#AContext`

[Koz98] Kozen, D.: Typed Kleene Algebra. Technical Report 98-1669, Computer Science Department, Cornell University (1998)

[Löh12] Löh, A.: lhs2TeX (2012). `http://www.andres-loeh.de/lhs2tex/`

[Mos13] Moshier, M.A.: A Relational Category of Polarities (unpublished draft) (2013)

[Nor07] Norell, U.: Towards a Practical Programming Language Based on Dependent Type Theory. PhD thesis, Department of Computer Science and Engineering, Chalmers University of Technology (2007). `http://wiki.portal.chalmers.se/agda/pmwiki.php`

[SS93] Schmidt, G., Ströhlein, T.: Relations and Graphs, Discrete Mathematics for Computer Scientists. EATCS-Monographs on Theoret. Comput. Sci. Springer (1993)

[Wil05] Wille, R.: Formal Concept Analysis as Mathematical Theory of Concepts and Concept Hierarchies. In: Ganter, B., Stumme, G., Wille, R. (eds.) Formal Concept Analysis. LNCS (LNAI), vol. 3626, pp. 1–33. Springer, Heidelberg (2005)

Reasoning about Computations and Programs

Metaphorisms in Programming

José N. Oliveira

High Assurance Software Laboratory
INESC TEC and University of Minho
Braga, Portugal
jno@di.uminho.pt

Abstract. This paper introduces the *metaphorism* pattern of relational specification and addresses how specification following this pattern can be refined into recursive programs.

Metaphorisms express input-output relationships which preserve relevant information while at the same time some intended optimization takes place. Text processing, sorting, representation changers, etc., are examples of metaphorisms.

The kind of metaphorism refinement proposed in this paper is a strategy known as *change of virtual data structure*. It gives sufficient conditions for such implementations to be calculated using relation algebra and illustrates the strategy with the derivation of *quicksort* as example.

Keywords: Programming from specifications, Algebra of programming.

Politicians and diapers should be changed often and for the same reason.

(attributed to Mark Twain)

1 Introduction

The witty quote by 19th century author Mark Twain that provided inspiration for the title of this paper embodies a *metaphor* which the reader will surely appreciate. But, what do metaphors of this kind have to do with computer programming?

Programming theory has been structured around concepts such as *syntax*, *semantics*, *generative grammar* and so on, which have been imported from Chomskian linguistics. The basis is that syntax provides the *shape* of information and that semantics express information *contents* in a syntax-driven way (e.g. meaning of the whole dependent on the meaning of the parts).

Cognitive linguistics breaks with such a *generative* tradition in its belief that semantics are conveyed in a different way, just by juxtaposing concepts in the form of *metaphors* which let meanings permeate each other by an innate capacity of our brain to function metaphor-wise. Thus we are led to the *metaphors we live by*, quoting the classic textbook by Lakoff and Johnson [8]. If in a public

W. Kahl et al. (Eds.): RAMiCS 2015, LNCS 9348, pp. 171–190, 2015.
DOI: 10.1007/978-3-319-24704-5_11

discussion one of the opponents is said to have *counterattacked* with a *winning* argument, the underlying metaphor is *argument is war*; metaphor *time is money* underlies everyday phrases such as *wasting time*, *investing time* and so on; Twain's quote lives in the metaphor *politics is dirt*, the same that would enable one to say that somebody might need to *clean his/her reputation*, for instance.

In his *Philosophy of Rhetoric* [14], Richards finds three kernel ingredients in a metaphor, namely a *tenor* (e.g. *politicians*), a *vehicle* (e.g. *diapers*) and a shared *attribute* (e.g. ... left for the reader to guess). The *flow of meaning* is from vehicle to tenor, through the (as a rule left unspecified) common attribute.

In [11] the author sketched a brief characterization of this construction in the form of a "cospan"

$$\begin{array}{ccc} \mathsf{T} & & \mathsf{V} \\ & {\scriptstyle f}\searrow \quad \swarrow{\scriptstyle g} & \\ & A & \end{array} \qquad (1)$$

where $f : \mathsf{T} \to A$ and $g : \mathsf{V} \to A$ are functions extracting a common attribute (A) from both tenor (T) and vehicle (V). The cognitive, æsthetic, or witty power of a metaphor is obtained by *hiding* A, thereby establishing a *composite*, binary relationship[1] $\mathsf{T} \xleftarrow{f^\circ \cdot g} \mathsf{V}$ between tenor and vehicle — the "T is V" metaphor — which leaves A implicit.

It turns out that, in the field of program specification, many problem statements are *metaphorical* in the same (formal) sense: they are characterized as input-output relationships in which the *preservation* of some kernel information is kept implicit, possibly subject to some form of optimization.

An example of this is *text formatting*, a relationship between formatted and unformatted text whose metaphor consists in preserving the sequence of words of both, while the output text is optimized wrt. some visual criteria.[2] Other examples could have been given:

- Change of base of numeric representation — the number represented in the source is the same represented by the result, cf. the 'representation changers' of [5].
- Conversion of finite lists into balanced search trees — the information preserved is the set of elements of the source list; the optimization is the invariant induced on the output tree, making it adequate for searching, etc.

[1] Given a binary relation R, writing $b\ R\ a$ ($\equiv$ "b is related to a by R") means the same as $a\ R^\circ\ b$, where R° is said to be the *converse* of R. So R° corresponds to *passive voice*, check e.g. *John loves Mary* compared to *Mary is loved by John*: $(loves)^\circ = (is\ loved\ by)$.

[2] It is the privilege of those who don't work with WYSIWYG text processors to feel the rewarding (if not æsthetic) contrast between the window where source text is edited and that showing the corresponding, nice-looking PDF output.

- Source code refactoring — the meaning of the source program is preserved, the target code being better styled wrt. coding conventions and best practices.
- Sorting — the bag (multiset) of elements of the source list is preserved, the optimization consisting in obtaining an ordered output.

The *optimization* implicit in all these examples can be expressed by reducing the *vagueness* of relation $f^\circ \cdot g$ in (1) according to some criterion telling which outputs are better than others. This can be achieved by adding such criteria in the form of a relation R which "shrinks" $f^\circ \cdot g$,

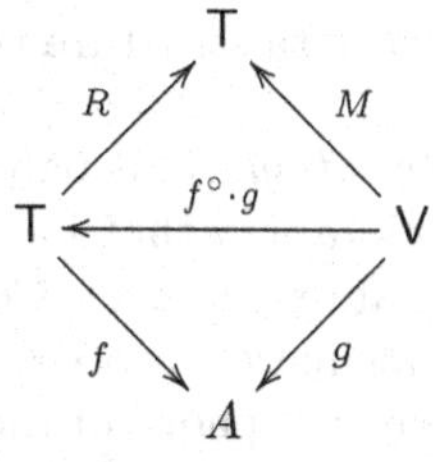

$$M = (f^\circ \cdot g) \upharpoonright R \qquad (2)$$

using the "shrinking" operator of [9] for reducing non-determinism, see the diagram above. By unfolding the meaning of this relational operator, the relationship established by M (2) is the following:

$$t \; M \; v \equiv (f \; t = g \; v) \wedge \langle \forall \; t' \; : \; f \; t' = g \; v : \; t \; R \; t' \rangle$$

In words: for each input v, choose among all outputs t' with the same (hidden) attribute of v those which are better than any other with respect to R, if any.

We will refer to construction (2) as a *metaphorism* wherever V and T are inductive types and functions f and g are recursive on such types. A *metaphorism* $M = (f^\circ \cdot g) \upharpoonright R$ therefore involves two functions and an optimization criterion. In the text formatting metaphorism, for instance,

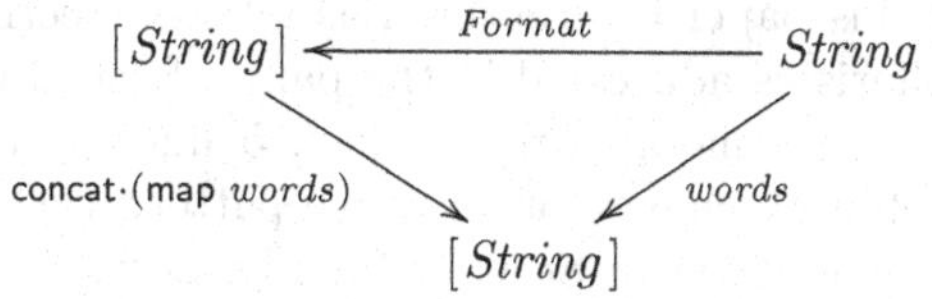

arrow *Format* relates a string (source text) to a list of strings (output text lines) such that the original sequence of words is preserved when white space is discarded. Formatting consists in (re)introducing white space evenly throughout the output text lines. For economy of presentation, the diagram omits the optimization part,

$$Format = (\mathsf{map}\; words^\circ \cdot \mathsf{concat}^\circ \cdot words) \upharpoonright R \qquad (3)$$

where $R : [String] \rightarrow [String]$ should capture the formatting criterion on lines of text, e.g. even spaced lines better than ill-spaced ones, etc. Metaphorism (3) also relies on a well-known property of relational converse, $(R \cdot S)^\circ = S^\circ \cdot R^\circ$.

Formally, nothing impedes f and g from being the same attribute function, in which case types V and T are also the same. Although less interesting from the strict (cognitive) metaphorical perspective, metaphorisms of this instance of (2) are very common in programming — take *sorting* as example, where V

and T are inhabited by finite sequences of the same type. Interestingly, some sorting algorithms actually involve *another* data-type, but this is hidden and kept implicit in the whole algorithmic process. Quicksort, for instance, unfolds recursively in a binary fashion which makes its use of the run-time heap look like a binary search tree — a pattern found in any *divide & conquer* algorithm. Because such a tree is not visible from outside, some authors refer to it as a *virtual* data structure [15].

Contribution. This paper addresses a generic process of implementing metaphorisms in a way that introduces *divide & conquer* strategies and the implicit virtual data structures. Conditions for the semantics of (2) to be preserved along the calculation process are discussed. Altogether, the reasoning shows how the "outer metaphor" of the specification (2) disappears and is replaced by a more implicit but more interesting "inner metaphor" which is at the heart of the implementation. We will restrict to a special case of (2) which is described in the next section and will use quicksort as running example.

Related Work. This paper follows the line of research of reference [9] in investigating relational specification patterns which involve the "shrinking" combinator for controling vagueness and non-determinism. It also relates to previous work on representation changers [5] and on the relational algebra of programming, in general [1, 10]. Our calculation of sufficient conditions for implementing metaphorisms via change of virtual data-structure, illustrated with quicksort, can be regarded as a generalization and expansion of the derivation of the same algorithm in [1], where it is given in a rather brief and terse style.

Paper Structure. The remainder of this paper is structured as follows. Sections 2 and 4 identify the class of metaphorisms addressed in the paper. Sect. 3 discusses implementation strategies for such metaphorisms. Sect. 5 finds generic conditions for these to be implemented by change of recursive pattern (virtual data-structure), an example of which is given in Sect. 6. Finally, Sect. 7 concludes. Some background on relation algebra and proofs of auxiliary results are given in appendices A and B, respectively.

2 Shrunken Equivalence Relations as Metaphorisms

Wherever $f = g$ in (2) we get $M = (f^\circ \cdot f) \upharpoonright R$, a "shrunken" equivalence relation because $f^\circ \cdot f$ is an equivalence, known as the *kernel* of f, $\mathsf{ker}\, f = f^\circ \cdot f$:

$$M = (\mathsf{ker}\, f) \upharpoonright R \tag{4}$$

So $y\ M\ x$ means not only that $f\ y = f\ x$ (this is the information to be preserved), but also that y is "best" among all other y' such that $f\ y' = f\ x$ holds, as expressed by the meaning of the shrinking combinator [9, 13], see property (37) in the appendix: $S \upharpoonright R$ is the largest sub-relation X of S such that, for all $b', b \in B$, if there exists some $a \in A$ such that $b' X a \wedge b S a$ holds, then $b' R b$ holds.

Example: take $\mathsf{V} = \mathsf{T} = [A]$ parametric on type A and $f = bag$, the function that extracts the bag of elements of a finite list. The equivalence relation is $Perm = \mathsf{ker}\ bag$, that is $y\ Perm\ x$ means that y is a *permutation* of x. What about R? If sorting is the intended optimization, one might want to specify that $y\ R\ x$ holds wherever y has less "out-of-order" entries than x, something like e.g. (in Haskell concrete syntax)

$$\begin{array}{l} y\ R\ x = oo\ y \leqslant oo\ x\ \mathbf{where} \\ \quad oo\ s = \mathsf{length}\ [n \mid n \leftarrow [0\mathinner{..}\mathsf{length}\ s], n+1 < \mathsf{length}\ s, s\,!!\,n > s\,!!\,(n+1)] \end{array}$$

where oo is the function that counts "out-of-order" entries.

For the calculational theory of [1, 9] to be applicable to metaphorism (4), one needs to express either $\mathsf{ker}\,f$ or R (or both) as relational (un)folds, also referred to as ana/catamorphisms in the literature [1]. This makes perfect sense since, in many situations, T will be an inductive (initial, tree-like) data-type and f a *fold* which recursively extracts information from T using some function k for this. The popular notation $f = (\!|k|\!)$ will be used to express (relational) folds, see Appendix A for the basic properties of such a combinator.

It turns out that, if f is surjective, then the equivalence relation $\mathsf{ker}\,f$ will be a fold too, this time relational

$$\mathsf{ker}\,f = (\!|\mathsf{ker}\,f \cdot \mathsf{in}|\!) \tag{5}$$

where $\mathsf{T} \xleftarrow{\mathsf{in}} \mathsf{F\,T}$ is the initial algebra of type T, for some functor F. (The proof of (5) is given in Appendix B.) So

$$\mathsf{ker}\,f \cdot \mathsf{in} = \mathsf{ker}\,f \cdot \mathsf{in} \cdot \mathsf{F}\,(\mathsf{ker}\,f) \tag{6}$$

holds, by fold-cancellation (28). In the case of lists, $\mathsf{F}\,X = 1 + A \times X$ and $\mathsf{in} = [\mathsf{nil}\ , \mathsf{cons}]$, where $\mathsf{nil}\ x = [\,]$ is the constant function which yields the empty list and $\mathsf{cons}\ (a, s) = a : s$ adds a to the front of s. For $f = bag$, the fold which extracts the multiset of elements of a given list, $\mathsf{ker}\,f = Perm$ and we have the following property of the list permutation equivalence relation:

$$Perm \cdot \mathsf{in} = Perm \cdot \mathsf{in} \cdot (\mathsf{F}\,Perm) \tag{7}$$

The useful part of (7) is

$$Perm \cdot \mathsf{cons} = Perm \cdot \mathsf{cons} \cdot (id \times Perm) \tag{8}$$

where we use notation $R \times S$ to express the (Kronecker) *product* of two relations: $(b, d)\ (R \times S)\ (a, c)$ holds iff both $b\ R\ a$ and $d\ S\ c$ hold. Thus (8) is the same as

$$y\ Perm\ (a : x) = \langle \exists\ z\ :\ z\ Perm\ x :\ y\ Perm\ (a : z) \rangle$$

which means that permuting a sequence with at least one element is the same as adding it to the front of a permutation of the tail and permuting again.

The usefulness of (5, 6) is that the inductive definition of an equivalence relation $\mathsf{ker}\,f$ generated by a surjective fold f is such that the recursive branch $\mathsf{F}\,(\mathsf{ker}\,f)$ in the unfolding of $\mathsf{ker}\,f$ can be removed if convenient.

Another meaning of (6) is that $\mathsf{ker}\, f$ is a *congruence* for the initial algebra in, cf. the following theorem.

Theorem 1. *Let R be a congruence for an algebra $h : \mathsf{F}\, A \to A$ of functor F, that is*

$$h \cdot (\mathsf{F}\ R) \subseteq R \cdot h \tag{9}$$

holds and R is an equivalence relation. Then this is the same as stating:

$$R \cdot h = R \cdot h \cdot (\mathsf{F}\ R) \tag{10}$$

(Proof: see Appendix B.) □

3 Calculating Metaphorisms

Given a metaphorism M (4) such that $f = (\!|k|\!)$, it can immediately be shown that

$$M = (\mathsf{ker}\ (\!|k|\!)) \upharpoonright R = ((\!|k|\!)^\circ \upharpoonright R) \cdot (\!|k|\!) \tag{11}$$

by this law of shrinking: $(S \cdot f) \upharpoonright R = (S \upharpoonright R) \cdot f$ [9]. Thus we have two main ways of calculating metaphorisms:

- either we shrink $\mathsf{ker}\ (\!|k|\!)$ as a whole — a relational fold (5), as we have seen, or
- we shrink $(\!|k|\!)^\circ$ and then fuse the outcome with $(\!|k|\!)$ (11).

There is still a third way, known as *changing the virtual data structure* [15]. Given any *surjective* function $f : A \to B$, its image $\mathsf{img}\, f = f \cdot f^\circ$ — the converse-dual of $\mathsf{ker}\ f = f^\circ \cdot f$ — is such that $\mathsf{img}\ f = id$, where function $id\ x = x$ is the identity function, i.e. the equality relation on its type. So $\mathsf{img}\ f : B \to B$ can be pasted anywhere it typechecks, i.e. where type B is present. Suppose another $(\!|h|\!) : \mathsf{W} \to \mathsf{T}$ is given which is surjective. Then

$$\begin{aligned} M &= (\mathsf{ker}\ (\!|k|\!)) \upharpoonright R \\ &= (\mathsf{img}\ (\!|h|\!) \cdot \mathsf{ker}\ (\!|k|\!)) \upharpoonright R \\ &= (\!|h|\!) \cdot (N \upharpoonright R') \ \mathbf{where}\ N = (\!|h|\!)^\circ \cdot \mathsf{ker}\ (\!|k|\!) \end{aligned} \tag{12}$$

for some R' to be calculated. Using type diagrams, the strategy starts from

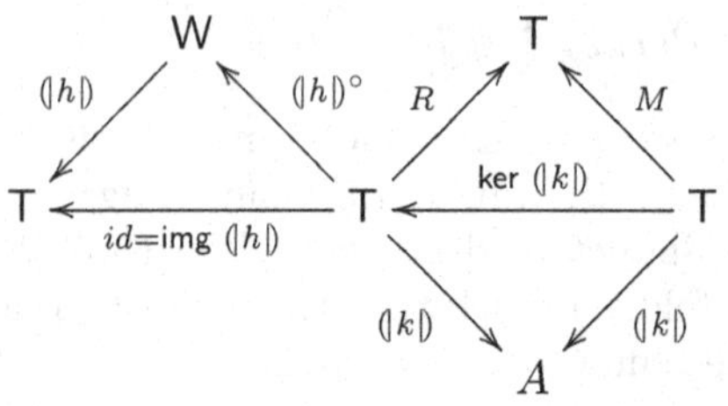

(13)

and then shifts the "ictus" of algorithmic control from type T to type W:

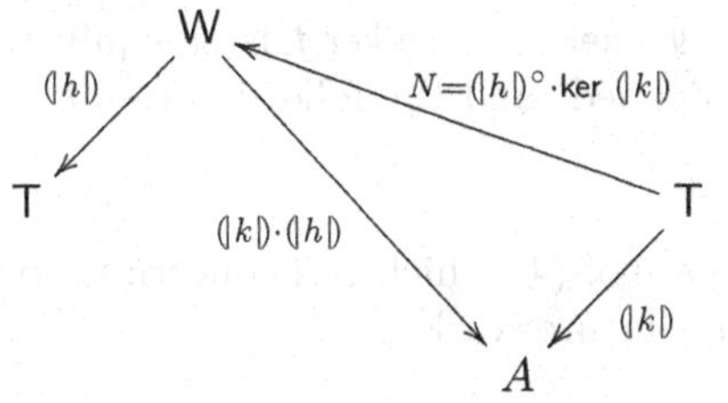

(14)

In this way, the starting, "outer" metaphor involving only T disappears and gives place to an "inner" metaphor between inductive types W and T, moving the optimization inside in the form of a relation R', which needs to be calculated:

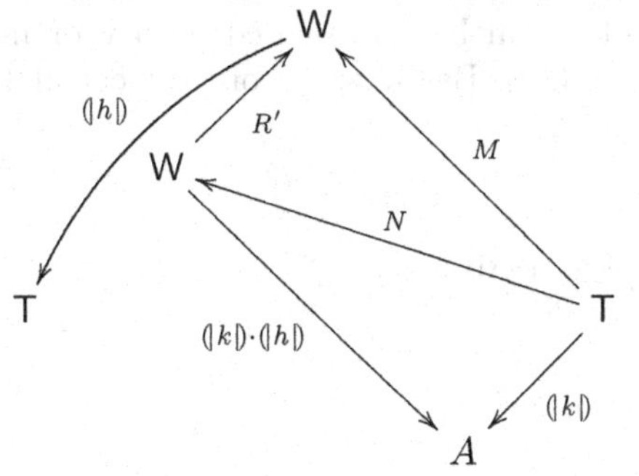

(15)

W is the (virtual) data type chosen to command the *divide & conquer* algorithmic control. It is usually a binary or n-ary tree structure and is regarded as virtual because, as mentioned above, it is doomed to disappear once the two-step composition process is fused into a single step.

In summary, finding a generic *divide & conquer* version of metaphorism $M = (\mathsf{ker}\ (\!|k|\!)) \upharpoonright R$ relying on virtual type W as *representation* of the original type T amounts to finding a function that implements the *divide* step, $(N \upharpoonright R')$ where $N = (\!|h|\!)^\circ \cdot \mathsf{ker}\ (\!|k|\!)$ and $(\!|h|\!)$ is an *abstraction* function. Finding R' is the hard part of the exercise, as we shall soon see.

4 Special Case of Shrinking

R in (2,4) is in general a *metric* indicating which structures are better than others, usually in the form of a preorder $R = \leqslant_h$ where h is the metric attribute to be compared and $\leqslant_h$ abbreviates $h^\circ \cdot (\leqslant) \cdot h$, that is: $y \leqslant_h x \equiv (h\ y) \leqslant (h\ x)$. For instance, trees can be compared by measuring their depth; programs under refactoring compared by counting LoC, and so on.

However, R can also take the form $R = \Psi \cdot \top$ in (4), where $\top$ is the "topmost" relation of its type (32) — $b \top a$ is true for every a and b — and $\Psi \subseteq id$ is a *partial identity* specifying some form of *selection*.[3] This indicates that only the outputs satisfying Ψ are regarded as good enough.

In case $R = \Psi \cdot \top$, (4) reduces to $M = \Psi \cdot \mathsf{ker}\, f$, since $\mathsf{ker}\, f$ is an equivalence relation and therefore entire (i.e. totally defined) and the following result holds

$$R \upharpoonright (\Psi \cdot \top) = \Psi \cdot R \quad \Leftarrow \quad R \text{ is entire} \tag{16}$$

(Proof in Appendix B.) It is this special case of (4) which will concern us in the sequel, leaving the full generality of (4) for future work.

5 Shrinking Metaphorisms into Hylomorphisms

Consider metaphorisms of form $M = \Psi \cdot \mathsf{ker}\ (\!|k|\!)$ which, as we have seen above, are special cases of (4). Suppose $(\!|h|\!) : \mathsf{W} \to \mathsf{T}$ is an abstraction function (surjective) which ensures that every inhabitant of T can be represented by one or more inhabitants of W, as in diagrams (13) to (15). Below we record the calculation implicit in such diagrams:

$$\begin{array}{ll}
& M = \Psi \cdot \mathsf{ker}\ (\!|k|\!) \\
\equiv & \quad \{\ \mathsf{img}\ (\!|h|\!) = id \text{ because } (\!|h|\!) \text{ is surjective }\ \} \\
& M = \mathsf{img}\ (\!|h|\!) \cdot \Psi \cdot \mathsf{ker}\ (\!|k|\!) \\
\equiv & \quad \{\ \text{inline image}\ \} \\
& M = (\!|h|\!) \cdot (\!|h|\!)^{\circ} \cdot \Psi \cdot \mathsf{ker}\ (\!|k|\!) \\
\equiv & \quad \{\ \text{hint: assume } \Phi \text{ such that } (\!|h|\!) \cdot \Phi = \Psi \cdot (\!|h|\!)\ ;\ \text{converses; } \Psi^{\circ} = \Psi\ \} \\
& M = (\!|h|\!) \cdot \Phi \cdot \underbrace{(\!|h|\!)^{\circ} \cdot \mathsf{ker}\ (\!|k|\!)}_{N}
\end{array}$$

The goals are, therefore: (a) to find Φ such that

$$(\!|h|\!) \cdot \Phi = \Psi \cdot (\!|h|\!) \tag{17}$$

holds, and (b) to convert $\Phi \cdot (\!|h|\!)^{\circ} \cdot \mathsf{ker}\ (\!|k|\!)$ into the converse of a fold, which we denote as usual by $[\!(g)\!]$, for some g.[4] Then the original metaphorism will be converted into a so-called *hylomorphism* [1] $(\!|h|\!) \cdot [\!(g)\!]$ with a "change of data-structure".

As W and T are inductive types, the two partial identities (coreflexives) will take the shape (say) $\Phi = (\!|\mathsf{in}_{\mathsf{W}} \cdot \Omega|\!)$ and $\Psi = (\!|\mathsf{in}_{\mathsf{T}} \cdot \Theta|\!)$, where in_{W} and in_{T} are the initial algebras of types W and T, respectively.

---

[3] We use uppercase Greek letters (e.g. Ψ, Φ, ...) to denote *partial identities*, also known as *coreflexives*, *monotypes* or *tests* [2, 3, 7]. Every partial identity Ψ is such that $\Psi \subseteq id$ and is in one-to-one correspondence with some predicate q. As in [9] we write $\Psi = q?$ wherever we want to indicate that q is the predicate captured by Ψ. Thus $\Psi = q?$ has the pointwise meaning $b\ \Psi\ a \equiv b = a \wedge q\ a$.

[4] Converses of folds are usually termed *unfolds* or anamorphisms. Notation $[\!(R)\!]$ means $(\!|R^{\circ}|\!)^{\circ}$.

Calculation of (17) proceeds by fusion (27), aiming to reduce both $(\!|h|\!) \cdot \Phi$ and $\Psi \cdot (\!|h|\!)$ to some fold $(\!|R|\!)$ over W. On the one side,

$$\Psi \cdot (\!|h|\!) = (\!|R|\!) \Leftarrow \Psi \cdot h = R \cdot (\mathsf{F}\,\Psi) \tag{18}$$

On the other side:

$$\begin{aligned}
& (\!|h|\!) \cdot \Phi = (\!|R|\!) \\
\equiv \quad & \{ \text{ inline } \Phi = (\!|\mathsf{in_W} \cdot \Omega|\!) \ \} \\
& (\!|h|\!) \cdot (\!|\mathsf{in_W} \cdot \Omega|\!) = (\!|R|\!) \\
\Leftarrow \quad & \{ \text{ fusion (27) } \} \\
& (\!|h|\!) \cdot \mathsf{in_W} \cdot \Omega = R \cdot \mathsf{F}\,(\!|h|\!) \\
\equiv \quad & \{ \text{ cancellation of } (\!|h|\!) \text{ (28) } \} \\
& h \cdot \mathsf{F}\,(\!|h|\!) \cdot \Omega = R \cdot \mathsf{F}\,(\!|h|\!) \\
\equiv \quad & \{ \text{ assume } \Lambda \text{ such that } \mathsf{F}\,(\!|h|\!) \cdot \Omega = \Lambda \cdot \mathsf{F}\,(\!|h|\!) \ \} \\
& h \cdot \Lambda \cdot \mathsf{F}\,(\!|h|\!) = R \cdot \mathsf{F}\,(\!|h|\!) \\
\Leftarrow \quad & \{ \text{ Leibniz } \} \\
& h \cdot \Lambda = R
\end{aligned}$$

Replacing this in $\Psi \cdot h = R \cdot \mathsf{F}\Psi$, the side condition of (18), we get: $\Psi \cdot h = h \cdot \Lambda \cdot (\mathsf{F}\,\Psi)$. Let us summarize both calculations in the form of a theorem.

Theorem 2. *Let $(\!|h|\!) : \mathsf{W} \to \mathsf{T}$ be an abstraction of inductive type T by W, and $\Psi = (\!|\mathsf{in_T} \cdot \Theta|\!)$ and $\Phi = (\!|\mathsf{in_W} \cdot \Omega|\!)$ be partial identities representing inductive predicates over such types.*

For $(\!|h|\!) \cdot \Phi = \Psi \cdot (\!|h|\!)$ (17) to hold, search for the existence of $\Lambda : \mathsf{F\,T} \to \mathsf{F\,T}$ such that

$$\Psi \cdot h = h \cdot \Lambda \cdot \mathsf{F}\Psi \tag{19}$$

$$\mathsf{F}\,(\!|h|\!) \cdot \Omega = \Lambda \cdot \mathsf{F}\,(\!|h|\!) \tag{20}$$

hold, where F is the base functor of W, that is, $\mathsf{in_W} : \mathsf{F\,W} \to \mathsf{W}$.
□

Note that condition (20) establishes Ω as *weakest precondition* for $\mathsf{F}\,(\!|h|\!)$ to ensure Λ on its output, cf. (35) in Appendix A. Likewise, (19) establishes Λ as weakest precondition for h to maintain invariant Ψ.

Searching for the Anamorphism. Thus far, the starting metaphor $\mathsf{ker}\ (\!|k|\!)$ has been left aside. Going back to

$$M = (\!|h|\!) \cdot \Phi \cdot \underbrace{(\!|h|\!)^\circ \cdot \mathsf{ker}\ (\!|k|\!)}_{N}$$

our aim is to convert $N = \Phi \cdot (\!|h|\!)^{\circ} \cdot \mathsf{ker}\ (\!|k|\!)$ into $[\!(R)\!]$ for some R. Below we shall need the extra condition that $\mathsf{ker}\ (\!|k|\!)$ is a congruence for h, that is,

$$h \cdot \mathsf{F}\ \mathsf{ker}\ (\!|k|\!) \subseteq \mathsf{ker}\ (\!|k|\!) \cdot h \tag{21}$$

holds, equivalent to

$$\mathsf{ker}\ (\!|k|\!) \cdot h = \mathsf{ker}\ (\!|k|\!) \cdot h \cdot (\mathsf{F}\ \mathsf{ker}\ (\!|k|\!)) \tag{22}$$

by Theorem 1. Another alternative to state (21) is

$$(\!|k|\!) \cdot h \leqslant \mathsf{F}\ (\!|k|\!) \tag{23}$$

meaning that $(\!|k|\!) \cdot h$ should be *less injective* (39) than $\mathsf{F}\ (\!|k|\!)$, see Appendix B. We shall also need the assumption:

$$\mathsf{F}\,(\mathsf{ker}\ (\!|k|\!)) \cdot \Lambda = \Lambda \cdot \mathsf{F}\,(\mathsf{ker}\ (\!|k|\!)) \tag{24}$$

We calculate:

$$\begin{array}{ll}
& \Phi \cdot (\!|h|\!)^{\circ} \cdot \mathsf{ker}\ (\!|k|\!) = [\!(R)\!] \\
\equiv & \quad \{ \text{ converses } \} \\
& \mathsf{ker}\ (\!|k|\!) \cdot (\!|h|\!) \cdot \Phi = (\!|R^{\circ}|\!) \\
\equiv & \quad \{ \ (\!|h|\!) \cdot \Phi = \Psi \cdot (\!|h|\!)\ (17), \text{ Theorem 2} \} \\
& \mathsf{ker}\ (\!|k|\!) \cdot \Psi \cdot (\!|h|\!) = (\!|R^{\circ}|\!) \\
\Leftarrow & \quad \{ \text{ fusion (27) } \} \\
& \mathsf{ker}\ (\!|k|\!) \cdot \Psi \cdot h = R^{\circ} \cdot \mathsf{F}\ (\mathsf{ker}\ (\!|k|\!) \cdot \Psi) \\
\Leftarrow & \quad \{ \text{ (19); functor F; Leibniz } \} \\
& \mathsf{ker}\ (\!|k|\!) \cdot h \cdot \Lambda = R^{\circ} \cdot \mathsf{F}\ \mathsf{ker}\ (\!|k|\!) \\
\equiv & \quad \{ \text{ (22) } \} \\
& \mathsf{ker}\ (\!|k|\!) \cdot h \cdot (\mathsf{F}\ \mathsf{ker}\ (\!|k|\!)) \cdot \Lambda = R^{\circ} \cdot \mathsf{F}\ \mathsf{ker}\ (\!|k|\!) \\
\Leftarrow & \quad \{ \text{ (24) ; Leibniz ; converses } \} \\
& R = \Lambda \cdot h^{\circ} \cdot \mathsf{ker}\ (\!|k|\!)
\end{array}$$

□

In summary, note how the original metaphorism $\Psi \cdot \mathsf{ker}\ (\!|k|\!)$ gets converted into a hylomorphism whose *divide* step is another metaphorism:

$$R = \Lambda \cdot ((\!|k|\!) \cdot h)^{\circ} \cdot (\!|k|\!) \tag{25}$$

That is, the "outer" metaphor which we started from (involving only T) disappears and gives place to an "inner" metaphor between inductive types W and T, whereby the optimization is internalized.

This "inner" metaphor is more interesting, as we can see by looking at an example of this reasoning.

6 Example: Quicksort

This section shows how the derivation of *quicksort* as given in e.g. [1] corresponds to the implementation strategy for metaphorisms given above, under the following instantiations:

- T is the usual finite list datatype with constructors (say) nil and cons, that is, $\mathsf{in_T} = [\mathsf{nil}\ , \mathsf{cons}]$.
- W is the binary tree data type whose base is $\mathsf{F}\,f = id + id \times (f \times f)$ and whose initial algebra is (say) $\mathsf{in_W} = [\mathsf{empty}\ , \mathsf{fork}]$.
- $(\!|k|\!) = bag$, the function which converts a list into the bag (multiset) of its elements.
- $\mathsf{ker}\ bag = Perm$, the list permutation relationship (the metaphor we start from).
- $(\!|h|\!) = \mathit{flatten}$, for $h = [\mathsf{nil}\ , inord]$ where $inord\ (a, (x, y)) = x +\!\!+ [a] +\!\!+ y$; that is, *flatten* is the binary tree into finite list surjection.
- Ψ filters ordered lists, $\Psi = (\!|[\mathsf{nil}\ , \mathsf{cons}] \cdot (id + \Theta)|\!)$ where $\Theta = mn?$ for $mn\ (x, xs) = \langle \forall\ x'\ :\ x' \in_{\mathsf{T}} xs :\ x' \geqslant x \rangle$, where $\in_{\mathsf{T}}$ denotes list membership; that is, predicate $mn\ (x, xs)$ ensures that list $x : xs$ is such that x is at most the minimum of xs, if it exists.

As seen in Sect. 5, we have to search for some partial identity $\Lambda = id + \Upsilon : id + id \times (\mathsf{T} \times \mathsf{T}) \to id + id \times (\mathsf{T} \times \mathsf{T})$ which, following (19), should be the weakest precondition for $[\mathsf{nil}\ , inord]$ to preserve ordered lists (Ψ):

$$
\begin{array}{ll}
& \Psi \cdot [\mathsf{nil}\ , inord] = [\mathsf{nil}\ , inord] \cdot (id + \Upsilon) \cdot (id + id \times (\Psi \times \Psi)) \\
\equiv & \quad \{\ \text{coproducts; } \Psi \cdot \mathsf{nil} = \mathsf{nil}\text{, since the empty list is trivially ordered}\ \} \\
& \Psi \cdot inord = inord \cdot \Upsilon \cdot (id \times (\Psi \times \Psi))
\end{array}
$$

Let *ord* and *wpl* be the predicates represented by partial identities Ψ and Υ, respectively, that is $\Psi = ord?$ and $\Upsilon = wpl?$. Unfolding *inord* we get the following pointwise calculation of weakest pre-condition *wpl*:

$$
\begin{array}{ll}
& ord\ (x +\!\!+ [a] +\!\!+ y) \\
\equiv & \quad \{\ \text{pointwise definition of ordered lists}\ \} \\
& (ord\ x) \wedge (ord\ y) \wedge \underbrace{\langle \forall\ b\ :\ b \in_{\mathsf{T}} x :\ b \leqslant a \rangle \wedge \langle \forall\ b\ :\ b \in_{\mathsf{T}} y :\ a \leqslant b \rangle}_{wpl\ (a,(x,y))}
\end{array}
$$

From this we get the following relational definition of the *divide* step (25) of the implementation,

$$
\begin{array}{l}
R : [A] \to 1 + A \times ([A] \times [A]) \\
R = (id + wpl?) \cdot (bag \cdot [\mathsf{nil}\ , inord])^{\circ} \cdot bag
\end{array}
\qquad (26)
$$

which we unfold as follows, by letting $R^\circ = [R_1^\circ , R_2^\circ]$ and using the converse of (26):

$$
\begin{array}{ll}
& [R_1^\circ , R_2^\circ] = bag^\circ \cdot (bag \cdot [\mathsf{nil} , inord]) \cdot (id + wpl?) \\
\equiv & \{ \ bag^\circ \cdot bag = Perm; \ Perm.\mathsf{nil} = \mathsf{nil}; \text{ converses } \} \\
& \begin{cases} R_1 = \mathsf{nil}^\circ \\ R_2 = wpl? \cdot inord^\circ \cdot Perm \end{cases}
\end{array}
$$

In summary, $y \ R \ x$ has the following meaning: either $x = [\,]$ and R yields the unique inhabitant of singleton type 1 (cf. R_1) or x is non-empty and R splits a permutation of x into two halves y and z separated by a "pivot" a, cf.

$$
(a, (y, z)) \ R_2 \ x = wpl \ (a \ (y, z)) \wedge (y +\!\!+ [a] +\!\!+ z) \ Perm \ x
$$

where wpl was calculated above. Pivot a can be taken from any position in the list. In the standard version, a is the head of x. There is, still, a check-list of proofs to discharge.

Ensuring Bi-ordered (virtual) Intermediate Trees. We start from the instantiation of (20) for this exercise,

$$
\mathsf{F}\, \mathit{flatten} \cdot (id + wp'?) = (id + wpl?) \cdot \mathsf{F}\, \mathit{flatten}
$$

where the goal is to find another weakest precondition wp' which is basically wpl "passed along" $\mathsf{F}\, \mathit{flatten}$ from lists to trees:

$$
\begin{array}{ll}
& (id \times (\mathit{flatten} \times \mathit{flatten})) \cdot wp'? = wpl? \cdot (id \times (\mathit{flatten} \times \mathit{flatten})) \\
\equiv & \{ \ (35) \ \} \\
& wp' = \mathsf{wp}(id \times (\mathit{flatten} \times \mathit{flatten}), wpl) \\
\equiv & \{ \text{ go pointwise } \} \\
& wp' \ (a, (t_1, t_2)) = wpl \ (a, (\mathit{flatten} \ t_1, \mathit{flatten} \ t_2)) \\
\equiv & \{ \text{ definition of } wpl \ \} \\
& wp' \ (a, (t_1, t_2)) = \begin{cases} \langle \forall \ b \ : \ b \in_{\mathsf{T}} (\mathit{flatten} \ t_1) : \ b \leqslant a \rangle \\ \langle \forall \ b \ : \ b \in_{\mathsf{T}} (\mathit{flatten} \ t_2) : \ a \leqslant b \rangle \end{cases} \\
\equiv & \{ \text{ define } \in_{\mathsf{W}} = \in_{\mathsf{T}} \cdot \mathit{flatten} \ \} \\
& wp' \ (a, (t_1, t_2)) = \langle \forall \ b \ : \ b \in_{\mathsf{W}} t_1 : \ b \leqslant a \rangle \wedge \langle \forall \ b \ : \ b \in_{\mathsf{W}} t_2 : \ a \leqslant b \rangle)
\end{array}
$$

Recall that $\Omega = id + wp'?$. In words, wp' in $\Phi = (\!|\mathsf{in}_{\mathsf{W}} \cdot \Omega|\!) = (\!|\mathsf{in}_{\mathsf{W}} \cdot (id + wp'?)|\!)$ ensures that the first part of the implementation, controlled by the *divide step* coalgebra R calculated above (26) yields trees which are *bi-ordered*. Trees with this property are known as *binary search trees* [6].

Preserving the Metaphor. Next we consider side condition (23), which instantiates to:

$$
\begin{array}{ll}
& bag \cdot [\mathsf{nil}\ , inord] \leqslant id + id \times (bag \times bag) \\
\Leftarrow & \quad \{\ \text{coproducts; (40)}\ \} \\
& bag \cdot \mathsf{nil} + bag \cdot inord \leqslant id + id \times (bag \times bag) \\
\equiv & \quad \{\ \text{(41) ; any } f \leqslant id \text{ [12] ; let } bag' = bag \cdot inord\ \} \\
& bag' \leqslant id \times (bag \times bag) \\
\equiv & \quad \{\ bag' \text{ loses more information than } id \times (bag \times bag)\ \} \\
& true
\end{array}
$$

In the last step we can easily observe that, while from $(a, (bag\ x, bag\ y))$ we can obtain $bag'\ (a, (x, y))$, the converse is false: bag' merges the multisets of x and y too quickly. Thus bag' is less injective than $id \times (bag \times bag)$.

Downto the Multiset Level. Finally, we have to check (24), for $\Lambda = id + \Upsilon = id + wpl?$:

$$
\begin{array}{ll}
& \mathsf{F}\, Perm \cdot \Lambda = \Lambda \cdot \mathsf{F}\, Perm \\
\equiv & \quad \{\ Perm = \mathsf{ker}\ bag\ ;\ \mathsf{F}\,(R^{\circ}) = (\mathsf{F}\, R)^{\circ}\ \} \\
& \mathsf{ker}\ (\mathsf{F}\ bag) \cdot \Lambda = \Lambda \cdot \mathsf{ker}\ (\mathsf{F}\ bag) \\
\equiv & \quad \{\ \mathsf{F}\, R = id + id \times (R \times R)\ ;\ \text{kernel of the sum (42)};\ \Lambda = id + wpl?\ \} \\
& \mathsf{ker}\ (id \times (bag \times bag)) \cdot wpl? = wpl? \cdot \mathsf{ker}\ (id \times (bag \times bag)) \\
\Leftarrow & \quad \{\ \text{(36), assuming that condition } q \text{ exists}\ \} \\
& wpl = \mathsf{wp}(id \times (bag \times bag), q)
\end{array}
$$

Thus we have to find post-condition q ensured by $id \times (bag \times bag)$ with wpl as weakest-precondition. We proceed as before:

$$
\begin{array}{ll}
& wpl\ (a, (x, y)) = q\ (a, (bag\ x, bag\ y)) \\
\equiv & \quad \{\ \text{unfold } wpl\ \} \\
& q\ (a, (bag\ x, bag\ y)) = \left\{ \begin{array}{l} \langle \forall\ b\ :\ b \in_{\mathsf{T}} x :\ b \leqslant a \rangle \\ \langle \forall\ b\ :\ b \in_{\mathsf{T}} y :\ a \leqslant b \rangle \end{array} \right. \\
\equiv & \quad \{\ \text{assume } \in_{\mathsf{B}} \text{ such that } \in_{\mathsf{T}} = \in_{\mathsf{B}} \cdot bag\ \} \\
& q\ (a, (bag\ x, bag\ y)) = \left\{ \begin{array}{l} \langle \forall\ b\ :\ b \in_{\mathsf{B}} (bag\ x) :\ b \leqslant a \rangle \\ \langle \forall\ b\ :\ b \in_{\mathsf{B}} (bag\ y) :\ a \leqslant b \rangle \end{array} \right. \\
\Leftarrow & \quad \{\ \text{substitution}\ \} \\
& q\ (a, (b_1, b_2)) = \left\{ \begin{array}{l} \langle \forall\ b\ :\ b \in_{\mathsf{B}} b_1 :\ b \leqslant a \rangle \\ \langle \forall\ b\ :\ b \in_{\mathsf{B}} b_2 :\ a \leqslant b \rangle \end{array} \right.
\end{array}
$$

□

Finally, multiset membership $\epsilon_{\mathsf{B}} = \in \cdot\, support$ can be obtained by taking multiset *supports*, whereby we land in standard set membership ($\in$). Thus we have a chain of memberships, from sets, to multisets, to finite lists and finally to binary (search) trees.

Note how this last proof of the check-list goes down to the very essence of sorting as a metaphorism: the attribute of a finite list which any sorting function is bound to preserve is the multiset (bag) of its elements.

7 Conclusions and Future Work

This paper identifies a pattern of relational specification, termed *metaphorism*, in which some kernel information of the input is preserved at the same time some form of optimization takes place towards the output. Text processing, sorting and representation changers are given as examples of metaphorisms. It then addresses the problem of refining metaphorisms into recursive programs.

The kind of metaphorism refinement proposed is known as *changing the virtual data structure*, whereby *divide & conquer* strategies can be introduced. The paper gives sufficient conditions for such implementations to be calculated in general, and gives the derivation of *quicksort* as example. This derivation can be regarded as a generalization of the reasoning about the same algorithm given in [1].

Altogether, the paper shows how such *divide & conquer* refinement strategies consist of replacing the "outer metaphor" of the starting specification (metaphorism) by a more implicit but more interesting "inner metaphor", which is at the heart of the implementation. In the quicksort example, the "outer metaphor" relates lists which permute each other, while the "inner metaphor" relates lists with binary search trees.

This research can be framed into the area of investigating how to manage or refine specification vagueness (non-determinism) by means of the "shrinking" combinator proposed in references [9, 13]. The pattern of shrinking addressed in the current paper is, however, far too restrictive: what is expected in general is shrinking over *preorders* which measure *progress* with respect to some other attribute, e.g. reducing the number of "out-of-order" entries in sorting, as presented in the introduction. Note how such metaphorisms expose the *variant/invariant* duality essential to program correctness and termination proofs, in their own way: there are two main attributes in the game, one is to be preserved (the essence of the metaphor, cf. *invariant*) while the other is to be mini(maxi)mized (the essence of the optimization, cf. *variant*).

This paper is intended as starting point for future work in exploiting the metaphorism concept in program derivation. Candidate case studies in program refactoring or text processing already pose significant challenges when compared to the sorting example given in the current paper. Comparative work is also welcome, in particular checking what benefits can be expected from regarding representation changers [5] from the metaphorism perspective, or (back to sorting) checking how the ideas of this paper combine with the work on parametric permutation functions by Henglein [4].

From the linguistics perspective, metaphorisms are *formal* metaphors and not exactly *cognitive* metaphors. But computer science is full of these as well, as its terminology (e.g. "stack", "pipe", "memory", "driver") amply shows. If a picture is worth a thousand words, perhaps a good metaphor is worth a thousand axioms?

Acknowledgements. The author wishes to thank the anonymous referees for their comments and suggestions. This work is funded by ERDF - European Regional Development Fund through the COMPETE Programme (operational programme for competitiveness) and by National Funds through the *FCT - Fundação para a Ciência e a Tecnologia* (Portuguese Foundation for Science and Technology) within project FCOMP-01-0124-FEDER-020537.

References

[1] Bird, R., de Moor, O.: Algebra of Programming. Series in Computer Science. Prentice-Hall International (1997)
[2] Doornbos, H., Backhouse, R., van der Woude, J.: A calculational approach to mathematical induction. TCS 179(1–2), 103–135 (1997)
[3] Freyd, P.J., Scedrov, A.: Categories, Allegories. Mathematical Library, vol. 39. North-Holland (1990)
[4] Henglein, F.: What is a sorting function? J. Logic and Algebraic Programming (JLAP) 78(5), 381–401 (2009)
[5] Hutton, G., Meijer, E.: Back to basics: Deriving representation changers functionally. Journal of Functional Programming 6(1), 181–188 (1996)
[6] Knuth, D.E.: The Art of Computer Programming, 2nd edn. Addison/Wesley (1997/1998); 3 volumes. First edition's dates: 1968 (volume 1), 1969 (volume 2) and 1973 (volume 3)
[7] Kozen, D.: Kleene algebra with tests. ACM Trans. Program. Lang. Syst. 19(3), 427–443 (1997)
[8] Lakoff, G., Johnson, M.: Metaphors we live by. University of Chicago Press, Chicago (1980)
[9] Mu, S.-C., Oliveira, J.N.: Programming from Galois connections. JLAP 81(6), 680–704 (2012)
[10] Oliveira, J.N.: Extended Static Checking by Calculation using the Pointfree Transform. In: Bove, A., Barbosa, L.S., Pardo, A., Pinto, J.S. (eds.) LerNet 2008. LNCS, vol. 5520, pp. 195–251. Springer, Heidelberg (2009)
[11] Oliveira, J.N.: On the 'A' that links the 'M's of maths, music and maps. Contributed talk to the 2013 CEHUM Autumn Colloquium XV(Maths and Comp. Science Panel), U. Minho, Braga, November 21-23 (2013)
[12] Oliveira, J.N.: A relation-algebraic approach to the "Hoare logic" of functional dependencies. JLAP 83(2), 249–262 (2014)
[13] Oliveira, J.N., Ferreira, M.A.: Alloy meets the algebra of programming: A case study. IEEE Trans. Soft. Eng. 39(3), 305–326 (2013)
[14] Richards, I.A.: The Philosophy of Rhetoric. Oxford University Press (1936)
[15] Swierstra, D., de Moor, O.: Virtual data structures. In: Möller, B., Partsch, H., Schuman, S. (eds.) Formal Program Development. LNCS, vol. 755, pp. 355–371. Springer, Heidelberg (1993)

A Background — Basic Definitions and Results of Relation Algebra

Relational Folds: this paper relies on basic properties of relational folds over a type T defined by initial algebra $\mathsf{T} \xleftarrow{\mathsf{in}} \mathsf{F\,T}$ on functor F, namely *fusion*

$$S \cdot (\!|R|\!) = (\!|Q|\!) \;\Leftarrow\; S \cdot R = Q \cdot \mathsf{F}\, S \tag{27}$$

and *cancellation*,

$$(\!|R|\!) \cdot \mathsf{in} = R \cdot \mathsf{F}\,(\!|R|\!) \tag{28}$$

both stemming from *universal property*:

$$X = (\!|R|\!) \;\equiv\; X \cdot \mathsf{in} = R \cdot \mathsf{F}\, X \tag{29}$$

Shunting rules for function f, where R, S are arbitrary binary relations:

$$f \cdot R \subseteq S \equiv R \subseteq f^{\circ} \cdot S \tag{30}$$

$$R \cdot f^{\circ} \subseteq S \equiv R \subseteq S \cdot f \tag{31}$$

Top relation — the topmost relation of its type can be defined by

$$!^{\circ} \cdot ! = \top \tag{32}$$

where $! : A \rightarrow 1$ is the constant function which maps every argument to the unique element of singleton type 1.

Pre/post restrictions where Φ and Ψ are partial identities:

$$R \cdot \Phi = R \cap \top \cdot \Phi \tag{33}$$

$$\Psi \cdot R = R \cap \Psi \cdot \top \tag{34}$$

Weakest Pre-conditions: let $p?$ and $q?$ be the partial identities for predicates p and q, respectively, and $\mathsf{wp}(f, q)$ denote the *weakest precondition* for function f to ensure post-condition q, that is: $\mathsf{wp}(f, q)\ x = q\ (f\ x)$. Then the following properties hold (proofs in Appendix B):

$$f \cdot p? = q? \cdot f \;\equiv\; p = \mathsf{wp}(f, q) \tag{35}$$

$$\mathsf{ker}\, f \cdot p? = p? \cdot \mathsf{ker}\, f \;\Leftarrow\; p = \mathsf{wp}(f, q) \tag{36}$$

"Shrinking" — let $B \xleftarrow{X,S} A$ and $B \xleftarrow{R} B$ be binary relations in universal property [9]:

$$X \subseteq S \upharpoonright R \;\equiv\; X \subseteq S \wedge X \cdot S^{\circ} \subseteq R \tag{37}$$

Coproducts: coproduct notation $C \xleftarrow{[R\,,S]} A + B$ denotes the junction of relations $C \xleftarrow{R} A$ and $C \xleftarrow{S} B$ (coproduct). Direct sum $R + S$ is the same as $[i_1 \cdot R\,, i_2 \cdot S]$, where i_1 and i_2 are the *injections* associated to datatype sums.

Injectivity Preorder: the kernel of a relation R,

$$\mathsf{ker}\ R \stackrel{\mathrm{def}}{=} R^{\circ} \cdot R \tag{38}$$

measures the *injectivity* of R. As in [12] we capture this by introducing a preorder on relations which compares their *injectivity*

$$R \leqslant S \equiv \mathsf{ker}\ S \subseteq \mathsf{ker}\ R \tag{39}$$

and satisfies, among many others, the following properties:

$$[R\ , S] \leqslant R + S \tag{40}$$

$$R + S \leqslant P + Q \equiv R \leqslant P \wedge S \leqslant Q \tag{41}$$

Moreover:

$$\mathsf{ker}\ (R + S) = \mathsf{ker}\ R + \mathsf{ker}\ S \tag{42}$$

$$\mathsf{ker}\ (R \times S) = \mathsf{ker}\ R \times \mathsf{ker}\ S \tag{43}$$

B Proofs of Auxiliary Results

Proof of (5), where $f = (\!| k |\!)$:

$$
\begin{array}{ll}
& \mathsf{ker}\ f = (\!| \mathsf{ker}\ f \cdot \mathsf{in} |\!) \\
\equiv & \quad \{ \text{ inline definition } f = (\!| k |\!)\ ;\ \mathsf{ker}\ f = f^{\circ} \cdot f \ \} \\
& (\!| k |\!)^{\circ} \cdot (\!| k |\!) = (\!| (\!| k |\!)^{\circ} \cdot (\!| k |\!) \cdot \mathsf{in} |\!) \\
\Leftarrow & \quad \{ \text{ fusion (27) } \} \\
& (\!| k |\!)^{\circ} \cdot k = (\!| k |\!)^{\circ} \cdot (\!| k |\!) \cdot \mathsf{in} \cdot \mathsf{F}\,(\!| k |\!)^{\circ} \\
\equiv & \quad \{ \text{ cancellation (28) } \} \\
& (\!| k |\!)^{\circ} \cdot k = (\!| k |\!)^{\circ} \cdot k \cdot \mathsf{F}\,(\!| k |\!) \cdot \mathsf{F}\,(\!| k |\!)^{\circ} \\
\Leftarrow & \quad \{ \text{ factor } (\!| k |\!)^{\circ} \cdot k \text{ out (Leibniz) ; functor } \mathsf{F} \ \} \\
& id = \mathsf{F}\,((\!| k |\!) \cdot (\!| k |\!)^{\circ}) \\
\equiv & \quad \{ f = (\!| k |\!)\ ;\ \mathsf{img}\ f = f \cdot f^{\circ} = id \text{ assuming } f \text{ surjective } \} \\
& id = \mathsf{F}\, id
\end{array}
$$

$$\equiv \quad \{ \text{ functor } \mathsf{F}\text{: } \mathsf{F}\, id = id \ \}$$

$$true$$

□

Proof of Theorem 1:

$$R \cdot h = R \cdot h \cdot (\mathsf{F}\ R)$$

$$\equiv \quad \{ \ R \cdot h \subseteq R \cdot h \cdot (\mathsf{F}\ R) \text{ holds by } id \subseteq \mathsf{F}\ R\text{, since } id \subseteq R \ \}$$

$$R \cdot h \cdot (\mathsf{F}\ R) \subseteq R \cdot h$$

$$\equiv \quad \{ \text{ the lower } R \text{ can be cancelled, since } R \text{ is an equivalence (see below) } \}$$

$$h \cdot (\mathsf{F}\ R) \subseteq R \cdot h$$

□

The last step can be justified by assuming the function k_R which maps every object to its equivalence class, as dictated by R. Then $R = \mathsf{ker}\ k_R$ and, for any suitably typed relations X and Y:

$$R \cdot X \subseteq R \cdot Y$$

$$\equiv \quad \{ \text{ inline } R = \mathsf{ker}\ k_R \ \}$$

$$\mathsf{ker}\ k_R \cdot X \subseteq \mathsf{ker}\ k_R \cdot Y$$

$$\equiv \quad \{ \ \mathsf{ker}\ k_R = k_R^\circ \cdot k_R \text{ ; shunting (30) } \}$$

$$k_R \cdot k_R^\circ \cdot k_R \cdot X \subseteq k_R \cdot Y$$

$$\equiv \quad \{ \ f \cdot f^\circ \cdot f = f \text{ (difunctionality) } \}$$

$$k_R \cdot X \subseteq k_R \cdot Y$$

$$\equiv \quad \{ \text{ shunting (30) ; } R = \mathsf{ker}\ k_R \ \}$$

$$X \subseteq R \cdot Y$$

□

Proof of (16):

$$X \subseteq R \upharpoonright (\Phi \cdot \top)$$

$$\equiv \quad \{ \ (37) \ \}$$

$$X \subseteq R \wedge X \cdot R^\circ \subseteq \Phi \cdot \top$$

$$\equiv \quad \{ \ (32) \text{ ; shunting (31) ; converses } \}$$

$$X \subseteq R \wedge X \cdot (! \cdot R)^\circ \subseteq \Phi \cdot !^\circ$$

$$\equiv \quad \{ \text{ assume } R \text{ entire } \}$$

$$X \subseteq R \wedge X \cdot !^\circ \subseteq \Phi \cdot !^\circ$$

$$
\begin{array}{ll}
\equiv & \{ \text{ shunting (31) ; (32) } \} \\
& X \subseteq R \wedge X \subseteq \Phi \cdot \top \\
\equiv & \{ \text{ (34) } \} \\
& X \subseteq \Phi \cdot R
\end{array}
$$

□

Proof that (23) is equivalent to (21), where g abbreviates $(\!|k|\!)$:

$$
\begin{array}{ll}
& h \cdot \mathsf{F}\,(\mathsf{ker}\ g) \subseteq \mathsf{ker}\ g \cdot h \\
\equiv & \{ \ \mathsf{F}\,(R^\circ) = (\mathsf{F}\,R)^\circ\text{; shunting (30) ; kernel (38) } \} \\
& \mathsf{ker}\,(\mathsf{F}\ g) \subseteq h^\circ \cdot g^\circ \cdot g \cdot h \\
\equiv & \{ \text{ kernel (38) ; injectivity preorder (39) } \} \\
& g \cdot h \leqslant \mathsf{F}\,g
\end{array}
$$

□

Proof of (35): abbreviating $\mathsf{wp}(f,q)$ by w, $p = \mathsf{wp}(f,q)$ is the same as $p? = w? = f^\circ \cdot q? \cdot f \cap id = \mathsf{dom}\,(q? \cdot f)$, where $\mathsf{dom}\,R$ denotes the *domain* of definition of relation R.

Step (⇒): $f \cdot p? = q? \cdot f$ is stronger than $f \cdot p? \subseteq q? \cdot f$ which immediately grants $p? \subseteq w?$. So we only have to ensure $w? \subseteq p?$:

$$
\begin{array}{ll}
& w? \subseteq p? \\
\equiv & \{ \ w? = f^\circ \cdot q? \cdot f \cap id \ \} \\
& f^\circ \cdot q? \cdot f \cap id \subseteq p? \\
\equiv & \{ \ f \cdot p? = q? \cdot f \text{ assumed } \} \\
& f^\circ \cdot f \cdot p? \cap id \subseteq p? \\
\equiv & \{ \text{ trivia } \} \\
& (f^\circ \cdot f \cap id) \cdot p? \subseteq p? \\
\Leftarrow & \{ \text{ monotonicity } \} \\
& f^\circ \cdot f \cap id \subseteq id \\
\equiv & \{ \ R \cap S \subseteq S \ \} \\
& \mathit{true}
\end{array}
$$

□

Step (⇐): $p? \subseteq w?$ is equivalent to $f \cdot p? \subseteq q? \cdot f$. We are left with:

$$
\begin{array}{ll}
& q? \cdot f \subseteq f \cdot p? \Leftarrow p? = w? \\
\equiv & \{ \text{ substitution } \} \\
& q? \cdot f \subseteq f \cdot w? \\
\equiv & \{ R \cdot \mathsf{dom}\, R = R \} \\
& (q? \cdot f) \cdot \mathsf{dom}\,(q? \cdot f) \subseteq f \cdot w? \\
\equiv & \{ w? = \mathsf{dom}\,(q? \cdot f) \} \\
& q? \cdot f \cdot w? \subseteq f \cdot w? \\
\Leftarrow & \{ q? \subseteq id; \text{ monotonicity } \} \\
& true
\end{array}
$$

□

Proof of (36):

$$
\begin{array}{ll}
& \mathsf{ker}\, f \cdot p? \\
= & \{ \text{ kernel (38) ; (35) since } p = \mathsf{wp}(f, q) \text{ is assumed } \} \\
& f^{\circ} \cdot q? \cdot f \\
= & \{ \text{ converses ; partial identities } \} \\
& (q? \cdot f)^{\circ} \cdot f \\
= & \{ \text{ again (35) ; converses ; kernel (38) } \} \\
& p? \cdot \mathsf{ker}\, f
\end{array}
$$

□

Relational Mathematics for Relative Correctness*

Jules Desharnais[1], Nafi Diallo[2], Wided Ghardallou[3], Marcelo F. Frias[4], Ali Jaoua[5], and Ali Mili[2]

[1] Université Laval, Québec City, Canada
[2] New Jersey Institute of Technology, Newark, NJ, USA
[3] University of Tunis El Manar, Tunis, Tunisia
[4] Instituto Tecnológico de Buenos Aires (ITBA), and CONICET, Argentina
[5] Qatar University, Qatar

jules.desharnais@ift.ulaval.ca, ncd8@njit.edu, wided.ghardallou@gmail.com, mfrias@itba.edu.ar, jaoua@qu.edu.qa, ali.mili@njit.edu

Abstract. In earlier work, we had presented a definition of software fault as being any feature of a program that admits a substitution that would make the program more-correct. This definition requires, in turn, that we define the concept of relative correctness, i.e., what it means for a program to be more-correct than another with respect to a given specification. In this paper we broaden our earlier definition to encompass non-deterministic programs, or non-deterministic representations of programs; also, we study the mathematical properties of the new definition, most notably its relation to the refinement ordering, as well as its algebraic properties with respect to the refinement lattice.

Keywords: Absolute correctness, relative correctness, refinement ordering, refinement lattice, faults, fault removal.

1 Introduction

1.1 What Is a Program Fault?

Our work stems from trying to define what is a software fault; usually we characterize a fault at some location in a program as a feature of the program that differs from what we believe it should be at that location. But this characterization presumes that we know with great precision and great certainty what the program ought to be doing at every location throughout its source code. Needless to say, such a presumption is unrealistic, since it is difficult in general to have a precise, complete, vetted specification of the overall software product, much less a specification of every small part thereof. Also, it is very common to find cases

* Acknowledgement: This publication was made possible by a grant from the Qatar National Research Fund, NPRP04-1109-1-174. Its contents are solely the responsibility of the authors and do not necessarily represent the official views of the QNRF.

W. Kahl et al. (Eds.): RAMiCS 2015, LNCS 9348, pp. 191–208, 2015.
DOI: 10.1007/978-3-319-24704-5_12

where the same faulty behavior of the program can be traced back to more than one possible feature, involving more than one location in the source code. In [9] we had defined a fault in a software product as any feature (be it a lexical token, a statement, a condition, a contiguous block, a set of non-contiguous statements, etc.) that admits a substitute that would make the program strictly more-correct, in a sense to be defined. Such a definition, once we decide what it means to be more-correct, has the advantage that it does not depend on a detailed knowledge of the design of the software product, and that it characterizes faults without making any assumption about whether other parts of the program are, or are not, correct. It is worth noting that this definition of a fault, like any definition we could think of, is based on an implicit level of granularity of the program; this level of granularity corresponds to the degree of precision with which we want to isolate faults. At one extreme in the scale of granularity, we could consider lexical tokens; at the opposite extreme, we could consider the whole program as a monolith; most programmers think of faults at the granularity level of an assignment statement or equivalent syntactic units.

1.2 Deterministic and Non-deterministic Programs

In [2] we briefly discuss the properties of relative correctness, and its implications for software engineering processes, such as software testing, software repair, software faultiness analysis and in [3] we discuss the implication of relative correctness for software design. In all of our discussions in [2, 9, 3], we consider deterministic programs. In this paper we wish to lift the hypothesis of determinacy, and define relative correctness in the broader context of possibly non-deterministic programs. One may want to ask: why do we need to define relative correctness for non-deterministic programs if most programming languages of interest are deterministic? There are several reasons why we may want to do so:

- We may want to apply the concept of relative correctness, not only to finished software products, but also to partially defined intermediate designs (as appear in a stepwise refinement process).
- Non-determinacy is a convenient tool to model deterministic programs whose detailed behavior is difficult to capture, unknown, or irrelevant to a particular analysis.
- We may want to reason about the relative correctness of programs without having to compute their functions is all their minute details.

As an illustration, we consider the space S defined by the following declarations:

```
a:  array [1..N] of itemtype; x:  itemtype;
low, high: 0..N+1;  found:  boolean;
```

where `itemtype` is some data type that represents an ordered set, and we consider the following specification R and program P:

$$R = \{(s, s')|\mathit{found}' \Leftrightarrow (\exists i : 1 \leq i \leq N : x = a[i])\}, \quad (1)$$

```
P:  {low=2; high=N;  found=false;
     while (low<=high)
        {indextype m=(low+high)/2;
         if (x<a[m]) {high=m-1;}
         else  if (x>a[m]) {low=m+1;}
               else {found=true; low=m+1; high=m-1;}}}.
```

We would like to think of the statement `low=2` as a fault, and that replacing this statement by `low=1` would produce a more-correct program; but to prove these claims using the original definition of relative correctness, we would have to compute the function of this program, i.e. determine the final values of all the program variables as a function of the initial values. But computing the final values of variables `low` and `high` is at the same time very difficult (as they depend on the position of `x` with respect to the array cells) and rather irrelevant (as they play an auxiliary role with respect to the function of the program). The interest of non-deterministic relations is that they enable us to focus on relevant functional aspects of a program, at the exclusion of complex and/or uninteresting details.

In section 2 we introduce some relational definitions and notations, which we use in section 3 to introduce a definition of relative correctness for non-deterministic programs; and in sections 4 and 5 we explore the properties of relative correctness, most notably its relation to the refinement ordering (section 4) and its relation to the refinement lattice (section 5). Finally, in section 6 we summarize our findings, compare them to related work, and sketch directions for future research. All propositions of the article were additionally proved with the theorem prover Prover9 [8].

2 Mathematics for Program Analysis

2.1 Relational Notations

In this section, we introduce some elements of relational mathematics that we use in the remainder of the paper to carry out our discussions. We assume the reader familiar with relational algebra [11, 12]. Dealing with programs, we represent sets using a programming-like notation, by introducing variable names and associated data type (sets of values). For example, if we represent set S by the variable declarations

$$x : X; y : Y; z : Z,$$

then S is the Cartesian product $X \times Y \times Z$. Elements of S are denoted in lower case s, and are triplets of elements of X, Y, and Z. Given an element s of S, we represent its X-component by $x(s)$, its Y-component by $y(s)$, and its Z-component by $z(s)$. When no risk of ambiguity exists, we may write x to represent $x(s)$, and x' to represent $x(s')$, letting the references to s and s' be implicit.

A relation on S is a subset of the Cartesian product $S \times S$; given a pair (s, s') in R, we say that s' is an *image* of s by R. Special relations on S include the *universal* relation $L = S \times S$, the *identity* relation $I = \{(s, s')|s' = s\}$, and the *empty* relation $\phi = \{\}$. Operations on relations (say, R and R') include the set theoretic operations of *union* ($R \cup R'$), *intersection* ($R \cap R'$), *difference* ($R \setminus R'$) and *complement* ($\overline{R}$). They also include the *relational product*, denoted by $R \circ R'$ (or RR', for short) and defined by

$$RR' = \{(s, s')|\exists s'' : (s, s'') \in R \wedge (s'', s') \in R'\}.$$

The *power* of relation R is denoted by R^n, for a natural number n, and defined by $R^0 = I$, and for $n > 0$, $R^n = R \circ R^{n-1}$. The *reflexive transitive closure* of relation R is denoted by R^* and defined by $R^* = \{(s, s')|\exists n \geq 0 : (s, s') \in R^n\}$. The *converse* of relation R is the relation denoted by $\widehat{R}$ and defined by

$$\widehat{R} = \{(s, s')|(s', s) \in R\}.$$

Finally, the *domain* of a relation R is defined as the set $dom(R) = \{s|\exists s' : (s, s') \in R\}$, and the *range* of relation R is defined as the domain of $\widehat{R}$.

A relation R is said to be *reflexive* if and only if $I \subseteq R$, *antisymmetric* if and only if $R \cap \widehat{R} \subseteq I$, *asymmetric* if and only if $R \cap \widehat{R} = \phi$, and *transitive* if and only if $RR \subseteq R$. A relation is said to be a *partial ordering* if and only if it is reflexive, antisymmetric, and transitive. Also, a relation R is said to be *total* if and only if $I \subseteq R\widehat{R}$, and is said to be *deterministic* (or: a *function*) if and only if $\widehat{R}R \subseteq I$. In this paper we use a property to the effect that two functions f and f' are identical if and only if $f \subseteq f'$ and $f'L \subseteq fL$. A relation R is said to be a *vector* if and only if $RL = R$; a vector on space S is a relation of the form $R = A \times S$, for some subset A of S; we use vectors to represent subsets of S, and we may by abuse of notation write $s \in R$ to mean $s \in A$; in particular, we use the product RL as a relational representation of the domain of R.

The following laws will be used in the forthcoming proofs. The first one is a special case of the Dedekind rule [11, 12].

$$PQ \cap R \subseteq P(Q \cap \widehat{P}R) \tag{2}$$

$$(PL \cap Q)R = PL \cap QR \tag{3}$$

2.2 A Refinement Calculus

Throughout this paper, we interpret relations as program specifications or as programs and we may use the same symbol to refer to a program and to the relation that the program defines on its space. Given two relations R and R', we say that R' *refines* R (abbrev: $R' \sqsupseteq R$) if and only if $RL \cap R'L \cap (R \cup R') = R$. We find that this condition is equivalent to $RL \subseteq R'L \ \wedge \ RL \cap R' \subseteq R$. We also find that the refinement relation is a partial ordering and that it has lattice-like properties, in the following sense [1]:

- Any two relations R and R' have a greatest lower bound, which we denote by $R \sqcap R'$ and to which we refer as the *meet* of R and R'. Also, we find: $R \sqcap R' = RL \cap R'L \cap (R \cup R')$.
- Given two relations R and R', we define the *join* of R and R' (denoted by $R \sqcup R'$) as:
$$(\overline{RL} \cap R') \cup (\overline{R'L} \cap R) \cup (R \cap R').$$
- Two relations R and R' admit a least upper bound if and only if they satisfy the condition $RL \cap R'L = (R \cap R')L$, which we call the *consistency condition.*
- The least upper bound of two relations that satisfy the consistency condition is their join. In other words, the join of two relations always exists, but it equals their least upper bound only if they meet the consistency condition.

Let R and P be two relations on space S; we say that P (interpreted as a possibly non-deterministic program) is *correct* with respect to R (interpreted as a specification) if and only if P refines R. We have a proposition (due to [10]) to the effect that if P is deterministic, then P is correct with respect to R if and only if $(R \cap P)L = RL$.

3 Relative Correctness of Non-deterministic Programs

3.1 Background

In this section, we briefly summarize our main findings with regards to deterministic programs, so as to convey our expectations with respect to non-deterministic programs. All our discussions about correctness, relative correctness, and faults refer to a relational specification, which we usually denote by R. We denote candidate programs by P and P', and for the sake of convenience we make no distinction between a program (as a syntactic representation) and the function or relation that the program defines on its space. Given a program P and a specification R, we find that the domain of $R \cap P$ represents the set of initial states for which P delivers an output that is considered correct with respect to R; we refer to this set as the *competence domain* of P with respect to R. A (deterministic) program P' is said to be more-correct than a (deterministic) program P with respect to specification R if and only if it has a larger competence domain; we denote this by $P' \sqsupseteq_R P$. Then we define a fault f in a program P as any feature of the program that admits a substitute that would make the program *strictly* more-correct (i.e. yield a strictly larger competence domain). Among the most salient properties we have found for the property of relative correctness, we cite:

- *Relative correctness is reflexive and transitive but not antisymmetric.* Reflexivity and transitivity stem from the reflexivity and transitivity of set inclusion, as it applies to competence domains. Relative correctness is not antisymmetric because programs may have the same competence domain and still be distinct, due to the non-determinacy of specifications. This property holds for non-deterministic programs (and the non-deterministic version of relative correctness), as we see in proposition 3.3.

- *Relative correctness culminates in absolute correctness.* A correct program with respect to specification R is more-correct with respect to R than any candidate program. Indeed, since we find (section 2.2) that a correct P satisfies the condition $dom(R \cap P) = dom(R)$, then a correct program has a maximal competence domain. This property holds for non-deterministic programs (and the non-deterministic version of relative correctness), as we see in proposition 4.2.
- *Relative correctness logically implies enhanced reliability.* We find that if a program is more-correct than another, then it is necessarily more reliable. Indeed, if we measure reliability by the probability of successful execution modulo some probability distribution θ of input states, then the probability of successful execution of a program P modulo probability distribution θ is the integral (or for discrete models, the sum) of θ over the competence domain of P; clearly, the larger the competence domain the higher the probability. We do not prove that this property survives the transition to non-deterministic programs, though we suspect that it does, modulo an angelic interpretation of competence domain (whereby a non-deterministic program is considered to behave correctly as soon as it provides at least one correct outcome with respect to R).
- *Relative Correctness and Refinement.* One of the most interesting properties we have found about relative correctness is its relationship to refinement. In [9], we find that a program P' refines a program P if and only if P' is more-correct than P with respect to any specification. Formally,

$$P' \sqsupseteq P \Leftrightarrow (\forall R : P' \sqsupseteq_R P).$$

This property does not hold for non-deterministic programs (and the non-deterministic version of relative correctness), but we have an interesting substitute in proposition 4.3.

3.2 Definitions

The purpose of this section is to define the concept of relative correctness for arbitrary programs, that are not necessarily deterministic.

In seeking to generalize the property of relative correctness to non-deterministic programs, we consider two requirements: first, the formula for non-deterministic programs must be equivalent to the formula we already have for deterministic programs when the programs are deterministic; second, we wish to preserve the properties we have listed above, most notably the relation between relative correctness and refinement. We submit the following definition.

Definition 3.1. *Let R, P and P' be relations on space S. We say that P' is more-correct than P with respect to R (abbrev: $P' \sqsupseteq_R P$) if and only if:*

$$(R \cap P)L \subseteq (R \cap P')L \ \wedge \ (R \cap P)L \cap \overline{R} \cap P' \subseteq P.$$

Interpretation: P' is more-correct than P with respect to R if and only if it has a (n equal or) larger competence domain, and for the elements in the competence domain of P, program P' has (the same or) fewer images that violate R than P does. Even though a more appropriate name for this relation is *at-least-as-correct-as*, we use the shorter version *more-correct-than.* As an illustration, we consider the set $S = \{0, 1, 2, 3, 4, 5, 6, 7\}$ and we let R, P and P' be defined by the following Boolean matrices:

$$\underset{R}{\begin{array}{c|cccccccc|} & 0&1&2&3&4&5&6&7 \\ \hline 0&1&1&0&0&0&0&0&0\\ 1&1&1&1&0&0&0&0&0\\ 2&0&1&1&1&0&0&0&0\\ 3&0&0&1&1&1&0&0&0\\ 4&0&0&0&1&1&1&0&0\\ 5&0&0&0&0&1&1&1&0\\ 6&0&0&0&0&0&1&1&1\\ 7&0&0&0&0&0&0&1&1 \end{array}} \quad \underset{P}{\begin{array}{c|cccccccc|} & 0&1&2&3&4&5&6&7 \\ \hline 0&0&0&1&1&0&0&0&0\\ 1&0&0&0&1&1&0&0&0\\ 2&1&1&0&0&0&0&0&0\\ 3&0&1&1&0&0&0&0&0\\ 4&0&0&1&1&0&0&0&0\\ 5&0&0&0&1&1&0&0&0\\ 6&0&0&0&1&1&0&0&0\\ 7&0&0&0&0&1&1&0&0 \end{array}} \quad \underset{P'}{\begin{array}{c|cccccccc|} & 0&1&2&3&4&5&6&7 \\ \hline 0&0&0&1&1&0&0&0&0\\ 1&0&0&1&1&0&0&0&0\\ 2&1&0&0&1&0&0&0&0\\ 3&0&1&0&0&1&0&0&0\\ 4&0&0&1&0&0&1&0&0\\ 5&0&0&0&1&0&0&1&0\\ 6&0&0&0&0&1&0&0&1\\ 7&0&0&0&0&1&1&0&0 \end{array}}$$

From these definitions, we compute:

$$\underset{(R \cap P)L}{\begin{array}{c|cccccccc|} & 0&1&2&3&4&5&6&7 \\ \hline 0&0&0&0&0&0&0&0&0\\ 1&0&0&0&0&0&0&0&0\\ 2&1&1&1&1&1&1&1&1\\ 3&1&1&1&1&1&1&1&1\\ 4&1&1&1&1&1&1&1&1\\ 5&1&1&1&1&1&1&1&1\\ 6&0&0&0&0&0&0&0&0\\ 7&0&0&0&0&0&0&0&0 \end{array}} \quad \underset{R \cap P'}{\begin{array}{c|cccccccc|} & 0&1&2&3&4&5&6&7 \\ \hline 0&0&0&0&0&0&0&0&0\\ 1&0&0&1&0&0&0&0&0\\ 2&0&0&0&1&0&0&0&0\\ 3&0&0&0&0&1&0&0&0\\ 4&0&0&0&0&0&1&0&0\\ 5&0&0&0&0&0&0&1&0\\ 6&0&0&0&0&0&0&0&1\\ 7&0&0&0&0&0&0&0&0 \end{array}} \quad \underset{(R \cap P')L}{\begin{array}{c|cccccccc|} & 0&1&2&3&4&5&6&7 \\ \hline 0&0&0&0&0&0&0&0&0\\ 1&1&1&1&1&1&1&1&1\\ 2&1&1&1&1&1&1&1&1\\ 3&1&1&1&1&1&1&1&1\\ 4&1&1&1&1&1&1&1&1\\ 5&1&1&1&1&1&1&1&1\\ 6&1&1&1&1&1&1&1&1\\ 7&0&0&0&0&0&0&0&0 \end{array}} \quad \underset{(R \cap P)L \cap \overline{R} \cap P'}{\begin{array}{c|cccccccc|} & 0&1&2&3&4&5&6&7 \\ \hline 0&0&0&0&0&0&0&0&0\\ 1&0&0&0&0&0&0&0&0\\ 2&1&0&0&0&0&0&0&0\\ 3&0&1&0&0&0&0&0&0\\ 4&0&0&1&0&0&0&0&0\\ 5&0&0&0&1&0&0&0&0\\ 6&0&0&0&0&0&0&0&0\\ 7&0&0&0&0&0&0&0&0 \end{array}}$$

We leave it to the reader to check that the two clauses of Definition 3.1 are satisfied. Program P' is more-correct than program P with respect to R because it has a larger competence domain ($\{1, 2, 3, 4, 5, 6\}$ vs $\{2, 3, 4, 5\}$) and because on the competence domain of P, program P' generates no incorrect output unless P also generates it.

As a second illustration, we consider the binary search program introduced above, and we capture an abstraction of its semantics by the following relation:

$$Q = \{(s, s')|a' = a \wedge x' = x \wedge (found' \Rightarrow (\exists i : 2 \leq i \leq N : x = a[i]))\}.$$

We let P' be the program obtained from P by replacing `low=2` by `low=1`, and we find as corresponding abstraction:

$$Q' = \{(s, s')|a' = a \wedge x' = x \wedge (found' \Rightarrow (\exists i : 1 \leq i \leq N : x = a[i]))\}.$$

If we knew that array a is sorted, the implications in Q and Q' could be strengthened to equivalences. We find that Q' is more-correct than Q with respect to R (see (1)). The first clause stems from

$$R \cap Q' = \{(s,s')|a' = a \wedge x' = x \wedge (found' \Leftrightarrow (\exists i : 1 \leq i \leq N : x = a[i]))\},$$

whence $(R \cap Q')L = L$. For the second clause, it suffices to show $\overline{R} \cap Q' \subseteq Q$, which follows from the definitions of Q, Q' and R.

3.3 Properties

The first property we want to check about this definition is that it generalizes the definition given in [9], which provides that a deterministic program P' is more-correct than a deterministic program P with respect to a specification R if and only if $(R \cap P)L \subseteq (R \cap P')L$. We have the following proposition.

Proposition 3.2. *Let R, P and P' be relations on S. If P' is deterministic then the conditions $(R \cap P)L \subseteq (R \cap P')L$ and $P' \sqsupseteq_R P$ are logically equivalent.*

Proof. The condition $P' \sqsupseteq_R P$ clearly implies $(R \cap P)L \subseteq (R \cap P')L$, hence we focus our attention on the reverse implication. We let P' be a function, we assume that P and P' satisfy the condition $(R \cap P)L \subseteq (R \cap P')L$, and we aim to prove the condition $(R \cap P)L \cap \overline{R} \cap P' \subseteq P$. We proceed as follows:

$$
\begin{array}{ll}
 & (R \cap P)L \cap \overline{R} \cap P' \subseteq P \\
\Leftarrow & \qquad \{\text{Since } (R \cap P)L \subseteq (R \cap P')L\} \\
 & (R \cap P')L \cap \overline{R} \cap P' \subseteq P \\
\Leftarrow & \qquad \{\text{Boolean algebra}\} \\
 & (R \cap P')L \cap P' \subseteq R \\
\Leftarrow & \qquad \{\text{By (2)}\} \\
 & (R \cap P')(L \cap \widehat{(R \cap P')}P') \subseteq R \\
\Leftarrow & \qquad \{\text{Boolean algebra}\} \\
 & (R \cap P')\widehat{(R \cap P')}P' \subseteq R \\
\Leftarrow & \qquad \{\text{Boolean algebra, monotonicity}\} \\
 & R\widehat{P'}P' \subseteq R \\
\Leftarrow & \qquad \{P' \text{ is deterministic, hence } \widehat{P'}P' \subseteq I\} \\
 & R \subseteq R,
\end{array}
$$

which is a tautology. **qed**

As we see, this proof assumes that P' is deterministic but imposes no condition on P: indeed the second clause in the definition of relative correctness imposes a condition restricting the possible incorrect behavior of P' on the competence domain of P, where P' is known to behave correctly (since it has a larger competence domain than P). Because P' is deterministic, it assigns only one image to any element of the competence domain of P, which is known to be a correct image; hence there is no scope for P' to associate an incorrect image. So that if P' satisfies the first clause of the definition of relative correctness and is deterministic, then it necessarily satisfies the second clause, regardless of relation P.

Proposition 3.3. *The relative correctness relation with respect to a given specification is reflexive and transitive.*

Proof. Reflexivity is trivial. To prove transitivity, we consider relations R, P, P' and P'', and we assume that P' is more-correct than P with respect to R, and that P'' is more-correct than P' with respect to R. The condition $(R \cap P)L \subseteq (R \cap P'')L$ stems readily from the hypothesis. We focus on the condition $(R \cap P)L \cap \overline{R} \cap P'' \subseteq P$, which we prove as follows:

$(R \cap P)L \cap \overline{R} \cap P'' \subseteq P$

$\Leftrightarrow$ {Hypothesis $P' \sqsupseteq_R P$, whence $(R \cap P)L \subseteq (R \cap P')L$, and Boolean algebra}

$(R \cap P)L \cap (R \cap P')L \cap \overline{R} \cap P'' \subseteq P$

$\Leftrightarrow$ {Hypothesis $P'' \sqsupseteq_R P'$, whence $(R \cap P')L \cap \overline{R} \cap P'' \subseteq P'$, and Boolean algebra}

$(R \cap P)L \cap (R \cap P')L \cap \overline{R} \cap P'' \cap P' \subseteq P$

$\Leftarrow$ {Boolean algebra}

$(R \cap P)L \cap \overline{R} \cap P' \subseteq P$,

which holds, by hypothesis. **qed**

Since $\sqsupseteq_R$ is reflexive and transitive, it is a preorder; we use this preorder to define an equivalence relation, as follows:

Definition 3.4. *Two relations P and P' are said to be* equally correct *with respect to specification R (abbrev: $P \equiv_R P'$) if and only if $P \sqsupseteq_R P'$ and $P' \sqsupseteq_R P$.*

For deterministic relations P and P', equal correctness simply means having the same competence domain; the following proposition characterizes equal correctness for arbitrary (not necessarily deterministic) relations.

Proposition 3.5. *Let R, P and P' be arbitrary relations on space S. Then*

$$(P \equiv_R P') \Leftrightarrow (R \cap P)L = (R \cap P')L \wedge (R \cap P)L \cap \overline{R} \cap P = (R \cap P')L \cap \overline{R} \cap P'.$$

Proof. From $P \equiv_R P'$ we infer readily that P and P' have the same competence domain with respect to R. Also, from $(R \cap P)L \cap \overline{R} \cap P' \subseteq P$ and $(R \cap P)L = (R \cap P')L$ we infer:

$$(R \cap P')L \cap \overline{R} \cap P' \subseteq (R \cap P)L \cap \overline{R} \cap P.$$

By interchanging P and P' and combining the two results, we find:

$$(R \cap P')L \cap \overline{R} \cap P' = (R \cap P)L \cap \overline{R} \cap P.$$

The converse implication is trivial, if we replace equality by inclusion, and note that an intersection is a subset of its terms. **qed**

Proposition 3.5 characterizes equivalence classes of relation $\equiv_R$ as being relations that share a common competence domain and a common set of incorrect outputs with respect to specification R. The following proposition singles out a representative element of each class, and shows that it is the least refined element of the class.

Proposition 3.6. *Let R and P be relations on space S. Then*

$$\rho_R(P) = (R \cap P)L \cap (R \cup P)$$

is in the same equivalence class of $\equiv_R$ as P and the domain of $\rho_R P$ is equal to the competence domain of P, i.e. $\rho_R(P)L = (R \cap P)L$. Furthermore, $\rho_R(P)$ is the least refined element of its equivalence class.

Proof. We check that P and $\rho_R(P)$ are equally correct.

$P \equiv_R \rho_R(P)$
$\Leftrightarrow$ {Proposition 3.5}
$(R \cap P)L = (R \cap (R \cap P)L \cap (R \cup P))L$
$\wedge\ (R \cap P)L \cap \overline{R} \cap P = (R \cap (R \cap P)L \cap (R \cup P))L \cap \overline{R} \cap (R \cap P)L \cap (R \cup P)$
$\Leftrightarrow$ {Boolean algebra, (3), monotonicity}
$(R \cap P)L = (R \cap P)L \wedge (R \cap P)L \cap \overline{R} \cap P = (R \cap P)L \cap \overline{R} \cap P$

which is a tautology. The equality $\rho_R(P)L = (R \cap P)L$ follows from (3), Boolean algebra and monotonicity. As for proving that $\rho_R(P)$ is the least refined element of its class, we let P' be an element in the equivalence class of P, and we show that P' refines $\rho_R(P)$.

$P' \equiv_R P$
$\Leftrightarrow$ {Proposition 3.5}
$(R \cap P')L = (R \cap P)L\ \wedge\ (R \cap P')L \cap \overline{R} \cap P' = (R \cap P)L \cap \overline{R} \cap P$
$\Rightarrow$ {Boolean algebra}
$(R \cap P')L = (R \cap P)L\ \wedge\ (R \cap P')L \cap \overline{R} \cap P' \subseteq P$
$\Leftrightarrow$ {Shunting}
$(R \cap P')L = (R \cap P)L\ \wedge\ (R \cap P')L \cap P' \subseteq R \cup P$
$\Leftrightarrow$ {P and P' have the same competence domain}
$(R \cap P')L = (R \cap P)L\ \wedge\ (R \cap P)L \cap P' \subseteq R \cup P$
$\Rightarrow$ {Boolean algebra, monotonicity}
$(R \cap P)L \subseteq P'L\ \wedge\ (R \cap P)L \cap P' \subseteq (R \cap P)L \cap (R \cup P)$
$\Leftrightarrow$ {Substitution of $\rho_R(P)$, and $\rho_R(P)L = (R \cap P)L$}
$\rho_R(P)L \subseteq P'L\ \wedge\ \rho_R(P)L \cap P' \subseteq \rho_R(P)$
$\Leftrightarrow$ {Definition of refinement}
$P' \sqsupseteq \rho_R(P)$. **qed**

It stems from this proposition that if P and P' are equally correct with respect to some specification R, then $\rho_R(P)$ and $\rho_R(P')$ are identical. Figure 1 shows an example of a specification R, two equally correct programs with respect to R, P and P', and the least refined relation of their shared equivalence class. The reader may check that P and P' are both refinements of $\rho_R(P)$ (=$\rho_R(P')$).

$$\begin{bmatrix}1&1&0&0&0&0\\1&1&1&0&0&0\\0&1&1&1&0&0\\0&0&1&1&1&0\\0&0&0&1&1&1\\0&0&0&0&1&1\end{bmatrix}\quad\begin{bmatrix}0&0&1&1&0&0\\0&1&1&1&0&0\\0&0&1&1&1&0\\0&0&0&1&1&1\\0&0&0&0&1&1\\0&0&1&1&0&0\end{bmatrix}\quad\begin{bmatrix}0&0&1&0&1&0\\1&1&0&1&0&0\\0&1&1&0&1&0\\0&0&1&1&0&1\\0&0&0&1&1&0\\1&0&0&1&0&0\end{bmatrix}\quad\begin{bmatrix}0&0&0&0&0&0\\1&1&1&1&0&0\\0&1&1&1&1&0\\0&0&1&1&1&1\\0&0&0&1&1&1\\0&0&0&0&0&0\end{bmatrix}$$

$$R \qquad\qquad P \qquad\qquad P' \qquad\qquad \rho_R(P) = \rho_R(P')$$

Fig. 1. Equal Correctness of Non-Deterministic Programs: $P' \equiv_R P$

4 Relative Correctness and Refinement

Because refinement plays a central role in the definition of (absolute) correctness, it is legitimate to explore the relationship between refinement and *relative* correctness; this is the subject of this section.

Proposition 4.1. *Let R, P and P' be relations on set S. Then P' is more-correct than P with respect to R if and only if $\rho_R(P')$ refines $\rho_R(P)$, i.e.*

$$P' \sqsupseteq_R P \Leftrightarrow \rho_R(P') \sqsupseteq \rho_R(P).$$

Proof. We proceed by equivalences:

$\rho_R(P') \sqsupseteq \rho_R(P)$

$\Leftrightarrow$ {Formula of refinement}

$\rho_R(P)L \subseteq \rho_R(P')L \wedge \rho_R(P)L \cap \rho_R(P') \subseteq \rho_R(P)$

$\Leftrightarrow$ {Substitution of $\rho_R(P)$, and $\rho_R(P)L = (R \cap P)L$}

$(R \cap P)L \subseteq (R \cap P')L$

$\wedge\ (R \cap P)L \cap (R \cap P')L \cap (R \cup P') \subseteq (R \cap P)L \cap (R \cup P)$

$\Leftrightarrow$ $\{A \subseteq B \Leftrightarrow A \cap B = A\}$

$(R \cap P)L \subseteq (R \cap P')L \ \wedge\ (R \cap P)L \cap (R \cup P') \subseteq (R \cap P)L \cap (R \cup P)$

$\Leftrightarrow$ $\{A \cap B \subseteq A \cap C \Leftrightarrow A \cap B \subseteq C\}$

$(R \cap P)L \subseteq (R \cap P')L \ \wedge\ (R \cap P)L \cap (R \cup P') \subseteq R \cup P$

$\Leftrightarrow$ {Shunting, Boolean algebra}

$(R \cap P)L \subseteq (R \cap P')L \ \wedge\ (R \cap P)L \cap P' \cap \overline{R} \subseteq P$

$\Leftrightarrow$ {Definition of $\sqsupseteq_R$}

$P' \sqsupseteq_R P.$ **qed**

The following proposition casts absolute correctness as the culmination of relative correctness, in the sense that a correct program is more-correct than (or as correct as) any candidate program.

Proposition 4.2. *Let R and P' be relations on set S. Then P' is correct with respect to R if and only if P' is more-correct with respect to R than any relation P on S, i.e.*

$$P' \sqsupseteq R \Leftrightarrow (\forall P : P' \sqsupseteq_R P).$$

Proof. Proof of $\Rightarrow$: Assume that P' refines R, i.e. $RL \subseteq P'L$ and $RL \cap P' \subseteq R$, and let P be an arbitrary relation on S. We must show that P' is more-correct than P with respect to R, i.e. that

$$(R \cap P)L \subseteq (R \cap P')L \ \wedge\ (R \cap P)L \cap \overline{R} \cap P' \subseteq P.$$

We write:

$$
\begin{array}{ll}
 & (R \cap P')L \\
\supseteq & \qquad \{\text{Hypothesis } RL \cap P' \subseteq R\text{, monotonicity}\} \\
 & (RL \cap P' \cap P')L \\
= & \qquad \{\text{Boolean algebra, (3)}\} \\
 & RL \cap P'L \\
= & \qquad \{\text{Hypothesis } RL \subseteq P'L\text{, Boolean algebra}\} \\
 & RL \\
\supseteq & \qquad \{\text{Boolean algebra, monotonicity}\} \\
 & (R \cap P)L.
\end{array}
$$

On the other hand,

$$
\begin{array}{ll}
 & (R \cap P)L \cap \overline{R} \cap P' \\
\subseteq & \qquad \{\text{Boolean algebra, monotonicity}\} \\
 & RL \cap \overline{R} \cap P' \\
\subseteq & \qquad \{\text{Hypothesis } RL \cap P' \subseteq R\text{, Boolean algebra}\} \\
 & R \cap \overline{R} \\
\subseteq & \qquad \{\text{Since } R \cap \overline{R} = \phi\} \\
 & P.
\end{array}
$$

Proof of $\Leftarrow$: We assume that P' is more-correct than any relation P on S with respect to specification R, and we write this property for $P = R$. This yields:

$$
\begin{array}{ll}
 & (R \cap R)L \subseteq (R \cap P')L \ \wedge\ (R \cap R)L \cap \overline{R} \cap P' \subseteq R \\
\Leftrightarrow & \qquad \{\text{Boolean algebra, shunting}\} \\
 & RL \subseteq (R \cap P')L \ \wedge\ RL \cap P' \subseteq R \cup R \\
\Rightarrow & \qquad \{\text{Boolean algebra, monotonicity}\} \\
 & RL \subseteq P'L \ \wedge\ RL \cap P' \subseteq R \\
\Leftrightarrow & \qquad \{\text{Definition of } \sqsupseteq\} \\
 & P' \sqsupseteq P.
\end{array}
$$

qed

In [9], we find that for deterministic relations P and P', P' refines P if and only if P' is more-correct than P with respect to any specification. This property can be interpreted as follows: if P' refines P, then whatever P does, P' can do as well or better; in particular, P' is more-correct than P with respect to any specification R. In other words, the only way for P' to be more-correct than P with respect to any specification R is for P' to merely refine P. When P and P' are not necessarily deterministic, we find that the condition $(\forall R : P' \sqsupseteq_R P)$ is too strong a sufficient condition for $P' \sqsupseteq P$, and too strong a necessary condition. We have the following proposition.

Proposition 4.3. *Let P and P' be relations on set S. P' refines P if and only if P' is more-correct than P with respect to P, i.e. $P' \sqsupseteq P \Leftrightarrow P' \sqsupseteq_P P$.*

Proof. Proof of $\Rightarrow$: This follows from Proposition 4.2 (renaming the bound variable) and elimination of the quantifier:

$$P' \sqsupseteq P \Leftrightarrow (\forall Q : P' \sqsupseteq_P Q) \Rightarrow P' \sqsupseteq_P P.$$

Proof of $\Leftarrow$: From $P' \sqsupseteq_P P$, we infer $(P \cap P)L \subseteq (P \cap P')L$ and $(P \cap P)L \cap \overline{P} \cap P' \subseteq P$, which imply $PL \subseteq P'L$ and $PL \cap P' \subseteq P$ by Boolean algebra, monotonicity and shunting. **qed**

In other words, according to this proposition, P' does not have to be more-correct than P with respect to any specification; it suffices that it be more-correct than P with respect to a single specification, namely P itself. The interpretation of this proposition is quite straightforward: The property of P' to be more-correct than P with respect to P can be interpreted to mean that P' beats P at its own game; this sounds like a good characterization of refinement. The following example disproves that $P' \sqsupseteq P$ implies $(\forall R : P' \sqsupseteq_R P)$. We take:

$$S = \{0, 1\}, \quad R = \{(0, 1)\}, \quad P = \{(0, 0), (0, 1)\}, \quad P' = \{(0, 0)\}.$$

Indeed, P' clearly refines P. Yet P' is not more-correct than P with respect to R, as we can check by observing that: $R \cap P = \{(0, 1)\}$, hence $(R \cap P)L = \{(0, 0), (0, 1)\}$ and $R \cap P' = \phi$, hence $(R \cap P')L = \phi$. While $P' \sqsupseteq P$ does not imply that P' is more-correct than P for any relation R, it does imply than P' is more-correct than P with respect to a single relation, namely P.

Hence while in [9] we have found that for deterministic relations P and P', $P' \sqsupseteq P$ is equivalent to $(\forall R : P' \sqsupseteq_R P)$, Proposition 4.3 provides that for relations P and P' that are not necessarily deterministic, $P' \sqsupseteq P$ is equivalent to $P' \sqsupseteq_P P$. This means in particular that for deterministic P and P', $P' \sqsupseteq_P P$ implies $(\forall R : P' \sqsupseteq_R P)$. This is an intriguing property, but one that we can understand intuitively: if we take two arbitrary programs P and P', then P' could conceivably be more-correct than P with respect to some specification, and less-correct with respect to other specifications; but if P' is more-correct than P with respect to P *itself*, then P' clearly dominates P, i.e. there is nothing P could do that P' could not; this conveys the same idea of subsumption that we associate with refinement.

To conclude this section, we consider the following question: Is it possible that if P' is more-correct than P with respect to some relation R, then it is more-correct than P with respect to any relation that R refines? Intuitively, it sounds like it should since refinement reflects the strength of a specification; the following example shows that this is not the case. We consider:

$$S = \{0, 1, 2\}, \quad P = \{(0, 0)\}, \quad P' = \{(0, 2)\}, \quad R = \{(0, 1)\}, \quad Q = \{(0, 0), (0, 1)\}.$$

We do have $R \sqsupseteq Q$, and we do have $P' \sqsupseteq_R P$ since $(R \cap P)L = \phi \subseteq (R \cap P')L$ and $(R \cap P)L \cap \overline{R} \cap P' = \phi \subseteq P$. Yet, P' is not more-correct than P with respect to Q, since $(Q \cap P)L = \{(0, 0), (0, 1), (0, 2)\}$ whereas $(Q \cap P')L = \phi$.

5 Relative Correctness and Refinement Lattice

In section 2.2 we have introduced some lattice-like operators including the least upper bound and the greatest lower bound of two specifications; we have found that when two specifications R and Q satisfy the consistency condition $RL \cap QL = (R \cap Q)L$ then they admit a least upper bound. In this section we first raise the question whether $P' \sqsupseteq_R P$ and $P' \sqsupseteq_Q P$ logically imply $P' \sqsupseteq_{R \sqcup Q} P$, where $R \sqcup Q$ is the least upper bound (modulo the refinement ordering) of R and Q. The following proposition gives a nuanced answer to this question.

Proposition 5.1. *Let P and P' be two programs (relations) on space S and let R and Q be two specifications on S. If P' is deterministic, and if it is more-correct than P with respect to R and with respect to Q then it is more-correct than P with respect to $R \sqcup Q$, i.e.*

$$\widehat{P'}P' \subseteq I \ \wedge \ P' \sqsupseteq_R P \ \wedge \ P' \sqsupseteq_Q P \ \Rightarrow \ P' \sqsupseteq_{R \sqcup Q} P.$$

Proof. We introduce a lemma that will be useful for our proof:

$$\widehat{P}P \subseteq I \ \wedge \ Q \subseteq P \ \Rightarrow \ (R \cap P)L \cap Q = R \cap Q.$$

Assume $\widehat{P}P \subseteq I$ and $Q \subseteq P$.

$(R \cap P)L \cap Q = R \cap Q$

$\Leftrightarrow$ $\{(R \cap P)L \cap Q \subseteq Q$, $Q \subseteq P$ hence $R \cap Q \subseteq (R \cap P)L$, $R \cap Q \subseteq Q\}$

$(R \cap P)L \cap Q \subseteq R$

$\Leftarrow$ {By (2)}

$(R \cap P)(L \cap \widehat{(R \cap P)}Q) \subseteq R$

$\Leftarrow$ {Hypothesis $Q \subseteq P$, Boolean algebra, monotonicity}

$(R \cap P)\widehat{(R \cap P)}P \subseteq R$

$\Leftarrow$ {Boolean algebra, monotonicity of converse and product}

$R\widehat{P}P \subseteq R$

$\Leftarrow$ {Monotonicity of product}

$\widehat{P}P \subseteq I$

$\Leftarrow$ {Hypothesis $\widehat{P}P \subseteq I$}

true.

Using this lemma, we now show the main theorem. Assume $\widehat{P'}P' \subseteq I$, $P' \sqsupseteq_R P$ and $P' \sqsupseteq_Q P$.

$P' \sqsupseteq_{R \sqcup Q} P$

$\Leftrightarrow$ {Hypothesis $\widehat{P'}P' \subseteq I$, Proposition 3.2}

$((R \sqcup Q) \cap P)L \subseteq ((R \sqcup Q) \cap P')L$

$\Leftrightarrow$ {Definition of $\sqcup$}

$(((\overline{QL} \cap R) \cup (\overline{RL} \cap Q) \cup (R \cap Q)) \cap P)L$
$\subseteq (((\overline{QL} \cap R) \cup (\overline{RL} \cap Q) \cup (R \cap Q)) \cap P')L$

$\Leftrightarrow$ {Boolean algebra, distributing L over $\cup$, (3)}

$(\overline{QL} \cap (R \cap P)L) \cup (\overline{RL} \cap (Q \cap P)L) \cup (R \cap Q \cap P)L$

$\subseteq (\overline{QL} \cap (R \cap P')L) \cup (\overline{RL} \cap (Q \cap P')L) \cup (R \cap Q \cap P')L$

$\Leftarrow$ {Boolean algebra, hypotheses $P' \sqsupseteq_Q P$ and $P' \sqsupseteq_R P$}

$(R \cap Q \cap P)L \subseteq (R \cap Q \cap P')L$

$\Leftarrow$ {For any relations A and B, $(A \cap B)L \subseteq AL \cap BL$}

$(R \cap P)L \cap (Q \cap P)L \subseteq (R \cap Q \cap P')L$

$\Leftarrow$ {Boolean algebra, hypotheses $P' \sqsupseteq_Q P$ and $P' \sqsupseteq_R P$}

$(R \cap P')L \cap (Q \cap P')L \subseteq (R \cap Q \cap P')L$

$\Leftrightarrow$ {By (3)}

$((R \cap P')L \cap Q \cap P')L \subseteq (R \cap Q \cap P')L$

$\Leftrightarrow$ {Lemma, using P', $R \cap P'$, Q for P, Q, R, and hypothesis $\widehat{P'}P' \subseteq I$}

$(R \cap Q \cap P')L \subseteq (R \cap Q \cap P')L$

$\Leftrightarrow$ {Tautology}

true. **qed**

This result holds regardless of whether R and Q satisfy the consistency condition: if they do, then this result pertains for their least upper bound; if not, then the result pertains for their join, which is not their least upper bound. To prove that the condition of determinacy of P' is a necessary condition in proposition 5.1, we consider the following (counter) example on set $S = \{0, 1, 2\}$ where P' is not deterministic, and we prove that then P' can be more-correct than P with respect to two specifications without being more-correct with respect to their join:

$$P = \{(0,0),(0,1),(0,2)\},\ P' = \{(0,1),(0,2)\},$$

$$R = \{(0,0),(0,2)\},\ Q = \{(0,0),(0,1)\}.$$

Indeed, we find $(R \cap P)L = (R \cap P')L = \{(0,0),(0,1),(0,2)\}$ and $(R \cap P)L \cap \overline{R} \cap P' = \{(0,1)\}$, which is a subset of P, hence $P' \sqsupseteq_R P$. On the other hand, we find $(Q \cap P)L = (Q \cap P')L = \{(0,0),(0,1),(0,2)\}$ and $(Q \cap P)L \cap \overline{Q} \cap P' = \{(0,2)\}$, which is a subset of P, hence $P' \sqsupseteq_Q P$. And yet, $R \sqcup Q = \{(0,0)\}$, whence $(P \cap (R \sqcup Q))L = \{(0,0),(0,1),(0,2)\} \not\subseteq \phi = (P' \cap (R \sqcup Q))L$; therefore P' is not more-correct than P with respect to $R \sqcup Q$.

Whereas proposition 5.1 elucidates how relative correctness distributes over the join, the following proposition explores the same property for the meet.

Proposition 5.2. *If P' is more-correct than P with respect to R and with respect to Q, then it is more-correct than P with respect to $R \sqcap Q$.*

Proof.

$P' \sqsupseteq_{R \sqcap Q} P$

$\Leftrightarrow$ {Definition of relative correctness}

$((R \sqcap Q) \cap P)L \subseteq ((R \sqcap Q) \cap P')L\ \wedge\ ((R \sqcap Q) \cap P)L \cap \overline{(R \sqcap Q)} \cap P' \subseteq P$

$\Leftrightarrow$ {Definition of meet}

$(RL \cap QL \cap (R \cup Q) \cap P)L \subseteq (RL \cap QL \cap (R \cup Q) \cap P')L$

$\wedge\ (RL \cap QL \cap (R \cup Q) \cap P)L \cap \overline{RL \cap QL \cap (R \cup Q)} \cap P' \subseteq P$

$\Leftrightarrow$ {Distribution, De Morgan}

$$\begin{aligned}
& (QL\cap R\cap P)L\cup(RL\cap Q\cap P)L\subseteq(QL\cap R\cap P')L\cup(RL\cap Q\cap P')L\\
& \wedge\ (RL\cap QL\cap(R\cup Q)\cap P)L\cap(\overline{RL}\cup\overline{QL}\cup(\overline{R}\cap\overline{Q}))\cap P'\subseteq P\\
\Leftarrow\quad & \qquad\qquad \{\text{By (3), Boolean algebra}\}\\
& (QL\cap(R\cap P)L)\cup(RL\cap(Q\cap P)L)\subseteq(QL\cap(R\cap P')L)\cup(RL\cap(Q\cap P')L)\\
& \wedge\ (RL\cap QL\cap(R\cup Q)\cap P)L\cap\overline{R}\cap\overline{Q}\cap P'\subseteq P\\
\Leftarrow\quad & \qquad\qquad \{\text{Boolean algebra, monotonicity}\}\\
& (R\cap P)L\subseteq(R\cap P')L\ \wedge\ (Q\cap P)L\subseteq(Q\cap P')L\\
& \wedge\ (R\cap P)L\cap\overline{R}\cap P'\subseteq P\ \wedge\ (Q\cap P)L\cap\overline{Q}\cap P'\subseteq P\\
\Leftarrow\quad & \qquad\qquad \{\text{Definition of relative correctness}\}\\
& P'\sqsupseteq_R P\wedge P'\sqsupseteq_Q P.
\end{aligned}$$

qed

6 Concluding Remarks

6.1 Summary

In [9] we have introduced the concept of relative correctness as it applies to deterministic programs, and have used it to define the concept of a fault in a program with respect to a specification. In this paper, we generalize the definition of relative correctness to non-deterministic programs, on the grounds that very often, even when we are dealing with deterministic programs, we may want to reason about relative correctness without having to compute the functions of candidate programs in all their minute details. To this effect, we introduce a definition, investigate its properties, and explore its relation to refinement as well as its algebraic properties with respect to lattice operations.

6.2 Prospects

One of the broadest venues of research that this paper opens pertains to the approximation of deterministic programs by non-deterministic relations. If we approximate program P by relation Π and program P' by relation Π', what relation must hold between P and Π, and between P' and Π', in order for a conclusion we draw on Π and Π' to carry over to P and P'. Interestingly, such a relation must necessarily involve R, the specification against which we define relative correctness. As an example, let P and P' be two programs on some space S defined by two variables, say x and y, and let R be the following specification on S:

$$R=\{(s,s')|y'=f(y)\},$$

for some function f. Clearly, we can reason about the relative correctness of P and P' by considering abstractions thereof, say Π and Π', that focus exclusively on variable y. We want to generalize this argument by characterizing the relation that must hold between P, P', Π, Π' and R so that we can analyze Π and Π' and infer conclusions about the relative correctness of P and P' with respect to R. This is currently under investigation.

6.3 Related Work

Several authors have introduced and studied concepts that are similar to relative correctness, and some refer to them by this exact name [7, 4, 6, 13, 14, 5]. In [7] Logozzo discusses a framework for ensuring that some semantic properties are preserved by program transformation in the context of software maintenance. In [4] Lahiri et al. present a technique for verifying the relative correctness of a program with respect to a previous version, where they represent specifications by means of executable assertions placed throughout the program, and they define relative correctness by means of inclusion relations between sets of successful traces and unsuccessful traces. Logozzo and Ball [6] take a similar approach to Lahiri et al. in the sense that they represent specifications by a network of executable assertions placed throughout the program, and they define relative correctness in terms of successful traces and unsuccessful traces of candidate programs; Logozzo and Ball distinguish between two categories of program failures, namely contract violations when functional requirements are violated and run-time errors, when operational requirements are violated. In [13], Nguyen et al. present an automated repair method based on symbolic execution, constraint solving, and program synthesis; they call their method SemFix, on the grounds that it performs program repair by means of semantic analysis. In [14], Weimer et al. discuss an automated program repair method that takes as input a faulty program, along with a set of positive tests (i.e. test data on which the program is known to perform correctly) and a set of negative tests (i.e. test data on which the program is known to fail) and returns a set of possible patches. In [5] Le Goues et al. survey existing technology in automated program repair and identify open research challenges; among the criteria for automated repair methods, they cite applicability (extent of real-world relevance), scalability (ability to operate effectively and efficiently for products of realistic size), generality (scope of application domain, types of faults repaired), and credibility (extent of confidence in the soundness of the repair tool).

Our work differs significantly from all these works in many ways:

- First, we use relational specifications that address the functional properties of the program as a whole, and have no cognizance of intermediate assertions that are expected to hold throughout the program; also, our relational specifications do not necessarily correspond to an abstraction of the assertions used in trace-based program analysis, because the initial and final assertions could be checking some local properties, whereas our specifications capture global input/ output properties.
- Second, our definition of relative correctness involves competence domains (for deterministic specifications) and the sets of states that candidate programs produce in violation of the specification (for non-deterministic programs).
- Third we conduct a detailed analysis of the relations between relative correctness and the property of refinement.
- Finally, we study how the property of relative correctness can be decomposed using lattice operators on the reference specification.

Acknowledgements. This publication was made possible by a grant from the Qatar National Research Fund, NPRP04-1109-1-174. Its contents are solely the responsibility of the authors and do not necessarily represent the official views of the QNRF.

The authors are very grateful to the anonymous reviewers for their valuable feedback, which has greatly improved the form and content of this paper.

References

[1] Boudriga, N., Elloumi, F., Mili, A.: On the Lattice of Specifications: Applications to a Specification Methodology. Formal Aspects of Computing 4, 544–571 (1992)
[2] Diallo, N., Ghardallou, W., Mili, A.: Correctness and relative correctness. In: Proceedings of the 37th International Conference on Software Engineering, Firenze, Italy, May 20–22 (2015)
[3] Diallo, N., Ghardallou, W., Mili, A.: Program derivation by correctness enhancements. In: Refinement 2015, Oslo, Norway (June 2015)
[4] Lahiri, S.K., McMillan, K.L., Sharma, R., Hawblitzel, C.: Differential assertion checking. In: Proceedings of the ESEC/SIGSOFT FSE, pp. 345–455 (2013)
[5] LeGoues, C., Forrest, S., Weimer, W.: Current challenges in automatic software repair. Software Quality Journal 21(3), 421–443 (2013)
[6] Logozzo, F., Ball, T.: Modular and verified automatic program repair. In: Proceedings of the OOPSLA, pp. 133–146 (2012)
[7] Logozzo, F., Lahiri, S., Faehndrich, M., Blackshear, S.: Verification modulo versions: Towards usable verification. In: Proceedings of the PLDI (2014)
[8] McCune, W.: Prover9 and Mace4, `http://www.cs.unm.edu/~mccune/prover9`
[9] Mili, A., Frias, M.F., Jaoua, A.: On faults and faulty programs. In: Höfner, P., Jipsen, P., Kahl, W., Müller, M.E. (eds.) RAMiCS 2014. LNCS, vol. 8428, pp. 191–207. Springer, Heidelberg (2014)
[10] Mills, H.D., Basili, V.R., Gannon, J.D., Hamlet, R.D.: Structured Programming: A Mathematical Approach. Allyn and Bacon, Boston (1986)
[11] Schmidt, G.: Relational Mathematics. Encyclopedia of Mathematics and Its Applications, vol. 132. Cambridge University Press (2010)
[12] Schmidt, G., Ströhlein, T.: Relations and Graphs – Discrete Mathematics for Computer Scientists. EATCS Monographs on Theoretical Computer Science. Springer (1988)
[13] Nguyen, H.D.T., Qi, D., Roychoudhury, A., Chandra, S.: Semfix: Program repair via semantic analysis. In: Proceedings of the ICSE, pp. 772–781 (2013)
[14] Weimer, W., Nguyen, T., Le Goues, C., Forrest, S.: Automatically finding patches using genetic programming. In: Proceedings of the ICSE 2009, pp. 364–374 (2009)

Encoding and Decoding in Refinement Algebra

Kim Solin

The University of Queensland, Brisbane, Australia

Abstract. Refinement algebras are axiomatic algebras for reasoning about programs in a total-correctness framework. We extend demonic and angelic refinement algebra with operators for encoding and decoding. Encoding gives one the least data refinement of a program with respect to a given data-refinement abstraction. Decoding gives one the greatest program that can be data refined into the decoded program with respect to a given abstraction statement. The resulting algebra is applied to reasoning about action systems.

1 Introduction: On Refinement Algebra

The axiomatic algebraisation of program refinement was initiated a decade ago by Joakim von Wright [22]. From a programming-theoretic perspective, the key difference between Kleene algebra and refinement algebra is that refinement algebras are algebras for total correctness, whereas Kleene algebra only captures partial correctness [12]. Reasoning in total correctness means that it is possible to reason about non-termination. In his seminal paper, von Wright started from Kleene algebra and proposed axioms for the demonic refinement algebra. He then applied this structure to reasoning about correctness rules, program transformation, basic data refinement and more. The author of this paper took von Wright's work further by extending the basic algebra to an algebra for both demonic and angelic nondeterminism [20], by adding enabledness and termination operators and reasoning about action systems [21], by deriving the while-loop normal form theorem in the algebra [19], and together with Larissa Meinicke by reasoning about probabilistic programs in a refinement algebra [16]. Meinicke and Ian Hayes added an explicit operator for probabilistic choice to the total-correctness algebra [15]. Viorel Preoteasa has also proposed an algebra similar to, but more abstract than, refinement algebra that is also a total-correctness algebra [18]. The advantage of these algebras is that one gets a simple and perspicuous way of reasoning that easily lends itself to mechanisation and automation [2, 11]. In addition to having provided an interesting reasoning tool for total-correctness program refinement, the advantage of the algebraic approach has also been given credence in a number of applications of related structures (see for instance [1, 5–7, 9, 10, 12–14] with references).

So one has basic proof of concept. The ensuing research around the refinement algebras can now take at least one of the following forms. One can use the algebras at hand for reasoning about larger case studies, thereby collecting

W. Kahl et al. (Eds.): RAMiCS 2015, LNCS 9348, pp. 209–224, 2015.
DOI: 10.1007/978-3-319-24704-5_13

data for further validating the claim that the algebraic approach makes reasoning simple, and, also, for empirically determining when and when not to use the algebraic methods; this is preferably supported by mechanisation. Then one can try to prove various more elaborate metamathematical results regarding the algebras, such as decidability, complexity, and completeness with respect to different models. And, thirdly, one can continue the incorporation of more of the traditional refinement calculus into the abstract-algebraic framework, proving that even more advanced concepts of program refinement are amenable to algebraic treatment. It is this last-mentioned research alternative that is our concern in this paper. In the paper we shall incorporate fully the data-refinement encoding and decoding operators of Back and von Wright's paper [4] into the abstract-algebraic framework. The approach in that paper is already quite algebraic in flavour, but not fully axiomatic, so some work must be done to put this piece of theory on a proper axiomatic basis.

The encoding of a program with an abstraction statement gives us the least data refinement of the program with respect to that abstraction statement. It is this operator that we shall consider axiomatically in an abstract algebra in this paper. We shall also consider its dual: decoding. The decoding of a program with an abstraction statement yields the greatest program that data refines into the decoded program with respect to the abstraction statement. One can view the least and the greatest data refinement as a refinement of a program that only changes the data representation: encoding into a more concrete data representation, decoding into a more abstract representation. This means that one can separate the data refinement from the algorithmic refinement, which makes the refinement process more modularised. This will be described in more detail in the main part of the paper.

Concretely speaking, the contributions of this paper are:

- determining the appropriate axiomatic algebra that is needed for handling encoding and decoding (it turns out to be demonic and angelic refinement algebra, DARA), and, in this algebra, characterising various programming-theoretic conditions and constructs that are essential for encoding and decoding,
- axiomatising encoding and decoding on the basis of, and in line with, Back and von Wright [4], axiomatically investigating the basic properties of those operators, and
- applying the axiomatisation to structural reasoning about action systems, a classical application area for refinement algebra.

The paper is structured as follows. The first section comprises the necessary background theory: the basic algebra, various algebraic characterisations and extensions to the basic algebra. The second section is the main part of the paper, and concerns the axiomatisation of the encoding operator. The third, and last, part deals with the decoding operator. To conclude, some thoughts about the results of the paper are presented together with ideas for future research.

2 Background Theory

The challenge when having decided to reason axiomatically about something is to choose which operators and axioms one actually needs. In order to be able to reason about encoding, decoding, and data refinement, we need to ensure that we have the possibility to capture both angelic and demonic nondeterminism. So we need the full power of demonic and angelic refinement algebra [20]. This algebra has monotone predicate transformers as a model. In this section, we outline the basic theory of this algebra, as well as its extension with guards and assertions, and with enabledness and termination operators. We also consider abstract characterisations of healthiness conditions and program inversion, and the addition of a special element, chaos. The reader already familiar with this algebra might prefer to move directly on to Section 3, perhaps stopping briefly at the sections about abstract characterisations, program inversion, and chaos.

2.1 Demonic and Angelic Refinement Algebra

Definition 1. A *demonic and angelic refinement algebra* (DARA) with carrier set R is an algebra

$$(R, \sqcap, \sqcup, ;, {}^*, {}^\omega, {}^\phi, {}^\dagger, \bot, \top, 1)$$

that satisfies the following axioms and rules for $x, y, z \in R$: [1]

$$x \sqcap (y \sqcap z) = (x \sqcap y) \sqcap z \qquad x \sqcup (y \sqcup z) = (x \sqcup y) \sqcup z \tag{1}$$
$$x \sqcap y = y \sqcap x \qquad x \sqcup y = y \sqcup x \tag{2}$$
$$x \sqcap \top = x \qquad x \sqcup \bot = x \tag{3}$$
$$x \sqcap x = x \qquad x \sqcup x = x \tag{4}$$
$$x \sqcap (y \sqcup z) = (x \sqcap y) \sqcup (x \sqcap z) \qquad x \sqcup (y \sqcap z) = (x \sqcup y) \sqcap (x \sqcup z) \tag{5}$$
$$x(yz) = (xy)z \tag{6}$$
$$1x = x = x1 \tag{7}$$
$$\top x = \top \qquad \bot x = \bot \tag{8}$$
$$x(y \sqcap z) \sqsubseteq xy \sqcap xz \qquad x(y \sqcup z) \sqsupseteq xy \sqcup xz \tag{9}$$
$$(x \sqcap y)z = xz \sqcap yz \qquad (x \sqcup y)z = xz \sqcup yz \tag{10}$$
$$x^* = xx^* \sqcap 1 \qquad x^\phi = xx^\phi \sqcup 1 \tag{11}$$
$$x \sqsubseteq yx \sqcap z \Rightarrow x \sqsubseteq y^*z \qquad yx \sqcup z \sqsubseteq x \;\Rightarrow\; y^\phi z \sqsubseteq x \tag{12}$$
$$x^\omega = xx^\omega \sqcap 1 \qquad x^\dagger = xx^\dagger \sqcup 1 \tag{13}$$
$$yx \sqcap z \sqsubseteq x \;\Rightarrow\; y^\omega z \sqsubseteq x \qquad x \sqsubseteq yx \sqcup z \;\Rightarrow\; x \sqsubseteq y^\dagger z \tag{14}$$

where the partial order $\sqsubseteq$ is defined by $x \sqsubseteq y \Leftrightarrow_{df} x \sqcap y = x$.

[1] The precedence of the operators is given by $\mathsf{prec}(\sqcap) = \mathsf{prec}(\sqcup) < \mathsf{prec}(;) < \mathsf{prec}(^*) = \mathsf{prec}(^\omega) = \mathsf{prec}(^\phi) = \mathsf{prec}(^\dagger)$ understood in the natural way, and $x; y$ is written xy when there is no risk for confusion.

The absorption laws $x \sqcap (y \sqcup x) = x$ and $x \sqcup (y \sqcap x) = x$ follow from the axioms, and it is easily seen that $\perp$ is the least element and $\top$ the greatest. The complete lattice of monotone predicate transformers form a total-correctness programming-theoretic model for the algebra. The axioms and rules all have natural interpretations from this perspective. Below, we only give the intuition for the basic constants and operators. It is easily seen that the left and the right hand side column form duals: so for every equality derived in the algebra, there is a dual equality that can be derived with the dual axioms. For the details of the model, for a richer explanation of the intuition behind the axioms and rules, and for more about the duality, refer to [20].

The intuition behind the basic components is given by the following. The constant 1 is the immediately terminating program that accomplishes nothing, it just leaves the state as it was. The constant $\perp$ is the always aborting program, and the constant $\top$ is the miraculous program, a program that can achieve anything. The binary choice operators, $\sqcap$ and $\sqcup$, are demonic and angelic choice, respectively. A demonic choice between two options is done nondeterministically by the system, whereas an angelic choice is done by the user. The operator ; is simply sequential composition. There are four different iteration operators. Considering first the terminating ones, weak iteration * iterates a program a finite number of times, nondeterministically chosen by the system, whereas angelic weak iteration $^\phi$ does the same but determined by the user. The strong iteration $^\omega$ is determined by the system, that can choose to iterate the statement infinitely, in which case it simply aborts. So here demonic nontermination equals abort. The dual angelic strong iteration †, is determined by the user and if the user manages to go on forever a miracle occurs; angelic nontermination equals magic.

2.2 Abstract Characterisations of Healthiness Conditions

Since the angelic and demonic refinement algebra is general enough to incorporate both kinds of nondeterminism it is often very useful to enforce the exclusion of either one of them, or even both. To do this we can use the following conditions. Given any x in the carrier set R, if for all $y, z \in R$

- $x(y \sqcap z) = xy \sqcap xz$ then x is *conjunctive*, if
- $x(y \sqcup z) = xy \sqcup xz$ then x is *disjunctive*, and if
- x is both conjunctive and disjunctive then it is *functional.*

A conjunctive element has demonic nondeterminism, but not angelic, and dually. A functional element has no nondeterminism at all. An element x is *strict* if $x\perp = \perp$ and *terminating* of $x\top = \top$. If an element is strict and disjunctive it is *universally* disjunctive, and if it is terminating and conjunctive it is *universally* conjunctive.

These characterisations, which can be given purely in terms of the algebra's operators, are usually enough for one to reason with sufficient power. However, when it comes to encoding and decoding, we will need the following characterisation to ensure the validity of some of the axioms with respect to the predicate-transformer model. Let R be the carrier set of a DARA, and let $\bigsqcap C$ and $\bigsqcup C$

denote the infimum and supremum of $C \subseteq R$ with respect to the partial order $\sqsubseteq$, respectively. We say that an element x is *sensu stricto* universally conjunctive and disjunctive, respectively, if for an arbitrary index set I,

- $x; (\bigsqcap_{i \in I} y_i) = \bigsqcap_{i \in I}(x; y_i)$, and
- $x; (\bigsqcup_{i \in I} y_i) = \bigsqcup_{i \in I}(x; y_i)$,

respectively. A universally conjunctive element is thus terminating (take $I = \emptyset$), and, likewise, a universally disjunctive element is strict. If I is finite, this definition reduces to the one in terms of the algebra's operators above.

2.3 Guards and Assertions

Guards are programs that skip if some predicate holds and otherwise establish a miracle. They are key when reasoning about programming constructs. Algebraically, they take the following form [20]. An element g of a DARA is a *guard* if

- g is functional,
- g has a functional complement $\bar{g}$ satisfying $g\bar{g} = \bar{g}g = \top$ and $g \sqcap \bar{g} = 1$, and
- for any g' also satisfying the first two conditions it holds that $gg' = g \sqcup g'$ and $\bar{g}g' = \bar{g} \sqcup g'$.

Assertions are programs that skip if some predicate holds and otherwise abort. Like guards, they are essential in programming theory. They are characterised dually to guards. An element p of a DARA is an *assertion* if

- p is functional,
- p has a functional complement $\bar{p}$ satisfying $p\bar{p} = \bar{p}p = \bot$ and $p \sqcup \bar{p} = 1$, and
- for any p' also satisfying the first two conditions it holds that $pp' = p \sqcap p'$ and $\bar{p}p' = \bar{p} \sqcap p'$.

One can show that both the set of guards and the set of assertions form Boolean algebras. Moreover, guards and assertions are each other's duals. For any guard g, the dual assertion g° is given by $\bar{g}\bot \sqcap 1$; and for any assertion p, the dual guard p° is given by $\bar{p}\top \sqcup 1$.

2.4 Enabledness and Termination

A program is enabled when it is feasible, that is, when it is possible to execute without having to resort to magic. The *enabledness operator* ϵ determines those states from which a program is enabled. It is axiomatised as follows for any x in the carrier set:

- ϵx is a guard, and
- $\epsilon x = x\bot \sqcup 1$.

A program terminates when it does not go on indefinitely, and the *termination operator* τ determines those states from which the program terminates. It is axiomatised as follows for any x in the carrier set:

- τx is an assertion, and
- $\tau x = x\top \sqcap 1$.

It can be shown that these definitions give us all the basic properties of enabledness and termination [21], which are very closely related to the operators of Kleene algebra with domain [6]. We will make use of the basic properties that $\epsilon x x = x$ and $\tau x x = x$.

2.5 Program Inversion

The basic idea of program inversion is that given a program x, one finds an *inverse program* $\breve{x}$ such that $\breve{x}$ returns to the state that x started from. So, in effect, one has $x; \breve{x} = \breve{x}; x = 1$. Now, this only holds when the program is deterministic and if each final state can only be reached from a unique initial state. If one wants to reason about nondeterministic programs and programs that can reach some final state from several different initial states, then one needs to generalise the notion of program inversion. Following Back and von Wright [4], one can do this generalisation as follows: assuming universal disjunctivity, the program $\breve{x}$ is the *inverse* of x iff

$$x\breve{x} \sqsubseteq 1 \sqsubseteq \breve{x}x.$$

A slightly different notion of program inversion, which is related to conjunctivity of programs, has figured in the abstract-algebraic setting earlier [22, 21].

2.6 Chaos

Chaos, in symbols C, is the least terminating program. Starting from a given state, chaos takes you to any state. In contrast to magic, it is always feasible, and it does not abort. It was called *havoc* by von Wright [22], and was axiomatised as follows:

$$\mathsf{C}\top = \top \tag{15}$$

$$x\top = \top \Rightarrow \mathsf{C} \sqsubseteq x \tag{16}$$

The first condition says that C terminates, and the second that it is the least such program. It could also be expressed using the termination operator, since it can be shown that $x\top = \top \Leftrightarrow \tau x = 1$ [21]. von Wright notes that the existence of C does not follow from the axioms of a demonic refinement algebra, and that, therefore, assuming that this element exists, will restrict the number of models [22]. For the remainder of this paper, it will be assumed that C exists.

2.7 Recapitulation

In sum, we have a demonic and angelic refinement algebra, extended with guards and assertions, and with operators for enabledeness and termination. Moreover, we have postulated that there is an element C, and it has been shown how to express different healthiness conditions and program inversion abstractly. On this basis, we can now move on to the main contribution of the paper: the axiomatic consideration of encoding and decoding.

3 Encoding Axiomatically

This section contains the main contribution of this paper. We axiomatise the encoding operator, investigate its basic properties in the context of DARA, and apply it to reasoning about action systems.

3.1 Axioms and Intuition

Data refinement means replacing some data representation of a program with another data representation that is more advantageous in some respect (such as being easier to implement, more efficient, or more secure) while preserving the intended functionality of the program. On an abstract level, data refinement can be understood as a commutativity property

$$dx \sqsubseteq yd,$$

where x is the program refined into y, and d is the abstraction statement that connects the two data representations. Data refinement was considered abstract-algebraically already in von Wright's seminal paper [22], and we shall here continue that work. The concern in this paper is the *least* data refinement of a program x with respect to an abstraction statement d. If one can single out the least data refinement of a program, this can be viewed as a refinement step that *only* changes the data representation. So in that sense, one can separate the pure data refinement from the ensuing algorithmic refinement. This separation makes the division of the refinement steps clearer and, hopefully, easier to handle. For more detail, see Back and von Wright [4].

For the remainder of this paper, we introduce the following abbreviation [4]:

$$x \sqsubseteq_d y \;\Leftrightarrow\; dx \sqsubseteq yd. \tag{17}$$

Again following Back and von Wright [4], we denote *the least data refinement of x with respect to d* by

$$x \downarrow d$$

and postulate the following two axioms in addition to the axioms of DARA:

$$x \sqsubseteq_d (x \downarrow d), \tag{18}$$

$$x \sqsubseteq_d y \Rightarrow (x \downarrow d) \sqsubseteq y. \tag{19}$$

It can be shown axiomatically (the proof in [4] can easily be replayed in DARA) that these axioms are equivalent to

$$x \downarrow d \sqsubseteq y \;\Leftrightarrow\; x \sqsubseteq_d y. \tag{20}$$

The binary operator $\downarrow$ is called *encoding.* It is easily seen that the above axioms are valid in the predicate-transformer model. We shall now turn to the basic properties of the encoding operator in the context of DARA extended with guards, assertions, enabledness and termination.

3.2 Basic Properties

Many of the basic properties can be given analogous proofs to those in the predicate-transformer model. In this section, we list all the basic properties of the encoding operator, but only write out the axiomatic proofs that are significantly different from the model-theoretic proofs. To begin with, it is easily seen that encoding is monotone in its first argument. It is not, however, monotone (or antitone) in its second argument. This can easily be justified along the lines of [4].

We next consider what Back and von Wright call structure-preserving encodings, that is refinements $x \downarrow d \sqsubseteq y$ such that x and y have the same overall structure. The first two theorems can be proved analogously to the proofs in the model [4].

Theorem 1. *Let d be an element in the carrier set of a* DARA *extended with the operators of the previous sections. Then*

$$\bot \downarrow d = \bot, \text{ if } d \text{ is strict,} \tag{21}$$

$$1 \downarrow d \sqsubseteq 1, \text{ and} \tag{22}$$

$$\top \downarrow d = \top, \text{ if } d \text{ is strict and terminating,} \tag{23}$$

all hold true.

Theorem 2. *Let x, y and d be elements in the carrier set of a* DARA *extended with the operators of the previous sections. Then*

$$xy \downarrow d = (x \downarrow d)(y \downarrow d), \tag{24}$$

$$(x \sqcap y) \downarrow d \sqsubseteq (x \downarrow d) \sqcap (y \downarrow d), \text{ and} \tag{25}$$

$$(x \sqcup y) \downarrow d \sqsubseteq (x \downarrow d) \sqcup (y \downarrow d), \tag{26}$$

all hold true.

For the next theorem, we will make use of Greg Nelson's pairing property. Nelson's property says that total correctness can be described as weak correctness plus an extra termination assumption [17]. Weak correctness can be seen as a form of partial correctness in a total-correctness conceptual framework. It was proved algebraically by Solin in [21], and has the following abstract form (g_1 is the precondition guard, x the program, g_2 the postcondition guard):

$$g_1 x \bar{g_2} = \top \;\Leftrightarrow\; (x g_2 \sqsubseteq g_1 x \;\wedge\; \tau(g_1 x) = 1). \tag{27}$$

We will also use the already mentioned fact that $x\top = \top \Leftrightarrow \tau x = 1$ [21]. In a sense, the following theorem says that least structure-preserving data refinement of guards and assertions reduces to proving correctness-like criteria for the abstraction statement.

Theorem 3. *Let p_1, p_2 be assertions, let g_1, g_2 be guards, and let d be an element in the carrier set of a* DARA *extended with the operators of the previous sections. Then*

$$g_1 \downarrow d \sqsubseteq g_2 \Leftrightarrow g_2 d \bar{g_1} = \top, \text{ if } d \text{ is universally conjunctive} \tag{28}$$

$$g_1 \downarrow d \sqsubseteq g_2 \Leftrightarrow g_2^\circ d \bar{g_1}^\circ = \bot, \text{ if } d \text{ is universally disjunctive, and} \tag{29}$$

$$p_1 \downarrow d \sqsubseteq p_2 \Leftrightarrow \bar{p_2} d p_1 = \bot \tag{30}$$

all hold true.

Proof. For the first statement, first note that by the assumption that d is universally conjunctive and basic guard properties, one has, for any guard g,

$$gd\top = g\top = g\bar{g}g = \top g = \top.$$

The first statement is then proved by

$$\begin{aligned}
& g_1 \downarrow d \sqsubseteq g_2 \\
\Leftrightarrow\ & \{(17,\ 20)\} \\
& dg_1 \sqsubseteq g_2 d \\
\Leftrightarrow\ & \{\text{above observation}\} \\
& dg_1 \sqsubseteq g_2 d \wedge g_2 d\top = \top \\
\Leftrightarrow\ & \{\text{Nelson's property (27)}\} \\
& g_2 d \bar{g_1} = \top.
\end{aligned}$$

The second one is immediate from duality. The third one follows from mutual implication as follows. First note that by (17) and (20), $p_1 \downarrow d \sqsubseteq p_2 \Leftrightarrow dp_1 \sqsubseteq p_2 d$. Then we show that $dp_1 \sqsubseteq p_2 d \Leftrightarrow \bar{p_2} d p_1 = \bot$. For the first direction (left to right), note that by the assumption and basic assertion properties, one has

$$\bar{p_2} d p_1 \sqsubseteq \bar{p_2} p_2 d = \bot d = \bot.$$

Since $\bot$ is the least element the first direction follows. For the other direction (right to left), derive

$$\begin{aligned}
& dp_1 \\
=\ & \{\text{axiom for 1, basic assertion property}\} \\
& (p_2 \sqcup \bar{p_2}) dp_1 \\
=\ & \{\text{distributivity axiom}\} \\
& p_2 dp_1 \sqcup \bar{p_2} dp_1 \\
=\ & \{\text{assumption, } \bot \text{ least element}\} \\
& p_2 dp_1 \\
\sqsubseteq\ & \{\text{for any assertion } p,\ p \sqsubseteq 1\} \\
& p_2 d.
\end{aligned}$$

This proves the theorem. □

The next theorem concerns the abstraction statement in the second argument of the encoding operator.

Theorem 4. *Let x be an element in the carrier set of a* DARA *extended with the operators of the previous sections. Then*

$$x \downarrow (y \sqcup z) \sqsubseteq (x \downarrow y) \sqcup (x \downarrow z) \tag{31}$$

$$x \downarrow \bot = \bot, \tag{32}$$

$$x \downarrow 1 = x, \text{ and} \tag{33}$$

$$x \downarrow \top = \mathsf{C} \tag{34}$$

all hold true.

Proof. The three first statements can be proved analogously to the proofs in [4]. We focus here on the fourth part, which involves C, and which has not been given a fully axiomatic proof. The proof is by mutual refinement:

$$\begin{aligned}
& x \downarrow \top \sqsubseteq \mathsf{C} \\
\Leftrightarrow\ & \{(17, 20)\} \\
& \top x \sqsubseteq \mathsf{C}\top \\
\Leftrightarrow\ & \{\mathsf{DARA}\text{ axioms}\} \\
& \top = \mathsf{C}\top \\
\Leftrightarrow\ & \{\mathsf{C}\text{ axiom}\} \\
& \mathsf{true}, \text{ and}
\end{aligned}$$

$$\begin{aligned}
& \mathsf{C} \sqsubseteq x \downarrow \top \\
\Leftarrow\ & \{\mathsf{C}\text{ axiom}\} \\
& (x \downarrow \top)\top = \top \\
\Leftrightarrow\ & \{\mathsf{DARA}\text{ axioms}\} \\
& \top x \sqsubseteq (x \downarrow \top)\top \\
\Leftrightarrow\ & \{(17)\} \\
& x \sqsubseteq_\top (x \downarrow \top) \\
\Leftrightarrow\ & \{\text{encoding axiom}\} \\
& \mathsf{true}.
\end{aligned}$$

This proves the statement. □

The following theorem can be proved along the lines of the proof in the model.

Theorem 5. *Let x, d_1 and d_2 be elements in the carrier set of a* DARA *extended with the operators of the previous sections. Then*

$$x \downarrow d_1 d_2 \sqsubseteq (x \downarrow d_2) \downarrow d_1$$

holds true.

It is an important theorem as motivated by the following. Suppose that we have $x \downarrow d_1 \sqsubseteq x_1$ and $x_1 \downarrow d_2 \sqsubseteq x_2$, then one can show directly from the definitions, using (20), that $x \downarrow d_2 d_1 \sqsubseteq x_2$. But from the same assumptions one can also show, using plain monotonicity reasoning, that $(x \downarrow d_1) \downarrow d_2 \sqsubseteq x$. So using Theorem 5, one can therefore separate the reasoning from the two assumptions into two encoding steps:

$$\begin{array}{l} \quad x \downarrow d_2 d_1 \sqsubseteq x_2 \\ \Leftarrow \{\text{Theorem 5}\} \\ \quad (x \downarrow d_1) \downarrow d_2 \sqsubseteq x_2 \\ \Leftarrow \{\text{assumption}\} \\ \quad (x \downarrow d_1) \downarrow d_2 \sqsubseteq x_1 \downarrow d_2 \\ \Leftarrow \{\text{monotonicity}\} \\ \quad x \downarrow d_1 \sqsubseteq x_1 \\ \Leftarrow \{\text{assumption}\} \\ \quad \mathsf{true} \end{array}$$

This makes the data-refinement process more structured.[2]

3.3 Forward and Backward Data Refinement

Forward data refinement can be seen as an encoding with a universally disjunctive (*sensu stricto*) abstraction, whereas backwards data refinement can be seen as an encoding with universally conjunctive (*sensu stricto*) abstraction.

The next two theorems can be proved analogously to the model-theoretic proofs [4]. For forward data refinement, encoding has an explicit characterisation.

Theorem 6. *Let x and d be elements in the carrier set of a* DARA *extended with the operators of the previous sections. Let d be universally disjunctive (sensu stricto). Then*

$$x \downarrow d = dx\breve{d}$$

holds true.

In connection to universally disjunctive abstraction statements, Theorem 5 can be strengthened to an equality.

Theorem 7. *Let x, d_1 and d_2 be elements in the carrier set of a* DARA *extended with the operators of the previous sections. Let either of d_1 and d_2 be universally disjunctive (sensu stricto). Then*

$$x \downarrow d_1 d_2 = (x \downarrow d_2) \downarrow d_1$$

holds true.

[2] I am grateful to Larissa Meinicke for help with clarifying the surprisingly foggy explanation of this in Back and von Wright's paper [4].

For backward data refinement we have not been able to prove any interesting properties on the abstraction level that DARA provides.[3]

3.4 Iteration Operators

We now consider the iteration operators in connection to the encoding operator. In order to capture the relationship between encoding and the iteration operators, we need to postulate the following four *conditional axioms* (the same technique was used in [20] when introducing a negation operator).

Assume that d is an element of the carrier set of a DARA that is universally disjunctive *sensu stricto*, then

$$x^* \downarrow d \sqsubseteq (x \downarrow d)^*, \text{ and} \tag{35}$$
$$x^\omega \downarrow d \sqsubseteq (x \downarrow d)^\omega. \tag{36}$$

Assume that d is an element of the carrier set of a DARA that is universally conjunctive *sensu stricto*, then

$$x^\phi \downarrow d \sqsubseteq (x \downarrow d)^\phi, \text{ and} \tag{37}$$
$$x^\dagger \downarrow d \sqsubseteq (x \downarrow d)^\dagger. \tag{38}$$

The assumption that d is universally conjunctive and disjunctive *sensu stricto*, respectively, allows one to prove the validity of the axioms with respect to the predicate-transformer model along the lines of Back and von Wright [4]. The validity proofs make use of the Fusion Theorem, so the conditions are essential, since they imply continuity and co-continuity, respectively. It might be possible to do abstract-algebraic characterisations of continuity and co-continuity in terms of the algebra's operators along the lines of Meinicke and Solin [16] and from those conditions prove the above axioms as theorems; this is, however, outside the scope of the current paper.[4]

3.5 Enabledness and Termination

We now consider some basic properties of enabledness and termination in connection to encoding. These properties are new, and were not considered in the model. The next theorems concern a basic transformation rule, and structure-preserving data refinements of enabledness and termination, respectively.

[3] Back and von Wright [4] are able to strengthen the properties of Theorems 1, 2 and 3, but the concepts and the conditions they use are not readily expressible in DARA with the current extensions. For the conditions to be expressible, one would have to introduce a subalgebra of demonic and angelic *relational* updates; a research topic well worth pursuing. One can conjecture that this subalgebra would be a certain type of axiomatic relational algebra.

[4] Note that the axioms for * and † are not independent of the other axioms introduced so far, but can be derived by applying the appropriate induction axiom.

Theorem 8. *Let x and y be elements in the carrier set of a* DARA *extended with the operators of the previous sections. Then*

$$x \downarrow \epsilon x \sqsubseteq y \Leftrightarrow x \sqsubseteq y\epsilon x \quad \textit{and} \tag{39}$$
$$x \downarrow \tau x \sqsubseteq y \Leftrightarrow x \sqsubseteq y\tau x \tag{40}$$

hold true.

Proof. The proof is immediate from (17) and (20) and basic properties of enabledness and termination. □

Theorem 9. *Let x, y and d be elements in the carrier set of a* DARA *extended with the operators of the previous sections. Then*

$$\epsilon x \downarrow d \sqsubseteq \epsilon y \Leftrightarrow d\epsilon x \sqsubseteq (\epsilon y)d \quad \textit{and} \tag{41}$$
$$\tau x \downarrow d \sqsubseteq \tau y \Leftrightarrow d\tau x \sqsubseteq (\tau y)d \tag{42}$$

hold true. Moreover, if $d = y$, then

$$\epsilon x \downarrow y \sqsubseteq \epsilon y \Leftrightarrow y\epsilon x \sqsubseteq y \quad \textit{and} \tag{43}$$
$$\tau x \downarrow y \sqsubseteq \tau y \Leftrightarrow y\tau x \sqsubseteq y \tag{44}$$

hold true.

Proof. The proof of the first claim is immediate from (20) and basic properties of enabledness and termination. The second claim is immediate from the first. □

3.6 An Application

Action systems are a formalism for reasoning about concurrent programs [3]. On an abstract level, an action system can be viewed as a possibly nonterminating iteration of a set of atomic actions $x_1, x_2, \ldots, x_n$, that terminates when none of the actions are enabled. There are also initialising and finalising actions, y and z, respectively. In refinement algebra, an action system can be formulated as [21]:

$$y; \mathsf{do}\, x_1 \| \ldots \| x_n\, \mathsf{od}; z \quad =_{df} \quad y(x_1 \sqcap \cdots \sqcap x_n)^{\omega}\overline{\epsilon x_1} \ldots \overline{\epsilon x_n} z.$$

Refinement algebra has earlier proved applicable to structural reasoning about action systems [22, 21, 16, 19, 20, 11]. We shall next consider a structural property of action systems in connection to the encoding operator, which has not earlier been studied in the literature. It is enough to consider an action system with two actions, x and y, and one can simply ignore the initialisation and the finalisation. Consider then the following, with d universally disjunctive *sensu stricto*:

$$
\begin{aligned}
& (x \sqcap y)^{\omega} \overline{\epsilon x}\ \overline{\epsilon y} \downarrow d \\
\sqsubseteq\ & \{\text{encoding property (24)}\} \\
& ((x \sqcap y)^{\omega} \downarrow d)(\overline{\epsilon x}\ \overline{\epsilon y} \downarrow d) \\
\sqsubseteq\ & \{\text{encoding property (36)}\} \\
& ((x \sqcap y) \downarrow d)^{\omega}(\overline{\epsilon x}\ \overline{\epsilon y} \downarrow d) \\
\sqsubseteq\ & \{\text{encoding property (25)}\} \\
& ((x \downarrow d) \sqcap (y \downarrow d))^{\omega}(\overline{\epsilon x}\ \overline{\epsilon y} \downarrow d) \\
\sqsubseteq\ & \{\text{encoding property (24)}\} \\
& ((x \downarrow d) \sqcap (y \downarrow d))^{\omega}(\overline{\epsilon x} \downarrow d)(\overline{\epsilon y} \downarrow d).
\end{aligned}
$$

The result easily generalises to action systems with any number of actions. It shows that if one wants to reason about the least data refinement of an action system, it is possible to reason only about the individual actions. Note that depending on the abstraction statement d, the ending enabledness statements can be refined further using Theorems 3 and 9. Note also that results of this kind can be derived even more directly when the explicit characterisation of encoding is used (forward data refinement). As usual, the abstract-algebraic framework makes the derivation much more perspicuous than if the result had been derived in the model.

4 Decoding Axiomatically

In this section the decoding is briefly considered axiomatically. Encoding is the least data refinement of a program x under the abstraction statement d. Decoding is the greatest program that can be data refined into y under the abstraction statement d. So the encoding is the least solution to the equation $dx \sqsubseteq \xi d$, whereas decoding is the greatest solution to the equation $d\xi \sqsubseteq yd$, when ξ represents the unknown. Unlike encoding, decoding does not always exist and must be axiomatised conditionally with the condition of universal disjunctivity *sensu stricto* [4]. It takes the following form: if d is universally disjunctive *sensu stricto*, then

$$x \uparrow d \sqsubseteq_d x, \tag{45}$$

$$y \sqsubseteq_d x \Rightarrow y \sqsubseteq (x \uparrow d). \tag{46}$$

It can again be shown axiomatically that these axioms are equivalent to

$$x \sqsubseteq y \uparrow d \Leftrightarrow x \sqsubseteq_d y. \tag{47}$$

And it is easy to see that decoding is monotone in the first argument.

The following theorem, which connects encoding and decoding, is immediate from (20) and (47).

Theorem 10. *Let x and d be elements in the carrier set of a* DARA *extended with the operators of the previous sections. Let d be universally disjunctive. Then*

$$x \sqsubseteq (x \downarrow d) \uparrow d \text{ and} \tag{48}$$

$$x \sqsubseteq (x \uparrow d) \downarrow d \tag{49}$$

hold true.

A number of structure-preserving refinements can be proved for decoding. Moreover, decoding can also be given an explicit characterisation like the one for forward data refinement ($x \uparrow d = \breve{d}xd$). For reason of space, the report of these properties is limited to the following highly interesting theorem, which follows axiomatically from the previous theorems together with basic properties of program inversion.

Theorem 11. *Let x, y and d be elements in the carrier set of a* DARA *extended with the operators of the previous sections. Let d be universally disjunctive. Then all the following,*

$$x \downarrow d \sqsubseteq y, \quad x \sqsubseteq y \uparrow d, \tag{50}$$

$$dx\breve{d} \sqsubseteq y, \quad x \sqsubseteq \breve{d}yd, \tag{51}$$

$$dx \sqsubseteq yd, \quad x\breve{d} \sqsubseteq \breve{d}y, \tag{52}$$

are equivalent.

5 Conclusion and Future Work

The contributions of this paper, listed in the Introduction, show that refinement algebra has reached a considerable level of maturity: the foundational work done in the earlier papers lets one approach the rather advanced theory of encoding and decoding abstract-algebraically. Nevertheless, some work still remains to be done when it comes to algebraisation of the refinement calculus; in particular, it would be pivotal to axiomatise the relational updates. Even more important would be to do larger case studies using refinement algebra, in order to properly evaluate its applicability and also in order to collect heuristic techniques that can be used in practice. For the case studies, making use of the available mechanisations (such as [2, 11]) would be of high importance.

Considering encoding and decoding in the elegant and very abstract framework of Walter Guttmann is likely to yield important insights into the operators' place in overall sequential program algebra [7]. It would also be interesting to consider the operators in connection to Preoteasa's work [18]. From a concurrent perspective, the operators could be considered in connection to concurrent Kleene algebra and in connection to rely/guarantee Kleene algebra [9, 1]. For the possibility to guarantee total correctness in concurrency, the addition of the operators to Ian Hayes's total-correctness algebra for rely/guarantee reasoning would be a very interesting route to take [8].

Acknowledgements. For helpful discussions on data refinement, I am grateful to Larissa Meinicke, Ian Hayes and Robert Colvin at The University of Queensland, and to Cliff Jones and Nisansala Yatapanage at Newcastle University. Anonymous referees helped improve the presentation. The work was supported by The Australian Research Council, Grant DP130102901.

References

1. Armstrong, A., Gomes, V.B.F., Struth, G.: Algebraic principles for rely-guarantee style concurrency verification tools. In: Jones, C., Pihlajasaari, P., Sun, J. (eds.) FM 2014. LNCS, vol. 8442, pp. 78–93. Springer, Heidelberg (2014)
2. Armstrong, A., Gomes, V.B.F., Struth, G.: Kleene algebra with tests and demonic refinement algebras. Archive of Formal Proofs (2014)
3. Back, R.-J., Kurki-Suonio, R.: Distributed cooperation with action systems. ACM Trans. Program. Lang. Syst. 10(4), 513–554 (1988)
4. Back, R.-J., von Wright, J.: Encoding, decoding and data refinement. Formal Asp. Comput. 12(5), 313–349 (2000)
5. Cohen, E.: Separation and reduction. In: Backhouse, R., Oliveira, J.N. (eds.) MPC 2000. LNCS, vol. 1837, pp. 45–59. Springer, Heidelberg (2000)
6. Desharnais, J., Möller, B., Struth, G.: Kleene algebra with domain. ACM Trans. Comput. Log. 7(4), 798–833 (2006)
7. Guttmann, W.: Algebras for correctness of sequential computations. Sci. Comput. Program. 85, 224–240 (2014)
8. Hayes, I.J.: Generalised rely-guarantee concurrency: An algebraic foundation. Unpublished manuscript, The University of Queensland (February 2015)
9. Hoare, T., Möller, B., Struth, G., Wehrman, I.: Concurrent Kleene algebra and its foundations. J. Log. Algebr. Program. 80(6), 266–296 (2011)
10. Höfner, P., Khédri, R., Möller, B.: An algebra of product families. Software and System Modeling 10(2), 161–182 (2011)
11. Höfner, P., Struth, G.: Automated reasoning in Kleene algebra. In: Pfenning, F. (ed.) CADE 2007. LNCS (LNAI), vol. 4603, pp. 279–294. Springer, Heidelberg (2007)
12. Kozen, D.: Kleene algebra with tests. ACM Trans. Program. Lang. Syst. 19(3), 427–443 (1997)
13. McIver, A., Gonzalia, C., Cohen, E., Morgan, C.C.: Using probabilistic Kleene algebra pKA for protocol verification. J. Log. Algebr. Program. 76(1), 90–111 (2008)
14. McIver, A., Meinicke, L., Morgan, C.: Hidden-Markov program algebra with iteration. Mathematical Structures in Computer Science 25(2), 320–360 (2015)
15. Meinicke, L., Hayes, I.: Probabilistic choice in refinement algebra. In: Audebaud, P., Paulin-Mohring, C. (eds.) MPC 2008. LNCS, vol. 5133, pp. 243–267. Springer, Heidelberg (2008)
16. Meinicke, L., Solin, K.: Refinement algebra for probabilistic programs. Formal Asp. Comput. 22(1), 3–31 (2010)
17. Nelson, G.: A generalization of Dijkstra's calculus. ACM Trans. Program. Lang. Syst. 11(4), 517–561 (1989)
18. Preoteasa, V.: Refinement algebra with dual operator. Sci. Comput. Program. 92, 179–210 (2014)
19. Solin, K.: Normal forms in total correctness for while programs and action systems. J. Log. Algebr. Program. 80(6), 362–375 (2011)
20. Solin, K.: Dual choice and iteration in an abstract algebra of action. Studia Logica 100(3), 607–630 (2012)
21. Solin, K., von Wright, J.: Enabledness and termination in refinement algebra. Sci. Comput. Program. 74(8), 654–668 (2009)
22. von Wright, J.: Towards a refinement algebra. Sci. Comput. Program. 51(1-2), 23–45 (2004)

Type Checking by Domain Analysis in Ampersand

Stef M.M. Joosten[1,2] and Sebastiaan J.C. Joosten[3,4]

[1] Open Universiteit Nederland, P.O. Box 2960, HEERLEN, The Netherlands
[2] Ordina NV, Nieuwegein
[3] Eindhoven University of Technology, P.O. Box 513, Eindhoven, The Netherlands
[4] Radboud University, Nijmegen, The Netherlands
stef.joosten@ou.nl

Abstract. In the process of incorporating subtyping in relation algebra, an algorithm was found to derive the subtyping relation from the program to be checked. By using domain analysis rather than type inference, this algorithm offers an attractive visualization of the type derivation process. This visualization can be used as a graphical proof that the type system has assigned types correctly. An implementation is linked to in this paper, written in Haskell. The algorithm has been tried and tested in Ampersand, a language that uses relation algebra for the purpose of designing information systems.

1 Introduction

In building information systems, the challenge is to translate business policy into a running system that can support that policy. According to the Business Rules Method [1], a business policy is best described by a set of agreements called business rules. In our view, creating an information system should *only* involve writing down these agreements in a formal language, a compiler should do the rest. Ampersand is a project in which such a compiler was developed [2]. It uses a variant of heterogeneous relation algebra [3] as formal language for three reasons: relation algebra is suited for symbolic manipulation, relation algebra is close to natural language, and relation algebra is easily implemented through SQL.

Specialization and overloading are used frequently in business rules, and Ampersand supports this. For this purpose, a slight modification of the heterogeneous relation algebra was implemented, and a new type system was developed. Van der Woude et al. describes this slightly modified algebra in [4]. In line with the typing rules of that paper, this paper describes a type system for Ampersand.

The type system has two purposes that are common for type systems. First, the type system should *assign one type* to every term in the script. This means that tools which use the script as input, have access to the type information, and can use it for its analysis. This way, a type system helps improve the output of such tools. Second, the type system should give *feedback to the user*. This includes alerting the user to type errors or warnings, and can even be feedback

W. Kahl et al. (Eds.): RAMiCS 2015, LNCS 9348, pp. 225–240, 2015.
DOI: 10.1007/978-3-319-24704-5_14

in the form of type derivations. This way, a type system helps the user to improve her code.

While type systems have been around for quite a while, we tried out a novel approach to typing. Our type system differs from conventional systems in two important ways:

- Even though type inference is quite common since the Hindley–Milner type system [5], rules for subtyping (aka generalization or specialization) are typically made explicit. This paper will show that type rules for subtyping can be derived too. This enabled a very uniform and appealing syntax in the Ampersand language, in which relations, domains and subdomains can be declared implicitly.
- Typically, type systems construct tree-shaped proofs in a type deduction system. Our system, however, uses a directed graph rather than a tree to derive types. This allows us to use simple and intuitive algorithms for doing type derivation. It also means that the type checker can explain its calculations by drawing the relevant portion of the graph.

The result appeals to our desire for simplicity and elegance. In order to validate this result beyond the usual toy examples, we have implemented the type system in the Ampersand compiler and tried it in practice. Soon, we found that the lack of distinction between rules and declarations led to unexpected results in scripts with errors. For practical reasons, it is imperative that erroneous scripts yield error messages that make sense to their authors. So we were forced to introduce the distinction between declarations (of relations), statements to specify generalization, and rules in Ampersand, which now has a more conventional type checker.

Still, the elegant properties remain. We present the type system in the hope that the reader finds joy in its elegance, finds another use for it, or finds the lessons learned from our attempt instructive.

2 Related Work

There is a large body of work on type systems. The rationale behind a good type system is to verify programs mechanically in order to prevent erroneous runs. Within the category of syntactically correct programs, a type system should distinguish those programs that can be interpreted semantically from those without such an interpretation. On top of that, a type system may forbid or warn against input that is unexpected.

The type system proposed in this article was designed to cater for subtyping in Ampersand. Already in the 90's it was recognized that subtyping increases the expressive power of a type system as well as the power of its mechanical verification [6–9]. However, this comes at a price. Subtyping can have performance penalties or it may incur problems with the decidability of type inference, for instance [10, 11]. All type systems we have studied assume there is, or define, an explicit subtype definition. The type system proposed in this paper derives the subtyping from the same program it analyses.

Undecidability of a type system does not mean it cannot be used in practice. Dynamically typed languages work around this issue by checking type correctness on run time, keeping track of type information at runtime. When some of that type information is calculated statically, the result is called hybrid type checking [12]. Our work takes the orthogonal approach: we derive type information, and refuse input for which this derivation failed. The user always has the option to add type information himself, in case the derivation was not powerful enough.

Ampersand is based on a variant of heterogeneous relation algebra described in [4]. Outside of the Ampersand project, we do not know of any type checker for this algebra. However, in the area of machine learning, the work of [13] seems to be working in the same setting. This work does not describe a type checker, but the notion of subclass coincides with ours, and applies to relations in the same way. In addition, the work - like ours - tries to derive subclass information without requiring additional input. Except for this work, and the scope of Ampersand, we could not find a setting in which the same variant of heterogeneous relation algebra would apply. Besides a shared notion of subsets in relation algebra, there are only differences: assigning a type to an entity for 'question answering', as solved in [13], at best gives a heuristic to perform dynamic type checking. Ampersand is designed for static type checking only.

3 Problem Definition

This paper focuses on the problem to assign to every term in Ampersand precisely one type, which is formulated in Equation 2. Ampersand interprets each concept[1] as a set and each relation as a set of pairs. Instances of a concept are called *atoms* and are elements in those sets. Ampersand lets its user model relations in combination with concepts. If, for example, the user defines a relation $account_{\langle Person,IBAN\rangle}$, she wants to be sure that the relation is populated with instances of those concepts, i.e. the relation contains pairs of persons and IBAN-numbers only. So, if ⟨`Peter`, `NL99BANK0123456789`⟩ is a pair within this relation, the type system must ensure that:

- `Peter` is an instance of *Person* and any concept that is more general (e.g. *LegalEntity*).
- `NL99BANK0123456789` is an instance of *IBAN* and any concept that is more general (e.g. *Accountnumber*).

The type system aims to:

- maintain the algebraic properties of homogeneous relation algebra (i.e. the axioms of Tarski).
- ensure that calculations with relational terms maintain type correctness.

[1] The notions *concept*, *atom*, *relation*, *term*, and *rule* are defined formally in section 4.

These first two requirements have been elaborated and published [4]. This study by van der Woude and Joosten analyses how heterogeneous relation algebra fails to maintain all of Tarski's axioms. It blames the complement operator and solves that problem by using the binary minus operator instead. That result is used in Equation 18 of the current paper. In everyday use, this means that one of Tarski's axioms is restricted in a way that does not inhibit practical applications.

Ampersand poses requirements to the type system. Some of these were trivial to implement, and have been omitted from the details in the paper. Others will be treated in the sections hereafter:

- By typing a term, we restrict the values it may have at run time. This can improve run time performance in many cases. In Ampersand, every term gets exactly one type. This requirement is called soundness, and is formalized at the end of this section in Equation 2.
- In order to make it easier to reuse names, the type system must allow overloading.
  For example, a user is free to declare both
  $$\mathit{account}_{\langle Person,IBAN\rangle} \quad \text{and}$$
  $$\mathit{account}_{\langle Purchase,IBAN\rangle}$$
  in the same script. This introduces two different relations, just because *Person* and *Purchase* are distinct concepts. Upon each use of a declaration, the type system must allow omission of the type in cases where no confusion can arise. In this paper, overloading of relations will not be discussed.
- In order to facilitate collaborative development and code reuse, the type system must cope with specialization.
  For example, if an apartment is a home, Ampersand allows the user to state: $\mathit{Apartment} \preceq \mathit{Home}$. This means that every instance of *Apartment* is an instance of *Home* as well. The consequence is that all relations that work with *Home* are applicable to *Apartment* as well. In this interpretation, the word generalization has the same meaning and may be used as a synonym. Specialization can be helpful for reusing code that was written for *Home*. Allowing specialization will be an emergent property of our type system
- The type system must allow intuitive explanation of results to users. We aimed to achieve this by using a graph as visualization of our type system.

Relation algebra uses binary relations. In Ampersand, each relation r has a type $\mathfrak{T}(r) = \langle A, B\rangle$, which is a tuple of concepts. As a shorthand for 'r with type $\langle A, B\rangle$', we write $r_{\langle A,B\rangle}$. A relation contains a set of pairs, the elements of which are called *atoms*. Each pair $\langle a, b\rangle$ has a source atom a and a target atom b. To distinguish between a relation and the set of pairs it contains, we use a function $\mathfrak{I}(\cdot)$. We will also use $\mathfrak{I}(\cdot)$ to indicate the set of atoms in concepts and types. This is defined formally in Definition 2.

Every atom is an element of a concept, which is called the type of that atom. Formally, the typing of two atoms a and b in a relation $r_{\langle A,B\rangle}$ is described by:

$$\langle a, b\rangle \in \mathfrak{I}(r_{\langle A,B\rangle}) \Rightarrow a \in \mathfrak{I}(A) \ \wedge \ b \in \mathfrak{I}(B) \tag{1}$$

The function $\mathfrak{I}(\cdot)$ is not available to the type system, so to ensure Property 1, the type system will reason with type terms rather than atoms. For each term t there are the type-terms $dom(t)$ and $cod(t)$[2]. For each concept C, there is the type-term $pop(C)$. The set of all source atoms in a relation r is indicated by $dom(r)$ (pronounced: domain of r) and the set of all target atoms in that relation is indicated by $cod(r)$ (codomain of r). The set of all atoms that are an instance of concept C can be indicated by $pop(C)$ (population of C). Note that $\mathfrak{I}(C)$ (the interpretation of the concept C) is equal to $\mathfrak{I}(pop(C))$ (the interpretation of the concept C interpreted as a type-term). Similar equalities will be used to established the soundness of our algorithm.

The problem for the type system to solve is to assign to every term t precisely one type $\mathfrak{T}(t)$ such that:

$$\mathfrak{T}(t) = \langle A, B\rangle \ \Rightarrow\ \mathfrak{I}(dom(t)) \subseteq \mathfrak{I}(pop(A)) \ \wedge\ \mathfrak{I}(cod(t)) \subseteq \mathfrak{I}(pop(B)) \qquad (2)$$

This equation specifies soundness. The task of the type algorithm is to ensure that every term t has precisely one type $\mathfrak{T}(t)$ and to decide whether a script can satisfy Property 2 at runtime.

4 Definitions

In order to describe the type system, we need definitions of the notions *atom*, *concept*, *relation*, *term*, and *rule*.

Since we describe a type system, it is not necessary for the reader to know what an Ampersand script does. Nevertheless, it might help with the intuitions, so we give those here: A script contains a set of rules. A rule is the equality between two terms, built from relations. Rules can be thought of as invariants throughout the execution of a program. Based on the relations used in them, Ampersand will generate a database and some interfaces. The phase in which Ampersand takes a script, and turns it into a database, is what we will refer to as compile time. The phase in which a (possibly different) user interacts with the database, is what we will refer to as runtime. The database can then be used to calculate which atom-pairs are in the relations, from which Ampersand can decide whether all rules are satisfied. If not, the last change is reverted, returning the database to a 'safe' state. The changes to the database are not specified at compile time, but given by the database user at runtime. This means that Ampersand scripts do not have a notion of execution, and that atoms are only a runtime concept.

Atoms are values that have no internal structure, meant to represent data elements in a database. From a business perspective, atoms are used to represent concrete items of the world, such as `Peter`, `1`, or `the king of France`. By convention throughout the remainder of this paper, variables a, b, and c are used to represent *atoms*. The set of all atoms is called $\mathbb{A}$. Each atom is an instance of a *concept*.

[2] Later, we will also introduce $inter(s, t)$ as a type-term.

Concepts are names we use to classify atoms in a meaningful way. For example, you might choose to classify `Peter` as a person, and `074238991` as a telephone number. We will use variables A, B, C, D to represent concepts. Let us call the set of all concepts in an Ampersand script $\mathbb{C}$. The expression $a \in A$ means that atom a is an instance of concept A. In the syntax of Ampersand, concepts form a separate syntactic category, allowing a parser to recognize them as concepts. The declaration of $A \preceq B$ (pronounce: A is a B) in an Ampersand script states that any instance of A is an instance of B as well. We call this *specialization*, but it is also known as *generalization* or *subtyping*. Specialization is needed to allow statements such as: "An orange is a fruit that ....".

Relations are used to represent sets of facts (i.e. statements that are true in a business context), to be stored and maintained as data in a computer. As data changes over time, so do the contents of these relations. In this paper relations are represented by variables r, s, and d. We represent the declaration of a relation r in an Ampersand script by $r_{\langle A,B\rangle}$, in which A is the source concept and B the target concept. The relation $\mathbb{I}_A$ represents the *identity relation* of concept A. The relation $\mathbb{V}_{A\times B}$ represents the *universal relation* over concepts A and B. The set of all identifiers that represent relations in an Ampersand script, is called $\mathbb{D}$. It is defined by:

$$r \in \mathbb{D} \text{ iff } r_{\langle A,B\rangle} \text{ occurs in the Ampersand script.} \tag{3}$$

The meaning of relations in Ampersand is defined by an interpretation function $\mathfrak{I}$. It maps each relation to a set of facts. The declaration of $r_{\langle A,B\rangle}$ implies $r \in \mathbb{D}$, and $A, B \in \mathbb{C}$. Furthermore, it is a runtime requirement that the pairs in r are contained in its type:

$$\langle a,b\rangle \in \mathfrak{I}(r) \Rightarrow\ a \in \mathfrak{I}(pop(A)) \wedge b \in \mathfrak{I}(pop(B)) \tag{4}$$

Terms are used to combine relations using operators. The set of terms is called $\mathbb{T}$. It is defined by:

Definition 1 (terms)
The set of terms, $\mathbb{T}$, is the smallest set that satisfies, for $r, s \in \mathbb{T}$, $d \in \mathbb{D}$ and $A, B \in \mathbb{C}$.

$$d \in \mathbb{T} \quad \textit{(every relation is a term)} \tag{5}$$
$$(r \cap s) \in \mathbb{T} \quad \textit{(intersection)} \tag{6}$$
$$(r - s) \in \mathbb{T} \quad \textit{(difference)} \tag{7}$$
$$(r; s) \in \mathbb{T} \quad \textit{(composition)} \tag{8}$$
$$r^{\smile} \in \mathbb{T} \quad \textit{(converse)} \tag{9}$$
$$\mathbb{I}_A \in \mathbb{T} \quad \textit{(identity)} \tag{10}$$
$$\mathbb{V}_{A\times B} \in \mathbb{T} \quad \textit{(full set)} \tag{11}$$

Throughout the remainder of this paper, terms are represented by variables r, s, d, and t.

The meaning of terms in Ampersand is an extension of interpretation function $\mathfrak{I}$. Let A and B be finite sets of atoms, then $\mathfrak{I}$ maps all terms to the set of facts for which that term stands.

Definition 2 (interpretation of terms)
For every $A, B \in \mathbb{C}$ and $r, s \in \mathbb{T}$

$$\mathfrak{I}(r \cap s) = \{\langle a,b\rangle \mid \langle a,b\rangle \in \mathfrak{I}(r) \text{ and } \langle a,b\rangle \in \mathfrak{I}(s)\} \tag{12}$$
$$\mathfrak{I}(r - s) = \{\langle a,b\rangle \mid \langle a,b\rangle \in \mathfrak{I}(r) \text{ and } \langle a,b\rangle \notin \mathfrak{I}(s)\} \tag{13}$$
$$\mathfrak{I}(r; s) = \{\langle a,c\rangle \mid \text{for some } b,\ \langle a,b\rangle \in \mathfrak{I}(r) \text{ and } \langle b,c\rangle \in \mathfrak{I}(s)\} \tag{14}$$
$$\mathfrak{I}(r^{\smile}) = \{\langle b,a\rangle \mid \langle a,b\rangle \in \mathfrak{I}(r)\} \tag{15}$$
$$\mathfrak{I}(\mathbb{I}_A) = \{\langle a,a\rangle \mid a \in A\} \tag{16}$$
$$\mathfrak{I}(\mathbb{V}_{A\times B}) = \{\langle a,b\rangle \mid a \in A, b \in B\} \tag{17}$$

In fact, Ampersand has even more operators: the complement (prefix unary $-$), Kleene closure operators (postfix $^+$ and *), left- and right residuals (infix $\backslash$ and $/$), relational addition (infix $\dagger$), and product (infix $\times$). These do not introduce new concepts, just more terms. We have constrained this exposition to the operators mentioned above, which is sufficient for explaining the type system.

To solve the problems with the complement operator in heterogeneous relation algebra [4], Ampersand uses a binary difference operator as in Equation 7. It is used to define a complement as a unary prefix operator $-$ for relations of which the type is known.

$$\mathfrak{T}(r) = \langle A, B\rangle \ \Rightarrow\ -r \ =\ \mathbb{V}_{A\times B} - r \tag{18}$$

After approval of the script by the type system, every term has a unique type. Since scripts with type errors cannot be executed, ordinary users of Ampersand never get to see any behavior of the complement other than they can predict with Tarski's axioms.

After defining concepts, relations and terms, let us now define rules. Rules are used to impose constraints on the data in relations.

A rule is a pair of terms $r, s \in \mathbb{T}$. We indicate the set of all rules in an Ampersand script by $\mathbb{R}$. To indicate that a pair of terms (r, s) is in $\mathbb{R}$, we will write:

$$\text{RULE } r = s$$

The rule RULE $r = s$ imposes the following restriction on the data in Ampersand:

$$\mathfrak{I}(r) = \mathfrak{I}(s)$$

For a user, this means that a rule restricts the possible populations in the database to those that satisfy the rule.

Note that a declaration $r_{\langle A,B\rangle}$ in Ampersand can be represented by the following rule:

$$\text{RULE } r = r \cap \mathbb{V}_{A\times B}$$

5 Domain Analysis

The core idea of the proposed type algorithm is an analysis of domains. A domain is a set of atoms. We introduce type-terms to represent such sets; our algorithm will act on type-terms. There are four functions that yield type-terms: $dom(\cdot)$, $cod(\cdot)$, $inter(\cdot,\cdot)$ and $pop(\cdot)$, with the following interpretation.

Definition 3 (Interpretation of type-terms)
For every $a, b \in \mathbb{A}$ *and* $r, s \in \mathbb{T}$

$$\mathfrak{I}(dom(r)) = \{a \mid \langle a, b\rangle \in \mathfrak{I}(r)\} \tag{19}$$

$$\mathfrak{I}(cod(r)) = \{b \mid \langle a, b\rangle \in \mathfrak{I}(r)\} \tag{20}$$

$$\mathfrak{I}(inter(r, s)) = \mathfrak{I}(cod(r)) \cap \mathfrak{I}(dom(s)) \tag{21}$$

$$\mathfrak{I}(pop(A)) = \mathfrak{I}(A) \tag{22}$$

Note that we use the word *type-term* to indicate the intermediate structures used by the type system. The word *type* is used for a pair of concepts.

Since $\mathfrak{I}(cod(r)) = \mathfrak{I}(dom(r^{\smile}))$, we can treat $cod(r)$ as a shorthand notation for $dom(r^{\smile})$. Similarly, $\mathfrak{I}(pop(A)) = \mathfrak{I}(A)$, so we can treat $pop(\cdot)$ as a shorthand notation, too. In fact, we could have avoided introducing $cod(\cdot)$ and $pop(\cdot)$, and rely solely on $dom(\cdot)$. However, the use of $cod(\cdot)$ and $pop(\cdot)$ makes it easier for the reader to keep track of the terms involved in these type terms. We will treat $inter(s, t)$ as is, without creating a shorthand: the obvious shorthand $dom(s \cap t^{\smile})$ contains a term that is not necessarily well typed.

Listing 1.1. A type correct Ampersand script

```
RELATION r[A*C]
RELATION s[A*B]
RELATION t[B*C]
RULE r = s;t
```

Listing 1.1 introduces three relations, `r[A*C]`, `s[A*B]`, and `t[B*C]` and one rule: "`r = s;t`". This rule has four terms: "`r`", "`s;t`", "`s`", and "`t`". Domain analysis can be used to derive the type of expressions. For example, in the context of listing 1.1, we can derive:

$$\begin{aligned} & \mathfrak{I}(dom(s;t)) \\ \subseteq\ & \mathfrak{I}(dom(s)) \\ =\ & \mathfrak{I}(dom(s_{A*B})) \\ \subseteq\ & \mathfrak{I}(pop(A)) \end{aligned}$$

$$\begin{aligned} & \mathfrak{I}(cod(s;t)) \\ \subseteq\ & \mathfrak{I}(cod(t)) \\ =\ & \mathfrak{I}(cod(t_{B*C})) \\ \subseteq\ & \mathfrak{I}(pop(C)) \end{aligned}$$

Together these two calculations prove that term $s;t$ has type $\langle A, C\rangle$.

The domain analysis introduces a relation *sub* between type terms. The intention is that *sub* is the counterpart of the subset relation $\subseteq$. It translates the observation that $\mathfrak{I}(dom(s;t)) \subseteq \mathfrak{I}(dom(t))$ to type terms by stating that $dom(s;t)$ *sub* $dom(t)$. By constructing the *sub* relation consistently for all operators in every term in listing 1.1, the algorithm from section 6 constructs a type-graph. The result is shown in figure 1. Every vertex (ellipse) in this graph represents a set of type-terms: in this particular example, the type system has distinguished 10 distinct sets. Type-terms that are proven to represent the same set of atoms are printed inside the same ellipse. A directed path from a vertex containing type-term t_1 to a vertex containing type-term t_2, indicates that atoms in $\mathfrak{I}(t_1)$ are also in $\mathfrak{I}(t_2)$. Each derivation such as the two above can be seen as a path in a type graph. In fact, the type graph even shows shorter derivations, when possible.

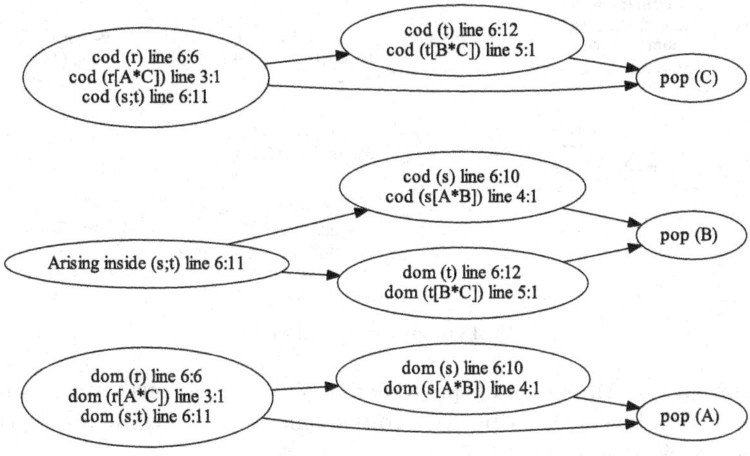

Fig. 1. Type graph for Listing 1.1

The type graph can be interpreted as a collection of derivations, that can be used to prove that an expression has a certain type. Because the graph itself contains all proofs that all terms have precisely one type, there is no need to write down all calculations. The graph itself can serve as a compact representation of all the necessary calculations.

Let a second example illustrate how mistakes are analyzed. Consider a script with a type error in Listing 1.2. The type graph of this script is represented in Figure 2. One of the errors is a mismatch between *cod*(s) and *dom*(t) (on line 6 position 11). Because Ampersand treats concepts `A` and `B` as sets with an empty intersection, this is treated as an error. We have highlighted that error in red dashed arrows in Figure 2. Another error is for example that the domain of `r` (on line 6 position 6) is *sub* of `A` and also of `B`. In general, every term that leads to more than one pop-vertex is erroneous, as well as the terms that lead to no

Listing 1.2. A type incorrect Ampersand script

```
3 RELATION r[A*C]
4 RELATION s[B*A]
5 RELATION t[B*C]
6 RULE r = s;t
```

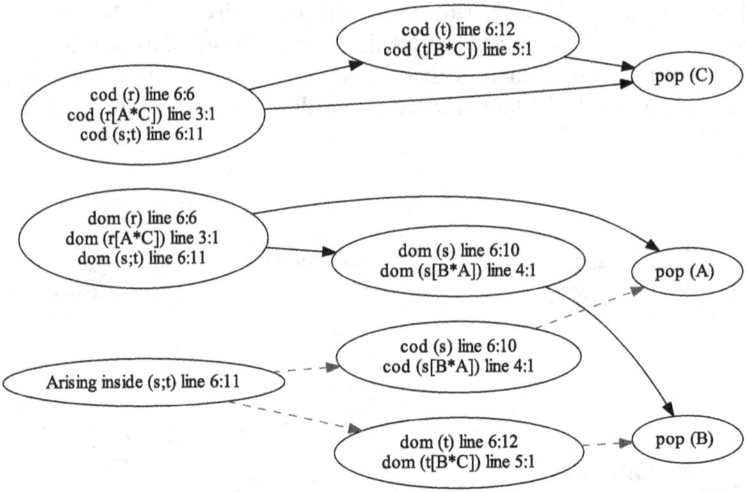

Fig. 2. Type graph for Listing 1.2

pop-vertex. In the Ampersand type system, a single erroneous term is sufficient reason to reject the entire script. As a consequence, every term in a type-correct script has precisely one type.

6 Algorithm

This section explains the algorithm of the proposed type system. A toy version of the domain analysis can be found online[3]. It is meant to communicate the idea and play with it in GHCi.

The algorithm for type checking is an algorithm on type graphs, for which we introduce a set of vertices $\mathbb{N}$ and a set of edges $sub_{\langle \mathbb{N},\mathbb{N} \rangle}$. Each vertex v represents a set of atoms. The relation sub represents edges in the type graph. So $v\ sub\ v'$ represents an edge from vertex v to vertex v', which means that the set of atoms v is a subset of v'.

[3] `http://cs.ru.nl/~B.Joosten/ampTypes/`

For each term in the script, edges are computed using the following rules.

$$dom(\mathbb{V}_{A\times B})\ sub\ pop(A) \wedge cod(\mathbb{V}_{A\times B})\ sub\ pop(B) \tag{23}$$

$$dom(r_{\langle A,B\rangle})\ sub\ pop(A) \wedge cod(r_{\langle A,B\rangle})\ sub\ pop(B) \tag{24}$$

$$dom(\mathbb{I}_{A})\ sub\ pop(A) \wedge cod(\mathbb{I}_{A})\ sub\ pop(A) \tag{25}$$

$$dom(r \cap s)\ sub\ dom(r) \wedge dom(r \cap s)\ sub\ dom(s) \tag{26}$$

$$cod(r \cap s)\ sub\ cod(r) \wedge cod(r \cap s)\ sub\ cod(s) \tag{27}$$

$$dom(r - s)\ sub\ dom(r) \wedge cod(r - s)\ sub\ cod(r) \tag{28}$$

$$dom(r; s)\ sub\ dom(r) \wedge cod(r; s)\ sub\ cod(s) \tag{29}$$

$$dom(r^{\smile})\ sub\ cod(r) \wedge cod(r^{\smile})\ sub\ dom(r) \tag{30}$$

$$dom(r)\ sub\ cod(r^{\smile}) \wedge cod(r)\ sub\ dom(r^{\smile}) \tag{31}$$

Note that we cannot state that $pop(A)\ sub\ dom(\mathbb{V}_{A\times B})$, because B can be empty. In that case, $\mathfrak{I}(dom(\mathbb{V}_{A\times B}))$ is a proper subset of $\mathfrak{I}(pop(A))$. The composition term $r; s$ needs a set of atoms beside the domain and codomain. The reason is that the interpretation $\mathfrak{I}(r; s)$ contains an existential quantifier, for which we require a type. For that purpose we introduce $inter(r, s)$ and the following rule:

$$inter(r, s)\ sub\ cod(r)\ \wedge\ inter(r, s)\ sub\ dom(s) \tag{32}$$

In addition, we add edges for every rule RULE $r = s$:

$$dom(r)\ sub\ dom(s) \wedge cod(r)\ sub\ cod(s) \wedge dom(s)\ sub\ dom(r) \wedge cod(s)\ sub\ cod(r) \tag{33}$$

Recall that RULE $r = s$ implies $\mathfrak{I}(r) = \mathfrak{I}(s)$. Using Definition 2, the reader should now be able to verify that Equations 23 to 33 all satisfy:

$$n_1\ sub\ n_2 \Rightarrow \mathfrak{I}(n_1) \subseteq \mathfrak{I}(n_2) \tag{34}$$

This means that an arrow connecting domains n_1 and n_2 in Figures 1 and 2 may be read as a subset relation that has been recognized (by the type algorithm) between these domains. In fact, Equations 23 through 33 describe the edges between vertices for all possible terms. As the compiler traverses the parse tree recursively, it visits all terms in the script and collects relevant edges on the way.

The type system must establish that each term gets precisely one type. It does so by taking all $pop(\cdot)$ vertices it encounters when traversing the graph from the term that is to be typed. For this, we introduce the relation of pre-types $P : \mathbb{T}\times\mathbb{C}$ using sub^* as the reflexive transitive closure of sub:

$$P = \{\langle x, C\rangle \mid x\ sub^*\ pop(C)\} \tag{35}$$

The relation P is total for terms: for every domain or subdomain x, there is a term $pop(C)$ such that $\langle x, C\rangle \in P$. This holds because we require every declaration to be declared with a type. So according to Equations 23 to 25, P is total for those. The only terms we can construct according to Definition 1 are terms that are smaller than the declarations they are made of. So for every term x:

$$(\exists_{C\in\mathbb{C}}\ \langle dom(x), C\rangle \in P) \wedge (\exists_{C\in\mathbb{C}}\ \langle cod(x), C\rangle \in P) \tag{36}$$

Using this property, we could easily make a type system that is complete in the sense that it assigns a type to every term, by picking any such concept arbitrarily. However, the fact that there is a choice often indicates a mistake of an Ampersand user. So, instead of choosing an arbitrary concept, the compiler emits an error message forcing the user to make that choice.

If one of these vertices is smallest with respect to sub^*, it is used as the type for that term. If not, a type error is shown to the user. In other words, the typing function is governed by these rules:

$$\mathfrak{T}(t) = \langle A, B\rangle \Rightarrow \langle dom(t), A\rangle \in P \wedge \langle cod(t), B\rangle \in P \quad (37)$$

$$\mathfrak{T}(t) = \langle A, B\rangle \wedge \langle dom(t), A'\rangle \in P \Rightarrow \langle pop(A), A'\rangle \in P \quad (38)$$

$$\mathfrak{T}(t) = \langle A, B\rangle \wedge \langle cod(t), B'\rangle \in P \Rightarrow \langle pop(B), B'\rangle \in P \quad (39)$$

In this manner, type checking involves computing a Kleene closure over the relation sub. The compiler uses the Warshall algorithm for computing the closure, giving it polynomial ($O(n^3)$ with n the number of type-terms) complexity.

7 Fulfillment of Requirements

In the previous sections, we have explained and illustrated how the type system works. The soundness of the type system is specified by Equation 2. Equations 35 and 37 yield:

$$\mathfrak{T}(t) = \langle A, B\rangle \Rightarrow dom(t)\ sub^*\ pop(A) \wedge cod(t)\ sub^*\ pop(B)$$

Together with Equation 34, and transitivity of $\subseteq$, this yields:

$$\mathfrak{T}(t) = \langle A, B\rangle \Rightarrow \mathfrak{I}(dom(t)) \subseteq \mathfrak{I}(pop(A)) \wedge \mathfrak{I}(cod(t)) \subseteq \mathfrak{I}(pop(B))$$

So we have established soundness, i.e. our type system satisfies Equation 2.

For sound scripts, P (Equation 35) is total. Since P associates one or more concepts to the domain or codomain of every term, P is a total relation. We have chosen to implement the algorithm by choosing the smallest concept instead of an arbitrary one. In the Example from Listing 1.2, the concepts P associates to $inter(s, t)$ are A and B: choosing an arbitrary one would be unsound. If the smallest concept is not unique, as in our example, the type algorithm emits an error message that forces the user to make a choice. By resolving such errors detected by Ampersand, the user can more easily be alerted to unintended scripts. In an Ampersand script without type errors, each term gets a unique type in a way that may be more predictable for the user.

The type graph can be used for a visual check: If for every vertex v there is a path to precisely one smallest $pop(\cdot)$ vertex, the script has no errors. This is precisely what was illustrated in figures 1 and 2. A reviewer of this paper suggested highlighting the relevant paths in the graph for each type and type error, which we think would be a great tool to assist Ampersand users.

The user has syntax to specialize concepts, e.g. $A \preceq B$ as defined in Section 4. The type system can already handle such statements, if we assume them to be syntax sugar for rules. The statement, $A \preceq B$ can be expressed as:

$$\text{RULE } \mathbb{I}_A = \mathbb{I}_B \cap \mathbb{I}_A \tag{40}$$

By Equation 33 the type algorithm calculates the right edges in the type graph, from which a pre-order of concepts is constructed.

The user is given control over intersection of concepts by insisting that every set must have a type. If the type algorithm computes an intersection set between two concepts, but that vertex is not associated a $pop(\cdot)$ of some concept, it is forbidden. This gives the user full control, because he can always add a statement of the same form as Equation 40 to make the type system accept what he wanted.

The type graph can be interpreted in term of good old sets, and it is easy to see that it represents type calculations as was done with figure 1. In our teaching practice, this has shown to appeal to the intuition of students, making it easy to explain.

The static typing is enabled by this type system because it reasons solely with concepts and not with atoms. For this reason, the implementation in the Ampersand compiler aborts after emitting type errors, or it proceeds if there are no mistakes.

8 Discussion

The authors appreciate domain analysis for visualizing the type system, as illustrated in figures 1 and 2. For large scripts, the type graph loses value because the user loses overview. But in smaller, yet complicated scripts, it gives an insightful outlook both on the correctness of the script as of its incorrectness.

In some cases, the type system might reject a script in which a user has correctly represented a desirable situation. For example, this script will be rejected:

$$\begin{array}{ll} \text{DECLARE} & r_{\langle A,B\rangle} \\ \text{DECLARE} & s_{\langle C,D\rangle} \\ \text{RULE} & \mathbb{I}_E \quad = \mathbb{I}_B \cap \mathbb{I}_C \\ \text{RULE} & r;s \quad = \mathbb{V}_{A\times D} \end{array}$$

In this script, the first two lines declare the relations r and s. The third line introduces a concept E, for which the type system will know that it is smaller than B and C. The $inter(r,s)$ vertex arising in the last rule is known to be a subset of both B and C. Even though this means that it must be a subset of $\mathbb{I}_E$, the type system does not discover this automatically. At this point, the Ampersand user is required to be more specific, and change the last line to:

$$\text{RULE } r;\mathbb{I}_E;s = \mathbb{V}_{A\times D}$$

This introduces $inter(\mathbb{I}_E, s)$ and $inter(r,(\mathbb{I}_E;s))$ as terms, which are both typed as $pop(E)$. For the Ampersand user, adding such terms should feel like type casting. We point out to users that type errors can often be resolved by adding type information to the script. The user should add type information to "help" the type system. This appears to be easy to explain and quite intuitive for users.

As indicated earlier, Ampersand has more operators, including the complement (unary $-$), product[4] ($\times$), and union ($\cup$). These can be implemented by using the previously introduced difference, intersection, and full relation, given that the $\cdot$ placeholders can be replaced by the correct concepts:

$$-r = \mathbb{V}_{\cdot \times \cdot} - r$$
$$(r \cup s) = -(-r \cap -s)$$
$$r \times s = r; \mathbb{V}_{\cdot \times \cdot}; s$$

These definitions require a type for the full relation $\mathbb{V}_{\cdot \times \cdot}$ to be given. The user may specify this type, but we do not require this. Separate heuristics try to infer the type, but require the user to specify the type if the type system cannot.

The type system allows the use of both $\mathbb{V}$ and $\mathbb{I}$ without type. As an example, consider a script with the relation $r_{\langle A,A \rangle}$. The union of r and $\mathbb{I}$ would be expressible in terms of the ambiguous $\mathbb{V}$:

$$(r \cup \mathbb{I}) = \mathbb{V} - ((\mathbb{V} - r) \cap (\mathbb{V} - \mathbb{I}))$$

In this case, the different occurrences of $\mathbb{V}$ may still refer to different relations. They are separated by their position in the script. After applying a heuristic to guess the type of each $\mathbb{V}$, the type system obtains an expression to which it can apply the methods discussed in this paper:

$$(r \cup \mathbb{I}_A) = \mathbb{V}_{A \times A} - ((\mathbb{V}_{A \times A} - r) \cap (\mathbb{V}_{A \times A} - \mathbb{I}_A))$$

For this heuristic, the type of any surrounding declared relation(s) is used.

Note that $(r \cup \mathbb{I}_A)$ and $(r \cup \mathbb{I}_B)$ may be terms with a very different interpretation. So, it is necessary to disambiguate the type of every expression in a script. In the above, the type system picks $(r \cup \mathbb{I}_A)$ as the intended meaning of $(r \cup \mathbb{I})$. In each case with multiple choices, the type system produces an error, alerting the user about the ambiguity.

So far, experience with the type algorithm shows that the amount of type information needed in an Ampersand script is reasonable in the eyes of users. In most cases where the type of a term can obviously be deduced by a user, the type system infers that type. There are limitations, of course. The constraint that every term should get a type is restrictive in the sense that some scripts will not be admissible. This often is desired behavior, as is the case in Listing 1.2. In such cases, the Ampersand compiler produces an error message (type error) to help the user identify and fix the mistake.

[4] The definition below is the one currently implemented in ampersand, even though the name 'product' and the symbol $\times$ might suggest another operator. We are open to suggestions for a better name.

A unique property of this type system is that the order of subtypes emerges from the script itself. In practice, however, this makes debugging an Ampersand script difficult, even when the user is presented with the partial order on concepts. Suppose for example that the user specifies:

$$\text{RULE } r \cap \mathbb{V}_{B\times A} = r$$
$$\text{RULE } r;r \cap \mathbb{I}_A = \mathbb{I}_A$$

The first line in this script says that r is of type $\langle B, A\rangle$ (or, to be precise, of a type $\langle B', A'\rangle$ with $B' \preceq B$ and $A' \preceq A$). The last line can be thought of as type incorrect: the user may have intended $\mathbb{I}_A \subseteq r^{\smile};r$. Given the current script, however, we can derive (and our algorithm derives):

$$\begin{aligned}
\mathfrak{I}(A) &= \mathfrak{I}(dom(A))\\
&= \mathfrak{I}(dom(r;r \cap A))\\
&\subseteq \mathfrak{I}(dom(r))\\
&= \mathfrak{I}(dom(r \cap \mathbb{V}_{B\times A}))\\
&\subseteq \mathfrak{I}(dom(\mathbb{V}_{B\times A}))\\
&\subseteq \mathfrak{I}(B)
\end{aligned}$$

This implies that $A \preceq B$, so no error message is produced. To make matters worse, even when $\mathbb{I}_B \cap \mathbb{I}_A = \mathbb{I}_B$ is added to intentionally imply $B \preceq A$, a type error does occur, saying that $A = B$ could be derived. At no point is the user alerted to the fact that $r;r$ contains an error at the ;. This shows that when errors are produced, they may not point to the problem. In a practical setting in industry, this leads to unacceptably confusing errors for users. Suppose, for example, that the user makes the mistake of omitting a converse operator $((\cdot)^{\smile})$. The consequence is that the compiler checks whether the type *Person* corresponds to *Account* (which is in the eyes of the user obviously not the case). The type checker finds an interpretation in which $Person \preceq Account$. The resulting error messages, if any, will be unintelligble in the eyes of the user. As a result, the type algorithm proposed here was *not* adopted in the Ampersand compiler.

9 Conclusion

This paper shows that domain analysis can be used as a mechanism for type checking. The type algorithm presented is unconventional: it does not work with proof trees. This results in an attractive type graph, which is useful for explaining the type-correctness or type-incorrectness of a script. The type graph is a comprehensive representation of all proofs that are needed to show that every term has precisely one type. Because of this, the diagram has demonstrated to be quite convincing in debates among Ampersand professionals in practice. The approach is not only simple, but it allows compilers with advanced features such as specialization and overloading with hardly any extra effort. The simplicity of the approach is illustrated by the Haskell code corresponding to this article, which contains the complete algorithm in under 200 lines of Haskell code, and can be found at: `http://cs.ru.nl/~B.Joosten/ampTypes/` The simplicity is also valued in the Ampersand project, because this yields code that is maintainable because of its simplicity.

This type algorithm adds a new feature to type checking: it uses information from the very rules it is checking to enhance the concept pre-order.

Acknowledgements. We thank the reviewers for their comments. A special thanks to the reviewer who tried our code on the examples in this paper.

References

1. The Business Rules Group: Business rules manifesto – the principles of rule independence (2003). `http://www.BusinessRulesGroup.org`
2. Michels, G., Joosten, S., van der Woude, J., Joosten, S.: Ampersand: Applying relation algebra in practice. In: de Swart, H. (ed.) RAMICS 2011. LNCS, vol. 6663, pp. 280–293. Springer, Heidelberg (2011)
3. Maddux, R.: Relation Algebras. Studies in Logic and the Foundations of Mathematics, vol. 150. Elsevier Science (2006)
4. van der Woude, J., Joosten, S.: Relational heterogeneity relaxed by subtyping. In: de Swart, H. (ed.) RAMICS 2011. LNCS, vol. 6663, pp. 347–361. Springer, Heidelberg (2011)
5. Milner, R.: A theory of type polymorphism in programming. Journal of Computer and System Sciences 17, 348–375 (1978)
6. Marlow, S., Wadler, P.: A practical subtyping system for erlang. SIGPLAN Not. 32, 136–149 (1997)
7. Liskov, B.H., Wing, J.M.: A behavioral notion of subtyping. ACM Transactions on Programming Languages and Systems (TOPLAS) 16, 1811–1841 (1994)
8. Mitchell, J.C.: Type inference with simple subtypes. Journal of Functional Programming 1, 245–285 (1991)
9. Amadio, R.M., Cardelli, L.: Subtyping recursive types. ACM Transactions on Programming Languages and Systems (TOPLAS) 15, 575–631 (1993)
10. Tiuryn, J., Urzyczyn, P.: The subtyping problem for second-order types is undecidable. In: IEEE Computer Society Symposium on Logic in Computer Science, pp. 74–74 (1996)
11. Wells, J.B.: The undecidability of mitchell's subtyping relationship. Technical report, Boston University Computer Science Department (1995)
12. Lindahl, T., Sagonas, K.: Practical type inference based on success typings. In: Proceedings of the 8th ACM SIGPLAN International Conference on Principles and Practice of Declarative Programming, pp. 167–178. ACM (2006)
13. Tomás, D., Vicedo, J.L.: Minimally supervised question classification on fine-grained taxonomies. Knowledge and Information Systems 36, 303–334 (2013)

Towards Interactive Verification of Programmable Logic Controllers Using Modal Kleene Algebra and KIV

Roland Glück and Florian Benedikt Krebs

German Aerospace Center
{roland.glueck,florian.krebs}@dlr.de

Abstract. In this paper we develop an approach to interactive verification of programmable logic controllers which often serve as controllers in safety critical systems and hence need thorough verification. As a verification tool we use the KIV system, whereas the formalization is done in modal Kleene algebra. We first prove a bunch of theorems from modal Kleene algebra in KIV, subsequently translate the desired properties of a program for a programmable logic controller in modal Kleene algebra, and finally prove these encoded properties interactively with KIV.

1 Introduction

1.1 Overview

Programmable Logic Controllers (PLCs) are widely used in automation systems as robots or machine tools like lathes, drill presses or milling machines. In particular, robots can become a safety threat to humans, so correctness of PLC programs is highly desired. Also in other safety critical scenarios, as e.g. control of nuclear power plants or airplanes, PLC programs have to work correctly. Failure of such a program can lead to economical or human damages for which reason verification of PLC programs is necessary both for financial and an ethical reasons.

Our approach is based on two concepts: Modal Kleene algebra (MKA) and interactive theorem proving with KIV (see [4]). MKA is known as an algebraic framework for temporal logics like LTL, CTL and CTL* (see e.g. [13,19]) and has already a long history in the context of automated reasoning and verification as in [10,15,16]. The KIV system has shown its ability and power for formal verification in [9,20,23] and on many other occasions. In our work we combine the algebraic elegance of MKA with the interactive proving strategy of KIV. This avoids the problems one has to face with automated theorem provers, and allows to use the full expressiveness of MKA. Whereas automated reasoning in semirings and Kleene algebra has already a long tradition (see again [12,16,15]), interactive verification using Kleene algebra is a rather new and rising area (as e.g. in [8]) with promising prospects. The reader should not expect new theoretical results but rather a preliminary demonstration of first steps towards interactive verification of PLC programs.

W. Kahl et al. (Eds.): RAMiCS 2015, LNCS 9348, pp. 241–256, 2015.
DOI: 10.1007/978-3-319-24704-5_15

The paper is organized as follows: In Section 2 we give a short introduction to programming PLCs. Section 3 provides a summary about Kleene algebra and its connections to temporal logic. The KIV system is introduced in Section 4. In Section 5 we show how to formalize PLC programs in MKA abstractly, and give two exemplary case studies in Section 6. A summary and an outlook on future work is the content of Section 7.

1.2 Related Work

There are some other approaches to PLC verification. [22] and [21] use model checking with NuSMV (see [5]) where the latter publication demonstrates how the state space obtained in the first paper can be reduced. Data flow analysis is a technique which is used in [17] also in the context of PLC verification. In [11] simulation of PLC programs serves as an auxiliary tool for testing.

In contrast to all these approaches, our concept has the potential to avoid the problem of state explosion which often occurs during the work with a model checker. This problem can be circumvented by a clever human proof strategy. Moreover, unsuccessful searching for a proof can lead the user not only to errors in the implementation but also to a correct solution.

2 A Short Introduction to Programmable Logic Controllers

2.1 Basics

Programmable logic controllers, or PLCs for short, are a common tool in industry for controlling machines or electrical devices of all kinds. A PLC has a set of inputs and outputs which may be of Boolean or numerical type. The inputs may stem from measuring results from sensors or Boolean values from switches activated by an operator or user whereas the outputs may go to actuators or signal devices for process monitoring. Moreover, a PLC has the capability to store values in internal variables. The type of a variable (input, output or internal) is specified in a separated variable table and can in general not be inferred from the bare program.

PLCs offer the possibility of performing the safety critical computations in a special part with doubly checked computation. Signals which are crucial for the safe operation of a plant like messages from a photoelectric sensor, a limit switch or an emergency stop are processed in a separate part of the hardware. Every computation is executed twice and the results are compared; if they differ the whole plant stops. In this part the instruction set is reduced to some Boolean functions, so we will concentrate here also on a subset of Boolean functions.

Most PLCs work in a cyclic way: in step n the inputs and the internal variables are read, and subsequently the outputs and the internal variables for the following step $n + 1$ are computed. Typical cycle times are in the dimension of 20 - 250 milliseconds. PLC programs can be executed in a synchronous and asynchronous way; in the first case an error occurs if the computation time exceeds a given cycle time, in the second case the computation of the outputs and internal variables is delayed till termination. Both concepts have their obvious advantages and drawbacks.

2.2 Programming Programmable Logic Controllers

The standard for PLCs is the norm IEC 61131 [2]. It defines five kinds of programming languages for PLCs:

1. Instruction List (IL), a kind of an assembly language with an operator stack and push and pop operations,
2. Ladder Diagram (LD) which reflects the origins of PLCs from electrical engineering since it resembles circuit diagrams,
3. Sequential Function Chart (SFC), a graphical method in the spirit of state diagrams,
4. Structured Text (ST) with a C-like syntax, and
5. Function Block Diagram (FBD), the most popular variant with an appearance of logic circuits.

Most PLC manufactures offer the possibility of mixing two or more languages in one PLC program. The standards also offer concepts known from traditional and object-oriented programming like modularization and encapsulation, but we consider here only the basic concepts.

2.3 A Function Block Diagram Crash Course

Each of the five languages has its advantages, but due to its widespread use we concentrate on FBD in the sequel. Our overview reclines on the syntax of Step7 as in [7] but can easily be transferred to other implementations. Furthermore, we do not describe the full extent of the language but restrict ourselves to negation, AND- and OR-gates as well as flip-flops.

In FBD, one can write Boolean functions in clearly arranged diagrams. There are predefined Boolean functions which are represented as rectangles with inputs at the left side and outputs at the right side as in Figure 1. The output of these elementary gates can serve as inputs to other elementary gates or can be forwarded to an output signal to the environment or stored in an internal variable. A program can consist of several connected components of elementary gates, and in each connected component the elementary gates are evaluated according to topological order. Moreover, the connected components are evaluated in order from top to bottom (this is done by the layout manager of the programming environment). So the evaluation order of the program from Figure 3 is `NEG1`, `OR1`, `OR2`, `AND1`, `OR3`, `AND2`, `OR4` and finally `AND3`. For the naming of this example we refer the reader to Section 5.

An AND-gate is symbolized by a rectangle containing an ampersand on its top; similarly an OR-gate is represented by a rectangle that contains the symbol `>=1`. Negation of a variable or result is denoted by a small circle on the associated line. So the FBD from Figure 1 stands for the function

$$(\texttt{IN1} \wedge \neg\texttt{IN2}) \vee \texttt{IN5} \vee \texttt{M10},$$

and the result is forwarded to the output OUT4. There are also other logic functions like NAND and EXOR which we skip for brevity.

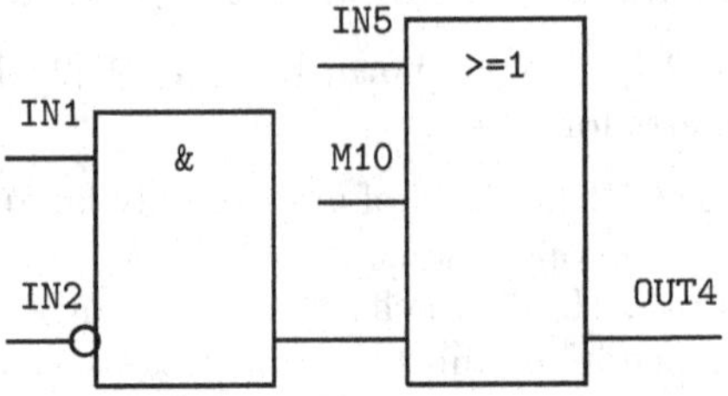

Fig. 1. AND, OR and Negation in FBD

In addition to these traditional Boolean functions PLC in general and FBD in particular provide dynamic Boolean functions whose results depend on the run of the input signals. As an example for this group we consider flip-flops.

Flip-flops have two inputs, one set input and one reset input, and one output. If a flip-flop receives a `true` signal on its set input the output is set to `true`. The output remains on `true` until the reset input receives a `true` signal which causes the output to switch to `false`. Depending on the behavior of the flip-flop receiving `true` signals at the same time both on the set and the reset input there are set dominant and reset dominant flip-flops: in this situation, a set dominant flip-flop sets it output to `true` whilst a reset dominant flip-flop sets it to `false`. For short, set dominant flip-flops are also called RS-flip-flops and reset dominant ones SR-flip-flops. By default, the initial output value of a flip-flop equals `false`.

The symbol for a flip-flop in FBD is a rectangle denoting its type (RS or SR) with two inputs, S for the set and R for the reset input. The output bears the letter Q and its recent value is stored in an internal variable written over the respective gate. So the flip-flop on the left side of Figure 2 is a reset dominant flip-flop with IN1 and IN2 on its set and reset input, resp., whose output is stored in the internal variable M0.1 and written to the output OUT4.

3 Kleene Algebra and Temporal Logic

In this section we first recapitulate, in Subsection 3.1, some basics about semirings and modal Kleene algebra before sketching the connections between modal Kleene algebra and linear temporal logic in Subsection 3.2. A more detailed overview of Subsection 3.1 offer e.g. [13,14], for Subsection 3.2 see also [19].

3.1 Modal Kleene Algebra

A central concept in algebraic system description are idempotent semirings:

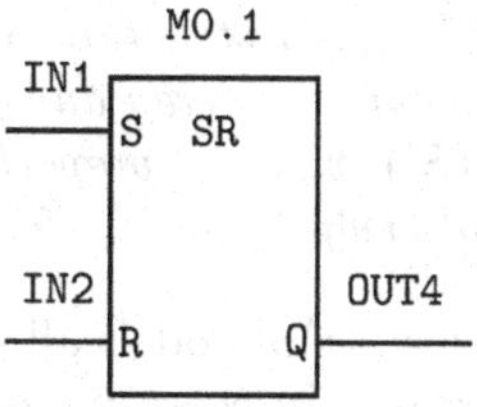

Fig. 2. A Reset Dominant Flip-flop

Definition 1. *An* idempotent semiring *is a structure* $S = (M, +, 0, \cdot, 1)$ *where* $(M, +, 0)$ *and* $(M, \cdot, 1)$ *are monoids and* $+$ *is commutative and idempotent. Moreover,* $\cdot$ *distributes both from the left and the right over* $+$ *and* 0 *is a left and right annihilator of* $\cdot$.

Given an idempotent semiring $S = (M, +, 0, \cdot, 1)$ we define the *natural order* $\leq \subseteq M \times M$ by $x \leq y :\Leftrightarrow x + y = y$. As suggested by its name, $\leq$ is an order with least element 0.

The operations $+$ and $\cdot$ are called *addition* and *multiplication* and model usually choice and composition, resp. A well-known instantiation of an idempotent semiring corresponding to this interpretation is the semiring $(\mathbf{Rel}_M, \cup, \emptyset, ;, \mathbf{id}_M)$ of the set of endorelations $\mathbf{Rel}_M$ over a set M with set union as addition (whose neutral element equals $\emptyset$) and relational composition ; as multiplication (with the identity relation $\mathbf{id}_M$ over M as neutral element).

Because of the associativity of addition and multiplication we are free to omit parentheses. Another convention we use is that multiplication binds stronger than addition. In order to increase readability we agree also on the possibility to omit the multiplication sign $\cdot$ if wanted; so $x(y + z)$ stands for $x \cdot (y + z)$.

This concept of a semiring suffices for modeling choice and composition but lacks the possibility of reasoning about subsets of M. To this effect, we introduce the concept of tests:

Definition 2. *Given an idempotent semiring* $S = (M, +, 0, \cdot, 1)$ *an element* $p \in M$ *is called a* test *iff an element* $\neg p$ *(the* complement *of* p*) exists with the properties* $p + \neg p = 1$ *and* $p \cdot \neg p = 0 = \neg p \cdot p$.

For an idempotent semiring S we write $\mathbf{test}(S)$ for the set of all tests of S. Usually, tests are denoted by p, q, r and s, and indexed and primed derivatives thereof. Later we deviate from this convention for good reasons. On $\mathbf{test}(S)$, infimum and multiplication coincide; in particular, multiplication of tests is idempotent and commutative. In every semiring 0 and 1 are tests, and they are the least and greatest element in $\mathbf{test}(S)$. An often used abbreviation for $p, q \in \mathbf{test}(S)$ is $p \to q$ instead of $\neg p + q$. The operator $\to$ has a weaker binding than $+$.

Using tests, we can also model analoga of image and preimage of a relation by the so called diamond and box operators:

Definition 3. *A* modal semiring *is a structure* $S = (M, +, 0, \cdot, 1, |\cdot\rangle, \langle\cdot|)$ *where* $S' = (M, +, 0, \cdot, 1)$ *is an idempotent semiring and* $|\cdot\rangle$ *and* $\langle\cdot|$ *are functions of the type* $(M \to \mathbf{test}(S')) \to \mathbf{test}(S')$ *with the properties* $|x\rangle p \leq q \Leftrightarrow \neg qxp \leq 0 \Leftrightarrow \langle x|p \leq \neg q$, $|xy\rangle p = |x\rangle|y\rangle p$ *and* $\langle xy|p = \langle y|\langle x|p$ *for all* $x \in M$ *and* $p, q \in S'$.

From an abstract point of view, $|x\rangle p$ models all states from where a transition via x into p is possible whereas $|x]p$ corresponds to all states from which a transition along x leads inevitably into p. A symmetric interpretation holds for the backward operators.

The operators $|\cdot\rangle$ and $\langle\cdot|$ are called the *forward* and *backward diamond*, resp. From these operators we can derive the *forward* and *backward box* operators $|\cdot]$ and $[\cdot|$ by the dualization $|x]p := \neg|x\rangle\neg p$ and $[x|p := \neg\langle x|\neg p$.

The diamond operators are distributive and hence isotone in both arguments. Moreover, we have $\langle x|0 = 0 = \langle 0|p$ for arbitrary $x \in M$ and $p \in \mathbf{test}(S')$ and the analogous properties for the backward diamond. The box operators are antitone in the first and isotone in the second argument. By dualization of the diamond properties we obtain $|x]1 = 1$, $|0]q = 1$ and $|1]q = q$ for every test q and the symmetric properties for the backward diamond.

Till now we are not able to model iteration in terms of semirings. This gap is filled by the Kleene star, introduced in [18] as follows:

Definition 4. *A* Kleene algebra K *is a structure* $K = (M, +, 0, \cdot, 1, {}^*)$ *where* $(M, +, 0, \cdot, 1)$ *is an idempotent semiring and* ${}^* : M \to M$ *is a unary operation in postfix notation with the following properties:*

$$1 + xx^* \leq x^* \qquad x + yz \leq z \Rightarrow y^*x \leq z$$
$$1 + x^*x \leq x^* \qquad x + yz \leq y \Rightarrow xz^* \leq y$$

As a unary operator, the Kleene star binds stronger than multiplication and addition.

Well-known properties of the Kleene star are its strictness ($x^* = 0 \Leftrightarrow x = 0$) and its isotony. Moreover, for all $n \in \mathbb{N}$ it fulfills the inequality $x^n \leq x^*$ where x^n is defined recursively in the standard way by $x^0 = 1$ and $x^{n+1} = x \cdot x^n$. For further properties see [13] or [14].

Finally, we call a structure $(M, +, 0, \cdot, 1, |\cdot\rangle, \langle\cdot|, {}^*)$ a *modal Kleene algebra*, or *MKA* for short, if $(M, +, 0, \cdot, 1, |\cdot\rangle, \langle\cdot|)$ is a modal semiring and $(M, +, 0, \cdot, 1, {}^*)$ is a Kleene algebra.

3.2 Modal Kleene Algebra and Linear Temporal Logic in a Nutshell

In this subsection we give a short overview of the results from [19] which are relevant for our paper. Our notation differs slightly from the one used there but the semantic remains the same.

Linear temporal logic, or LTL for short, allows reasoning about the temporal behavior of logic variables in traces of a transition system. A trace $s = s_0s_1s_2\ldots$

consists of a sequence of states s_0, s_1, s_2, ... of a transition system. For a trace $s = s_0 s_1 s_2 \ldots$ we define the trace s^i by $s^i = s_i s_{i+1} s_{i+2} \ldots$. Furthermore, we have a set $\Pi = \{\pi_1, \pi_2, \ldots, \pi_m\}$ of propositional variables which can be `true` or `false` in a state. To reason about the behavior of these variables along traces we use the set Ψ of LTL formulae, generated by the following (not minimal but expressive) grammar:

$$\Psi ::= \bot \mid \Pi \mid \neg\Psi; \mid \Psi \rightarrow \Psi \mid \Box\Psi \mid \Diamond\Psi \mid \circ\Psi \mid \Psi \,\mathsf{U}\, \Psi$$

For a formula $\pi \in \Pi$ we say that π is *valid* in a trace $s = s_0 s_1 s_2 \ldots$ iff π holds in s_0. Then the validity of an LTL formula with respect to s is defined as follows:

- $\bot$ is valid wrt. $s \Leftrightarrow$ `false`
- $\neg\psi$ is valid wrt. $s \Leftrightarrow \psi$ is not valid wrt. s
- $\psi_1 \rightarrow \psi_2$ is valid wrt. $s \Leftrightarrow$ validity of ψ_1 wrt. s implies validity of ψ_2 wrt. s
- $\Box\psi \Leftrightarrow \psi$ is valid wrt. s in all states of s
- $\Diamond\psi \Leftrightarrow \psi$ is valid wrt. s in some state of s
- $\circ\psi \Leftrightarrow \psi$ is valid wrt. s^1
- $\psi_1 \,\mathsf{U}\, \psi_2 \Leftrightarrow \exists i \geq 0 : \psi_2$ is valid wrt. s^i and ψ_1 is valid wrt. all s^j with $j < i$

We use the usual abbreviations $\neg\psi$ for $\psi \rightarrow \bot$, $\psi_1 \wedge \psi_2$ for $\neg(\psi_1 \rightarrow \neg\psi_2)$ and $\psi_1 \vee \psi_2$ for $\neg\psi_1 \rightarrow \psi_2$. A formula is said to be *valid* if it is valid with respect to all traces.

This can be extended to a set valued interpretation which maps each formula ψ to to a set of traces $[\![\psi]\!]$ with respect to which ψ is valid. These sets can be modeled by tests, and a general element a is going to be used for the transition relation which transforms a trace s into its successor s^1. Then we can define inductively:

$$\begin{aligned}
[\![\bot]\!] &= 0 \\
[\![\neg\psi]\!] &= \neg[\![\psi]\!] \\
[\![\psi_1 \rightarrow \psi_2]\!] &= \neg[\![\psi_1]\!]+[\![\psi_2]\!] \\
[\![\psi_1 \wedge \psi_2]\!] &= [\![\psi_1]\!]\cdot[\![\psi_2]\!] \\
[\![\psi_1 \vee \psi_2]\!] &= [\![\psi_1]\!]+[\![\psi_2]\!] \\
[\![\Box\psi]\!] &= |a^*][\![\psi]\!] \\
[\![\Diamond\psi]\!] &= |a^*\rangle[\![\psi]\!] \\
[\![\circ\psi]\!] &= [\![|a\rangle\psi]\!] \\
[\![\psi_1 \,\mathsf{U}\, \psi_2]\!] &= |([\![\psi_1]\!]\cdot a)^*\rangle[\![\psi_2]\!]
\end{aligned}$$

Note the twofold meaning of negation in the second line: the first negation is a logical operator, the second one a semiring's complement operator.

The relation a that transforms a trace s into its successor s^1 is a left total function. In the context of MKA, this leads to the property $|a\rangle p = |a]p$ for all tests p.

Now an LTL formula is valid iff its associated semiring term evaluates to 1. So for example, we have

$$\begin{array}{ll}
[\![\square\,(\psi \rightarrow \psi)]\!] & = \\
|a^*][\![(\psi \rightarrow \psi)]\!] & = \\
|a^*](\neg[\![\psi]\!]+[\![\psi]\!]) & = \\
|a^*]1 & = \\
1 &
\end{array}$$

due to the above rules and the properties of the complement and the forward box which shows the validity of the LTL formula $\square\,(\psi \rightarrow \psi)$ for every transition system. In Section 5 we describe how to obtain a concrete description of a for reasoning about FBD programs.

4 The KIV System

The KIV system from [4], developed mainly in Karlsruhe and Ulm and currently administrated in Augsburg is a tool for interactive proving and verification which was already successfully employed in verification of practical applications as e.g. in [9,20,23]. Our approach is to prove first a sufficient amount of theorems from the area of semirings and MKA, then to model PLC programs in MKA and finally to formalize and prove interactively desired properties of given PLC programs using the theorems from the first step.

KIV comes along as a plug-in for Eclipse and is easy to install. The data is organized in modules consisting of so called specification and associated sequents parts with a self-explanatory and easily understandable syntax. A specification part contains the sorts, constants, functions, predicates, variables and axioms used by the associated sequents part which contains the theorems to be proved. It is possible to enrich already existing specifications by additional axioms in a way similar to the inheritance concept of object-oriented programming. The theorems under consideration are entered in the associated sequents file and can be proved interactively.

In the basic settings the user has to insert all lemmata from other sequents as well as axioms from specifications and instanciate all quantifiers by hand. However, KIV offers a lot of automatically applicable heuristics which ease its use significantly. The use of rewrite lemmata allows even a kind of local automated reasoning.

The choice of an interactive prover instead of an automated prover like Prover9 (see [6]) was motivated by unsatisfactory experiences with automated provers. Given too many axioms they get lost in an inflated search space. The specification of the transition function of an FBD program even of moderate size generates an amount of axioms which swamps automated theorem provers. In contrast, using an interactive theorem prover one can select suitable lemmata for a prove step. Another advantage is that potential bugs can be discovered and fixed better: using an interactive theorem prover one can detect where the verification fails and change the program accordingly whereas the counterexamples produced by automated theorem provers are hardly readable and comprehensible for humans.

In some provers like Coq and Isabelle (see [3,1]), a lot of theorems from MKA are already implemented. However, we decided to use KIV because we need a lot of additional lemmata which we had to prove in either case, so the additional work was not overboarding.

5 Modeling Function Block Diagrams in Modal Kleene Algebra

The overarching idea of our approach for modeling FBD programs in MKA consists of four steps: first, we assign each elementary gate an MKA element. Second, we model each input and output signal as well as intermediate computation results and internal variables as tests by assigning to each such signal, result or variable `x` two tests `x_0` and `x_1` indicating a value of `true` or `false`, resp. Already existing names remain untouched. Of course, this modeling implies also $\texttt{x\_0} = \neg\texttt{x\_1}$. In the third step we define the behavior of each elementary gate with respect to its inputs and outputs and other signals by formulae from MKA. Finally, all the elementary gates are put together in a big product corresponding to the behavior of the overall program. Our running example illustrating this idea is the program from Figure 3.

In the first step we are free to choose the names of the elementary gates but it is useful to give meaningful names. Such names are written in gray at the bottom of each elementary gate in Figure 3 and are used throughout the further course. These descriptions are not part of the FBD syntax but serve only for better understanding.

As signals in our running example we have the inputs `IN1`, `IN2` and `IN3` and the outputs `OUT1`, `OUT2` and `OUT3` whereas intermediate computation results arise at `OR1`, `OR2`, `OR3`, `OR4` and `AND2`. We use the convention to denote gates by uppercase letters, the corresponding semiring element by lowercase letters and tests corresponding to the output of a gate by the gate's name in lowercase letters followed by `_0` and `_1`. Hence we have e.g. the constant tests `in1_0`, `in1_1`, `out3_0`, `out3_1`, `or3_0`, `or3_1` and the equations $\texttt{in2\_0} = \neg\texttt{in2\_1}$ and $\texttt{and2\_0} = \neg\texttt{and2\_1}$.

The behavior of an elementary gate with respect to its inputs and outputs depends of course of its type. We model this as follows:

- For an `AND`-gate `ANDK` with inputs `in1`, ..., `inn` we have the equations $\texttt{in1\_1} \cdot \cdots \cdot \texttt{inn\_1} \leq |\texttt{andk}\rangle\texttt{andk\_1}$ and $\texttt{in1\_0} + \cdots + \texttt{inn\_0} \leq |\texttt{andk}\rangle\texttt{andk\_0}$.
- For an `OR`-gate `ORK` with inputs `in1`, ..., `inn` we have the equations $\texttt{in1\_1} + \cdots + \texttt{inn\_1} \leq |\texttt{ork}\rangle\texttt{ork\_1}$ and $\texttt{in1\_0} \cdot \cdots \cdot \texttt{inn\_0} \leq |\texttt{ork}\rangle\texttt{ork\_0}$.
- For a set dominant flip-flop `RSK` with set input `s`, reset input `r` and output `q` whose value is stored in the internal variable `m` we have the following characterizations:
  - $\texttt{s\_1} + \texttt{m\_1} \cdot \texttt{r\_0} \leq |\texttt{rsk}\rangle\texttt{q\_1}$
  - $\texttt{s\_1} + \texttt{m\_1} \cdot \texttt{r\_0} \leq |\texttt{rsk}\rangle\texttt{m\_1}$
  - $\texttt{s\_0} \cdot \texttt{r\_1} + \texttt{m\_0} \cdot \texttt{s\_0} \leq |\texttt{rsk}\rangle\texttt{q\_0}$
  - $\texttt{s\_0} \cdot \texttt{r\_1} + \texttt{m\_0} \cdot \texttt{s\_0} \leq |\texttt{rsk}\rangle\texttt{m\_0}$

In the third and fourth equation one could factor out s_0 but it turned out to be more convenient to use the form given above. Reset dominant flip-flops are treated symmetrically.

- To handle the case of the negation of a signal x we simply swap x_0 and x_1. In general, we do not introduce a new gate and hence a new element of M for a negation, except the case if a negated signal is directly forwarded as in the topmost component of Figure 3. These considerations lead e.g. to the formulae $\texttt{in1\_0} \leq |\texttt{neg1}\rangle\texttt{out1\_1}$ or $\texttt{in1\_0}+\ \texttt{in2\_0} \leq |\texttt{OR2}\rangle\texttt{or2\_1}$.

These requirements are not enough for an adequate description of an FBD program since we do not take into account the fact that a gate does not overwrite signals from other gates. To this purpose we order the elementary gates according to their evaluation order. In our running example this would lead to the sequence OR_1, OR_2, AND_1, OR_3, AND_2, OR_4, AND_3. Given this sequence for an FBD program we have to ensure that all signals produced before a gate GAT and used by another gate after GAT remain unchanged by GAT. The same holds for signals which are used as final output or are stored in an internal variable. So for each such gate GAT and every such signal x we introduce the formulae $\texttt{x\_0} \leq |\texttt{gat}\rangle\texttt{x\_0}$ and $\texttt{x\_1} \leq |\texttt{gat}\rangle\texttt{x\_1}$. In the FBD program from Figure 1 this would lead e.g. to the formulae $\texttt{in1\_1} \leq |\texttt{or1}\rangle\texttt{in1\_1}$ (because in1 is used e.g. by AND2) or $\texttt{out2\_0} \leq |\texttt{or4}\rangle\texttt{out2\_0}$ (because out4 is used as a final output).

The last step of our construction puts all gates together in a big product consisting of the sequence of the previous step. This product plays the role of the transition relation a from Subsection 3.2. In our example we get as description of the overall system the term

$$\texttt{neg1} \cdot \texttt{or1} \cdot \texttt{or2} \cdot \texttt{and1} \cdot \texttt{or3} \cdot \texttt{and2} \cdot \texttt{or4} \cdot \texttt{and3}.$$

6 Case Studies

6.1 A 3-Bit Incrementer

As already suggested by its caption, the example from Figure 3 implements the increment function modulo 8 on three bits in little endian representation. The modeling of this FBD program was already shown in Section 5, and we adopt the naming from there. As an abbreviation we introduce the constant inc, given by

$$\texttt{inc} = \texttt{neg1} \cdot \texttt{or1} \cdot \texttt{or2} \cdot \texttt{and1} \cdot \texttt{or3} \cdot \texttt{and2} \cdot \texttt{or4} \cdot \texttt{and3}.$$

A possible approach to verification of this program is to prove its behavior on the eight inputs $\texttt{in1\_b1} \cdot \texttt{in2\_b2} \cdot \texttt{in3\_b3}$ with $\texttt{b1}, \texttt{b2}, \texttt{b3} \in \{0, 1\}$. This leads e.g. to proof obligations like

$$|\texttt{inc}^*](\ \texttt{in1\_0} \cdot \texttt{in2\_0} \cdot \texttt{in3\_1} \rightarrow |\texttt{inc}\rangle(\texttt{out1\_1} \cdot \texttt{out2\_0} \cdot \texttt{out3\_1})) = 1$$

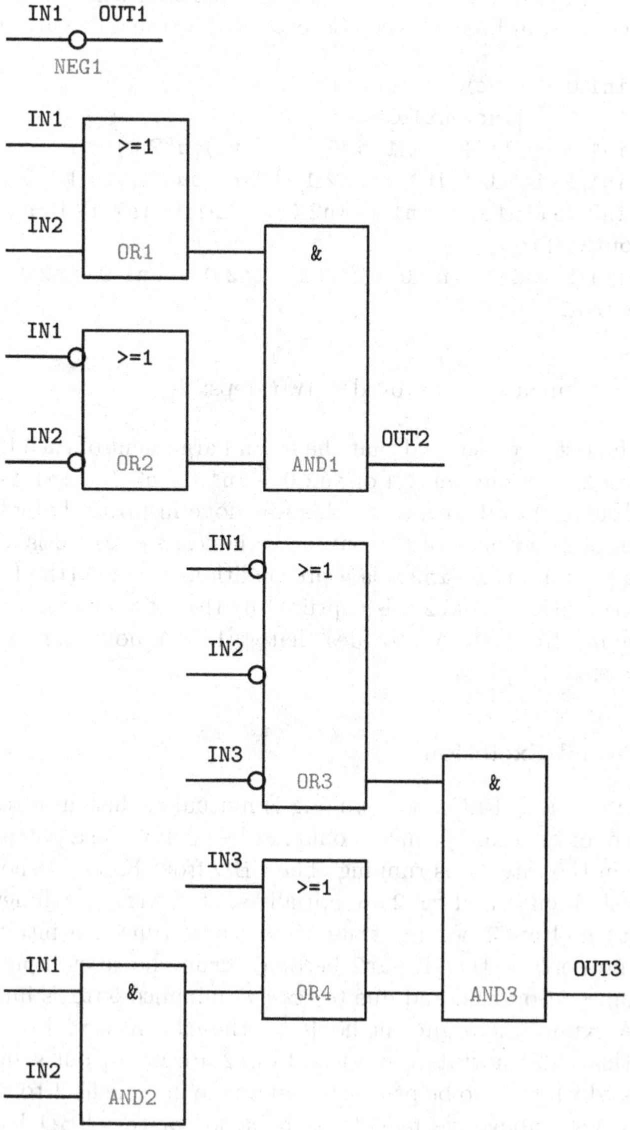

Fig. 3. A 3-bit Incrementer

or

$$|\mathtt{inc}^*](\ \mathtt{in1\_1} \cdot \mathtt{in2\_1} \cdot \mathtt{in3\_0} \rightarrow |\mathtt{inc}\rangle(\mathtt{out1\_0} \cdot \mathtt{out2\_0} \cdot \mathtt{out3\_1})) = 1.$$

It turned out that these formulae are very tedious to prove (and there are eight of them!) so we switched to another approach which verifies the behavior of the single output bits. This can be expressed by the following proof obligations:

$$|\mathtt{inc}^*](\mathtt{in1\_0} \rightarrow |\mathtt{inc}\rangle\mathtt{out1\_1}) = 1$$
$$|\mathtt{inc}^*](\mathtt{in1\_1} \rightarrow |\mathtt{inc}\rangle\mathtt{out1\_0}) = 1$$
$$|\mathtt{inc}^*](\mathtt{in1\_0} \cdot \mathtt{in2\_0} + \mathtt{in1\_1} \cdot \mathtt{in2\_1} \rightarrow |\mathtt{inc}\rangle\mathtt{out2\_0}) = 1$$
$$|\mathtt{inc}^*](\mathtt{in1\_1} \cdot \mathtt{in2\_0} + \mathtt{in1\_0} \cdot \mathtt{in2\_1} \rightarrow |\mathtt{inc}\rangle\mathtt{out2\_1}) = 1$$
$$|\mathtt{inc}^*](\mathtt{in3\_0} \cdot (\mathtt{in1\_0} + \mathtt{in1\_1} \cdot \mathtt{in2\_0}) + \mathtt{in1\_1} \cdot \mathtt{in2\_1} \cdot \mathtt{in3\_1} \rightarrow |\mathtt{inc}\rangle\mathtt{out3\_0}) = 1$$
$$|\mathtt{inc}^*](\mathtt{in1\_1} \cdot \mathtt{in2\_1} \cdot \mathtt{in3\_0} + \mathtt{in3\_1} \cdot (\mathtt{in2\_0} + \mathtt{in1\_0} \cdot \mathtt{in2\_1})) \rightarrow |\mathtt{inc}\rangle\mathtt{out3\_1}) = 1$$

These statements were proved in two steps:

- In a first step we showed that the second argument of each forward diamond (e.g. $\mathtt{in1\_0} \rightarrow |\mathtt{inc}\rangle\mathtt{out1\_1}$ or $\mathtt{in2\_0} + \mathtt{in1\_0} \cdot \mathtt{in2\_1} + \mathtt{in1\_1} \cdot \mathtt{in2\_1} \cdot \mathtt{in3\_1} \rightarrow |\mathtt{inc}\rangle\mathtt{out3\_0}$) evaluates to 1. This was done in principle by backward tracking of preconditions of the output variables, e.g. or1_0 is a precondition of out2_0, and $\mathtt{in1\_0} \cdot \mathtt{in2\_0}$ is a precondition of or1_0 (the fact that or1_0 is a precondition of out2_0 is captured by the axiom $\mathtt{or1\_0} \leq |\mathtt{and1}\rangle\mathtt{out2\_0}$).
- A lemma from MKA provides that $|x]1 = 1$ holds for arbitrary x which completes the proof.

6.2 Mutual Exclusion

A frequent task in PLC programming is mutual exclusion of resources. For example, a door on a safety fence should not be open while a potentially dangerous machine in the interior is running. The FBD from Figure 4 shows a solution of such a task: if out1 and out2 are initially set to false it will never happen that both out1 and out2 become true at the same time. An intuitive explanation for this property is that if out2 becomes true the reset input of the flip-flop SR1 becomes also true, and due to reset dominance out1 is immediately set to false. A symmetric argument holds for the case if out1 becomes true. Note that in this implementation out1 and out2 are no output signals but internal variables which have to be processed before having an effect to the environment.

Similarly to above we model the behavior of this FBD by the expression cycle, given by

$$\mathtt{cycle} = \mathtt{or1} \cdot \mathtt{sr1} \cdot \mathtt{or2} \cdot \mathtt{sr2}$$

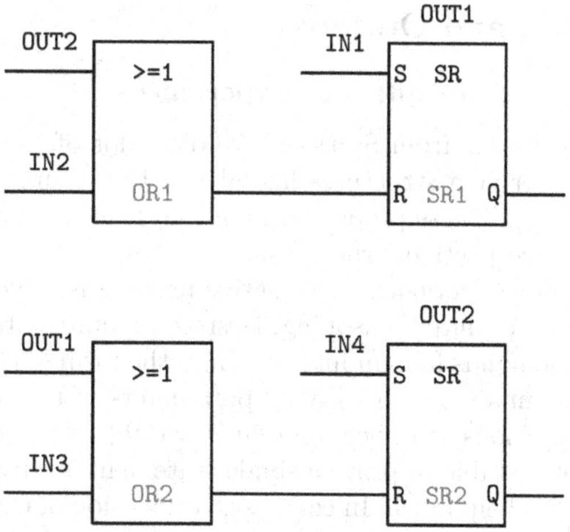

Fig. 4. Mutual Exclusion

where the characterizations of the right hand terms were given in Section 5. Now the desired behavior can be described by the following MKA expressions:

$$\texttt{out1\_0} \cdot \texttt{out2\_0} \rightarrow |\texttt{cycle}^*](\texttt{out1\_1} \rightarrow \texttt{out2\_0}) = 1$$
$$\texttt{out1\_0} \cdot \texttt{out2\_0} \rightarrow |\texttt{cycle}^*](\texttt{out2\_1} \rightarrow \texttt{out1\_0}) = 1$$

Of course these two formulations can be merged into one but in this form the formalization is both easier to read and to prove. The proofs of these claims were harder to do than the ones of the seemingly more complicated expressions from the previous subsection due to the deeper nesting of implications. The rough idea of the proof of the first claim is as follows:

- First we showed that $\texttt{out1\_0} \cdot \texttt{out2\_0} + \texttt{out1\_0} \cdot \texttt{out2\_1} + \texttt{out1\_1} \cdot \texttt{out2\_0}$ is an invariant of $\texttt{cycle}$.
- Due to a lemma from MKA $\texttt{out1\_0} \cdot \texttt{out2\_0} + \texttt{out1\_0} \cdot \texttt{out2\_1} + \texttt{out1\_1} \cdot \texttt{out2\_0}$ is an invariant of $\texttt{cycle}^*$, too.
- Finally, also a theorem of MKA states the implication

$$p \leq q \wedge qx\neg q = 0 \wedge q \leq r \Rightarrow p \rightarrow |x]r = 1$$

for all tests p, q, r and arbitrary x which shows the claim due to $\texttt{out1\_0} \cdot \texttt{out2\_0} \leq \texttt{out1\_0} \cdot \texttt{out2\_0} + \texttt{out1\_0} \cdot \texttt{out2\_1} + \texttt{out1\_1} \cdot \texttt{out2\_0}$ and $\texttt{out1\_0} \cdot \texttt{out2\_0} + \texttt{out1\_0} \cdot \texttt{out2\_1} + \texttt{out1\_1} \cdot \texttt{out2\_0} \leq \texttt{out1\_1} \rightarrow \texttt{out2\_0}$ (note that $qx\neg q = 0$ means that q is an invariant of x).

To show te second claim we only have to show the additional inequality $\texttt{out1\_0} \cdot \texttt{out2\_0} + \texttt{out1\_0} \cdot \texttt{out2\_1} + \texttt{out1\_1} \cdot \texttt{out2\_0} \leq \texttt{out2\_1} \rightarrow \texttt{out1\_0}$.

7 Conclusion and Outlook

7.1 Experimental Results and Experiences

Before doing the proofs from Section 6 we did a lot of proofs from the area of MKA and its underlying structures like idempotent semirings, Kleene algebras and modal semirings. These proofs were built up from scratch, i.e. assuming only the axioms of the respective structures.

According to our experiences, interactive proving is a more powerful tool for verification than automated reasoning. Despite its undisputed power automated reasoning often demands human interaction in the form of choosing an appropriate axiom and lemma set or modifying parameters of the search space. We did the same formalizations as in Section 6 in Prover9 (see [6]) with a disappointing result: it was not possible to prove a single statement, even selection of lemmata and axioms did not help much. In contrast, verification of the same formalization with KIV became after some time of familiarization a pleasant and comfortable task.

7.2 Future Work

This paper presented only the first step towards interactive verification of PLC programs. There are several lines of further research we plan to investigate. A first idea is automated generation of the specification and sequents files on the base of a given PLC program (the examples presented here were built by hand). Another direction of research is to expand our approach to other languages for programming PLCs besides FBD. Basically, this can be achieved by a suitable generator for KIV files which does not take FBD programs as input but PLC programs in another language. Moreover, it is consequent to tackle other Boolean functions but the ones considered here (especially flank evaluators seem to be a challenging task) and numerical functions. The second intention may be be eased by the fact that KIV offers already an implementation of the naturals. As usual in programming, PLC programs contain a lot of frequently used constructions like the mutual exclusion gadget from Figure 4. Here it can help shorten proofs if predefined properties of such structures as additional lemmata are already at the user's disposal.

The crucial test for our approach is the verification of real world instances. A typical robot cell with one robot has about 32 up to 64 safety-critical signals and two up to four doors with about four signals each. The associated PLC program consists of 50 op to 100 gates which are in general no primitive gates as in our examples but more complex ones with predefined behavior whose correctness is guaranteed by the producer. Their behavior needs to be modeled in MKA analogously to Section 5 which will not pose any problems. After doing so, our approach is expected to work without difficulty, eased by the fact that each signal is processed only in a small fraction of all gates. In such cases with a large number of gates and signals it would be interesting to investigate the influence the state explosion if one uses an approach based on model checking. This will show whether the considerations from Subsection 1.2 are correct.

Acknowledgments. We are grateful to Gerhard Schellhorn for his patient help with the KIV system and to the anonymous reviewers for pointing out a lot of minor and major mistakes.

References

1. Coq. https://coq.inria.fr/ (Online; accessed July 7, 2015)
2. IEC61131. http://webstore.iec.ch/webstore/webstore.nsf/artnum/048541opendocument (Online; accessed April 1, 2015)
3. Isabelle. https://isabelle.in.tum.de/ (Online; accessed July 7, 2015)
4. The KIV system. http://www.informatik.uni-augsburg.de/lehrstuehle/swt/se/kiv/ (Online; accessed November 5, 2014)
5. NuSMV. http://nusmv.fbk.eu/ (Online; accessed July 7, 2015)
6. Prover9. https://www.cs.unm.edu/~mccune/mace4/ (Online; accessed July 7, 2015)
7. Step7. http://w3.siemens.com/mcms/simatic-controller-software/en/step7/ (Online; accessed April 1, 2015)
8. Armstrong, A., Struth, G., Weber, T.: Program analysis and verification based on kleene algebra in isabelle/hol. In: Blazy, S., Paulin-Mohring, C., Pichardie, D. (eds.) ITP 2013. LNCS, vol. 7998, pp. 197–212. Springer, Heidelberg (2013)
9. Balser, M., Reif, W., Schellhorn, G., Stenzel, K., Thums, A.: Formal system development with KIV. In: Maibaum, T. (ed.) FASE 2000. LNCS, vol. 1783, pp. 363–366. Springer, Heidelberg (2000)
10. Berghammer, R., Höfner, P., Stucke, I.: Automated verification of relational while-programs. In: Höfner, P., Jipsen, P., Kahl, W., Müller, M.E. (eds.) RAMiCS 2014. LNCS, vol. 8428, pp. 173–190. Springer, Heidelberg (2014)
11. Carlsson, H., Svensson, B., Danielson, F., Lennartson, B.: Methods for reliable simulation-based PLC code verification. IEEE Trans. Industrial Informatics 8(2), 267–278 (2012)
12. Dang, H., Höfner, P.: Automated higher-order reasoning about quantales. In: Schmidt, R.A., Schulz, S., Konev, B. (eds.) Proceedings of the 2nd Workshop on Practical Aspects of Automated Reasoning, PAAR 2010. EPiC Series, vol. 9, pp. 40–51. EasyChair, Edinburgh (2010)
13. Desharnais, J., Möller, B., Struth, G.: Modal kleene algebra and applications - a survey. Journal on Relational Methods in Computer Science 1, 93–131 (2004)
14. Desharnais, J., Möller, B., Struth, G.: Kleene algebra with domain. ACM Transactions on Computational Logic 7, 798–833 (2006)
15. Höfner, P.: Automated reasoning for hybrid systems - two case studies -. In: Berghammer, R., Möller, B., Struth, G. (eds.) RelMiCS/AKA 2008. LNCS, vol. 4988, pp. 191–205. Springer, Heidelberg (2008)
16. Höfner, P., Struth, G.: Automated reasoning in Kleene algebra. In: Pfenning, F. (ed.) CADE 2007. LNCS (LNAI), vol. 4603, pp. 279–294. Springer, Heidelberg (2007)
17. Jee, E., Yoo, J., Cha, S.D., Bae, D.: A data flow-based structural testing technique for FBD programs. Information & Software Technology 51(7), 1131–1139 (2009)
18. Kozen, D.: A completeness theorem for kleene algebras and the algebra of regular events. Information and Computation 110(2), 366–390

19. Möller, B., Höfner, P., Struth, G.: Quantales and temporal logics. In: Johnson, M., Vene, V. (eds.) AMAST 2006. LNCS, vol. 4019, pp. 263–277. Springer, Heidelberg (2006)
20. Ortmeier, F., Schellhorn, G., Thums, A., Reif, W., Hering, B., Trappschuh, H.: Safety analysis of the height control system for the elbtunnel. Rel. Eng. & Sys. Safety 81(3), 259–268 (2003)
21. Pavlovic, O., Ehrich, H.: Model checking PLC software written in function block diagram. In: Third International Conference on Software Testing, Verification and Validation, ICST 2010, Paris, France, April 7-9. CEUR Workshop Proceedings. IEEE Computer Society (2010)
22. Pavlovic, O., Pinger, R., Kollmann, M.: Automation of formal verification of PLC programs written in IL. In: Beckert, B. (ed.) Proceedings of 4th International Verification Workshop in connection with CADE-21, Bremen, Germany, July 15-16. CEUR Workshop Proceedings, vol. 259. CEUR-WS.org (2007)
23. Schmitt, J., Hoffmann, A., Balser, M., Reif, W., Marcos, M.: Interactive verification of medical guidelines. In: Misra, J., Nipkow, T., Sekerinski, E. (eds.) FM 2006. LNCS, vol. 4085, pp. 32–47. Springer, Heidelberg (2006)

Investigating and Computing Bipartitions with Algebraic Means

Rudolf Berghammer[1], Insa Stucke[1], and Michael Winter[2,*]

[1] Institut für Informatik, Universität Kiel, 24098 Kiel, Germany
[2] Department of Computer Science, Brock University, St. Catharines, ON, Canada

Abstract. Using Dedekind categories as an algebraic structure for (binary) set-theoretic relations without complements, we present purely algebraic definitions of "to be bipartite" and "to possess no odd cycles" and prove that both notions coincide. This generalises D. König's well-known theorem from undirected graphs to abstract relations, and, hence, to models such as L-relations that are different from set-theoretic relations. One direction of this generalisation is shown by specifying a bipartition for the relation in question in form of a pair of disjoint relational vectors. For set-theoretic relations this immediately leads to relational programs for computing bipartitions. We also discuss how the algebraic proofs can be mechanised using theorem proving tools.

1 Introduction

The modern axiomatic investigation of the calculus of (binary) relations started with the seminal paper [18] of A. Tarski. For many years this calculus has been widely used by mathematicians, computer scientists and engineers as a conceptual and methodological base for investigating fundamental notions and problem solving. A lot of examples and references to relevant literature can be found, for instance, in the textbooks [16,17] and the proceedings of the international conferences RAMiCS.

Relation algebra, the axiomatic algebraic structure underlying the calculus of relations, has been applied to many concrete examples, particularly to graph-theoretic problems. This is mainly due to the fact that a directed graph can be seen as a binary relation on the vertex set. Other kinds of graphs can also be modeled easily using relation algebra as, e.g., demonstrated in [16]. These investigations have been accompanied by tool support. The latter concerns the mechanisation of relation algebra and the execution of relational programs in tools like RelView (see [1,25]) as well as theorem proving in the context of relation algebra (see [3,4,6,9]).

In this paper we continue this line of research. Primarily, we prove some results concerning the bipartition of graphs with purely relation-algebraic means, that is, without any reference to the fact that relations are sets of pairs over

* The author gratefully acknowledges support from the Natural Sciences and Engineering Research Council of Canada.

W. Kahl et al. (Eds.): RAMiCS 2015, LNCS 9348, pp. 257–274, 2015.
DOI: 10.1007/978-3-319-24704-5_16

certain carrier sets. We even avoid the use of complements which, algebraically, means that we do not work with relation algebra in the sense of [18,19] (homogeneous approach) or [16,17] (heterogeneous approach), but with the more general algebraic structure of a Dedekind category. This algebraic structure has been introduced in [12] and is, for example, used in [8] to investigate crispness of L-relations and in [21] to model processes. L-relations generalise fuzzy relations by replacing the unit interval $[0, 1] \subseteq \mathbb{R}$ as the domain of membership by an arbitrary lattice L. In the well-known matrix model of relations fuzzy relations are matrices with entries from the unit interval $[0, 1] \subseteq \mathbb{R}$, whereas L-relations are matrices with entries from a suitable lattice L. Since L-relations form a Dedekind category, our results also apply to this generalisation of set-theoretic relations.

Considering abstract relations as morphisms of a Dedekind category, we present purely algebraic definitions of "to be bipartite" and "to possess no odd cycles". For set-theoretic relations these notions coincide with the corresponding notions from graph theory. Then we algebraically prove for all relations that they are bipartite if and only if they do not possess odd cycles. This generalises D. König's well-known theorem (published in [10]) from undirected graphs to Dedekind categories. One direction of this generalisation is shown by specifying for the input relations bipartitions in form of pairs of disjoint relational vectors. This is done by means of algebraic expressions. When using RelView, in case of symmetric set-theoretic relations (i.e., undirected graphs) these are based on the algebraic construction of a splitting, that generalises projections of set-theoretic equivalence relations and in RelView is not available as a pre-defined operation. But splittings easily can be computed by means of a simple relational program.

Algebraic calculations concerning relations are extremely formal. This not only minimises the danger of making errors within proofs, but also allows the use of theorem provers and proof assistant tools. We have used the automated theorem prover Prover9 (see [24]) and the proof assistant tool Coq (see [23]) to check our results and we also report on our experience in respect thereof.

2 Relation-Algebraic Preliminaries

In the following we recall the algebraic preliminaries we will need in this paper. For more details on relation algebra see e.g., [16,17], and for more details on Dedekind categories see e.g., [8,21]. Especially the proofs concerning all unproven basic facts of this section can be found there.

We assume the reader is familiar with the basic operations on set-theoretic relations, viz. R^{T} (transposition), $\overline{R}$ (complement), $R \cup S$ (union), $R \cap S$ (intersection), $R;S$ (composition), the predicates $R \subseteq S$ (inclusion) and $R = S$ (equality), and the special relations O (empty relation), L (universal relation) and I (identity relation). Restricted to relations of the same type, the Boolean operations $\overline{}$, $\cup$ and $\cap$, the ordering $\subseteq$ and the constants O and L lead to a complete Boolean lattice. Some further well-known properties of set-theoretic relations are $\overline{R^{\mathsf{T}}} = \overline{R}^{\mathsf{T}}$, $(R \cup S)^{\mathsf{T}} = R^{\mathsf{T}} \cup S^{\mathsf{T}}$, $(R \cap S)^{\mathsf{T}} = R^{\mathsf{T}} \cap S^{\mathsf{T}}$, $(R^{\mathsf{T}})^{\mathsf{T}} = R$,

$(R;S)^{\mathsf{T}} = S^{\mathsf{T}};R^{\mathsf{T}}$, that from $R \subseteq S$ it follows $R^{\mathsf{T}} \subseteq S^{\mathsf{T}}$ and that also union, intersection and composition are monotonic in both arguments.

The theoretical framework for these rules (and much others) to hold is that of a (heterogeneous) *relation algebra.* This algebraic structure is a specific category with typed relations $R : X \leftrightarrow Y$ as morphisms, where the object X is the source and the object Y is the target of R. Most of the complement-free relation-algebraic rules already hold in a *Dedekind category.* This algebraic structure generalises relation algebra by using *left residuals* S/R instead of complements. Hence, the constants and operations of a Dedekind category are those of set-theoretic relations, except complementation, and additionally left residual. As usual, we overload the symbols O, L and I, i.e., avoid the binding of types to them. The axioms of a Dedekind category are

(a) the axioms of a complete distributive lattice for all relations of the same type under union, intersection, the ordering, empty and universal relation,
(b) the associativity of composition and that identity relations are neutral elements w.r.t. composition,
(c) the monotonicity of transposition and $(R^{\mathsf{T}})^{\mathsf{T}} = R$ and $(R;S)^{\mathsf{T}} = S^{\mathsf{T}};R^{\mathsf{T}}$, for all $R : X \leftrightarrow Y$ and $S : Y \leftrightarrow Z$,
(d) the *modular law* saying that $Q;R \cap S \subseteq Q;(R \cap Q^{\mathsf{T}};S)$, for all $Q : X \leftrightarrow Y$, $R : Y \leftrightarrow Z$ and $S : X \leftrightarrow Z$,
(e) that $Q;R \subseteq S$ if and only if $Q \subseteq S/R$, for all $Q : X \leftrightarrow Y$, $R : Y \leftrightarrow Z$ and $S : X \leftrightarrow Z$.

From the modular law we obtain the *dual modular law* $Q;R \cap S \subseteq (Q \cap S;R^{\mathsf{T}});R$, for all $Q : X \leftrightarrow Y$, $R : Y \leftrightarrow Z$ and $S : X \leftrightarrow Z$. In the following we always will assume that expressions and formulas are well-typed and in the proofs we will mention only the modular laws and the "non-obvious" consequences of the axioms. Well-known rules like those presented above remain unmentioned. During the entire paper residuals will not be applied.

The basic operations and constants mentioned above can be used for defining specific classes of relations in a purely algebraic way. In the following we introduce the classes which will be used in the remainder of this paper.

A relation R is *univalent* if $R^{\mathsf{T}};R \subseteq \mathsf{I}$, and *total* if $R;\mathsf{L} = \mathsf{L}$, which is equivalent to $\mathsf{I} \subseteq R;R^{\mathsf{T}}$. A *mapping* is a univalent and total relation. Relation R is *injective* if R^{T} is univalent and *surjective* if R^{T} is total.

A relation R is *homogeneous* if $R \cup R^{\mathsf{T}}$ is defined, i.e., source and target coincide. Let R be homogeneous. Then R is *reflexive* if $\mathsf{I} \subseteq R$, *irreflexive* if $R \cap \mathsf{I} = \mathsf{O}$, *symmetric* if $R = R^{\mathsf{T}}$, *antisymmetric* if $R \cap R^{\mathsf{T}} \subseteq \mathsf{I}$, and *transitive* if $R;R \subseteq R$. A symmetric and transitive relation is a *partial equivalence relation*; reflexive partial equivalence relations are *equivalence relations.* Finally, a reflexive, antisymmetric and transitive relation R is a *partial order relation*; if additional $R \cup R^{\mathsf{T}} = \mathsf{L}$ holds, then it is a *linear order relation.*

Assuming R as homogeneous set-theoretic relation, the least transitive relation containing R is its *transitive closure* R^+ and the least reflexive and transitive relation containing R is its *reflexive-transitive closure* R^*. In [11] K.C. Ng and

A. Tarski added the Kleene star as an additional operation to (abstract) relation algebras to denote the reflexive-transitive closure R^*. We do the same for Dedekind categories and specify the Kleene star as an additional operation for such structures by the following two laws (known from Kleene algebra) to hold for all relations Q, R, and S:

$$\mathsf{I} \cup R;R^* = R^* \qquad\qquad R;Q \cup S \subseteq Q \Rightarrow R^*;S \subseteq Q$$

Equivalently, for all relations Q, R and S we could demand $\mathsf{I} \cup R^*;R = R^*$ and that $Q;R \cup S \subseteq Q$ implies $S;R^* \subseteq Q$. The Kleene plus for the transitive closure is then reduced to the Kleene star by $R^+ = R;R^*$ (or, equivalently, $R^+ = R^*;R$), for all relations R. With these specifications well-known properties can be proved by purely algebraic means. In this paper we will need

$$\begin{array}{ccc} \mathsf{I} \subseteq R^* & R^*;R^* \subseteq R^* & R^* = \mathsf{I} \cup R^+ \\ (R^*)^{\mathsf{T}} = (R^{\mathsf{T}})^* & R^+;R^+ \subseteq R^+ & R^* = (R;R)^* \cup R;(R;R)^* \\ & R;(R;R)^* = (R;R)^*;R & \end{array}$$

and that $R^+ = R$ if R is transitive, for all relations R, as well as that the operations $R \mapsto R^+$ and $R \mapsto R^*$ are extensive, monotonic and idempotent, i.e., closure operators in the order-theoretic sense. Our introduction of the Kleene star via additional axioms means no restriction. Since in Dedekind categories the set $[X \leftrightarrow X]$ of relations of type $X \leftrightarrow X$ forms a complete lattice, for $R : X \leftrightarrow X$ we could alternatively define $R^* = \bigcap\{S \in [X \leftrightarrow X] \mid \mathsf{I} \subseteq S \wedge S;S \subseteq S \wedge R \subseteq S\}$, which equals $R^* = \mu(f_R)$, where $\mu(f_R)$ is the least fixed point of the monotonic function $f_R : [X \leftrightarrow X] \to [X \leftrightarrow X]$ with $f_R(S) = \mathsf{I} \cup S;S \cup R$.

The algebraic approach offers different ways for describing sets. We use only *vectors*, i.e., relations v with $v = v;\mathsf{L}$. Usually vectors are denoted by lower-case letters. For $v : X \leftrightarrow Y$ being a set-theoretic relation the condition $v = v;\mathsf{L}$ means that v can be written in the form $v = Z \times Y$ with a subset Z of X. Then we say that v *describes the subset* Z of X. For this application the target of a set-theoretic vector is irrelevant and, therefore, we always will use the singleton set $\mathbf{1}$. A vector is a *point* if it is injective and surjective. In the set-theoretic case this means that the point $v : X \leftrightarrow \mathbf{1}$ describes a singleton subset $\{x\}$ of X and then we say that it *describes the element* x of X. In the well-known Boolean matrix model of relations a vector is a row-constant matrix and a point is a matrix where exactly one row consists of ones only.

For vectors v and w and a relation R, also $R;v$, $v \cup w$ and $v \cap w$ are vectors. In case of set-theoretic relations and $R : X \leftrightarrow Y$ the vector $R;\mathsf{L}$ describes the subset $\{x \in X \mid \exists y \in Y : x\,R\,y\}$ of X, i.e., the *domain* of R. In the following lemma basic properties of disjoint vectors are collected.

Lemma 2.1. *Let v and w be vectors with $v \cap w = \mathsf{O}$. Then we have:*

(1) $v^{\mathsf{T}};w = \mathsf{O}$ and $w^{\mathsf{T}};v = \mathsf{O}$.
(2) $v;v^{\mathsf{T}} \cup w;w^{\mathsf{T}}$ is transitive.
(3) $(v;w^{\mathsf{T}} \cup w;v^{\mathsf{T}});(v;w^{\mathsf{T}} \cup w;v^{\mathsf{T}}) \subseteq v;v^{\mathsf{T}} \cup w;w^{\mathsf{T}}$.

Proof. (1) Using the modular law, that v is a vector and $v \cap w = \mathsf{O}$. the first equation is shown as follows and the second equation is shown analogously:

$$v^{\mathsf{T}}; w = v^{\mathsf{T}}; w \cap \mathsf{L} \subseteq v^{\mathsf{T}}; (w \cap (v^{\mathsf{T}})^{\mathsf{T}}; \mathsf{L}) = v^{\mathsf{T}}; (w \cap v) = \mathsf{O}$$

(2) The following calculation shows that $v; v^{\mathsf{T}} \cup w; w^{\mathsf{T}}$ is transitive:

$$\begin{aligned}
&(v; v^{\mathsf{T}} \cup w; w^{\mathsf{T}}); (v; v^{\mathsf{T}} \cup w; w^{\mathsf{T}}) \\
&= v; v^{\mathsf{T}}; v; v^{\mathsf{T}} \cup v; v^{\mathsf{T}}; w; w^{\mathsf{T}} \cup w; w^{\mathsf{T}}; v; v^{\mathsf{T}} \cup w; w^{\mathsf{T}}; w; w^{\mathsf{T}} \\
&= v; v^{\mathsf{T}}; v; v^{\mathsf{T}} \cup w; w^{\mathsf{T}}; w; w^{\mathsf{T}} && \text{by (1)} \\
&\subseteq v; \mathsf{L}; v^{\mathsf{T}} \cup w; \mathsf{L}; w^{\mathsf{T}} \\
&= v; v^{\mathsf{T}} \cup w; w^{\mathsf{T}} && v, w \text{ vectors}
\end{aligned}$$

(3) Similar to (2) we obtain the claim as follows:

$$\begin{aligned}
&(v; w^{\mathsf{T}} \cup w; v^{\mathsf{T}}); (v; w^{\mathsf{T}} \cup w; v^{\mathsf{T}}) \\
&= v; w^{\mathsf{T}}; v; w^{\mathsf{T}} \cup v; w^{\mathsf{T}}; w; v^{\mathsf{T}} \cup w; v^{\mathsf{T}}; v; w^{\mathsf{T}} \cup w; v^{\mathsf{T}}; w; v^{\mathsf{T}} \\
&= v; w^{\mathsf{T}}; w; v^{\mathsf{T}} \cup w; v^{\mathsf{T}}; v; w^{\mathsf{T}} && \text{by (1)} \\
&\subseteq v; \mathsf{L}; v^{\mathsf{T}} \cup w; \mathsf{L}; w^{\mathsf{T}} \\
&= v; v^{\mathsf{T}} \cup w; w^{\mathsf{T}} && v, w \text{ vectors}
\end{aligned}$$ □

3 Bipartite Relations do not Possess Odd Cycles

We assume the reader is familiar with the fundamental facts of graph theory; otherwise we refer to the textbook [5], for example. In graph theory a graph G is called *bipartite* if its vertex set X can be divided into two disjoint sets V and W such that each edge of G only connects vertices from different sets. Then the pair (V, W) is called a *bipartition* of G.

In this paper we investigate bipartitions of arbitrary homogeneous relations with algebraic means. If such a relation $R : X \leftrightarrow X$ is the *adjacency relation* of a *directed graph* $G = (X, R)$ (that is, R is a set-theoretic relation and consists of the directed edges of G) and the set-theoretic vectors $v : X \leftrightarrow \mathbf{1}$ and $w : X \leftrightarrow \mathbf{1}$ describe the subsets V and W of X in the sense of Section 2, then $R \subseteq v; w^{\mathsf{T}} \cup w; v^{\mathsf{T}}$ specifies that edges of G either start in V and end in W or start in W and end in V. Hence, the pair (V, W) is a bipartition of G if and only if $v \cap w = \mathsf{O}$, $v \cup w = \mathsf{L}$ and $R \subseteq v; w^{\mathsf{T}} \cup w; v^{\mathsf{T}}$. Generalising this to abstract relations, we define:

Definition 3.1. *Given a relation R and vectors v and w we say that the pair (v, w) is a* bipartition *of R if $v \cap w = \mathsf{O}$ and $R \subseteq v; w^{\mathsf{T}} \cup w; v^{\mathsf{T}}$. If there exists a bipartition, then R is called* bipartite.

Since in our algebraic proofs only the properties $v \cap w = \mathsf{O}$ and $R \subseteq v; w^{\mathsf{T}} \cup w; v^{\mathsf{T}}$ will play a role, we have dropped the demand $v \cup w = \mathsf{L}$ that, in graph-theoretic terminology means that $V \cup W = X$. Note, however, that if R, v and w are elements of a relation algebra (or set-theoretic relations), i.e., complements may be formed, then $v \cap w = \mathsf{O}$ and $R \subseteq v; w^{\mathsf{T}} \cup w; v^{\mathsf{T}}$ imply

$$R \subseteq v; w^{\mathsf{T}} \cup w; v^{\mathsf{T}} \subseteq v; \overline{v}^{\mathsf{T}} \cup \overline{v}; v^{\mathsf{T}}$$

such that $(v, \overline{v})$ is a bipartition of R if (v, w) is a bipartition of R. We start with a theorem that, assuming $v \cap w = \mathsf{O}$, provides an equivalent description of the second demand $R \subseteq v; w^{\mathsf{T}} \cup w; v^{\mathsf{T}}$ of a bipartition.

Theorem 3.1. *Let R be a relation and assume v and w to be vectors such that $v \cap w = \mathsf{O}$. Then we have $R \subseteq v; w^{\mathsf{T}} \cup w; v^{\mathsf{T}}$ if and only if $R; v \subseteq w$ and $R; w \subseteq v$ and $R \subseteq (v \cup w); (v \cup w)^{\mathsf{T}}$.*

Proof. "⇒": Using Lemma 2.1(1) and that w is a vector we obtain

$$R; v \subseteq (v; w^{\mathsf{T}} \cup w; v^{\mathsf{T}}); v = v; w^{\mathsf{T}}; v \cup w; v^{\mathsf{T}}; v = w; v^{\mathsf{T}}; v \subseteq w; \mathsf{L} = w.$$

The inclusion $R; w \subseteq v$ follows analogously. Finally, we have

$$R \subseteq v; w^{\mathsf{T}} \cup w; v^{\mathsf{T}} \subseteq v; v^{\mathsf{T}} \cup v; w^{\mathsf{T}} \cup w; v^{\mathsf{T}} \cup w; w^{\mathsf{T}} = (v \cup w); (v \cup w)^{\mathsf{T}}.$$

"⇐": First of all, we have as auxiliary result

$$R \cap v; v^{\mathsf{T}} = v; v^{\mathsf{T}} \cap R \subseteq (v \cap R; (v^{\mathsf{T}})^{\mathsf{T}}); v^{\mathsf{T}} = (v \cap R; v); v^{\mathsf{T}} \subseteq (v \cap w); v^{\mathsf{T}} = \mathsf{O},$$

where we use in the second step the dual modular law. By a similar computation we obtain $R \cap w; w^{\mathsf{T}} = \mathsf{O}$. We conclude

$$\begin{aligned} R &= R \cap (v \cup w); (v \cup w)^{\mathsf{T}} \\ &= R \cap (v; v^{\mathsf{T}} \cup v; w^{\mathsf{T}} \cup w; v^{\mathsf{T}} \cup w; w^{\mathsf{T}}) \\ &= (R \cap v; v^{\mathsf{T}}) \cup (R \cap w; w^{\mathsf{T}}) \cup (R \cap (v; w^{\mathsf{T}} \cup w; v^{\mathsf{T}})) \\ &= R \cap (v; w^{\mathsf{T}} \cup w; v^{\mathsf{T}}), \end{aligned}$$

using $R \subseteq (v \cup w); (v \cup w)^{\mathsf{T}}$ in the first step and the auxiliary results in the last step, and, hence, get the desired result $R \subseteq v; w^{\mathsf{T}} \cup w; v^{\mathsf{T}}$. □

Now, assume again $R : X \leftrightarrow X$ to be the adjacency relation of a directed graph $G = (X, R)$ and the set-theoretic vectors $v : X \leftrightarrow \mathbf{1}$ and $w : X \leftrightarrow \mathbf{1}$ to describe subsets V and W of its vertex set X. If we suppose besides $v \cap w = \mathsf{O}$ also $v \cup w = \mathsf{L}$, then we get $w = \overline{v}$, such that W is the complement $\overline{V}$ of V relative to X. From $w = \overline{v}$ we obtain that $R \subseteq (v \cup w); (v \cup w)^{\mathsf{T}}$ is true, $R; v \subseteq w$ is equivalent to $R; v \subseteq \overline{v}$ and $R; w \subseteq v$ is equivalent to $R; \overline{v} \subseteq v$ The inclusion $R; v \subseteq \overline{v}$ specifies V as *independent set* (or *stable set*) of G in the sense that no pair of vertices from V is connected by a directed edge (cf. [16]) and $R; \overline{v} \subseteq v$ does the same for $\overline{V}$. Under this point of view, hence, Theorem 3.1 relation-algebraically describes that the pair $(V, \overline{V})$ is a bipartition of G if and only if V and $\overline{V}$ are independent sets of G.

By means of Lemma 2.1 and Theorem 3.1 we are in the position to prove the first main result of the paper.

Theorem 3.2. *Let R be a relation. If there exists a bipartition (v, w) of R, then we have $R; (R; R)^* \cap \mathsf{I} = \mathsf{O}$.*

Proof. Using the modular law in the first step, we immediately obtain

$$v; w^{\mathsf{T}} \cap \mathsf{I} \subseteq v; (w^{\mathsf{T}} \cap v^{\mathsf{T}}; \mathsf{I}) = v; (v \cap w)^{\mathsf{T}} = \mathsf{O},$$

and $w; v^{\mathsf{T}} \cap \mathsf{I} = \mathsf{O}$ follows similarly. Now, we calculate as given below:

$$\begin{array}{lll}
R; (R; R)^* \cap \mathsf{I} & \subseteq R; ((v; w^{\mathsf{T}} \cup w; v^{\mathsf{T}}); (v; w^{\mathsf{T}} \cup w; v^{\mathsf{T}}))^* \cap \mathsf{I} & \text{assumption} \\
& \subseteq R; (v; v^{\mathsf{T}} \cup w; w^{\mathsf{T}})^* \cap \mathsf{I} & \text{Lemma 2.1(3)} \\
& = R; (\mathsf{I} \cup (v; v^{\mathsf{T}} \cup w; w^{\mathsf{T}})^+) \cap \mathsf{I} & \text{property closures} \\
& = R; (\mathsf{I} \cup v; v^{\mathsf{T}} \cup w; w^{\mathsf{T}}) \cap \mathsf{I} & \text{Lemma 2.1(2)} \\
& = (R \cup R; v; v^{\mathsf{T}} \cup R; w; w^{\mathsf{T}}) \cap \mathsf{I} & \\
& \subseteq (R \cup w; v^{\mathsf{T}} \cup v; w^{\mathsf{T}}) \cap \mathsf{I} & \text{Theorem 3.1 “}\Rightarrow\text{”} \\
& = (w; v^{\mathsf{T}} \cup v; w^{\mathsf{T}}) \cap \mathsf{I} & \text{assumption} \\
& = (w; v^{\mathsf{T}} \cap \mathsf{I}) \cup (v; w^{\mathsf{T}} \cap \mathsf{I}) & \\
& = \mathsf{O} & \text{aux. results}
\end{array}$$

□

So, for bipartite relations we have $R; (R; R)^* \cap \mathsf{I} = \mathsf{O}$. If again $R : X \leftrightarrow X$ is a set-theoretic relation and the adjacency relation of a directed graph $G = (X, R)$, then for all vertices $x, y \in X$ we have $x\,(R; (R; R)^*)\,y$ if and only if there is a (directed) path from x to y with odd length. As a consequence, the property $R; (R; R)^* \cap \mathsf{I} = \mathsf{O}$ holds if and only if in G there is no (directed) cycle with odd length. Summing up, Theorem 3.2 is the first direction of the generalisation of D. König's theorem from undirected graphs to Dedekind categories: *All bipartite relations do not posses odd cycles.*

4 Relations without Odd Cycles are Bipartite

In this section we show the remaining direction of the generalisation of D. König's theorem from undirected graphs to Dedekind categories, viz. that all relations without odd cycles are bipartite.

In graph theory the notion "bipartite" usually is studied for undirected graphs only – despite of the fact that directed bipartite graphs have a lot of applications, for instance, in game theory, as signature diagrams of algebraic specifications or as static parts of Petri nets. An undirected graph is of the form $G = (X, E)$, where each undirected edge e from the edge set E is a subset of X such that $|e| = 2$. Undirected graphs and irreflexive and symmetric relation are essentially the same, since $x\,R\,y$ if and only if $\{x, y\} \in E$, for all $x, y \in E$, defines an irreflexive and symmetric relation $R : X \leftrightarrow X$ and this correspondence between undirected graphs over a vertex set and irreflexive and symmetric relations on the same set is even one-to-one. But for all relations R, from $R; (R; R)^* \cap \mathsf{I} = \mathsf{O}$ we get irreflexivity $R \cap \mathsf{I} = \mathsf{O}$, such that we may neglect irreflexivity as pre-condition for the theorem we want to prove.

The next theorem shows that for a relation R and its *symmetric closure* $R \cup R^{\mathsf{T}}$ the sets of their bipartitions coincide. As a consequence it suffices to prove that all symmetric relations R with $R; (R; R)^* \cap \mathsf{I} = \mathsf{O}$ are bipartite to get that all relations without odd cycles are bipartite.

Theorem 4.1. *Let R be a relation and v, w be vectors. Then $R \subseteq v; w^{\mathsf{T}} \cup w; v^{\mathsf{T}}$ is equivalent to $R \cup R^{\mathsf{T}} \subseteq v; w^{\mathsf{T}} \cup w; v^{\mathsf{T}}$.*

Proof. "$\Rightarrow$": If $R \subseteq v; w^{\mathsf{T}} \cup w; v^{\mathsf{T}}$, then $R \cup R^{\mathsf{T}} \subseteq v; w^{\mathsf{T}} \cup w; v^{\mathsf{T}}$ follows from

$$R^{\mathsf{T}} \subseteq (v; w^{\mathsf{T}} \cup w; v^{\mathsf{T}})^{\mathsf{T}} = (v; w^{\mathsf{T}})^{\mathsf{T}} \cup (w; v^{\mathsf{T}})^{\mathsf{T}} = w; v^{\mathsf{T}} \cup v; w^{\mathsf{T}}.$$

"$\Leftarrow$": A proof of this direction is trivial. □

The Theorem 4.2 given below is the key of our proof that all symmetric relations without odd cycles are bipartite. We prepare its proof by the following fact.

Lemma 4.1. *Let R be a relation such that R^* is symmetric. Then for all relations U, from $R; (R; R)^*; U \cap U = \mathsf{O}$ it follows $(R; R)^*; U \cap R; (R; R)^*; U = \mathsf{O}$.*

Proof. Let E abbreviate the reflexive-transitive closure $(R; R)^*$. Then we have

$$E; U \subseteq (E \cup R; E); U = R^*; U = (R^*)^{\mathsf{T}}; U$$

using the property $R^* = (R; R)^* \cup R; (R; R)^*$ of reflexive transitive closures and that the relation R^* is symmetric, and

$$\begin{aligned} (R; E)^{\mathsf{T}}; U \cap E; U &\subseteq (R; E)^{\mathsf{T}}; (U \cap R; E; E; U) && \text{modular law} \\ &= (R; E)^{\mathsf{T}}; (R; E; U \cap U) && \text{property refl.-trans. closure} \\ &= \mathsf{O} && \text{assumption.} \end{aligned}$$

With these auxiliary results we get

$$\begin{aligned} E; U &= E; U \cap (R^*)^{\mathsf{T}}; U && \text{first auxiliary result} \\ &= E; U \cap (E \cup R; E)^{\mathsf{T}}; U && \text{property refl.-trans. closure} \\ &= (E; U \cap E^{\mathsf{T}}; U) \cup (E; U \cap (R; E)^{\mathsf{T}}; U) && \\ &= E; U \cap E^{\mathsf{T}}; U && \text{second auxiliary result} \end{aligned}$$

such that $E; U \subseteq E^{\mathsf{T}}; U$. Now, we are able to conclude the proof as follows:

$$\begin{aligned} E; U \cap R; E; U &\subseteq E^{\mathsf{T}}; U \cap R; E; U && \text{as } E; U \subseteq E^{\mathsf{T}}; U \\ &\subseteq E^{\mathsf{T}}; (U \cap E; R; E; U) && \text{modular law} \\ &= E^{\mathsf{T}}; (U \cap R; E; E; U) && \text{property refl.-trans. closure} \\ &= E^{\mathsf{T}}; (U \cap R; E; U) && \text{property refl.-trans. closure} \\ &= \mathsf{O} && \text{assumption} \end{aligned}$$

□

Because of $\mathsf{I} \subseteq (R; R)^*$, for all relations R with symmetric R^* and all relations U the properties $R; (R; R)^*; U \cap U = \mathsf{O}$ and $(R; R)^*; U \cap R; (R; R)^*; U = \mathsf{O}$ even are equivalent. For the proof of the following Theorem 4.2, however, we only need the direction stated in the lemma.

Theorem 4.2. *Let R be a relation and R^* be symmetric. If u is a vector such that $R; (R; R)^*; u \cap u = \mathsf{O}$ and $R \subseteq R^*; u; u^{\mathsf{T}}; R^*$ and we define $v := (R; R)^*; u$ and $w := R; v = R; (R; R)^*; u$, then v as well as w are vectors, $v \cap w = \mathsf{O}$ and $R \subseteq v; w^{\mathsf{T}} \cup w; v^{\mathsf{T}}$. I.e., the pair (v, w) is a bipartition of R.*

Proof. That v and w are vectors is obvious. From Lemma 4.1, $R^* = (R^*)^\mathsf{T}$ and $R;(R;R)^*;u \cap u = \mathsf{O}$ we get $(R;R)^*;u \cap R;(R;R)^*;u = \mathsf{O}$ such that

$$v \cap w = (R;R)^*;u \cap R;(R;R)^*;u = \mathsf{O}$$

by the definition of v and w. To prove $R \subseteq v;w^\mathsf{T} \cup w;v^\mathsf{T}$ we use direction "$\Leftarrow$" of Theorem 3.1 and have to verify three conditions. For the first verification

$$R;v = R;(R;R)^*;u = w$$

we use the definition of v and w. Also the second verification

$$R;w = R;R;(R;R)^*;u = (R;R)^+;u \subseteq (R;R)^*;u = v$$

follows from the definition of v and w. For the third verification we start with

$$v \cup w = (R;R)^*;u \cup R;(R;R)^*;u = ((R;R)^* \cup R;(R;R)^*);u = R^*;u,$$

using the definition of v and w once more, which, in combination with the assumptions $R \subseteq R^*;u;u^\mathsf{T};R^*$ and $R^* = (R^*)^\mathsf{T}$, implies

$$R \subseteq R^*;u;u^\mathsf{T};R^* = (R^*;u);(R^*;u)^\mathsf{T} = (v \cup w);(v \cup w)^\mathsf{T}. \qquad \Box$$

In graph-theoretic terminology this theorem reads as follows: Let a (directed or undirected) graph G be given such that for each pair $x, y \in X$ of vertices x is reachable from y if and only if y is reachable from x. Furthermore, let U be a subset of the vertex set X of G such that (1) no pair of vertices of U is connected via a path of odd length and (2) for each edge of G its vertices are reachable from vertices of U. If we define V as the set of vertices which can be reached from a vertex of U via a path of even length and W as the set of vertices which can be reached from a vertex of U via a path of odd length, then V and W are disjoint and for each edge of G one vertex is in V and the other one is in W.

It is remarkable that also the converse of Theorem 4.2 is valid such that, in general, we have the following characterisation of bipartite relations.

Theorem 4.3. *Let R be a relation and R^* be symmetric. Then there exists a vector u such that $R;(R;R)^*;u \cap u = \mathsf{O}$ and $R \subseteq R^*;u;u^\mathsf{T};R^*$ if and only if there exists a bipartition (v, w) of R.*

Namely, if the pair (v, w) is a bipartition of R, then $R;(R;R)^*;v \cap v = \mathsf{O}$ and $R \subseteq R^*;v;v^\mathsf{T};R^*$, such that v can be taken as u. Since, however, in the remainder of the paper only the direction expressed by Theorem 4.2 is applied, we have shifted the proof of these facts into the appendix. Instead, we present now an immediate consequence of Theorem 4.2. It shows how to get for *strongly connected* relations (that is, relations R with $R^* = \mathsf{L}$) without odd cycles a bipartition.

Corollary 4.1. *Let R be a relation such that $R^* = \mathsf{L}$ and $R;(R;R)^* \cap \mathsf{I} = \mathsf{O}$ and let p be a homogeneous point. If we define $v := (R;R)^*;p$ and $w := R;v = R;(R;R)^*;p$, then v and w are vectors such that $v \cap w = \mathsf{O}$ and $R \subseteq v;w^\mathsf{T} \cup w;v^\mathsf{T}$. I.e., the pair (v, w) is a bipartition of R. Furthermore, $v \cup w = \mathsf{L}$.*

Proof. To prove that the pair (v, w) is a bipartition of R, we verify the pre-conditions of Theorem 4.2 for u being the point p. From $R^* = \mathsf{L}$ we get that R^* is symmetric. As a point, p is also a vector, Next, we verify

$$R;(R;R)^*;p \cap p = R;(R;R)^*;p \cap \mathsf{I};p = (R;(R;R)^* \cap \mathsf{I});p = \mathsf{O}$$

using that p is injective and $R;(R;R)^* \cap \mathsf{I} = \mathsf{O}$. The last pre-condition

$$R \subseteq \mathsf{L} = \mathsf{LL} = \mathsf{L};p;p^\mathsf{T};\mathsf{L} = R^*;p;p^\mathsf{T};R^*$$

holds because p is surjective. The additionally stated property is shown by the calculation

$$v \cup w = (R;R)^*;p \cup R;(R;R)^*;p = ((R;R)^* \cup R;(R;R)^*);p = R^*;p = \mathsf{L};p = \mathsf{L}$$

that uses the definition of v and w and the surjectivity of the point p. □

Note that in case of three non-homogeneous universal relations it may happen that $\mathsf{L};\mathsf{L} \subset \mathsf{L}$ as demonstrated in [20]. However, if at least one of the universal relations on the left-hand side is homogeneous, then $\mathsf{L};\mathsf{L} = \mathsf{L}$ follows from the fact that $\mathsf{I} \subseteq \mathsf{L}$.

In order to show that all symmetric relations without odd cycles in an arbitrary Dedekind category are bipartite we are going to generalise the previous corollary. We use Theorem 4.2 again and prove the existence of a vector u with the required properties by algebraic means. Before we start we want to provide the idea behind this generalisation using regular graph/set theory. For an undirected graph G without odd cycles, let $R : X \leftrightarrow X$ be the corresponding symmetric relation such that $R;(R;R)^* \cap \mathsf{I} = \mathsf{O}$. Then R^* is an equivalence relation so that we may consider the set of equivalence classes of R^*, i.e., the set of *connected components* of G. We select from each connected component a single vertex and combine all these vertices to a subset U of X. If $u := U \times \mathbf{1}$ is the vector that describes U as subset of X, then each vertex of G is reachable from a vertex of some connected component, such that $R \subseteq R^*;u;u^\mathsf{T};R^*$ follows. The remaining assumption $R;(R;R)^*;u \cap u = \mathsf{O}$ is a consequence of the absence of odd cycles. In order to verify this, assume there are $x, y \in U$ which are connected via a path of odd length. Then this implies $x = y$ since x and y have to be in the same connected component and U contains from each connected component precisely one vertex. This is a contradiction to the absence of odd cycles.

The proof outlined above is based on set-theoretic arguments, i.e., refers to a relation as a set of pairs. In the following we want to show these results in context of abstract Dedekind categories. In doing so, a specific non-homogeneous relation will play a decisive role. This is the reason for using heterogeneous relations, i.e, a categorical approach to relations, instead of classical relation algebras in the sense of [18,19] that model homogeneous relations only.

Since we do not assume sets to be finite or countable, the selection of the single vertices from the connected components corresponds to an application of the *Axiom of Choice.* We use the following relation-algebraic variant of this axiom of set theory. E.g., it can be found as property **AC 4** in [15], with $\mathcal{D}(R)$ as notation for the domain of a relation R instead of our vector description $R;\mathsf{L}$.

Axiom 4.1. *(Relational Axiom of Choice) For all relations R there exists a univalent relation F such that $F \subseteq R$ and $F;\mathsf{L} = R;\mathsf{L}$.*

For all symmetric set-theoretic relations $R : X \leftrightarrow X$ there exists the *projection* (or canonical epimorphism) $\pi : X \to X/R^*$ that maps an element of X to its equivalence classes w.r.t. the equivalence relation R^* in the set of all equivalence classes X/R^* of R^*. If we regard π as a relation of type $X \leftrightarrow X/R^*$, that relates $x \in X$ and $Y \in 2^X$ if and only if Y is the equivalence class $[x]_{R^*}$, then it fulfills the following properties: $\pi;\pi^\mathsf{T} = R^*$ and $\pi^\mathsf{T};\pi = \mathsf{I}$. I.e., the relation π is a splitting of the equivalence relation R^* in the following sense.

Definition 4.1. *A relation S is called a* splitting *of a partial equivalence relation P if $S;S^\mathsf{T} = P$ and $S^\mathsf{T};S = \mathsf{I}$.*

This definition stems from [2]. Also in [7,20] the notion of a splitting for partial equivalence relations is introduced, thereby, compared with the above definition, changing source and target. The reason of our typing is that we want to keep projections as models of splittings. So, each set-theoretic partial equivalence relation possesses a splitting. When relations are considered as morphisms of Dedekind categories, then the above equations specify splittings up to isomorphism. However, it may happen that a partial equivalence relation P of a Dedekind category does not has a splitting. Nevertheless, then a splitting of P exists in a Dedekind category that extends the given one, as shown in [7,20]. Hence, we have:

Lemma 4.2. *Assume Axiom 4.1 to be true. Then for all relations R such that R^* is symmetric there exists a splitting S of R^* and a mapping F with $F \subseteq S^\mathsf{T}$.*

Proof. Since R^* is symmetric, it is a partial equivalence relation. From [7,20] it follows that R^* possesses a splitting S, possibly in an extension of the Dedekind category under consideration. Axiom 4.1 implies that there exists a univalent relation F such that $F \subseteq S^\mathsf{T}$ and $F;\mathsf{L} = S^\mathsf{T};\mathsf{L}$. Next, $\mathsf{I} \subseteq S^\mathsf{T};S$ shows that S is surjective. From this we get totality of S^T, which in turn yields $F;\mathsf{L} = S^\mathsf{T};\mathsf{L} = \mathsf{L}$. Thus, F is also total. □

After these preparations we are able to prove the existence of the vector u with purely algebraic means. In case of an undirected graph G without odd cycles and $R : X \leftrightarrow X$ as its corresponding relation and the projection relation $\pi : X \leftrightarrow X/R^*$ as splitting S, the mapping F of Theorem 4.4 is a choice function for the set of sets X/R^* in the usual mathematical sense.

Theorem 4.4. *Assume Axiom 4.1 to be true. Furthermore, let R be a relation with symmetric R^* and $R;(R;R)^* \cap \mathsf{I} = \mathsf{O}$, S be a splitting of R^* and F be a mapping such that $F \subseteq S^\mathsf{T}$. With L homogeneous, $F^\mathsf{T};\mathsf{L}$ is a vector and we have $R;(R;R)^*;F^\mathsf{T};\mathsf{L} \cap F^\mathsf{T};\mathsf{L} = \mathsf{O}$ and $R \subseteq R^*;F^\mathsf{T};\mathsf{L};(F^\mathsf{T};\mathsf{L})^\mathsf{T};R^*$.*

Proof. Since $\mathsf{L};\mathsf{L} = \mathsf{L}$ if one of the universal relations on the left-hand side of the equation is homogeneous, $F^\mathsf{T};\mathsf{L}$ is a vector. To prove its first property we start with an auxiliary result using that S is a splitting, $F \subseteq S^\mathsf{T}$ and F is total:

$$F;R;(R;R)^*;F^\mathsf{T} \subseteq F;R^*;F^\mathsf{T} = F;S;S^\mathsf{T};F^\mathsf{T} \subseteq S^\mathsf{T};S;S^\mathsf{T};S = \mathsf{I};\mathsf{I} \subseteq F;F^\mathsf{T}$$

Now, the claim can be shown as follows:

$$
\begin{aligned}
R;(R;R)^*;F^\mathsf{T};\mathsf{L} \cap F^\mathsf{T};\mathsf{L} &\subseteq F^\mathsf{T};(\mathsf{L} \cap F;R;(R;R)^*;F^\mathsf{T};\mathsf{L}) && \text{modular law}\\
&= F^\mathsf{T};F;R;(R;R)^*;F^\mathsf{T};\mathsf{L}\\
&= F^\mathsf{T};(F;R;(R;R)^*;F^\mathsf{T} \cap F;F^\mathsf{T});\mathsf{L} && \text{aux. result}\\
&= F^\mathsf{T};F;(R;(R;R)^* \cap \mathsf{I});F^\mathsf{T};\mathsf{L} && F \text{ univalent}\\
&= \mathsf{O} && \text{assumption}
\end{aligned}
$$

The following verification of the remaining property concludes the proof:

$$
\begin{aligned}
R &\subseteq S;\mathsf{I};S^\mathsf{T} && R \subseteq R^*,\ S \text{ splitting of } R^*\\
&\subseteq S;F;F^\mathsf{T};F;F^\mathsf{T};S^\mathsf{T} && F \text{ total}\\
&\subseteq S;S^\mathsf{T};F^\mathsf{T};F;S;S^\mathsf{T} && \text{as } F \subseteq S^\mathsf{T}\\
&= R^*;F^\mathsf{T};F;R^* && S \text{ splitting}\\
&\subseteq R^*;F^\mathsf{T};\mathsf{L};F;R^*\\
&= R^*;F^\mathsf{T};\mathsf{L};(F^\mathsf{T};\mathsf{L})^\mathsf{T};R^* && \mathsf{L} \text{ homogeneous}
\end{aligned}
$$

□

Due to Theorem 4.4, the definition $v := (R;R)^*;F^\mathsf{T};\mathsf{L}$ and $w := R;v$ leads to a bipartition (v, w) of the relation R. If R is symmetric, then R^* is also symmetric. Under the assumption of the relational Axiom of Choice, by Theorem 4.1 to 4.4 we, therefore, have completed the proof of the second direction of the generalisation of D. König's theorem from undirected graphs to Dedekind categories: *All relations without odd cycles are bipartite.* As already mentioned in Section 3, if complements may be formed, then from the bipartion (v, w) we get the specific bipartition $(v, \overline{v})$ depending on one vector only.

5 An Example from Fuzzy Relations

In the following, we present an example using L-relations. This example demonstrates Theorem 4.2 and 4.3 and provides a situation in which Lemma 4.2 and Theorem 4.4 cannot be applied. The set of L-relations for a complete distributive lattice $(L, \vee, \wedge, 0, 1)$ with component-wise defined meet and join also forms a complete distributive lattice. Together with regular converse and sup-meet composition, i.e., $(Q;R)(x,z) = \bigvee_{y \in L} Q(x,y) \wedge R(y,z)$, we obtain a Dedekind category as already mentioned in the introduction.

We consider the persons Alan, Betty, Chris, Dave, Eve, Frank, Gwen, i.e., the set $P = \{A, B, C, D, E, F, G\}$, and certain relationship among them. As far as this example is concerned, a relationship between two persons has two aspects. First, a person might like another persons house or apartment and, second, a person might like the car of the another person. Both aspects are rated by either "no" (n), "somewhat" (s), or "yes" (y). The two criteria and the three level scale for each lead to the lattice L_9 the order of which is depicted below:

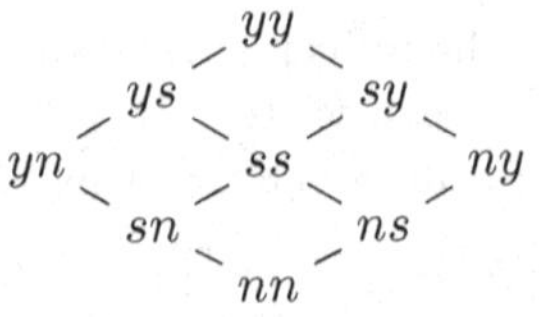

Now, the relationship among the persons from P is given as a L_9 fuzzy relation R on P, i.e., as a function from $P \times P$ to L_9. In the next two pictures we visualise this relation by the 7×7 matrix on the left, where we assume that a person corresponds to a row/column in order they are listed in P, e.g., the entry ys in the first row and sixth column indicates that Alan definitely likes Frank's apartment but likes his car only somewhat.

$$R = \begin{pmatrix} nn & ns & yn & ns & nn & ys & nn \\ ns & nn & ny & nn & yy & nn & ss \\ yn & ns & nn & ny & nn & ns & nn \\ ns & nn & ns & nn & ss & nn & yy \\ nn & ss & nn & yy & nn & ns & nn \\ ys & nn & ns & nn & ns & nn & ny \\ nn & yy & nn & ss & nn & ny & nn \end{pmatrix} \qquad R;(R;R)^* = \begin{pmatrix} nn & ns & ys & ns & nn & ys & nn \\ ns & nn & ny & nn & yy & nn & yy \\ yn & ny & nn & ny & nn & ny & nn \\ ns & nn & ny & nn & yy & nn & yy \\ nn & yy & nn & yy & nn & ny & nn \\ ys & nn & ny & nn & ny & nn & ny \\ nn & yy & nn & yy & nn & ny & nn \end{pmatrix}$$

Note, that R is not symmetric because Gwen likes both Dave's house and car only somewhat but Dave definitely likes both Gwen's apartment and car. On the other hand, the relation R^* is symmetric but different from the universal relation, i.e., the graph is not strongly connected. The relation also does not admit odd cycles which can be seen from the fact that the diagonal of the matrix $R;(R;R)^*$ above only contains the smallest element nn of the lattice L_9. Finally, we consider the three 7×1 vectors u, v and w that are presented below:

$$u = \begin{pmatrix} yn \\ yn \\ ny \\ nn \\ nn \\ nn \\ nn \end{pmatrix}, \quad v = \begin{pmatrix} ys \\ yn \\ ny \\ yn \\ ny \\ nn \\ ny \end{pmatrix}, \quad w = \begin{pmatrix} nn \\ ny \\ yn \\ ny \\ yn \\ yy \\ yn \end{pmatrix}$$

The vector u satisfies the assumptions of Theorem 4.2 and produces the bipartition given by (v, w). Note, that v is not the complement of w. However, if we change the first entry in u from yn to yy, then w remains unchanged and v becomes the complement of w. In either case, if we project to the first resp. second aspect of the relationships, then we obtain bipartitions for each of them. For instance, if we only consider the entries in (v, w) that have a non-zero first component (not equal to n), then we obtain the two sets $\{A, B, D\}$ and $\{C, E, F, G\}$, a bipartition with respect to liking each others home.

Since L_9 is not a Boolean lattice, the set of partial identities on a set, i.e., L_9 fuzzy relations smaller or equal than the identity, do not form a Boolean lattice either. It is shown in [22] that this implies that Axiom 4.1 is not valid, i.e., we cannot apply Lemma 4.2 and Theorem 4.4.

6 Relational Programs for Computing Bipartitions

Given a set-theoretic relation $R : X \leftrightarrow X$ on a finite set X with symmetric R^* and $R;(R;R)^* \cap \mathsf{I} = \mathsf{O}$, in this section we This brings it down to 18 pages. present relational programs that allow to compute from R a vector v such that the pair $(v, \overline{v})$ is a bipartition of R.

As $v := (R;R)^*;F^\mathsf{T};\mathsf{L}$ solves the task if F is a mapping contained in the transpose of a splitting S of R^*, we start with the following relational program *splitting* for computing a splitting of a partial equivalence relation P:

splitting(P)
  $\{P = P^\mathsf{T} \wedge P;P \subseteq P\}$
  $v := point(P;\mathsf{L})$;
  while $P;v \neq P;\mathsf{L}$ **do**
    $v := v \cup point(\overline{P;v} \cap P;\mathsf{L})$ **od**
  $\{v = v;\mathsf{L} \wedge P \cap v;v^\mathsf{T} \subseteq \mathsf{I} \wedge v;\mathsf{L} \subseteq P;\mathsf{L} \wedge P;v = P;\mathsf{L}\}$
  return $P;inj(v)^\mathsf{T}$

Here *point* selects a point from a non-empty vector. The call *inj*(v) of the **return**-clause of *splitting* computes the *embedding mapping* generated by $v : X \leftrightarrow \mathbf{1}$. If v describes the subset Y of X, then $inj(v) : Y \leftrightarrow X$ is nothing else than the identity function $id : Y \to X$, regarded as injective mapping in the sense of Section 2. In [2] the following relation-algebraic axiomatisation of embedding mappings is given, that specifies *inj*(v) for all vectors $v \neq \mathsf{O}$ up to isomorphism:

inj(v) is a mapping   *inj*(v) is injective $inj(v)^\mathsf{T};\mathsf{L} = v$

Furthermore, it is shown that for each vector $v \neq \mathsf{O}$ and each splitting S of the partial equivalence relation $\mathsf{I} \cap v;v^\mathsf{T}$ the transpose S^T fulfills these axioms. As a consequence of the results of [7,20] we, therefore, get that embedding mappings exist, even in case of abstract relations. In the programming language of RelView *point* and *inj* are available as pre-defined operations.

That the body of the relational program *splitting* is correct w.r.t. the annotated pre- and post-condition is shown in [2] by combining relation-algebraic calculations and the well-known invariant-based verification technique for while-programs. In [2] it is also verified that the post-condition of the body indeed implies that the relation $P;inj(v)^\mathsf{T}$ is a splitting of P.

Next, we modify the relational program *splitting* in such a way that it only yields the embedding mapping as result:

mapping(P)
  $\{P = P^\mathsf{T} \wedge P;P \subseteq P\}$
  $v := point(P;\mathsf{L})$;
  while $P;v \neq P;\mathsf{L}$ **do**
    $v := v \cup point(\overline{P;v} \cap P;\mathsf{L})$ **od**
  $\{v = v;\mathsf{L} \wedge P \cap v;v^\mathsf{T} \subseteq \mathsf{I} \wedge v;\mathsf{L} \subseteq P;\mathsf{L} \wedge P;v = P;\mathsf{L}\}$
  return *inj*(v)

Because of the axiomatisation of embedding mappings the result of the relational program *mapping* is a mapping. For an equivalence relation P this mapping is even contained in the transpose of the splitting $P;inj(v)^\mathsf{T}$ computed by the program *splitting*, as reflexivity and symmetry of P imply

$$inj(v) = inj(v); \mathsf{I} \subseteq inj(v); P^{\mathsf{T}} = (P; inj(v)^{\mathsf{T}})^{\mathsf{T}}.$$

So, *mapping*(R^*) is a candidate for F and this leads to the following relational program *bipartition* for computing the vector v of a bipartition $(v, \overline{v})$ of R:

$$\begin{array}{l} bipartition(R) \\ \quad \{ R = R^{\mathsf{T}} \wedge R; (R; R)^* \cap \mathsf{I} = \mathsf{O} \} \\ \quad v := (R; R)^*; mapping(R^*)^{\mathsf{T}}; \mathsf{L} \\ \quad \{ R \subseteq v; \overline{v}^{\mathsf{T}} \cup \overline{v}; v^{\mathsf{T}} \} \\ \quad \textbf{return } v \end{array}$$

Note, that now the program *splitting* is superfluous. We started our development with *splitting* for pedagogical reasons only.

7 Application of Theorem Provers and Proof Assistants

In particular, in the context of software and program verification the application of tools for theorem proving becomes more and more important. In this section, we show how automated theorem prover as well as proof assistants can be used for the verification of the already presented proofs of the Sections 2 to 4.

The formality of algebraic proofs and their primary use of rewriting is a vantage point for the use of tools for theorem proving as, for example, demonstrated in [6,9]. Based on this, in [3,4] the theorem prover Prover9 is used for the automated verification of the proof obligations occurring in the assertion-based verification of relational programs. Prover9 is an automated theorem prover for first-order equational logic. and, coupled with Mace4, a tool for searching models and counterexamples. Its handling is very straightforward because of its quite natural syntax. For more details we refer to [24].

However, Prover9 does not provide the opportunity of typing. So, we restricted us to homogeneous relations with one type only. For the encoding of the axiomatisation of Dedekind categories presented in Section 2 we had to consider that completeness is not a first-order property. To overcome this difficulty we, therefore, additionally weakened Axiom (a) of a Dedekind category by demanding only that relations form a distributive lattice with greatest element L and least element O, i.e., a bounded distributive lattice. This requires $\mathsf{O}; R = \mathsf{O}$, for all R, as additional axiom, since in Dedekind categories $\mathsf{O}; R = \mathsf{O}$ follows from completeness and distributivity. By these adaptations we precisely obtained the axioms of a *division allegory* with a singleton set of objects; see [7]. With this new axiomatisation and a few additionally added auxiliary facts Prover9 found proofs of the three statements of Lemma 2.1 in 142.62, 5.06, and 0.32 seconds, respectively. However, the tool was not able to verify Theorem 3.1 without any user interaction. Even if we transformed the equivalence in two implications a proof of only one of them was found by the tool. We encoded all presented lemmata and theorems and obtained similar negative results. These restrictions of automated theorem proving and the necessity of interactions became so serious that the change to a proof assistant was virtual essential.

The idea behind proof assistants is to check proofs mechanically, by the usage of so-called tactics. Two popular tools in this area are Isabelle/HOL and Coq. We have decided on Coq because of an already existing library (see [14]) which provides a large number of algebraic structures and, in particular, Dedekind categories. Furthermore, the library includes tactics to automate specific proofs about relation algebra and Kleene algebra (see [13]).

In general, the library is divided into several modules. For the derivation of our proofs we imported the modules `monoid`, `kleene`, `normalisation`, and `kat_tac`. By the import of the former two modules we provided the operations and axioms which are necessary for defining the desired algebraic structure, in our case a Dedeking category. The exact hierarchy and dependencies of the possible structures are managed in the module `level`. The latter two modules include the specific tactics `ra` for relation algebra and `ka` for Kleene algebra, respectively. For example, the tactic `ka` can be used to prove automatically properties like $R;(R;R)^* \cup (R;R)^* = R^*$ since it proves all universally true equations about Kleene algebra. Besides proving many auxiliary results, in this way we reproduced all presented proofs with Coq.

The Prover9 input files for the mentioned lemma and theorem as well as the proof scripts for all Coq proofs can be found in the web (see [26]).

8 Concluding Remarks

Abstracting set-theoretic relations to morphisms of a Dedekind category, we have shown that D. König's well-known characterisation of bipartite graphs via the absence of odd cycles also holds in this general algebraic setting. For one direction we had to assume the relational Axiom of Choice to hold; for set-theoretic relations this direction immediately led to relational programs for computing bipartitions. Without using the relational Axiom of Choice we have proved this direction for strongly-connected relations and without it we also have proved the characterisation of bipartite relations of Theorem 4.3. We also have reported on our experience with automated theorem provers and proof assistants.

In [22] it is shown that in the context of Dedekind categories the relational Axiom of Choice implies the existence of complements such that all morphism sets are Boolean lattices. Strictly speaking, thus, we have shown one direction of the generalisation of D. König's theorem for heterogeneous relation algebras only – but without using complements. The example of Section 5 demonstrates that bipartitions also may exist in the non-Boolean case. For the future we plan to weaken Axiom 4.1 in such a way that Lemma 4.2 and Theorem 4.4 remain valid, but from the weakening the existence of complements does not follow.

Acknowledgement. We thank the unknown referees for carefully reading the paper and their valuable remarks. We also thank D. Pous for his support concerning the use of Coq.

References

1. Berghammer, R., Neumann, F.: RELVIEW – An OBDD-based Computer Algebra system for relations. In: Gansha, V.G., Mayr, E.W., Vorozhtsov, E. (eds.) Computer Algebra in Scientific Computing. LNCS, vol. 3718, pp. 40–51. Springer, Heidelberg (2005)
2. Berghammer, R., Winter, M.: Embedding mappings and splittings with applications. Acta Informatica 47, 77–110 (2010)
3. Berghammer, R., Struth, G.: On automated program construction and verification. In: Bolduc, C., Desharnais, J., Ktari, B. (eds.) MPC 2010. LNCS, vol. 6120, pp. 22–41. Springer, Heidelberg (2010)
4. Berghammer, R., Höfner, P., Stucke, I.: Automated verification of relational while-programs. In: Widłak, W. (ed.) Molecular Biology - Not Only for Bioinformatics. LNCS, vol. 8248, pp. 309–326. Springer, Heidelberg (2013)
5. Diestel, R.: Graph theory, 3rd edn. Springer (2005)
6. Foster, S., Struth, G., Weber, T.: Automated engineering of relational and algebraic methods in Isabelle/HOL. In: de Swart, H. (ed.) Relational and Algebraic Methods in Computer Science. LNCS, vol. 6663, pp. 52–67. Springer, Heidelberg (2011)
7. Freyd, P., Scedrov, A.: Categories, allegories. North-Holland (1990)
8. Furusawa, H., Kawahara, Y., Winter, M.: Dedekind categories with cutoff operators. Fuzzy Sets and Systems 173, 1–24 (2011)
9. Höfner, P., Struth, G.: On automating the calculus of relations. In: Armando, A., Baumgartner, P., Dowek, G. (eds.) IJCAR 2008. LNCS (LNAI), vol. 5195, pp. 50–66. Springer, Heidelberg (2008)
10. König, D.: Über Graphen und ihre Anwendung in der Determinantentheorie und Mengenlehre. Mathematische Annalen 77, 453–465 (1916)
11. Ng, K.C., Tarski, A.: Relation algebras with transitive closure. Abstract 742-02-09, Notices Amer. Math. Soc. 24, A29-A30 (1977)
12. Oliver, J.P., Serrato, D.: Categories de Dedekind: Morphismes dans les categories de Schröder. C.R. Acad. Sci.Paris 290, 939–941 (1980)
13. Pous, D.: Kleene algebra with tests and Coq tools for while programs. In: Blazy, S., Paulin-Mohring, C., Pichardie, D. (eds.) ITP 2013. LNCS, vol. 7998, pp. 180–196. Springer, Heidelberg (2013)
14. Pous, D.: Relation algebra and KAT in Coq. http://perso.ens-lyon.fr/damien.pous/ra/
15. Rubin, H., Rubin, J.E.: Equivalents of the Axiom of Choice. North-Holland (1970)
16. Schmidt, G., Ströhlein, T.: Relations and graphs, Discrete mathematics for computer scientists, EATCS Monographs on Theoretical Computer Science, Springer (1993)
17. Schmidt, G.: Relational mathematics. Encyclopedia of Mathematics and its Applications, vol. 132. Cambridge University Press (2010)
18. Tarski, A.: On the calculus of relations. Journal of Symbolic Logic 6, 73–89 (1941)
19. Tarski, A., Givant, S.: A formalization of set theory without variables. Colloquium Publications 41. American Mathematical Society (1987)
20. Winter, M.: Strukturtheorie heterogener Relationenalgebren mit Anwendung auf Nichtdetermismus in Programmiersprachen. Dissertation, Fakultät für Informatik, Universität der Bundeswehr München, Dissertationsverlag NG Kopierladen GmbH (1998)
21. Winter, M.: An ordered category of processes. In: Berghammer, R., Möller, B., Struth, G. (eds.) RelMiCS/AKA 2008. LNCS, vol. 4988, pp. 367–381. Springer, Heidelberg (2008)

22. Winter, M.: Complements in distributive allegories. In: Berghammer, R., Jaoua, A.M., Möller, B. (eds.) RelMiCS 2009. LNCS, vol. 5827, pp. 337–350. Springer, Heidelberg (2009)
23. Coq-homepage: `http://coq.infia.fr`
24. Prover9-homepage: `http://www.prover9.org`
25. RelView-homepage: `http://www.informatik.uni-kiel.de/~progsys/relview/`
26. Input files, proof scripts: `http://media.informatik.uni-kiel.de/Ramics2015/`

Appendix

Let R be a relation, R^* be symmetric and assume that the pair (v, w) is a bipartition of R. As mentioned in the paper after Theorem 4.3, we are able to prove $R; (R; R)^*; v \cap v = \mathsf{O}$ as well as $R \subseteq R^*; v; v^\mathsf{T}; R^*$, such that v can be taken as vector u. A proof of the first property $R; (R; R)^*; v \cap v = \mathsf{O}$ looks as follows:

$$\begin{aligned}
R; (R; R)^*; v \cap v &\subseteq (v; w^\mathsf{T} \cup w; v^\mathsf{T}); v \cap v && \text{see proof Theorem 3.2}\\
&= (v; w^\mathsf{T}; v \cup w; v^\mathsf{T}; v) \cap v\\
&= w; v^\mathsf{T}; v \cap v && \text{Lemma 2.1(1)}\\
&\subseteq w; \mathsf{L} \cap v\\
&= w \cap v && w \text{ vector}\\
&= \mathsf{O} && (v, w) \text{ bipartition}
\end{aligned}$$

In the following proof of the second property $R \subseteq R^*; v; v^\mathsf{T}; R^*$ we use R^s as abbreviation of the symmetric closure $R \cup R^\mathsf{T}$ of R. First of all, we compute

$$R = R; \mathsf{I} \cap R \subseteq R; (\mathsf{I} \cap R^\mathsf{T}; R) \subseteq R; R^\mathsf{T}; R$$

as auxiliary result, using the modular law. From $R \subseteq v; w^\mathsf{T} \cup w; v^\mathsf{T}$ and the proof of Theorem 4.1 we get $R^\mathsf{T} \subseteq v; w^\mathsf{T} \cup w; v^\mathsf{T}$, such that Theorem 3.1 "$\Rightarrow$" yields

$$R^\mathsf{T} \subseteq (v \cup w); (v \cup w)^\mathsf{T}$$

as second auxiliary result. Finally, similar to $R; w \subseteq v$ in Theorem 3.1 "$\Rightarrow$" we can calculate

$$R^\mathsf{T}; w \subseteq (v; w^\mathsf{T} \cup w; v^\mathsf{T})^\mathsf{T}; w = w; v^\mathsf{T}; w \cup v; w^\mathsf{T}; w = v; w^\mathsf{T}; w \subseteq v; \mathsf{L} = v,$$

so that we conclude

$$R^s; w \subseteq v$$

as third auxiliary result. Now, we obtain the desided result as follows:

$$\begin{aligned}
R &\subseteq R; R^\mathsf{T}; R && \text{first auxiliary result}\\
&\subseteq R^s; R^\mathsf{T}; R^s\\
&= R^s; (v \cup w); (v \cup w)^\mathsf{T}; R^s && \text{second auxiliary result}\\
&= (R^s; v \cup R^s; w); (R^s; v \cup R^s; w)^\mathsf{T} && R^s \text{ symmetric}\\
&\subseteq (R^s; v \cup v); (R^s; v \cup v)^\mathsf{T} && \text{third auxiliary result}\\
&= (\mathsf{I} \cup R^s); v; v^\mathsf{T}; (\mathsf{I} \cup R^s) && R^s \text{ symmetric}\\
&\subseteq R^*; v; v^\mathsf{T}; R^* && R^s \subseteq R^* \text{ as } R^* \text{ is symmetric}
\end{aligned}$$

Tool-Based Verification of a Relational Vertex Coloring Program

Rudolf Berghammer[1], Peter Höfner[2,3], and Insa Stucke[1]

[1] Institut für Informatik, Christian-Albrechts-Universität zu Kiel, Germany
[2] NICTA, Australia
[3] Computer Science and Engineering, University of New South Wales, Australia

Abstract. We present different approaches of using a special purpose computer algebra system and theorem provers in software verification. To this end, we first develop a purely algebraic while-program for computing a vertex coloring of an undirected (loop-free) graph. For showing its correctness, we then combine the well-known assertion-based verification method with relation-algebraic calculations. Based on this, we show how automatically to test loop-invariants by means of the RelView tool and also compare the usage of three different theorem provers in respect to the verification of the proof obligations: the automated theorem prover Prover9 and the two proof assistants Coq and Isabelle/HOL. As a result, we illustrate that algebraic abstraction yields verification tasks that can easily be verified with off-the-shelf theorem provers, but also reveal some shortcomings and difficulties with theorem provers that are nowadays available.

1 Introduction

Provably correct programs can be obtained in different ways. Formal program verification is one of them. It means to prove with mathematical rigor that a given program meets a given formal specification of the problem. In case of imperative programs the use of pre- and post-conditions as specifications and intermediate assertions for the verification is a widely accepted and frequently used technique. Besides proof rules for the control structures of the programming language used it requires formal specifications for the data types on which the programs are applied. Experience has shown that algebraic/axiomatic specifications or modeling by algebraic structures are most suitable for that. In the present paper we consider a graph-theoretic problem and use relation algebra for modeling undirected graphs, single vertices, sets of vertices as well as functions which assign values to vertices.

In the present paper we consider a graph-theoretic problem and use relation algebra for modeling undirected graphs. The axiomatization of relation-algebraic calculus started with [26]. The calculus is widely used and many examples in the context of program verification can be found in the literature, e.g., [2,3,4,6,7]. For the use of relation algebra in graph theory we refer again to [24,25].

W. Kahl et al. (Eds.): RAMiCS 2015, LNCS 9348, pp. 275–292, 2015.
DOI: 10.1007/978-3-319-24704-5_17

Relation-algebraic proofs are precise and hence allow formal first-order reasoning, often even equational reasoning. This is a vantage point for the use of theorem provers as, for instance, demonstrated in [15,17,18]. Based on these positive experiences, in [8,9] the automated theorem prover Prover9 [20] is used for the automated verification of proof obligations appearing in the assertion-based verification of relational programs. This paper is a continuation of as well as a step further in this work. We consider a well-known graph theoretical problem, viz. vertex coloring. However, we do not restrict ourselves to the verification of the proof obligations via an automated theorem prover. We aim to gain more experience with tool support in formal verification of relational programs. Therefore, we also investigate the use of two different proof assistants tools, viz. Coq [11] and Isabelle/HOL [21], and of a specific purpose computer algebra system for relation algebra, viz. RelView [5,30]. The paper illustrates that algebraic abstraction yields verification tasks that can be verified with off-the-shelf theorem provers, but also reveals some shortcomings and difficulties with tools that are nowadays available.

One aim of the paper is to provide a guideline on how to get started with different tools with different approaches and possibilities when computations and mechanical proofs in relation algebra are desired or required. For that reason we restrict ourselves to a single and not too difficult problem. By this the general approach is easily visible and is not hidden by complex technical details. All input files and proof scripts can be found in the web [32].

2 Relation-Algebraic Preliminaries

To model undirected graphs, single vertices, sets of vertices and colorings, we will use binary relations and manipulate and calculate with such objects in a purely algebraic manner. Therefore, we recall the fundamentals of relation algebra based on the homogeneous approach of [26], its developments in [13,19,27] and the generalization to heterogeneous relation algebra in [24,25].

Set-theoretic relations form the standard model of relation algebras. We assume the reader to be familiar with the basic operations on them, viz. R^{T} (transposition), $\overline{R}$ (complementation), $R \cup S$ (union), $R \cap S$ (intersection), RS (composition), the predicates $R \subseteq S$ (inclusion) and $R = S$ (equality), and the special relations O (empty relation), L (universal relation), and I (identity relation). The three Boolean operations $\overline{}$, $\cup$ and $\cap$, the order $\subseteq$ and the two constants O and L form Boolean lattices. Well-known properties of set-theoretic relations are $\overline{R}^{\mathsf{T}} = \overline{R^{\mathsf{T}}}$, $(R \cup S)^{\mathsf{T}} = R^{\mathsf{T}} \cup S^{\mathsf{T}}$, $(R \cap S)^{\mathsf{T}} = R^{\mathsf{T}} \cap S^{\mathsf{T}}$, $(R^{\mathsf{T}})^{\mathsf{T}} = R$, $(RS)^{\mathsf{T}} = S^{\mathsf{T}} R^{\mathsf{T}}$, and the monotonicity of the transposition operation. Furthermore, union, intersection and composition are monotonic in both arguments.

The theoretical framework for these rules (and many others) to hold is that of a (heterogeneous) *relation algebra* with typed relations as elements. Typing means that each relation has a source and a target and we write $R : X \leftrightarrow Y$ to express that X is the source and Y is the target of R. We call $X \leftrightarrow Y$ the type of R. As constants and operations of a relation algebra we have those of set-theoretic relations, where we (as usual) overload the symbols O, L and I, i.e., avoid the binding of types to them. The axioms of a relation algebra are

(1) the axioms of a Boolean lattice for all relations of the same type under the Boolean operations, the order, empty relation and universal relation,
(2) the associativity of composition and that identity relations are neutral elements w.r.t. composition,
(3) that $QR \subseteq S$, $Q^{\mathsf{T}}\overline{S} \subseteq \overline{R}$ and $\overline{S}\,R^{\mathsf{T}} \subseteq \overline{Q}$ are equivalent, for all relations Q, R, S (with appropriate types),
(4) that $R \neq \mathsf{O}$ is equivalent to $\mathsf{L}R\mathsf{L} = \mathsf{L}$, for all relations R and all universal relations (with appropriate types).

We do not require the Boolean lattice to be complete, as in [26]. In [24] the equivalences of (3) are called the *Schröder equivalences* and direction '⇒' of (4) is called the *Tarski rule*. Our variant of the Tarski rule is motivated by the fact that it avoids the degenerated case of a Boolean lattice with one element only. In the relation-algebraic proofs of this paper we will mention only applications of the Schröder equivalences, the Tarski rule and 'non-obvious' consequences of the axioms. Furthermore, we will assume that complementation and transposition bind stronger than composition, composition binds stronger than union and intersection, and that all expressions and formulas are well-typed. Since types are helpful for the understanding, they frequently are presented in the text surrounding the corresponding formulae.

In this paper we make use of the following classes of relations. A relation R is *univalent* if $R^{\mathsf{T}}R \subseteq \mathsf{I}$ and *total* if $R\mathsf{L} = \mathsf{L}$. As usual, a univalent and total relation is a *function*. A relation R is *injective* if R^{T} is univalent and *surjective* if R^{T} is total. Finally, a relation R is *irreflexive* if $R \subseteq \overline{\mathsf{I}}$ and *symmetric* if $R = R^{\mathsf{T}}$. In case of set-theoretic relations the equivalence of these relation-algebraic specifications and the common logical specifications can easily be derived.

Relation algebra provides different ways to model subsets and single elements of sets. In the present paper we use vectors, a special class of relations introduced in [24], and usually denoted by lower-case letters. A relation v is a *vector* if $v = v\mathsf{L}$. For a set-theoretic relation $v : X \leftrightarrow Y$ the condition $v = v\mathsf{L}$ means that v is (as set of pairs) of the specific form $V \times Y$, with a subset V of X, i.e., for all $x \in X$ and $y \in Y$ we have $(x, y) \in v$ if and only if $x \in V$. We may consider v as relational model of the subset V of its source X. For modeling an element $x \in X$ we identify the singleton set $\{x\}$ with the only element x it contains. This leads to a specific class of vectors. A *point* p is an injective and surjective vector. In the set-theoretic case and if the point $p : X \leftrightarrow Y$ is of the specific form $p = P \times Y$ with $P \subseteq X$, then injectivity of p means that P contains at most one element and surjectivity of p means that P contains at least one element. Next, we prove properties of points which are consequences of our variant of the Tarski rule.

Lemma 2.1. *If p is a point, then we have $p \neq \mathsf{O}$, and if p and q are points, then we have $pq^{\mathsf{T}} \neq \mathsf{O}$.*

Proof. Using the Tarski rule and that the point p is a surjective vector, we get

$$p \neq \mathsf{O} \iff \mathsf{L}p\mathsf{L} = \mathsf{L} \iff \mathsf{L}p = \mathsf{L} \iff \mathsf{L} = \mathsf{L},$$

that is, the first claim, and using the Tarski rule twice, surjectivity of p and non-emptiness of points (i.e., the first claim), the second claim follows from

$$pq^\mathsf{T} \neq \mathsf{O} \iff \mathsf{L}pq^\mathsf{T}\mathsf{L} = \mathsf{L} \iff \mathsf{L}q^\mathsf{T}\mathsf{L} = \mathsf{L} \iff q^\mathsf{T} \neq \mathsf{O} \iff q \neq \mathsf{O}. \qquad \Box$$

In the context of algorithms the choice of an element from a non-empty set is frequently used. In the same way the choice of a point from a non-empty vector is fundamental for relational programming. Therefore, we assume a corresponding operation *point* to be at hand – as in the programming language of RelView; see [30] – such that $point(v)$ is a point and $point(v) \subseteq v$, for all non-empty vectors v. Note that *point* is a (deterministic) operation in the usual mathematical sense, such that each call $point(v)$ yields the same point in v. However, the above requirements allow different realizations. The specific implementation of *point* in RelView uses the fact that RelView deals only with relations on finite sets, which are linearly ordered by an internal enumeration. A call $point(v)$ then chooses that point which describes the least element of the set described by v.

3 A Relational Program for Vertex Coloring

Graph coloring in general and vertex coloring in particular is one of the most important and most studied concepts in graph theory. It leads to many interesting applications in mathematics and computer science, e.g., in the construction of timetables. In this section we develop a relational program to compute a vertex coloring of a given undirected graph, i.e., a labeling of the vertices with colors such that two adjacent vertices are labeled with different colors.

Assume G to be an undirected (loop-free) graph with vertex set X. We model G by the *adjacency relation* $E : X \leftrightarrow X$ such that for all $x, y \in X$ it holds $(x, y) \in E$ if and only if x and y are adjacent. Since G is assumed to be undirected (and loop-free), E is symmetric and irreflexive. E is the input of the relational program we want to develop and to prove as correct. Since we tent to a while-program and the use of the inductive assertion method, this leads to

$$\mathrm{Pre}(E) \;:\iff\; E = E^\mathsf{T} \wedge E \subseteq \overline{\mathsf{I}} \qquad \text{(Pre)}$$

as pre-condition. The output of our relational program should be a vertex coloring of G. Usually natural numbers are taken as colors and, thus, a vertex coloring of G would be a function $C : X \to \mathbb{N}$ such that $C(x) = C(y)$ implies $(x, y) \notin E$, for all $x, y \in X$. Functions are specific relations and so vertex colorings are relations as well. We want to stay as abstract as possible and do not want to use natural numbers as colors, but elements of an abstract set F of colors. As a consequence, a vertex coloring of G is a relation $C : X \leftrightarrow F$ that is univalent, total, and for all $x, y \in X$ if there exists $f \in F$ such that $(x, f) \in C$ and $(y, f) \in C$ this implies $(x,y) \in \overline{E}$. It is easy to show that the third requirement is equivalent to $CC^{\mathsf{T}} \subseteq \overline{E}$. This yields

$$\mathrm{Post}(C,E) \;:\Longleftrightarrow\; C^{\mathsf{T}}C \subseteq \mathsf{I} \wedge C\mathsf{L} = \mathsf{L} \wedge CC^{\mathsf{T}} \subseteq \overline{E} \tag{Post}$$

as post-condition. We call the formula $CC^{\mathsf{T}} \subseteq \overline{E}$ of $\mathrm{Post}(C,E)$ the *coloring property* of C w.r.t. E.

To develop a relational while-program with input E and output C which is correct w.r.t. the pre-condition $\mathrm{Pre}(E)$ and the post-condition $\mathrm{Post}(C,E)$, it seems to be reasonable to follow a greedy approach. Using a loop, the program assigns to each vertex an available color that is not already used for one of its neighbors. Such an approach means that we work with *partial colorings*. Formally, that means we use

$$\mathrm{Inv}(C,E) \;:\Longleftrightarrow\; C^{\mathsf{T}}C \subseteq \mathsf{I} \wedge CC^{\mathsf{T}} \subseteq \overline{E} \tag{Inv}$$

as loop-invariant, and want to extend C in each run through the loop by coloring an uncolored vertex with an allowed color in the above described manner until C is total. Summing up, we have

$$\{\,\mathrm{Pre}(E)\,\}\ \ldots;\ \{\,\mathrm{Inv}(C,E)\,\}\ \textbf{while}\ C\mathsf{L} \neq \mathsf{L}\ \textbf{do} \ldots \textbf{od}\ \{\,\mathrm{Post}(C,E)\,\}$$

as program outline. Because of the definition of the loop-invariant and the post-condition we immediately obtain the implication

$$\mathrm{Inv}(C,E) \wedge C\mathsf{L} = \mathsf{L} \;\Longrightarrow\; \mathrm{Post}(C,E) \tag{PO1}$$

to be valid. Hence, by (PO1) we have the first proof obligation of program verification, viz. that the loop-invariant in conjunction with the exit-condition of the loop implies the post-condition. It remains to develop an initialization that establishes the loop-invariant and a loop-body that maintains the loop-invariant as long as $C\mathsf{L} \neq \mathsf{L}$ holds. Obviously, we have:

Lemma 3.1. *The empty relation* $\mathsf{O} : X \leftrightarrow F$ *is univalent and fulfills the coloring property w.r.t.* E.

As an immediate consequence of this lemma we get that the implication

$$\mathrm{Pre}(E) \;\Longrightarrow\; \mathrm{Inv}(\mathsf{O},E) \tag{PO2}$$

is valid. If we, guided by this fact, change the above program outline by concretizing the initialization to $C := \mathsf{O}$, then (PO2) is the second proof obligation of program verification and says for the new program outline that the loop-invariant is established by the initialization if the pre-condition holds.

To develop a loop-body, we use the fact that the vector $C\mathsf{L}$ models the domain of the univalent relation $C : X \leftrightarrow F$, i.e., the set of vertices of G which are already colored. If $C\mathsf{L} \neq \mathsf{L}$, then the call $point(\overline{C\mathsf{L}})$ selects a point, say p, with $p \subseteq \overline{C\mathsf{L}}$ that models an uncolored vertex, say $x \in X$. Guided by the above mentioned greedy approach, we now consider the vector Ep. A little componentwise reflection shows that it models the set of neighbors of x and that the derived vector $C^{\mathsf{T}}Ep$ models the image of the set of neighbors of x under the univalent relation C, that is, the set of colors already assigned to a neighbor of x. As a consequence, the complemented vector $\overline{C^{\mathsf{T}}Ep}$ models the set of colors that are allowed to be assigned to x without contradicting the coloring property w.r.t. E. If we define a point q as $q := \mathit{point}(\overline{C^{\mathsf{T}}Ep})$, then q models one of these colors, say $f \in F$, and the union $C \cup pq^{\mathsf{T}}$ extends the relation C by additionally assigning f to x. This yields the following complete program outline:

$$\begin{array}{l} C := \mathsf{O};\\ \textbf{while } C\mathsf{L} \neq \mathsf{L} \textbf{ do}\\ \quad \textbf{let } p = \mathit{point}(\overline{C\mathsf{L}});\\ \quad \textbf{let } q = \mathit{point}(\overline{C^{\mathsf{T}}Ep});\\ \quad C := C \cup pq^{\mathsf{T}} \textbf{ od} \end{array} \tag{VC}$$

To improve readability of (VC), we use two **let**-clauses for assigning the above mentioned points p and q.

We have already verified two out of the three proof obligations needed to prove partial correctness of the relational program (VC) w.r.t. the above pre- and post-condition specification. It remains to verify the third proof obligation

$$\mathrm{Inv}(C, E) \wedge C\mathsf{L} \neq \mathsf{L} \implies \mathrm{Inv}(C \cup pq^{\mathsf{T}}, E) \tag{PO3'}$$

for partial correctness, where p and q are defined as in the relational program (VC). In case programs do not change the input and the precondition Pre remains unchanged, the pre-condition can be added to the loop-invariant. For the relational program (VC) this is the case and hence it suffices to show

$$\mathrm{Pre}(E) \wedge \mathrm{Inv}(C, E) \wedge C\mathsf{L} \neq \mathsf{L} \implies \mathrm{Inv}(C \cup pq^{\mathsf{T}}, E) \tag{PO3}$$

We prove (PO3) in two steps. First, we show that enlarging a univalent relation by the product of two points as done in line 5 of the relational program (VC) yields again a univalent relation.

Lemma 3.2. *Let C, p and q be relations such that C is univalent, p and q are points, $C\mathsf{L} \neq \mathsf{L}$, and $p \subseteq \overline{C\mathsf{L}}$. Then $C \cup pq^{\mathsf{T}}$ is univalent.*

Proof. Because of the equation

$$(C \cup pq^{\mathsf{T}})^{\mathsf{T}}(C \cup pq^{\mathsf{T}}) = C^{\mathsf{T}}C \cup qp^{\mathsf{T}}C \cup C^{\mathsf{T}}pq^{\mathsf{T}} \cup qp^{\mathsf{T}}pq^{\mathsf{T}}$$

it suffices to show the following four inclusions:

$$(1)\ C^{\mathsf{T}}C \subseteq \mathsf{I} \quad (2)\ qp^{\mathsf{T}}C \subseteq \mathsf{I} \quad (3)\ C^{\mathsf{T}}pq^{\mathsf{T}} \subseteq \mathsf{I} \quad (4)\ qp^{\mathsf{T}}pq^{\mathsf{T}} \subseteq \mathsf{I}$$

Inclusion (1) holds as C is univalent. Since $qp^{\mathsf{T}}C = (C^{\mathsf{T}}pq^{\mathsf{T}})^{\mathsf{T}}$ and $\mathsf{I} = \mathsf{I}^{\mathsf{T}}$, inclusion (2) is equivalent to inclusion (3) and, thus, it suffices to show that one of them holds. To prove inclusion (3), we calculate

$$p \subseteq \overline{C\mathsf{L}} \iff C\mathsf{L} \subseteq \overline{p} \iff C^{\mathsf{T}}p \subseteq \mathsf{O},$$

where we apply one of the Schröder equivalences in the second step. So, we have $C^{\mathsf{T}}p = \mathsf{O}$ and this implies $C^{\mathsf{T}}pq^{\mathsf{T}} = \mathsf{O} \subseteq \mathsf{I}$. Using the vector property and the injectivity of the point q, inclusion (4) is shown by

$$qp^{\mathsf{T}}pq^{\mathsf{T}} \subseteq q\mathsf{L}q^{\mathsf{T}} = qq^{\mathsf{T}} \subseteq \mathsf{I}\,. \qquad \square$$

The following lemma states the second fact we have to prove for verifying proof obligation (PO3). We show that the enlargement maintains the coloring property.

Lemma 3.3. *Let E, C, p and q be relations such that E is symmetric and irreflexive, p and q are points, C fulfills the coloring property w.r.t. E, $C^{\mathsf{T}}Ep \neq \mathsf{L}$, and $q \subseteq \overline{C^{\mathsf{T}}Ep}$. Then $C \cup pq^{\mathsf{T}}$ fulfills the coloring property w.r.t. E.*

Proof. We follow exactly the proof of Lemma 3.2 and start with

$$(C \cup pq^{\mathsf{T}})(C \cup pq^{\mathsf{T}})^{\mathsf{T}} = CC^{\mathsf{T}} \cup pq^{\mathsf{T}}C^{\mathsf{T}} \cup Cqp^{\mathsf{T}} \cup pq^{\mathsf{T}}qp^{\mathsf{T}},$$

such that it suffices to show the following four inclusions:

$$(1)\ CC^{\mathsf{T}} \subseteq \overline{E} \qquad (2)\ pq^{\mathsf{T}}C^{\mathsf{T}} \subseteq \overline{E} \qquad (3)\ Cqp^{\mathsf{T}} \subseteq \overline{E} \qquad (4)\ pq^{\mathsf{T}}qp^{\mathsf{T}} \subseteq \overline{E}$$

Inclusion (1) holds since it is assumed that C fulfills the coloring property. Because of $pq^{\mathsf{T}}C^{\mathsf{T}} = (Cqp^{\mathsf{T}})^{\mathsf{T}}$ and $\overline{E} = \overline{E}^{\mathsf{T}}$ the inclusions (2) and (3) are again equivalent. To prove inclusion (3), we calculate

$$\begin{aligned} q \subseteq \overline{C^{\mathsf{T}}Ep} &\iff C^{\mathsf{T}}Ep \subseteq \overline{q} \iff Cq \subseteq \overline{Ep} \\ &\iff Ep \subseteq \overline{Cq} \quad\iff Cqp^{\mathsf{T}} \subseteq \overline{E}\,, \end{aligned}$$

where we apply the Schröder equivalences in the second and the fourth step. Using the vector property and the injectivity of the point q and the irreflexivity of E, inclusion (4) is shown by

$$pq^{\mathsf{T}}qp^{\mathsf{T}} \subseteq p\mathsf{L}p^{\mathsf{T}} = pp^{\mathsf{T}} \subseteq \mathsf{I} \subseteq \overline{E}\,. \qquad \square$$

Combining the Lemmata 3.2 and 3.3, we immediately obtain (PO3) and, thus, altogether the partial correctness of the relational program (VC) w.r.t. the pre-condition $\mathrm{Pre}(E)$ and the post-condition $\mathrm{Post}(C, E)$. Note that only for the maintenance of the coloring property the pre-condition is required.

We are not only interested in partial correctness, but also in total correctness. Therefore, it remains to prove the proof obligation

$$\mathrm{Pre}(E) \implies \text{the relational program (VC) yields a defined value}\,. \qquad \text{(PO4)}$$

To verify (PO4), we have to verify two facts: first, we have to prove that the loop of the relational program (VC) terminates, and, secondly, that the partial operation *point* is only applied to non-empty vectors (i.e., yields a defined value). The following lemma shows that the relation C is strictly enlarged in each execution of the loop-body.

Lemma 3.4. *Let C, p and q be relations such that p and q are points, $C\mathsf{L} \neq \mathsf{L}$, and $p \subseteq \overline{C\mathsf{L}}$. Then $C \subseteq C \cup pq^\mathsf{T}$ and $C \neq C \cup pq^\mathsf{T}$.*

Proof. Inclusion $C \subseteq C \cup pq^\mathsf{T}$ is trivial. Next, we show $pq^\mathsf{T} \subseteq \overline{C}$ by

$$\begin{aligned} Cq \subseteq C\mathsf{L} &\iff C^\mathsf{T}\overline{C\mathsf{L}} \subseteq \overline{q} && \text{Schröder equivalences} \\ &\implies C^\mathsf{T}p \subseteq \overline{q} && \text{as } p \subseteq \overline{C\mathsf{L}} \\ &\iff p^\mathsf{T}C \subseteq \overline{q}^\mathsf{T} \\ &\iff pq^\mathsf{T} \subseteq \overline{C} && \text{Schröder equivalences}. \end{aligned}$$

Using $pq^\mathsf{T} \subseteq \overline{C}$, the second claim $C \neq C \cup pq^\mathsf{T}$ now can be shown by contradiction: $C = C \cup pq^\mathsf{T}$ would imply $pq^\mathsf{T} \subseteq C$, such that $pq^\mathsf{T} \subseteq C \cap \overline{C} = \mathsf{O}$ follows. But the latter fact contradicts Lemma 2.1. □

From this lemma we obtain that the loop of the relational program (VC) terminates if $E : X \leftrightarrow X$ is a relation on a finite set X, i.e., if the graph G is finite. However, to verify (PO4) we also have to ensure that the partial operation *point* is only applied to non-empty vectors. In case of the call *point*($\overline{C\mathsf{L}}$) non-emptiness of $\overline{C\mathsf{L}}$ follows from the loop-condition. However, since we do not assume specific properties for the set F of colors, in case of the call *point*($\overline{C^\mathsf{T}Ep}$) it may happen that $\overline{C^\mathsf{T}Ep}$ is empty, viz. if there are too few colors and each color is already assigned to a neighbor of the vertex modeled by the point p. This situation can not appear if there are enough colors. Obviously $|X|$ colors suffice. So, we have the following result:

Theorem 3.1 *If E is a relation on a finite set X and F consists of at least $|X|$ colors, then the relational program (VC) is totally correct w.r.t. the pre-condition* Pre(E) *and the post-condition* Post(C, E).

The assumptions of this theorem and its proof confirm again the experience we have made so far with the assertion-based verification of relational programs: algebra is an ideal base to verify the proof obligations for partial correctness, but for showing total correctness non-algebraic arguments are necessary, typically. Usually, they concern the sizes of the carrier sets of the relations in question.

In the following sections we demonstrate how the presented proofs can be automated, or at least supported by the tools mentioned in the introduction.

4 Invariant Testing Using RelView

Relation algebra has a fixed and small set of constants and operations which (in the case of finite carrier sets) can be implemented very efficiently. At the University of Kiel we have developed RelView, a special purpose computer algebra system for relation algebra. It uses BDDs for implementing relations and makes full use of a graphical user interface. Details can be found in [5,30].

Translating the relational program (VC) into the programming language of RelView yields the following code:

```
color(E)
  DECL C, p, q
  BEG  C = O(E);
       WHILE -eq(C*L(C),L(C)) DO
         ASSERT(Inv, incl(C^*C,I(C)) & incl(C*C^,-E));
         p = point(-(C*L(C)));
         q = point(-(C^*E*p));
         C = C | p*q^ OD
       RETURN C
  END.
```

In this RelView-program the symbols `-`, `^`, `|`, `&` and `*` denote the operations for complementation, transposition, union, intersection and composition, respectively. Furthermore, `eq` and `incl` are base-operations for testing the equality and inclusion of relations, respectively. All tests yield relations on a specific singleton set $\mathbf{1}$ as result, where $\mathsf{L} : \mathbf{1} \leftrightarrow \mathbf{1}$ models 'true' and $\mathsf{O} : \mathbf{1} \leftrightarrow \mathbf{1}$ models 'false'. A call of the base-operation `O` generates an empty relation, with the same type as the argument. The operations `L` and `I` perform the same for the universal relation and the identity relation, respectively. Due to the initialization of the variable `C` in `color` by the empty relation of the same type as the input `E`, hence, the vertex set X of the graph G is taken as set F of colors, implicitly. As a consequence, there are enough colors and the RelView-program `color` is totally correct w.r.t. the pre-condition $\mathrm{Pre}(E)$ and the post-condition $\mathrm{Post}(C, E)$.

Within the RelView-program `color` we also use the `ASSERT`-statement for testing the loop-invariant. If the second part of `ASSERT` (a relation-algebraic formula formulated as RelView-expression) is true, then the statement is without effect, otherwise the execution stops and RelView allows to inspect the values of the variables via the debug window. Combining the specification of the loop-invariant in the program via `ASSERT` with RelView's feature for generating relations randomly (also with specific properties like, in our case, symmetry and irreflexivity) has the general advantage that no invariant-tests have to be done by hand (which takes time and is vulnerable to mistakes) and a lot of tests can be done in a very short time. Consequently, one gets a good feeling if a loop-invariant was chosen correctly. Because of the specific modeling of truth-values in RelView, furthermore, on the two relations $\mathsf{L} : \mathbf{1} \leftrightarrow \mathbf{1}$ and $\mathsf{O} : \mathbf{1} \leftrightarrow \mathbf{1}$ the Boolean operations $\overline{}$, $\cup$ and $\cap$ precisely correspond to the logical connectives $\neg$, $\vee$ and $\wedge$, respectively. This allows to formulate all Boolean combinations over inclusions of relations as RelView-expressions and to test them via `ASSERT`. Experience has shown that this suffices for most practical applications. It also has shown that stepwise execution and visualization via RelView are frequently helpful if invariants are not correct, e.g., too weak.

5 Verification of Proof Obligations Using Prover9

Along the lines of [8,9] we now show how the correctness proof of the relational program (VC) can be supported by an automated theorem prover, i.e., we automate the proofs of Section 3 as far as possible. Intending a user-optimized approach we choose Prover9 as verification tool. This choice is based on an evaluation that shows that Prover9 performs best for automated reasoning in the context of relation algebra; see [9] for details. A further reason for the choice of Prover9 is the positive experience made in [8,9] in the automated verification of relational programs with this tool.

Prover9 [20] is a resolution- and paramodulation-based automated theorem prover for first-order and equational logic. However, it does not include a type system. Of course, types can be realized using predicates. Since this is a bit cumbersome, we have decided to restrict our experiments to homogenous relation algebra in the sense of [13,26,27], with untyped relations. This algebraic structure axiomatizes the algebra of relations on one set (the universe) and its axioms are obtained from those of Section 2 if all demands concerning types are removed. A consequence of our decision is that the sets of vertices and colors coincide, as in case of the RelView-program `color` of Section 4.

For each result of Section 3 we want to prove, we create one input file. Each file consists of three parts, where the first two parts of each file coincide. The first part contains the language options, in particular the list of operations of relation algebra. We use the symbols `^`, `'`, `\/`, `/\` and `*` for transposition, complement, union, intersection and composition, respectively, with the binding strengths of Section 2. The second part is a list of assumptions and contains the axioms of homogeneous relation algebra, some auxiliary facts which turned out to be well suited for proving relation-algebraic results, and predicates for defining properties of relations. The constants L, O, I and the inclusion of relations are implicitly defined via the axioms, in symbols `L`, `0`, `I` and `<=`. The encoding of the axiomatization in Prover9 is straightforward. For example, the distributivity laws can be formulated as follows:

```
x /\ (y \/ z) = (x /\ y) \/ (x /\ z).
x \/ (y /\ z) = (x \/ y) /\ (x \/ z).
```

To give another example in the notation of Prover9 the Schröder equivalences look as follows:

```
x*y <= z <-> x^*z' <= y'.
x*y <= z <-> z'*y^ <= x'.
```

Although Prover9 accepts capital letters as variable names, such as `Q`, `R` and `S`, we use the small letters `x`, `y` and `z` for variables, since variables which are denoted by these letters are automatically assumed as universally quantified. The set of auxiliary facts only lists statements which are already proven by Prover9 (e.g., in [16,17]). The predicates to specify, for instance, univalent relations or relations with the coloring property can be encoded as follows:

```
univalent(x)            <-> x^*x <= I.
coloringProperty(x,z) <-> x*x^ <= z'.
```

The goal to be proven by Prover9 is specified in the third part of the file. As mentioned, we apply algebra only for proving facts, where arguments concerning sizes of sets etc. are not necessary. This means that, besides the auxiliary lemma about points of Section 2, we apply Prover9 only for proving the lemmata of Section 3. Doing so, we use the variables `p` and `q` for the general points p and

q of Lemma 2.1 as well as for the specifically selected points $p := point(\overline{C\mathsf{L}})$ and $q := point(\overline{C^\mathsf{T}Ep})$ of Section 3 and the (again automatically universally quantified) variables x for the relation C and z for the relation E, respectively. Then, e.g., the statement of Lemma 3.3 can be encoded as follows:

```
all p all q (symmetric(z) & irreflexive(z) & point(p) & point(q) &
            coloringProperty(x,z) & (x^*z)*p != L & q <= ((x^*z)*p)'
            -> coloringProperty(x \/ p*q^,z)).
```

Prover9 has no problems to derive a proof of Lemma 2.1 and requires only a couple of milliseconds. For the input files for the Lemmata 3.1, 3.2 and 3.4 Prover9 generates output files containing their proofs instantaneously as well. However, in case of Lemma 3.3 Prover9 is not able to find a proof in an appropriate time (we stopped the execution after one hour). Guided by our experience gained by previous case studies we know that in such a situation the unfolding of definitions, the subdivision of the entire task into appropriate subtasks and the removal of laws may help, since these steps reduce the size of the search space, frequently even dramatically. In the present case replacing `coloringProperty(x\/p*q^,z)` by its definition is not sufficient. Also the removal of formulae seems not to be helpful. If, however, the proof of Lemma 3.3 is divided into, first, showing that its conclusion is equivalent to the conjunction of the inclusions (1) to (4) of its proof and, secondly, that from its assumptions this conjunction follows, then for each of these tasks Prover9 needs again no time.

If Prover9 fails to find a proof, besides the unfolding of definitions, the manual change of the goal and the removal of axioms or auxiliary facts, one can use that the tool allows a weighting of formulae to specify on them an order of significance in view of the present problem. Because of our experiments with different weightings, in the case of Lemma 3.3 we believe that also a weighting of formulae does not lead to a proof in an appropriate time.

Summing up, Prover9 was able to prove the desired results and requires in one case a small user interaction only. As all automatic theorem provers, if the goals are appropriately formulated, then no interaction is needed and, hence, no deeper knowledge about (relation-)algebraic reasoning is required from the user.

6 Verification of Proof Obligations Using Coq

Now we change the paradigm from 'automated' to 'user-controlled' and demonstrate how to verify Lemma 2.1 and the lemmata of Section 3 by use of the proof assistant Coq. More information about this tool can be found in [11,29].

Since the functionality of Coq is based on the *predicative calculus of inductive constructions*, each object has a type. Thus, heterogeneous relation algebra can be modeled, and it has already been done within a relation-algebra library, presented in [22] and available via the web (see [23]). This library does not only include a model for heterogeneous relation algebra but also for a large number of other algebraic structures. For this purpose, sets of operations and laws are provided, mainly in the modules `lattice` (for lattice theory), `monoid` (for preordered monoids) and `kat` (for Kleene algebra with tests). The dependencies of the structures w.r.t. the operations and laws are managed in the module level, i.e., one can choose which kind of structure should be used by providing the required operations and laws.

Since we want to derive the proofs of Lemma 2.1 and those of Section 3, we assume a heterogeneous relation algebra, where the constants, operations, predicates and laws are defined in the mentioned modules. At the moment, the Axioms (3) and (4), i.e., the Schröder equivalences and the Tarski rule, are not formulated in the library of [22], yet.

The Schröder equivalences can be derived via the two so-called modular laws of a Dedekind category, that is, via the law named capdotx and its dual law capxdot, which are defined in the module monoid.v. One of the Schröder equivalences can be encoded as follows:

```
Lemma schroe1 `{laws} `{BL+STR+CNV<<l} n m p(Q:X n m)(R:X m p)(S:X n p):
  Q*R <== S <-> Q`*!S <== !R.
```

With `{laws} and `{BL+STR+CNV<<l}, respectively, we provide the operations and axioms of relation algebra. The symbols `, !, +, ^ and * are used for transposition, complement, union, intersection and composition, respectively.

The missing Tarski rule has to be added since it is necessary for the proofs of Lemma 2.1 and (implicitly) of Lemma 3.4. In contrast to the Schröder equivalences, the Tarski rule is not a consequence of the given laws, i.e., we have to provide it as an additional axiom. In Section 2, we specify the Tarski rule by the equivalence of $R \neq \mathsf{O}$ and $\mathsf{L}R\mathsf{L} = \mathsf{L}$ for all relations R and universal relations with appropriate types. Of course, constants are typed objects in Coq, too. But, if Coq can infer the type from the context, then it is not necessary to specify it. In such a case the universal relation, empty relation and identity relation are denoted with top, 0 and 1, respectively. In case of non-inferable types we have to specify them by, for instance, top' X Y for the universal relation $\mathsf{L} : X \leftrightarrow Y$. We have a universal quantification over the three occurring universal relations in the Tarski rule. For its formulation within Coq, besides the type of R we have to specify the types of the two universal relations of the left-hand side of $\mathsf{L}R\mathsf{L} = \mathsf{L}$ only. The type of O in $R \neq \mathsf{O}$ and the type of the right-hand side of $\mathsf{L}R\mathsf{L} = \mathsf{L}$ then can be inferred from the context. Considering this typing and using a Coq-definition, the Tarski rule can be encoded as follows:

```
Definition Tarski_rule `{laws} : Prop :=
(forall a b c d (R:X b c),(top' a b)*R*(top' c d) == top <-> ~(R == 0)).
```

Note that we omit the assumption `{BL+STR+CNV<<l} about the level, because such definitions can be written without having a structure satisfying any law. We assume the definition Tarski_rule in each lemma whose relation-algebraic proof uses the Tarski rule. As already mentioned, this concerns Lemma 2.1 of Section 2 and Lemma 3.4 of Section 3.

As in the previous section, we define predicates specifying, e.g., relations as points or relations with the coloring property. In Coq, this can be done as follows:

```
Definition coloringProperty `{laws} {n} {m}: X n m -> X m m -> Prop :=
  fun x y => x*x`<== !y.
Definition point `{laws} {n} {m}: X n m -> Prop :=
  fun p => vector p /\ p*p` <== 1 /\ (forall a, top' a m == top*p).
```

Here `vector` is yet another predicate for describing vectors. Such predicates improve the readability of the encoding.

Using the defined predicates and the definition specifying the Tarski rule, the first statement of Lemma 2.1 can be encoded as follows:

```
Lemma lemma_2_1_1 `{laws} `{BL+STR+CNV<<l} m n:
tarski_rule -> forall (p:X m n), point p -> ~(p == 0).
```

Its second statement and the lemmata of Section 3 can be formulated in a similar way. For example, the Coq-version of Lemma 3.3 is given below; it looks rather similar to the version in Prover9 with typed relations though (note that the conjunction symbol /\ of Coq corresponds to the symbol & in Prover9):

```
Lemma lem3_3 `{laws} `{BL+STR+CNV<<l} v f (C:X v f)(E p:X v v)(q:X f v):
  symmetric E /\ irreflexive E /\ point p /\ point q /\
  coloringProperty C E /\ ~(C`*E*p == top) /\ q <== !(C`*E*p)
  -> coloringProperty (C + p*q`) E.
```

Usually, the development of proofs in Coq is done via various tactics. The proofs of the mentioned lemmata, apart from Lemma 3.4, can be managed with only a few basic tactics, such as `intro` for introducing new variables or hypotheses, `unfold` for unfolding upcoming predicates in the goal as well as in the hypotheses and `rewrite` for replacing terms. More interesting are the tactics defined in the module `normalisation.v` of the library described in [22]. This module includes three specific tactics called `ra`, `ra_simpl` and `ra_normalise` which can be used to automate parts of the proofs, for instance in case of universally quantified inclusions and equalities. The proof of Lemma 3.4 has to be handled in a different way since the occurring negated equality. For this purpose, we slightly change the proof presented in Section 3 and import a module for classical propositional logic, viz. the module `Coq.Logic.Classical_Prop`, to provide the required De Morgan's laws. The advantage of this approach is that we are able to prove the lemma with the already mentioned tactics, i.e., we avoid to deal with contradiction in Coq.

In summary, Coq offers a type system and allows to model heterogeneous relation algebra in a very natural way. Amongst others, the library we have used offers a typed model for heterogeneous relations, a large number of already proven algebraic theorems and very helpful tactics for reasoning about relation algebra.

7 Verification of Proof Obligations Using Isabelle/HOL

In this section we discuss the verification of the relational program (VC) by means of Isabelle/HOL, a proof assistant that additionally offers support for automated theorem proving via the Sledgehammer tool. For more details on Isabelle/HOL, see e.g., [21].

Similar to Coq, libraries can be included in Isabelle/HOL. The development of such libraries usually takes a long time and deep insights in the theorem prover at hand. Luckily, as in the case of Coq, relation algebra has already be formalized in Isabelle/HOL and is available via the web (see [1]). However, the library of [1] formalizes homogeneous relation algebra only. A consequence of its use with regard to the verification of the relational program (VC) is again that the sets of vertices and colors coincide.

The formalization [1] of homogeneous relation algebra follows the lines of [19,26]. Besides the basic constants, operations and predicates and the axioms it includes a number of further important relation-algebraic concepts such as subidentities, vectors and points, as well as various notions associated to functions – together with numerous proven facts. For example, all facts about relation algebra listed in Section 2 have been proven. As a consequence, it seems to be an ideal basis for the verification of relational programs such as (VC). However, the current implementation does not contain the Tarski-rule (similar to Coq), so we added this rule to the set of assumptions of a lemma when necessary.

The provided libraries are included by a simple `imports`-statement; encoding of the lemmata is easy and straightforward. For example, Lemma 3.3 is encoded as follows, where $\smile$ indicates transposition and the symbol `-` is used for both negation and complement:

```
lemma assumes "symmetric e"               and "irreflexive e"
          and "is_point p"                and "is_point q"
          and "coloringProperty x e"      and "q ≤ -(x⌣ ; e ; p)"
        shows "coloringProperty (x + p;q⌣) e"
```

It might be confusing that we use different symbols for the same operation (e.g., $^\mathsf{T}$, $\hat{}$, ‘ and $\smile$ for transposition). However, since we use different tools we decided to stick to the notation of these frameworks. Since the GUI of Isabelle/HOL allows non-ascii symbols, transposition can be encoded as $\smile$.

The used predicates are basically identical to the ones of Prover9 and Coq. For example, `coloringProperty` is defined as follows:

```
definition coloringProperty
     where "coloringProperty x e ≡ x ; x⌣ ≤ -e"
```

A straightforward approach would be the use of the Isabelle/Isar tool (see [28]), which basically replays the proofs given in Section 3. The advantage of this approach is that it provides a proof certificate and verifies the manual proofs. Moreover the generated proofs are easy to read. However, this strategy does not provide (much) automation and hence requires expert knowledge in relation-algebraic reasoning.

As mentioned in Section 5, automated reasoning within the relation-algebraic setting is successful if the proof-goals are appropriately formulated. In contrast to Coq, Isabelle/HOL offers support for (first-order) automated theorem provers via the integrated tool Sledgehammer (see [12] for more details), thus, allows to combine the ‘automated’ and ‘user-controlled’ paradigm. The Sledgehammer tool takes the given goal and proven facts available, feeds them to automated theorem provers, such as E and Z3, and awaits their output. In case one of the provers is successful in finding a proof, the proof is included in the Isabelle-file; in case all theorem provers fail, the GUI continues to assist in a manual proof derivation. That means that Isabelle/HOL provides both proof-assistance and proof-automation and it seems to be the perfect combination of interaction and automation. In fact, a proof of Lemma 3.3 becomes 'nearly' automatic: after a first manual step using Isabelle's unfolding and simplification mechanisms (`simp add: unfold_defs distrib_left distrib_right, safe`) we end up with the four subgoals (1) to (4) presented in the proof of Lemma 3.3. This command unfolds automatically all predicates (`unfold_defs`) and uses the built-in simplifier, which is manually extended by the two distributivity axioms of relation algebra (`distrib_left distrib_right`).

The derived subgoals can now all be proven automatically by Sledgehammer using not only the axioms of relation algebra, but also the facts provided by the theories of relation algebra of [1]. The fact `"x;y ≤ z ⇔ y ≤ -(x˘;-z)"`, for example, which was proven in the framework of [1], is automatically chosen to prove the second subgoal `(p;q˘;x˘≤ -e)` (in the proof $pq^{\mathsf{T}}C^{\mathsf{T}} \subseteq \overline{E}$). All other lemmata presented in the paper, except Lemma 3.4, can be proven in an identical way: first derive subgoals using the inbuilt simplifier, and then use Sledgehammer and the provided automated theorem provers to prove these subgoals automatically. Lemma 3.4 could not be proven by this strategy. In fact, we could only reply the proof by contradiction given in Section 3 – real proof automation was not possible.

8 Assessment and Concluding Remarks

In this paper we have developed a relational program for calculating a vertex coloring of an undirected graph, which is modeled by the adjacency relation. A relation-algebraic approach was chosen since case studies have shown that such an approach is not only very suitable for prototyping and testing programs by systems like RelView, but also for proof automation. The program verification was performed by classical reasoning about pre- and post-conditions, and loop-invariants. The proofs of the proof obligations were executed with the help of Prover9, Coq and Isabelle/HOL, which are prominent tools to support verification tasks. This repetition of mechanized proofs and the comparison with the original mathematical proofs allow us to compare these mentioned tools. As one might expect each and every tool has its pros and cons.

Prover9 does not include a type system such that the typing of relations in heterogeneous relation algebra would have to be realized by predicates. Such predicates make the encoding more complicated and decrease the readability. In our example types could be avoided, but if, e.g., incidence relations are used to model undirected graphs or hypergraphs, then types are mandatory. All but one theorems of Section 3 were proved full automatically by Prover9 in nearly no time. Unfortunately, in spite of weighting the significant rules such that they should be applied first, Prover9 was still not able to find a proof for Lemma 3.3 in an appropriate time frame. At the end, we decided not to change the weights of the assumptions but to split the goal into subgoals. Indeed, this approach requires a kind of interaction by the user, but it yields to uniform assumptions for all theorems of Section 3 as well as to short proving times. These results are not only based on the investigations presented in this paper but additionally coincide with those discussed in [8,9,14]. Since Prover9 is fully automatic it can be used as a black box and without having a deep understanding of its functionality. From a user's point of view a big advantage of Prover9 is that the encoding of the axiomatization of homogenous relation algebra, the definition of predicates and the formulation of theorems which have to be proved is very straightforward and also comprehensible for non-experts.

The proof assistant Coq is the very opposite of Prover9. It is completely user-controlled, i.e., a purely interactive theorem prover. Coq has a sophisticated type system with type inference, which is comparable with those of functional programming languages such as Haskell, ML and OCaml. For performing our proofs we used an already existing library for relation algebra. In this case the library implements homogenous as well as heterogeneous relation algebra. Due to this we were able to reproduce all proofs of Section 3 without any restrictions on types. The used library does also include tactics, that is, strategies for proof-finding and proof execution. They support relation-algebraic reasoning. Some of them implemented decision procedures for subsets of relation algebra, which we might use in future experiments. Furthermore, the library comes with a large number of algebraic structures related to relation algebras, e.g., Dedekind categories, Kleene algebras and Kleene algebras with tests. For this reason, the library can also be used in the context of reasoning about such structures as well. We refer to [10] for an application concerning Dedekind categories. The usage of Coq requires a lot of knowledge about the internals of the tool, such as the available tactics and the hierarchy and dependencies of the modules. Besides this, in our case the user needs expertise about relation-algebraic reasoning and the used library for relation algebra to be able to derive the proofs step by step. So, from a user's point of view Coq is far more complicated than Prover9; it is suitable for advanced users only.

For the proof verification with Isabelle/HOL we were also able to build on already existing theories. Since a library for heterogeneous relation algebra does not exist yet, we used a library that implements homogenous relation algebra. Using this library we again have to avoid typed relations, although Isabelle/HOL provides a type system similar to that of Coq. Concerning proof paradigms, Isabelle/HOL bridges the gap between interactive and automated reasoning via the Sledgehammer tool. Our experiments have shown that our strategy, i.e., first using the inbuilt simplifier for deriving subgoals and then Sledgehammer to prove these subgoals automatically, was successful with all theorems of Section 3 except Lemma 3.3, again. For the latter one, we have to derive new subgoals with two manual steps. However, more intrinsic proofs (here a proof by contradiction) requires again expert knowledge in relation-algebraic reasoning. Isabelle/HOL does not offer tactics nor decision procedures, yet, for relation-algebraic reasoning specifically. With regard to usage, Isabelle/HOL is powerful enough to support non-expert users with many (standard) tasks in case the problems in question are not very complex. But in case of more complex problems it requires experience and is then, like Coq, suitable for advanced users only.

As future work we plan to exhaust the capabilities of Prover9 w.r.t. the different options, e.g., the weighting of the given assumptions, to hopefully achieve best proving times for relation-algebraic theorems. Concerning Coq, we plan to explore the full power of the tactics and to investigate whether they are supportive in verification tasks. For Isabelle/HOL, we want to consider the implementation of tactics which are specific for relation-algebraic reasoning. Furthermore, an extension of the used library to heterogeneous relation algebra is desirable. We assume that by all this many verification tasks concerning programs on relations or related objects can be automated to a large extent – this can, however, only be verified by further and more complicated case studies. Finally, we plan to investigate in the future how tools for generating loop invariants (as Why3, see [31]) are applicable for our purposes.

Acknowledgement. We thank D. Pous for valuable remarks. We also thank the unknown referees for their constructive criticims and suggestions which helped to improve the paper. NICTA is funded by the Australian Government through the Department of Communications and the Australian Research Council through the ICT Centre of Excellence Program.

References

1. Armstrong, A., Foster, S., Struth, G., Weber, T.: Relation algebra. Archive of Formal Proofs (2014). `http://afp.sf.net/entries/Relation_Algebra.shtml`
2. Berghammer, R.: Combining relational calculus and the Dijkstra-Gries method for deriving relational programs. Inform. Sci. 119, 155–171 (1999)
3. Berghammer, R., Hoffmann, T.: Deriving relational programs for computing kernels by reconstructing a proof of Richardson's theorem. Sci. Comput, Prog. 38, 1–25 (2000)
4. Berghammer, R., Hoffmann, T.: Relational depth-first-search with applications. Inform. Sci. 139, 167–186 (2001)
5. Berghammer, R., Neumann, F.: RelView – An OBDD-based Computer Algebra system for relations. In: Ganzha, V.G., Mayr, E.W., Vorozhtsov, E.V. (eds.) CASC 2005. LNCS, vol. 3718, pp. 40–51. Springer, Heidelberg (2005)
6. Berghammer, R.: Applying relation algebra and RelView to solve problems on orders and lattices. Acta Inform. 45, 211–236 (2008)
7. Berghammer, R., Winter, M.: Embedding mappings and splittings with applications. Acta Inform. 47, 77–110 (2010)
8. Berghammer, R., Struth, G.: On automated program construction and verification. In: Bolduc, C., Desharnais, J., Ktari, B. (eds.) MPC 2010. LNCS, vol. 6120, pp. 22–41. Springer, Heidelberg (2010)
9. Berghammer, R., Höfner, P., Stucke, I.: Automated verification of relational while-programs. In: Höfner, P., Jipsen, P., Kahl, W., Müller, M.E. (eds.) RAMiCS 2014. LNCS, vol. 8428, pp. 309–326. Springer, Heidelberg (2014)
10. Berghammer, R., Stucke, I., Winter, M.: Investigating and computing bipartitions with algebraic means. In: Kahl, W., Oliviera, J.N., Winter, M. (eds.) Relational and Algebraic Methods in Computer Science (to appear)

11. Bertot, Y., Casteran, P.: Interactive theorem proving and program development. Coq'Art: The calculus of inductive constructions. Texts in Theoretical Computer Science. Springer (2004)
12. Blanchette, J.C., Böhme, S., Paulson, L.C.: Extending Sledgehammer with SMT solvers. In: Bjørner, N., Sofronie-Stokkermans, V. (eds.) CADE 2011. LNCS, vol. 6803, pp. 116–130. Springer, Heidelberg (2011)
13. Chin, L.H., Tarski, A.: Distributive and modular laws in the arithmetic of relation algebras. Univ. of California Publ. Math. (new series) 1, 341–384 (1951)
14. Dang, H.H., Höfner, P.: First-order theorem prover evaluation w.r.t. relation- and Kleene algebra. In: Berghammer, R., Möller, B., Struth, G. (eds.) Relations and Kleene Algebra in Computer Science – Ph.D. Programme at RelMiCS10/AKA05. Technical Report 2008-04, Institut für Informatik, Universität Augsburg, 48–52 (2008)
15. Foster, S., Struth, G., Weber, T.: Automated engineering of relational and algebraic methods in Isabelle/HOL (Invited tutorial). In: de Swart, H. (ed.) RAMICS 2011. LNCS, vol. 6663, pp. 52–67. Springer, Heidelberg (2011)
16. Höfner, P., Struth, G.: Automated reasoning in Kleene algebra. In: Pfenning, F. (ed.) CADE 2007. LNCS (LNAI), vol. 4603, pp. 279–294. Springer, Heidelberg (2007)
17. Höfner, P., Struth, G.: On automating the calculus of relations. In: Armando, A., Baumgartner, P., Dowek, G. (eds.) IJCAR 2008. LNCS (LNAI), vol. 5195, pp. 50–66. Springer, Heidelberg (2008)
18. Kahl, W.: Calculational relation-algebraic proofs in Isabelle/Isar. In: Berghammer, R., Möller, B., Struth, G. (eds.) RelMiCS 2003. LNCS, vol. 3051, pp. 179–190. Springer, Heidelberg (2004)
19. Maddux, R.: Relation algebras. Studies in Logic and the Foundations of Mathematics, vol. 150. Elsevier (2006)
20. McCune, W.W.: Prover9 and Mace4, `http://www.cs.unm.edu/~mccune/prover9`
21. Nipkow, T., Paulson, L.C., Wenzel, M.: Isabelle/HOL. LNCS, vol. 2283. Springer, Heidelberg (2002)
22. Pous, D.: Kleene algebra with tests and Coq tools for while programs. In: Blazy, S., Paulin-Mohring, C., Pichardie, D. (eds.) ITP 2013. LNCS, vol. 7998, pp. 180–196. Springer, Heidelberg (2013)
23. Pous, D.: Relation algebra and KAT in Coq, `http://perso.ens-lyon.fr/damien.pous/ra/`
24. Schmidt, G., Ströhlein, T.: Relations and graphs, Discrete mathematics for computer scientists. EATCS Monographs on Theoretical Computer Science. Springer (1993)
25. Schmidt, G.: Relational mathematics. Encyclopedia of Mathematics and its Applications. Cambridge University Press (2010)
26. Tarski, A.: On the calculus of relations. J. Symb. Logic 6(3), 73–89 (1941)
27. Tarski, A., Givant, S.: A formalization of set theory without variables. AMS Colloquium Publications, American Mathematical Society (1987)
28. Wenzel, M.: Isabelle/Isar – a versatile environment for human-readable formal proof documents. Dissertation, Technische Universität München (2002)
29. Coq-homepage: `https://coq.inria.fr`
30. RelView-homepage: `http://www.informatik.uni-kiel.de/~progsys/relview/`
31. Why3-homepage: `http://why3.lri.fr/`
32. Input files and proof scripts: `http://www.hoefner-online.de/ramics15/`

Applications of Relational and Algebraic Methods

L-Fuzzy Databases in Arrow Categories

Evans Adjei, Wazed Chowdhury, and Michael Winter*

Department of Computer Science,
Brock University,
St. Catharines, Ontario, Canada, L2S 3A1
{ea12gq,wc12ss,mwinter}@brocku.ca

Abstract. In this paper we present a query language for lattice-based (or L-fuzzy) databases. These databases store L-fuzzy sets in their attributes instead of (crisp) values in order to handle imprecise or incomplete information. A semantics for the language is defined using the abstract notion of an arrow category.

1 Introduction

Nowadays relational databases can be found almost everywhere starting with the contact list on a cell phone to a customer database of a big company. The common language to maintain a relational database and to retrieve information is the Structured Query Language (SQL). Even though relational databases can handle all kind of information, including missing information by using so-called *null* values, they are not very well suited to deal with imprecise data. For example, if a database of persons has a field for the height of a person, then any new entry is required to provide the height in centimeters or inches (or *null* if the height is unknown). If it is only known that Joe is *tall*, then we have only two alternatives. Either we use *null* because we do not know his height or we pick a random value that we consider to be *tall*. Both approaches do not reflect the information about Joe correctly. The first approach does not provide any information about Joe's height, i.e., the information that Joe is *tall* is dropped. The second approach provides information that is most likely wrong. In order to handle imprecise information such as *tall*, fuzzy relational databases and the query language Fuzzy Structured Query Language (FSQL) have been developed [5–7]. In such a database every field is allowed to store a fuzzy set instead of a single value. Fuzzy sets were introduced by Zadeh [22], and they constitute a generalization of regular sets. A fuzzy set is a set in which each element has a degree of membership from the unit interval $[0\ldots1]$ of the real numbers up to which the element is part of the set. An element with degree 0 is definitely not in the set and an element with degree 1 is definitely in the set. Formally, a fuzzy subset B of A is represented by a characteristic function $\chi_B : A \to [0\ldots1]$. Fuzzy sets are used to provide a mathematical interpretation of common language expressions, also called linguistic entities or labels, such as *tall*. For example, *tall* can be interpreted by the following fuzzy set:

* The author gratefully acknowledges support from the Natural Sciences and Engineering Research Council of Canada.

W. Kahl et al. (Eds.): RAMiCS 2015, LNCS 9348, pp. 295–311, 2015.
DOI: 10.1007/978-3-319-24704-5_18

$$\chi_{tall}(x) = \begin{cases} 0 & \text{iff } x \leqslant 160\text{cm}, \\ \frac{x-160}{20} & \text{iff } 160\text{cm} < x \leqslant 180\text{cm}, \\ 1 & \text{iff } 180\text{cm} < x \end{cases}$$

The result of a FSQL query is a list of entries of the database. Each entry comes with the degree up to which it satisfies the conditions of the query. The language FSQL adds to the regular SQL statements operations that are specific to fuzzy sets. For example, each comparison operations such as $=$, $\leqslant$ and $<$ are available in the form of a possibility $F=$, $F\leqslant$, $F<$ and a necessity operation $NF=$, $NF\leqslant$, $NF<$ (see also Section 3) computing the degree of the possibility resp. the necessity that the two fuzzy sets are in the corresponding relationship. In addition, the language also allows to specify thresholds, i.e., minimal degrees up to which the property must be true, and to use t-norms/t-conorms instead of min and max for computing the logical connective *and* and *or*. Linguistic labels are preceded by the symbol $ and their characteristic function is stored in a meta database. A typical example for a FSQL select statement is:

```
SELECT Name, Height, Age FROM Persons
    WHERE Height F= $Tall AND
          Age NF< $Old THOLD 0.5
```

In this paper we are going to introduce *L*-fuzzy databases and the query language *L*-fuzzy Structured Query Language (LFSQL). *L*-fuzzy sets were introduced by Goguen [8], and they generalize fuzzy sets even further. An *L*-fuzzy set is a set in which each element has a degree of membership from an arbitrary bounded lattice *L*, i.e., an ordered structure with a meet and a join operation and a least element 0 and a greatest element 1. As before an element with degree 0 is definitely not in the set and an element with degree 1 is definitely in the set. Formally, an *L*-fuzzy subset *B* of *A* is a characteristic function $\chi_B : A \to L$. The unit interval forms a lattice, i.e., fuzzy sets are $[0 \ldots 1]$-fuzzy sets. However, the unit interval is linearly ordered, i.e., for each pair $x, y \in [0 \ldots 1]$ of elements we have either $x \leqslant y$ or $y \leqslant x$. This property implies that we are always able to tell for any two elements *a* and *b* which is *more* in a given fuzzy set *B* by comparing $\chi_B(a)$ and $\chi_B(b)$. This might not be suitable to model certain situations. For example, we want to consider buying a TV and model the screen size of TV's as an *L*-fuzzy set. The degree of membership of a given screen size in the set of *good* sizes indicates how well-suited we consider this particular screen size. A screen size of 60in might be *good* because of the viewers experience but not so good because of the price of the corresponding TV. On the other hand, a screen size of 40in might be *good* because of the price but not so good in terms of the viewers experience. For these reasons both screen sizes should be in the *L*-fuzzy set of *good* sizes up to a certain degree. However, it seems hard, or even impossible or unwanted, to decide which screen size is better, i.e., we do not want that $\chi_{good}(40in) \leqslant \chi_{good}(60in)$ or vice versa. In this example we want to choose a lattice *L* that is not linearly ordered as the domain of membership values.

In addition to the syntax of LFSQL we present a formal semantics of the language in terms of abstract arrow categories. Arrow categories provide a suitable categorical/algebraic theory for *L*-fuzzy relations, and hence a general framework to interpret LFSQL. This semantics can be used in multiple ways. It was already used in an implementation of a prototype of LFSQL in the functional programming language Haskell

[1, 3]. This prototype is based on a concrete implementation of the arrow category of sets and (finite) L-fuzzy relations. The execution of a LFSQL statement uses a Haskell function that implements the semantics exactly as defined in Section 3.3. Furthermore, the semantics serves as the foundation for future investigation on dependencies, normal forms and data mining based on L-fuzzy databases.

The remainder of this paper is organized as follows. In Section 2 we recall the theory of arrow categories. Section 3 defines and discusses L-fuzzy databases and the language LFSQL. In addition, we define the semantics of LFSQL in terms of arrow categories [17, 20, 21]. A conclusion and future work is presented in Section 4.

2 Arrow Categories

In this section we want to recall the mathematical structures that we will be using in order to define the semantics of LFSQL. These structures include lattices, categories and arrow categories. For further details we refer to [2, 4, 20].

We will use the notation $R : A \to B$ to indicate that a morphism R of a category $\mathcal{R}$ has source A and target B. The collection of all morphisms with source A and target B is denoted by $\mathcal{R}[A, B]$. Composition is denoted by ; and has to be read from left to right, i.e., $R; S$ means R first, and then S. The identity morphism on A is written as $\mathbb{I}_A$.

A lattice L is called a complete Heyting algebra iff L is complete and the first infinite distributivity law $x \sqcap \bigsqcup M = \bigsqcup_{y \in M} (x \sqcap y)$ holds for all $x \in L$ and $M \subseteq L$. Therefore, a complete Heyting algebra is distributive and has relative pseudo complements, i.e., for each pair $x, y \in L$ there is an element $x \to y$ so that $z \leqslant x \to y$ is equivalent to $x \sqcap z \sqsubseteq y$.

Dedekind categories [11, 12] have been shown to be a suitable categorical/algebraic framework to describe binary relations.

Definition 1. *A Dedekind category $\mathcal{R}$ is a category satisfying the following:*

1. *For all objects A and B the collection $\mathcal{R}[A, B]$ is a complete Heyting algebra. Meet, join, the induced ordering, the least and the greatest element are denoted by $\sqcap, \sqcup, \sqsubseteq, \mathbb{L}_{AB}, \mathbb{T}_{AB}$, respectively.*
2. *There is a monotone operation $^{\smile}$ (called converse) mapping a relation $Q : A \to B$ to $Q^{\smile} : B \to A$ such that for all relations $Q : A \to B$ and $R : B \to C$ the following holds: $(Q; R)^{\smile} = R^{\smile}; Q^{\smile}$ and $(Q^{\smile})^{\smile} = Q$.*
3. *For all relations $Q : A \to B, R : B \to C$ and $S : A \to C$ the modular law $(Q; R) \sqcap S \sqsubseteq Q; (R \sqcap (Q^{\smile}; S))$ holds.*
4. *For all relations $R : B \to C$ and $S : A \to C$ there is a relation $S/R : A \to B$ (called the left residual of S and R) such that for all $X : A \to B$ the following holds: $X; R \sqsubseteq S \iff X \sqsubseteq S/R$.*

Notice that the axiom provided above are not independent. For example, the existence of residuals follows from the fact that each collection $\mathcal{R}[A, B]$ is a complete Heyting algebra.

The left residual also implies the existence of a right residual characterized by

$$Q;Y \sqsubseteq S \iff Y \sqsubseteq Q\backslash S.$$

In fact, we have $Q\backslash S = (S^{\smile}/Q^{\smile})^{\smile}$. Both residuals are monotone in one argument and antitone in the other. If $S \sqsubseteq S'$, $R' \sqsubseteq R$ and $Q' \sqsubseteq Q$, then $S/R \sqsubseteq S'/R'$ and $Q\backslash S \sqsubseteq Q'\backslash S'$.

The category **Rel** of binary relations between sets with the usual definition of the operations forms a Dedekind category. In addition, the collection of L-fuzzy relations, i.e., relations R given by their L-valued characteristic function $R : A \times B \to L$, between sets form a Dedekind category. Notice that **Rel** is actually a special case of L-fuzzy relations where L is the Boolean algebra of truth values. We will call relations in **Rel** regular relations in order to distinguish them from L-fuzzy relations.

We will often use a matrix representation in order to visualize finite examples in L-fuzzy or regular relations. For example, if $R : A \times B \to D_6$ with $A = \{r, s, t, u\}$ and $B = \{0, 1, 2, 3, 4\}$ is a D_6-fuzzy relation, then one such relation R can be visualized by the matrix

$$R := \begin{pmatrix} 1 & a & b & 1 & 0 \\ c & d & 0 & 0 & 1 \\ 0 & 1 & c & a & b \\ 1 & 0 & a & c & d \end{pmatrix} \qquad D_6 :=$$

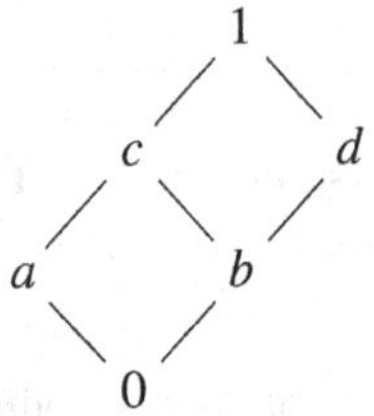

By assuming the order in which the elements are presented in A and B the matrix should be read as follows. The b in the third row of the matrix indicates that the third element of A, the element t, is in relation R to the fifth element of B, the element 4, by a degree b. Notice that D_6 is isomorphic to the product $\{0, 1\} \times \{0, m, 1\}$ of the two linear orderings $0 \leqslant 1$ and $0 \leqslant m \leqslant 1$. Therefore, D_6 can be used to model two aspects of membership similar to the TV example in the introduction. The first aspect is a *yes-no* and the second aspect a *yes-no-maybe* relationship. For example, $c = (1, m)$ represents the fact that c is *definitely* in the set with respect to the first aspect and *maybe* with respect to second aspect.

Before we introduce arrow categories we want to recall some important concepts and constructions within Dedekind categories. For more details on these constructions we refer to [13–16].

An order relation $E : A \to A$ is a relation that is reflexive, transitive, and antisymmetric, i.e., it satisfies $\mathbb{I}_A \sqsubseteq E$, $E;E \sqsubseteq E$, and $E \sqcap E^{\smile} \sqsubseteq \mathbb{I}_A$. For a given relation $X : B \to A$ it is possible to compute the upper bounds or lower bounds for each row of X, i.e., of the set of elements related to one $b \in B$, by using a residual

$$\mathrm{ubd}_E(X) = X^{\smile}\backslash E \quad \text{and} \quad \mathrm{lbd}_E(X) = X^{\smile}\backslash E^{\smile}.$$

Another important class of relations is given by maps. A map is a relation $Q : A \to B$ that is univalent (or partial function), i.e., $Q^{\smile};Q \sqsubseteq \mathbb{I}_B$, and total, i.e., $\mathbb{I}_A \sqsubseteq Q;Q^{\smile}$.

In the category **Rel** the empty set and singleton sets play an important role. The empty set is a zero object, i.e., initial and terminal, in **Rel**. This can be characterized by the fact that the smallest relation on the empty set is equal to the greatest relation. Therefore, we call an object 0 of a Dedekind category a zero object iff $\perp\!\!\!\perp_{00} = \mathbb{T}_{00}$. Singleton sets in **Rel** are terminal objects in the subcategory of maps. In **Rel** itself they can be characterized as so-called units. A unit 1 is an object of a Dedekind category for which $\mathbb{I}_1 = \mathbb{T}_{11}$ and $\mathbb{T}_{A1}$ is total for all objects A.

A relation $v : 1 \to A$ is called a vector. These relations represent a subset of A in an abstract manner. Similarly, a map $p : 1 \to A$ represents an element of A. Such a relation is called a point.

The abstract version of a cartesian product is given by a relational product.

Definition 2. *The relational product of two objects A and B is an object $A \times B$ together with two relations $\pi : A \times B \to A$ and $\rho : A \times B \to B$ so that the following equations hold*

$$\pi^{\smile};\pi \sqsubseteq \mathbb{I}_A, \quad \rho^{\smile};\rho \sqsubseteq \mathbb{I}_B, \quad \pi^{\smile};\rho = \mathbb{T}_{AB}, \quad \pi;\pi^{\smile} \sqcap \rho;\rho^{\smile} = \mathbb{I}_{A\times B}.$$

Another important construction is based on forming the disjoint union of sets.

Definition 3. *Let A and B be objects of a Dedekind category. An object $A + B$ together with two relations $\iota : A \to A + B$ and $\kappa : B \to A + B$ is called a relational sum of A and B iff*

$$\iota;\iota^{\smile} = \mathbb{I}_A, \quad \kappa;\kappa^{\smile} = \mathbb{I}_B, \quad \iota;\kappa^{\smile} = \perp\!\!\!\perp_{AB}, \quad \iota^{\smile};\iota \sqcup \kappa^{\smile};\kappa = \mathbb{I}_{A+B}.$$

Last but not least, the abstract version of subsets and/or the set of equivalence classes is given by splittings.

Definition 4. *Let A be an object of a Dedekind category and $\Xi : A \to A$ a partial equivalence relation, i.e., Ξ is symmetric $\Xi^{\smile} = \Xi$ and idempotent $\Xi;\Xi = \Xi$. An object B together with a relation $R : B \to A$ is called a splitting of Ξ iff $R;R^{\smile} = \mathbb{I}_A$ and $R^{\smile};R = \Xi$.*

In a Dedekind category one can identify the underlying lattice L of membership values by the scalar relations on an object.

Definition 5. *A relation $\alpha : A \to A$ is called a scalar on A iff $\alpha \sqsubseteq \mathbb{I}_A$ and $\mathbb{T}_{AA};\alpha = \alpha;\mathbb{T}_{AA}$.*

The notion of scalars was introduced by Furusawa and Kawahara [10] and is equivalent to the notion of ideals, i.e., relations $R : A \to B$ that satisfy $\mathbb{T}_{AA};R;\mathbb{T}_{BB} = R$, which were introduced by Jónsson and Tarski [9].

Notice that **Rel** has only two scalars $\perp\!\!\!\perp_{AA}$ and $\mathbb{I}_A$. This shows again that **Rel** seen as a category of fuzzy relations is based on the Boolean algebra of the truth values. In addition, these two scalars are available in every Dedekind category indicating that **Rel** is embedded in every Dedekind category of L-fuzzy relations. We call an L-fuzzy relation R crisp iff $R(x, y) \in \{0, 1\} \subseteq L$ for all x and y.

The next definition introduces arrow categories, i.e., the basic theory for L-fuzzy relations. Arrow categories add two operations to Dedekind categories. The relation $R^{\uparrow}$ is the smallest crisp relation that contains R, and $R^{\downarrow}$ is the greatest crisp relation included in R [17, 20, 21].

Definition 6. *An arrow category $\mathcal{A}$ is a Dedekind category with $\mathbb{T}_{AB} \neq \mathbb{L}_{AB}$ for all A, B and two operations $\uparrow$ and $\downarrow$ satisfying:*

1. *$R^\uparrow, R^\downarrow : A \to B$ for all $R : A \to B$.*
2. *$(^\uparrow, ^\downarrow)$ forms a Galois correspondence, i.e., $Q^\uparrow \sqsubseteq R$ iff $Q \sqsubseteq R^\downarrow$ for all $Q, R : A \to B$.*
3. *$(R^\smile; S^\downarrow)^\uparrow = R^{\uparrow\smile}; S^\downarrow$ for all $R : B \to A$ and $S : B \to C$.*
4. *$(Q \sqcap R^\downarrow)^\uparrow = Q^\uparrow \sqcap R^\downarrow$ for all $Q, R : A \to B$.*
5. *If $\alpha_A \neq \mathbb{L}_{AA}$ is a non-zero scalar then $\alpha_A^\uparrow = \mathbb{I}_A$.*

A relation that satisfies $R^\uparrow = R$, or equivalently $R^\downarrow = R$, is called crisp. Notice that the complete Heyting algebra of scalar relations on each object are isomorphic.

If α is scalar, then the relation $(\alpha \backslash R)^\downarrow$ is called the α-cut of R. If we identify the scalar α with its corresponding element in L, then the α-cut for L-fuzzy relations can be characterized by

$$(\alpha \backslash R)^\downarrow(x,y) = 1 \iff \alpha \leqslant R(x,y).$$

In fuzzy theory t-norms and t-conorms are essential for defining new operations for fuzzy sets or relations. In [8] a generalization of these operations for arbitrary complete lattices was introduced, called complete lattice-ordered semigroups. Given such an operation $* : L \times L \to L$ we may define a new meet or composition based operation on L-fuzzy relations $Q, R : A \to B$ and $S : B \to C$ by

$$(Q \sqcap_* R)(x,y) = Q(x,y) * R(x,y) \quad \text{and} \quad (Q;_* S)(x,z) = \bigsqcup_{y \in B} Q(x,y) * S(y,z).$$

In an abstract arrow category we require $*$ to be defined on the complete Heyting algebra of scalar elements. As shown in [19, 20] the corresponding operations on relations are defined as follows.

Definition 7. *Let Q, R be relations, $\otimes \in \{\sqcap, ;\}$ such that $Q \otimes R$ is defined, and $*$ the operation of a complete lattice-ordered semigroup on the set of scalar relations. Then we define*

$$Q \otimes_* R := \bigsqcup_{\alpha,\beta \text{ scalars}} (\alpha * \beta); ((\alpha \backslash Q)^\downarrow \otimes (\beta \backslash R)^\downarrow).$$

We distinguish two kinds of commutative complete lattice-ordered semigroup operations corresponding to either t-norms or t-conorms. If the neutral element of the semigroup is equal to 1 (greatest element of the lattice) we call the operation a t-norm like operation. t-norm like operations will be used together with $\sqcap$ and ; to form new operations on relations. If the neutral element of the semigroup is equal to 0 (smallest element of the lattice) we call the operation a t-conorm like operation. These operations will only be used together with $\sqcap$.

Notice that we also have residuals based on semigroup operations. They are defined as the left (resp. right) adjoint of $;_*$. For more details on these constructions we refer to [20].

3 L-Fuzzy Databases and LFSQL

In this section we are going to define L-fuzzy databases and the language LFSQL. The language can be used to create and update tables using a CREATE, INSERT and DELETE statement and to retrieve information using a SELECT statement.

3.1 L-Fuzzy Databases

Similar to a regular database, a table (or relation) in a L-fuzzy database has objects (or rows or tuples) and attributes (or columns). Each attribute, and, hence, each column, has a set of possible values, the so-called domain, assigned to it. Each domain provides some comparison operations. Unordered domains provide at least equality. Ordered domains will also provide an order $\leqslant$ and its corresponding strict order $<$ with $x \leqslant y$ iff $x < y$ or $x = y$. Finally, some domains may also provide one or more approximate equalities $\equiv$. This binary comparison operation returns a degree of membership from L indicating up to which degree two given elements are consider to be equal. For example, if we consider temperatures, we might consider two degree values as almost equal (high degree) if the differ by less than $0.1°$ and almost different (low degree) if they differ by more than $2°$. An approximate equality is required to be reflexive, i.e., $x \equiv x = 1$ and symmetric, i.e., $x \equiv y = y \equiv x$ but not transitive [20].

An entry in a table at a row and column is an L-fuzzy subset of the domain of the row. Notice that a single (crisp) value x is modeled by a fuzzy set that has degree 1 for x and 0 otherwise.

In addition to tables, an L-fuzzy database uses a meta-database in which the lattice L, any t-norm (resp. t-conorm) like operation on L, and the extend of the linguistic labels are stored. The extent of a linguistic label is an L-fuzzy set. The meta-database may also contain some pre-implemented characteristic functions, i.e., functions from certain domains into L. These functions can later be used to define L-fuzzy subsets explicitly. They can be parametric and depend heavily on the lattice L. For example, if L is the unit interval and A a linear ordered set, then the meta-database should include functions that generate triangular or trapezoidal fuzzy subsets B of A, i.e., the portion of the graph of the membership function that is not equal to 0 takes the form of a triangle or a trapezoid if visualized graphically.

Lingustic Labels and L-Fuzzy Sets: Besides L-fuzzy subsets that are stored in the meta-database we may define them either explicitly or by modifying already existing sets. Any of those sets can be used in statements where L-fuzzy subsets are required, or they can be stored in the meta-database under a new name.

An L-fuzzy subset B of a domain D can be defined explicitly by the syntactic notation $\{l_1/d_1, \ldots, l_n/d_n\}$. This defines B as the set with the following characteristic function:

$$\chi_B(x) = \begin{cases} l_1 \text{ iff } x = d_1, \\ \vdots \quad \vdots \\ l_n \text{ iff } x = d_n. \\ 0 \text{ otherwise} \end{cases}$$

Alternatively, an L-fuzzy set can be defined by using one of the pre-implemented functions stored in the meta-database. If f is the name of such a function, then $\#f$ defines the L-fuzzy set with characteristic function f.

There are two ways of obtaining new sets from previously defined sets. The first method is only available for ordered domains. We may compute the lower or upper bounds as already introduced in the previous section, i.e., $\mathrm{lbd}(m)$ (resp. $\mathrm{ubd}(m)$) is the set of lower bound (resp. upper bounds) of m with respect to the order of the domain. Notice that we are using these constructions in the context of L-fuzzy set and relations so that the set of lower or upper bounds is itself an L-fuzzy set. Both constructions can be based on a t-norm like operation already stored in the meta-database, i.e., we might write $\mathrm{lbd}(*, m)$.

The second method uses an approximate equality $\equiv$ defined on the domain. This relation can be used to intensify or weaken the notion given by the set m. We use $\mathrm{extremely}(\equiv, m)$ and $\mathrm{very}(\equiv, m)$ as intensifying and $\mathrm{more\_or\_less}(\equiv, m)$ and $\mathrm{roughly}(\equiv, m)$ as weakening modifiers. These sets satisfy the following chain of inclusions

$$\mathrm{extremely}(\equiv, m) \sqsubseteq \mathrm{very}(\equiv, m) \sqsubseteq m \sqsubseteq \mathrm{more\_or\_less}(\equiv, m) \sqsubseteq \mathrm{roughly}(\equiv, m).$$

As before, both operations, the residual and the composition, can be based on a t-norm like operation defined on L.

3.2 LFSQL

The language LFSQL is inspired by SQL resp. FSQL [5–7]. In this paper we will define and investigate the CREATE, the INSERT, the DELETE and a basic SELECT statement. But first we study different comparison operations for L-fuzzy sets induced by a binary comparator on elements.

***L*-Fuzzy Comparators:** A binary comparator C such as $=$, $\leqslant$ or $<$ compares two elements of a given set A, i.e., is a binary relation $C : A \to A$. If we want to compare L-fuzzy sets using C we have to lift the comparison from elements to sets. This can be done in at least two ways. Given a binary comparator C we will follow the notation in [7] and denote the possibility (fuzzy) comparison based on C by FC and the necessity (fuzzy) comparison based on C by NFC. We want to motivate both constructions by using regular sets instead of fuzzy sets. Assume that the age of John is modeled by the set $\{19, 20, 21\}$, i.e., John could be 19 or 20 or 21. Furthermore, assume that we consider any age 20 and below as young, i.e., the set *Young* is $\{0, 1, \ldots, 19, 20\}$. If we want to know whether John is young, we have to compare John's age with the set *Young*. A possibility comparison would ask the following question: Is it possible for John to have an age that is considered as *Young*? In our example this is true since John could be 19 (or 20) which is in the set *Young*. On the other hand, a necessity comparison would ask the question: Are all ages that John could possibly have considered as *Young*? This time the answer is no because John could be 21 which is not in the set *Young*. Notice that a necessity comparator need not to be symmetric even if the underlying binary comparator C is. In [5] it is mentioned that possibility comparators are more general than necessity comparators, i.e., a necessity comparison retrieves less tuples than a possibility comparison. We will show in the next section that this is true if all fuzzy sets are total. In FSQL all fuzzy sets are trapezoidal and, hence, normalized. The latter is equivalent to total in the case that $L = [0 \ldots 1]$.

The CREATE Statement: The CREATE statement creates a new (empty) table based on the attributes provided. With every attribute the user has to specify the domain used for this attribute. Notice that an empty table has no row. The general form of the create statement is

$$\text{CREATE TABLE } R(A_1 : D_1, \ldots, A_n : D_n);$$

where R is a new name for the table, $A_1, \ldots, A_n$ are attribute names and $D_1, \ldots, D_n$ are domains. Syntactic conditions for this statement are that the table name R must be new and that all domains are defined. The attributes $A_1, \ldots, A_n$ are local to the table. With $R.A_i$ we refer to the column of R corresponding to the attribute A_i. If A_i is unique in the context given, we may drop the prefix R.

The INSERT Statement: The INSERT statement adds a new row to an already existing table. The syntax is

$$\text{INSERT INTO } R \text{ VALUES } (m_1, \ldots, m_n);$$

where R is the name of a table already defined with attributes $A_1, \ldots, A_n$ and corresponding domains $D_1, \ldots, D_n$ and $m_1, \ldots, m_n$ are L-fuzzy subsets of $D_1, \ldots, D_n$, respectively. The L-fuzzy sets can be defined within the statement as introduced above or refer to an already defined set in the meta-database using a linguistic label.

The WHERE Clause: A basic comparison in the language is of the form $S\ LFC\ S'$ where S, S' are either references to attributes of the form $R.A$ or an L-fuzzy set and LFC is an L-fuzzy comparator of the form FC or NFC for a binary comparison C. As before an L-fuzzy set can be defined within the comparison or refer to the meta-database. Comparisons are only considered syntactically correct if the domain used by S and S' and of C are equal. A comparison always returns the degree up to which the statement is true.

Multiple comparisons can be combined by logical connectives. A logical connective is either the keyword AND or OR refereing to the meet and join operation of L. Alternatively, each version can be based on a t-norm like or t-conorm like operation, respectively. The result of such an expression is an element of L that is obtained by applying the logical connectives to the elements returned by the individual comparisons.

In addition, any comparison Com, either basic or combined, can optionally be equipped by a threshold, i.e., it can have the form Com THOLD l, where l is an element from L. This statement returns the degree of Com if it is greater than or equal to l, 0 otherwise.

A WHERE clause consists of the keyword WHERE followed by a comparison. Some examples are:

$$\text{WHERE } R.Age\ F{=}\ \$Young \text{ THOLD } l \text{ AND } R.Height\ NF{<}\ \$Tall$$
$$\text{WHERE } (R.Age\ F{=}\ S.Age \text{ OR}(*)\ R.Height\ F{>}\ \$Short) \text{ THOLD } l$$

The DELETE Statement: The DELETE statement deletes tuples from a table. It has the form

$$\text{DELETE FROM } R \text{ WHERE wh;},$$

where R is the name of a table already defined and wh is a WHERE clause. This statement deletes all tuples from R for which the degree returned by the WHERE clause wh is not zero.

The SELECT Statement: A basic SELECT statement returns a new table created from a number of given tables. From each table one row is selected in order to create a row of the result. Each column of this new row takes its value from the corresponding column in one of the old tables. From this new table only the rows that satisfies the conditions of the WHERE clause in the SELECT statement (degree not equal to 0) are kept. A basic SELECT statement has the form

$$\text{SELECT } S_1, \ldots, S_m \text{ FROM } R_1, \ldots, R_n \text{ WHERE wh;},$$

where $S_1, \ldots, S_m$ are selections of attributes from the tables $R_1, \ldots, R_n$. A SELECT statement is syntactically correct if the selections uniquely identify the table from which the attribute is selected. Notice that the tables need not to be table names. They can be tables recursively generated by nested SELECT statements.

Example: In this example, we will use the lattice D_6 from Section 2. Recall that D_6 can be used to model two aspects of membership. As outlined in the introduction we will rate features of TV's such as screen size with respect to the price (*yes-no*) and the viewer experience (*yes-no-maybe*) using D_6. Some information about TV's is known (or precise) which we indicate by a crisp value in the table as an abbreviation for the corresponding crisp D_6-fuzzy set. Some other information about certain TV's is unknown or imprecise. Notice that in the eaxmple we also use a weakening modifier on the crisp set given by the value 60. In the following we have listed the two D_6 fuzzy sets \$Big and \$Small of screen sizes contained in the meta-database, the table of TVs (left-side) and a SELECT statement and its resulting table (right-side):

$$\begin{aligned} \$\text{Big} &= \{0/30in, 0/32in, 0/39in, a/40in, c/42in, b/50in, b/55in, b/58in, d/60in\} \\ \$\text{Small} &= \{a/30in, a/32in, a/39in, a/40in, c/42in, b/50in, 0/55in, 0/58in, 0/60in\} \end{aligned}$$

Brand	Screen	Weight
Suni	\$Big	75
LB	50in	roughly(60)
SIMSANG	\$Small	60

Brand	Screen	Weight
Suni	\$Big	75
LB	50in	roughly(60)

SELECT Brand, Screen, Weight FROM TVs
WHERE Size $NF\geqslant$ 40in THOLD b AND Weight $F=$ roughly(60);

The first TV satisfies the size condition because all values in \$Big are included in the elements greater or equal 40in (with degree 1). The last TV does not satisfy this condition because its screen size can be 30in with degree a which is not included in the set or equal 40in (with degree 1), i.e., its screen size is not necessarily big.

A Note on Inner Joins: Joins are used to combine two table into one within a SELECT statement. A basic join clause has the form

$$R_1 \text{ INNER JOIN } R_2 \text{ ON } R_1.A_i = R_2.B_j,$$

where R_1 and R_2 are tables and A_i and B_j are attributes of R_1 and R_2, respectively, with the same domain. This clause is equivalent to a SELECT statement. If R_1 has attributes $A_1, \ldots, A_m$ and R_2 has attributes $B_1, \ldots, B_n$, then the statement

$$\begin{aligned} \text{SELECT } & R_1.A_1, \ldots, R_1.A_{i-1}, R_1.A_{i+1}, \ldots, R_1.A_m, \\ & R_2.B_1, \ldots, R_2.B_{j-1}, R_2.B_{j+1}, \ldots, R_2.B_n \\ \text{FROM } & R_1, R_2 \text{ WHERE } R_1.A_i \; F{=} \; R_2.B_j;, \end{aligned}$$

computes the inner join above.

3.3 Semantics of LFSQL

In this section we want to provide a semantics of LFSQL in an arrow category $\mathcal{A}$. We will require that all injections, projections, and splittings used are crisp relations. This does not constitute a major restriction since crisp versions of these relational construction do exist in most cases [18, 20]. In order to provide an adequate semantics, in particular, in order to model domains, the lattice L and the meta-database, we require the following items:

1. $\mathcal{A}$ is an arrow category with (crisp) relational products, sums, splittings, a zero object and a unit.
2. The complete Heyting algebra of scalar elements of $\mathcal{A}$ is isomorphic to L. In particular, for every $l \in L$ we have a scalar $I(l)$ in $\mathcal{A}$.
3. For every domain D we have an object $I(D)$ of $\mathcal{A}$, and for every element $d \in D$ we have a crisp point $I(d) : 1 \to I(D)$. In addition, we have:
   (a) If D is ordered, then we have a crisp order relation $I(\leqslant) : I(D) \to I(D)$ so that $d \leqslant d'$ iff $I(d); I(\leqslant); I(d')^{\smile} = \mathbb{T}_{11}$.
   (b) If D has an approximate equality $\equiv$, then we have a relation $I(\equiv) : I(D) \to I(D)$ so that $d_1 \equiv d_2 = l$ iff $I(d_1); I(\equiv); I(d_2)^{\smile} = I(l)$.

Notice that the isomorphism between L and the scalar element of $\mathcal{A}$ implies that for every t-norm $*$ (resp. t-conorm) like operation in the meta-database there is a corresponding operation on the scalars. In the remainder of this section we will identify both operations.

If n is a natural number we will write $I(n)$ for the object $\underbrace{1 + \cdots + 1}_{n-\text{times}}$. Notice that $I(0) = 0$ and that $I(m \cdot n)$ is isomorphic to $I(m) \times I(n)$ because products distribute over sums. We will identify these objects.

Semantics of *L*-Fuzzy Sets: If m is an L-fuzzy subset of D in LFSQL, then the semantics of m is a vector $[\![m]\!] : 1 \to I(D)$. We model the part of the meta-database storing L-fuzzy sets by a function σ_s that maps a name $\$m$ of an L-fuzzy subset of the domain D to a relation $\sigma_s(\$m) : 1 \to I(D)$. Given such a function we define the semantics of basic L-fuzzy sets by:

$$\begin{aligned}
[\![\$m]\!](\sigma_s) &= \sigma_s(\$m), \\
[\![\{l_1/d_1, \dots, l_n/d_n\}]\!](\sigma_s) &= \bigsqcup_{i=1}^{n} (I(l_i); I(d_i)), \\
[\![\#f]\!](\sigma_s) &= \bigsqcup_{d \in D} (I(f(d)); I(d)).
\end{aligned}$$

Relation algebraically intensifying modifiers are computed using residuals, and weakening modifiers are computed using composition [20]. Therefore, we define

$$\begin{aligned}
[\![\text{extremely}(\equiv, m)]\!](\sigma_s) &= ([\![m]\!](\sigma_s)/I(\equiv))/I(\equiv), \\
[\![\text{very}(\equiv, m)]\!](\sigma_s) &= [\![m]\!](\sigma_s)/I(\equiv), \\
[\![\text{more_or_less}(\equiv, m)]\!](\sigma_s) &= [\![m]\!](\sigma_s); I(\equiv), \\
[\![\text{roughly}(\equiv, m)]\!](\sigma_s) &= [\![m]\!](\sigma_s); I(\equiv); I(\equiv),
\end{aligned}$$

If a t-norm like operation is used, then the residual resp. the composition based on that operation is used instead.

Semantics of Tables and Databases: If a table R has r rows and attributes $A_1, \dots, A_n$ with domains $D_1, \dots, D_n$, then the semantics of a table is a relation $[\![R]\!] : I(r) \to I(D_1) + \cdots + I(D_n)$. Notice that n-ary sums are obtained by iterating binary sums. We will denote the injection from $I(D_i)$ into $I(D_1) + \cdots + I(D_n)$ by ι_i. This interpretation, and the target object of it in particular, needs some explanation. Let us first consider the non-fuzzy case. If R is a table in a regular (non-fuzzy) database, then R can be seen as a finite set of tuples, i.e., as a finite subset of $D_1 \times \cdots \times D_n$. Relation algebraically this can be modeled by either a point $[\![R]\!] : 1 \to \mathcal{P}(I(D_1) \times \cdots \times I(D_n))$, where $\mathcal{P}(X)$ is an abstract version of a power set construction, or a vector $[\![R]\!] : 1 \to I(D_1) \times \cdots \times I(D_n)$ or a function $[\![R]\!] : I(r) \to I(D_1) \times \cdots \times I(D_n)$ since we deal with finite sets. If the attributes store sets (fuzzy or non-fuzzy ones), then the target object within the last option becomes $\mathcal{P}(I(D_1)) \times \cdots \times \mathcal{P}(I(D_n))$. It is well-known that this object is isomorphic to $\mathcal{P}(I(D_1) + \cdots + I(D_n))$ [15, 16]. Last but not least, having a function of the form $[\![R]\!] : I(r) \to \mathcal{P}(I(D_1) + \cdots + I(D_n))$ is equivalent to having a relation of the form $[\![R]\!] : I(r) \to I(D_1) + \cdots + I(D_n)$.

With the interpretation above, projecting to an attribute A_i in a table R becomes the converse of an injection, i.e., we have $[\![R.A_i]\!] = [\![R]\!]; \iota_i^{\smile}$.

The semantics of a database is given by a function σ_t that maps table names to the semantic of the table, i.e., with the conventions above we have $\sigma_t(R) : I(r) \to I(D_1) + \cdots + I(D_n)$. As usual, we denote by $\sigma_t[Q/R]$ the update of σ_t at R by the relation Q.

Semantics of L-Fuzzy Comparators: It is well-known that comparators as described in the previous section can be computed using composition (possibility) and the residual (necessity). We want to illustrate this again by the previous example. If we denote the vectors *John.Age* by J and *Young* by Y, then we have:

$$
\begin{aligned}
&John.Age\ F{=}\ Young\\
&= J;Y^{\smile}\\
&= \overset{\begin{matrix}0 & 1 & \cdots & 18 & 19 & 20 & 21 & 22 & \cdots\end{matrix}}{\begin{pmatrix}0 & 0 & \cdots & 0 & 1 & 1 & 1 & 0 & \cdots\end{pmatrix}}; \overset{\begin{matrix}0 & 1 & \cdots & 18 & 19 & 20 & 21 & 22 & \cdots\end{matrix}}{\begin{pmatrix}1 & 1 & \cdots & 1 & 1 & 1 & 0 & 0 & \cdots\end{pmatrix}}^{\smile}\\
&= (1)\\
&John.Age\ NF{=}\ Young\\
&= (Y/J)^{\smile}\\
&= \left(\overset{\begin{matrix}0 & 1 & \cdots & 18 & 19 & 20 & 21 & 22 & \cdots\end{matrix}}{\begin{pmatrix}1 & 1 & \cdots & 1 & 1 & 1 & 0 & 0 & \cdots\end{pmatrix}} / \overset{\begin{matrix}0 & 1 & \cdots & 18 & 19 & 20 & 21 & 22 & \cdots\end{matrix}}{\begin{pmatrix}0 & 0 & \cdots & 0 & 1 & 1 & 1 & 0 & \cdots\end{pmatrix}}\right)^{\smile}\\
&= (0)
\end{aligned}
$$

As already shown in [20] the operations generalize to the L-fuzzy case as expected. The composition (possibility) operation will compute the least upper bound of all values obtained as the degree of an element belonging to both sets, i.e., we have

$$v;w^{\smile} = \bigsqcup_{x\in A}(v(x) \sqcap w(x)).$$

The residual (necessity) operation computes the greatest lower bound of all values obtained as the maximal degree of which an element belongs to the first set implies that it belongs also to the second set, i.e., we have

$$w/v = \bigsqcap_{x\in A}(v(x) \to w(x)).$$

If the underlying binary comparison is not =, then the corresponding relation has to be added in the composition resp. in the residual (see below).

We now define the semantics of a comparison $S\ LFC\ S'$. Recall that LFC is either a possibility or a necessity comparator based on C and that S, S' are either selections of the form $R.A_i$ or an L-fuzzy sets. We define the semantics $[\![S]\!](\sigma_s,\sigma_t) : I(r) \to I(D_i)$ of a selection S by

$$
\begin{aligned}
[\![R.A_i]\!](\sigma_s,\sigma_t) &= [\![R.A_i]\!](\sigma_t) = \sigma_t(R);\iota_i^{\smile},\\
[\![m]\!](\sigma_s,\sigma_t) &= \pi_{I(n)1};[\![m]\!](\sigma_s).
\end{aligned}
$$

Based on this definition the semantics $[\![S\ LFC\ S']\!](\sigma_s,\sigma_t) : I(r) \to I(r)$ of a comparison $S\ LFC\ S'$ is defined by

$$
\begin{aligned}
[\![S\ FC\ S']\!](\sigma_s,\sigma_t) &= [\![S]\!](\sigma_s,\sigma_t);I(C);[\![S]\!](\sigma_s,\sigma_t)^{\smile} \sqcap \mathbb{I}_{I(r)},\\
[\![S\ NFC\ S']\!](\sigma_s,\sigma_t) &= (([\![S']\!](\sigma_s,\sigma_t);I(C)^{\smile})/[\![S]\!](\sigma_s,\sigma_t))^{\smile} \sqcap \mathbb{I}_{I(r)}.
\end{aligned}
$$

Notice that the semantics of a comparison is a partial identity, i.e., a relation smaller or equal that $\mathbb{I}_{I(r)}$.

In addition, from $X \sqsubseteq ((R;C^{\smile})/Q)^{\smile}$ we obtain $X^{\smile};Q \sqsubseteq R;C^{\smile}$, and, hence, $X \sqsubseteq Q;Q^{\smile};X \sqsubseteq Q;C;R^{\smile}$ if Q is total. This verifies $((R;C^{\smile})/Q)^{\smile} \sqsubseteq Q;C;R^{\smile}$, i.e., that possibility comparisons are more general than necessity comparisons as mentioned in Section 3.

Semantics of the WHERE Clause: Combining the semantics of individual comparisons by AND or OR is simply based on $\sqcap$ and $\sqcup$, respectively. If a t-norm like or a t-conorm like operation $*$ is used for AND resp. OR, we use $\sqcap_*$. Notice that even in the case of OR we use $\sqcap_*$ (and not $\sqcup_*$). The usage of the basic relation algebraic operation $\sqcap$ in Definition 7 just guarantees a component-wise application of $*$. In fact, if we restrict $\sqcup$ to scalar relations and use it as a t-conorm like operation, then we obtain $\sqcap_\sqcup = \sqcup$ [20].

Finally, a threshold is modeled by an α-cut, i.e.,

$$[\![Com \text{ THOLD } l]\!](\sigma_s,\sigma_t) = (I(l)\backslash[\![Com]\!](\sigma_s,\sigma_t))^{\downarrow}.$$

Since partial identities are closed under meets, joins (including the t-norm and/or t-conorm based versions) and α-cuts, the semantics of a WHERE clause is also a partial identity.

Semantics of Statements: The semantics of an L-fuzzy database is given by a function σ_t that maps table names to the semantics of the table as described in the previous section. As usual we denote by $\sigma_t[Q/R]$ the update of σ_t at table name R by the relation Q.

The semantics of the create statement modifies σ_t by adding a new (empty) relation:

$$[\![\text{CREATE TABLE } R(A_1 : D_1,\ldots,A_n : D_n);]\!](\sigma_t) = \sigma_t[\perp\!\!\!\perp_{0I(D_1)+\cdots+I(D_n)}/R].$$

If $(m_1,\ldots,m_n)$ are L-fuzzy sets we can define the semantics of the tuple given by those sets by $[\![(m_1,\ldots,m_n)]\!](\sigma_s) = \bigsqcup_{i=1}^{n}[\![m_i]\!](\sigma_s);\iota_i$. With this notion we obtain from the following diagram the semantics $[\![R(m_1,\ldots,m_n)]\!](\sigma_s,\sigma_t)$ of the table in which the tuple $(m_1,\ldots,m_n)$ has been added to R, i.e., we have $[\![R(m_1,\ldots,m_n)]\!](\sigma_s,\sigma_t) = \iota^{\smile};\sigma_t(R) \sqcup \kappa^{\smile};[\![(m_1,\ldots,m_n)]\!](\sigma_s)$.

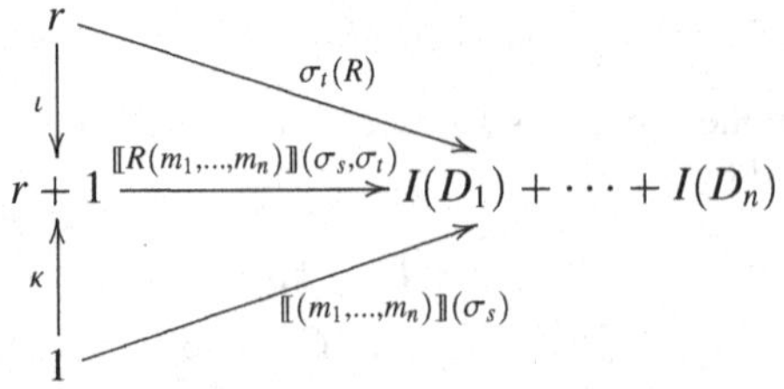

Based on these definitions the semantics of an INSERT statement is as follows:

$$\begin{aligned}&[\![\text{INSERT INTO } R \text{ VALUES } (m_1,\dots,m_n);]\!](\sigma_s,\sigma_t)\\&= \sigma_t[[\![R(m_1,\dots,m_n)]\!](\sigma_s,\sigma_t)/R].\end{aligned}$$

In the semantics of the DELETE statement we have to filter the rows that satisfy the condition imposed by the WHERE clause. This can be done by using the splitting of its semantics $[\![\mathrm{wh}]\!](\sigma_s,\sigma_t)^{\uparrow}$. In order to see that the object of this splitting is again the interpretation of a natural number define $A = \{i \in \{1,\dots,r\} \mid \iota_i;X;\iota_i^{\smile} \neq \mathbb{⊥}_{11}\}$ where X is an abbreviation for $[\![\mathrm{wh}]\!](\sigma_s,\sigma_t)^{\uparrow}$. Now we want to show that the object $I(|A|)$, i.e., the interpretation of the cardinality of A, together with the relation $S = \bigsqcup_{i\in A}\iota_i^{\smile};\iota_i : I(|A|) \to I(r)$ is a splitting of X. The property $S;S^{\smile} = \mathbb{I}_{I(|A|)}$ follows immediately from the properties of relational sums. In order to show that $S^{\smile};S = X$ we first notice that $\iota_i;X;\iota_i^{\smile}$ is crisp since ι_i and X are crisp, and hence either equal to $\mathbb{⊥}_{11}$ or $\mathbb{⊤}_{11}$ because all relations with source and target 1 are scalars. Furthermore, for $i \neq j$ we have $\iota_i;X;\iota_j^{\smile} \sqsubseteq \iota_i;\iota_j^{\smile} = \mathbb{⊥}_{11}$. We conclude

$$\begin{aligned}X &= (\bigsqcup_{i=1}^{n}\iota_i^{\smile};\iota_i);X;(\bigsqcup_{i=1}^{n}\iota_i^{\smile};\iota_i) = \bigsqcup_{i,j=1}^{n}\iota_i^{\smile};\iota_i;X;\iota_j^{\smile};\iota_j\\
&= \bigsqcup_{i=1}^{n}\iota_i^{\smile};\iota_i;X;\iota_i^{\smile};\iota_i && \text{see above}\\
&= \bigsqcup_{i\in A}\iota_i^{\smile};\iota_i && \iota_i;X;\iota_i^{\smile} = \begin{cases}\mathbb{⊥}_{11} & \text{iff } i \notin A\\ \mathbb{⊤}_{11} = \mathbb{I}_1 & \text{iff } i \in A\end{cases}\\
&= \bigsqcup_{i,j\in A}\iota_i^{\smile};\iota_i;\iota_j^{\smile};\iota_j && \text{relational sums}\\
&= (\bigsqcup_{i\in A}\iota_i^{\smile};\iota_i);(\bigsqcup_{j\in A}\iota_j^{\smile};\iota_j) = S;S^{\smile}.\end{aligned}$$

Now, we define

$$[\![\text{DELETE FROM } R \text{ WHERE wh};]\!](\sigma_s,\sigma_t) = \sigma_t[S;\sigma_t(R)/R].$$

In order to provide the semantics of a SELECT statement of the form

$$\text{SELECT } S_1,\dots,S_m \text{ FROM } R_1,\dots,R_n \text{ WHERE wh};$$

we first have to generate the new table that consists of the attributes $S_1,\dots,S_n$ from the tables $R_1,\dots,R_m$. Suppose that $S_{j_1},\dots,S_{j_{k_i}}$ are the selections that are associated with table R_i. If we denote the semantics of the table R_i reduced to the attributes $S_{j_1},\dots,S_{j_{k_i}}$ by Q_i, then we have

$$Q_i = \bigsqcup_{l=1}^{k_i}[\![R_i.S_{j_l}]\!](\sigma_s,\sigma_t);\iota_l : I(r_i) \to I(D_{j_1}) + \cdots + I(D_{j_{k_i}}).$$

With this abbreviation we obtain the semantics of the intermediate table as

$$Q = \bigsqcup_{i=1}^{n} \pi_i; Q_i; \iota_i : I(\prod_{i=1}^{n} r_i) \to \sum_{i=1}^{n} I(D_{j_1}) + \cdots + I(D_{j_{i_1}}).$$

The semantics of the SELECT statement filters the rows that satisfy the WHERE clause from this intermediate table similar to the DELETE statement. If $S : I(r') \to I(\prod_{i=1}^{n} r_i)$ splits the relation $[\![\mathrm{wh}]\!](\sigma_s, \sigma_t)^{\uparrow}$, then we define the semantics of the SELECT statement by

$$[\![\text{SELECT } S_1, \ldots, S_m \text{ FROM } R_1, \ldots, R_n \text{ WHERE wh; }]\!](\sigma_s, \sigma_t) = S; Q.$$

4 Conclusion and Future Work

In this paper we have introduced *L*-fuzzy databases and the query language LFSQL. An abstract semantics in terms of arrow categories was also provided. A prototype implementation using the programming language Haskell that is based on the formal semantics has been developed. Further details on the language and its implementation can be found in [1, 3].

There are multiple ways to extend the work presented in this paper. We want to outline three of them. The language can be extended to include more features that are useful in practical applications. For example, FSQL has a special function CDEG that applies to attributes and computes the compatibility degree of conditions involving these attributes. This value can also be added to the list of selections in a SELECT statement. The semantics of this component is immediate since the compatibility degree is already available in the semantics of the WHERE clause. Further extensions include fuzzy quantifiers, joins, additional comparisons and ORDER, HAVING and GROUP clauses.

Another option is a thorough investigation of functional dependencies based on the semantics provided here. The fuzzy case allows for several generalizations of the notion of a functional dependency. Each of them can be investigated including correct and complete set of axioms for them.

Last but not least, several methods of data mining could be explored in the context of *L*-fuzzy databases.

References

1. Adjei, E.: L-Fuzzy Structural Query Language. Brock University (M.Sc. Thesis), tbp (2015)
2. Birkhoff, G.: Lattice Theory, 3rd edn., vol. XXV. American Mathematical Society Colloquium Publications (1967)
3. Chowdhury, W.: An Abstract Algebraic Theory of L-Fuzzy Relations in Relational Databases. Brock University (M.Sc. Thesis), tbp (2015)
4. Freyd, P., Scedrov, A.: Categories, Allegories. North-Holland (1990)

5. Galindo, J., Medina, J.M., Pons, O., Cubero, J.C.: A Server for Fuzzy SQL Queries. In: Andreasen, T., Christiansen, H., Larsen, H.L. (eds.) FQAS 1998. LNCS (LNAI), vol. 1495, pp. 164–174. Springer, Heidelberg (1998)
6. Galindo, J.: New Characteristics in FSQL, a Fuzzy SQL for Fuzzy Databases. WSEAS Transactions on Information Science and Applications 2(2), 161–169 (2005)
7. Galindo, J., Urrutia, A., Piattini, M.: Fuzzy Databases: Modeling, Design and Implementation. Idea Group Publishing Hershey, USA (2006)
8. Goguen, J.A.: L-fuzzy Sets. J. Math. Anal. Appl.1̃8, 145–157 (1967)
9. Jónsson, B., Tarski, A.: Boolean Algebras with Operators, I, II. Amer. J. Math.7̃3, 891–939 (1951); 74, 127–162 (1952)
10. Kawahara, Y., Furusawa, H.: Crispness and Representation Theorems in Dedekind Categories. DOI-TR 143, Kyushu University (1997)
11. Olivier, J.P., Serrato, D.: Catégories de Dedekind. Morphismes dans les Catégories de Schröder. C.R. Acad. Sci. Paris 290, 939–941 (1980)
12. Olivier, J.P., Serrato, D.: Squares and Rectangles in Relational Categories - Three Cases: Semilattice, Distributive lattice and Boolean Non-unitary. Fuzzy Sets and Systems 72, 167–178 (1995)
13. Schmidt, G., Hattensperger, C., Winter, M.: Heterogeneous Relation Algebras. In: Brink, C., Kahl, W., Schmidt, G. (eds.) Relational Methods in Computer Science. Advances in Computer Science. Springer, Vienna (1997)
14. Schmidt, G., Ströhlein, T.: Relationen und Graphen. Springer (1989); English version: Relations and Graphs. Discrete Mathematics for Computer Scientists, EATCS Monographs on Theoret. Comput. Sci., Springer (1993)
15. Schmidt, G.: Relational Mathematics. Encyplopedia of Mathematics and Its Applications 132 (2011)
16. Winter, M.: Strukturtheorie heterogener Relationenalgebren mit Anwendung auf Nichtdetermismus in Programmiersprachen. Dissertationsverlag NG Kopierladen GmbH, München (1998)
17. Winter, M.: A new Algebraic Approach to *L*-Fuzzy Relations Convenient to Study Crispness. INS Information Science 139, 233–252 (2001)
18. Winter, M.: Relational Constructions in Goguen Categories. In: de Swart, H. (ed.) RelMiCS 2001. LNCS, vol. 2561, pp. 212–227. Springer, Heidelberg (2002)
19. Winter, M.: Derived Operations in Goguen Categories. TAC Theory and Applications of Categories 10(11), 220–247 (2002)
20. Winter, M.: Goguen Categories - A Categorical Approach to *L*-fuzzy relations. Trends in Logic, vol. 25. Springer (2007)
21. Winter, M.: Arrow Categories. Fuzzy Sets and Systems 160, 2893–2909 (2009)
22. Zadeh, L.A.: Fuzzy Sets. Information and Control 8, 338–353 (1965)

Text Categorization Using Hyper Rectangular Keyword Extraction: Application to News Articles Classification

Abdelaali Hassaine, Souad Mecheter, and Ali Jaoua

Computer Science and Engineering Department
College of Engineering, Qatar University, Doha, Qatar
{hassaine,200656336,jaoua}@qu.edu.qa

Abstract. Automatic text categorization is still a very important research topic. Typical applications include assisting end-users in archiving existing documents, or helping them in browsing existing corpus of documents in a hierarchical way. Text categorization is usually composed of two main steps: keyword extraction and classification. In this paper, a corpus of documents is represented by a binary relation linking each document to the words it contains. From this relation, the Hyper Rectangle Algorithm extracts the list of the most representative words in a hierarchical way. A hyper-Rectangle associated to an element of the range of a binary relation is the union of all non-enlargeable rectangles containing it. The extracted keywords are fed into the random forest classifier in order to predict the category of each document. The method has been validated on the popular Reuters 21578 news articles database. Results are very promising and show the effectiveness of the Hyper Rectangular method in extracting relevant keywords.

Keywords: Hyper Rectangular Coverage, Text categorization, Random forests.

1 Introduction

With the exponential increase of Internet content, automatic text classification has become a very active research area. Automatic text classification has many applications including news articles classification, structuring of large online corpora, spam emails filtering, webpage classification, anomaly detection and authentication.

In this paper, a corpus is initially represented by a binary relation $\mathcal{R}$ linking each document (i.e. objects) to the words it contains (i.e. attributes). Naturally, words are associated to several concepts with respect to some context represented here by the binary relation $\mathcal{R}$. As for example, in the working environment the word "good" may be associated to the concept of students, instructor, or administrators. In this paper, we assimilated a real concept to a non-enlargeable rectangle in a relation (i.e. maximal rectangle or formal concept). Therefore, a

W. Kahl et al. (Eds.): RAMiCS 2015, LNCS 9348, pp. 312–325, 2015.
DOI: 10.1007/978-3-319-24704-5_19

word w is represented by the union $HR(w)$ of the set of non-enlargeable rectangles involving it. The relation $HR(w)$ is called hyper-rectangle associated to word w. The weight of a word w is calculated by a metric applied on the hyper-rectangle associated to w. Very often, few hyper-rectangles may cover the entire domain. A multilevel browsing tree of words is derived from each corpus. Different coverage levels of the relation $\mathcal{R}$ are obtained from most generic levels to most specific ones. The last level matches with a rectangular coverage of relation $\mathcal{R}$ (i.e. union of non-enlargeable rectangles covering relation $\mathcal{R}$). Therefore, most relevant representative keywords may be found using the hyper rectangular segmentation method.

The remainder of this paper is structured as follows: Section 2 presents existing methods related to text classification. Section 3 describes the hyper rectangle keyword extraction method. Section 4 describes the classification method used to predict the text categories. Section 5 presents experiments we conducted and section 6 concludes this article and presents some future work perspectives.

2 Related Work

Several studies related to text categorization exist in the literature. In this section we give an overview of the recent approaches and advances in this field.

Jiang et al. proposed a text categorization method based on a modified K-nearest-neighbor algorithm which is combined with a constrained one pass clustering algorithm [13]. The proposed algorithm is also incremental and scalable, but the validation has only been done on a proprietary Chinese database and may need some further validation on public databases in order to be comparable with other approaches. Uğuz et al. proposed an information gain feature selection method for reducing the number of features. Subsequently, genetic algorithm and principal component analysis are applied for feature selection. The authors then apply a k-nearest neighbor and decision trees algorithm to classify the documents [19]. The method is efficient but can be improved by the introduction of some more powerful classifiers or ensemble of classifiers. Yang et al. also proposed a feature selection method in which the importance of each word is comprehensively measured both in inter-category and intra-category [20]. Azam et al. [2] compared term frequency and document frequency based feature selection metrics, they concluded that term frequency metrics are useful especially for small feature sets. Yoshikawa et al. proposed a kernel-based discriminative algorithm for the classification of bag-of-words data [21]. The algorithm achieves 87% on the WebKB database, 94% on the Reuters-21578 database and 60% on the 20 Newsgroups database. Jia et al. proposed a feature voting scheme for text classification [12]. The algorithm achieves 84.6% on the WebKB-4 database and 95.75% on the Reuters-R8 database. Lee et al. proposed an enhanced support vector machine classification framework that uses an Euclidean distance function which is reported to have low impact on the implementation of kernel function and soft margin parameter C [15]. Kurian et al. showed that feature reduction leads generally to an improved performance and that Latent Semantic Indexing is generally preferred over Principal Component Analysis [14]. Cardoso-Cachopo et al. also showed that the combination of several classifiers generally leads to an improved performance over a single classifier [7]. Li et al. proposed an improved version of the back propagation neural network algorithm in order to prevent it from being trapped into a local minimum [17]. A comparative study by Aphinyanaphongs et al. [1] concluded that the optimal classifier depends on the corpora and that there is no global optimal classifier.

To sum up, we can say that existing methods focus either on feature selection approaches or classification approaches. Each developed method needs to be adapted to a specific database in order to reach optimal results. In this study, we propose a feature selection approach based on formal concept analysis which extracts keywords in a hierarchical ordering of importance.

3 Hyper Rectangle Feature Extraction

In this section, first we briefly introduce Relation Algebra and Formal Concept Analysis. For more in-depth introduction about theoretical foundations one may refer to [18,8]. Second, we present the hyper rectangular feature extraction method.

A corpus of documents can be represented by a binary relation R as a subset of the Cartesian product of the set of documents by the set of words. The cardinality of a relation is defined by the number of pairs it contains. The domain of R is the list of documents (or objects) and the codomain of R is the list of words (or attributes). A relation can be characterized by its cardinality r, the cardinality of its domain d and the cardinality of its codomain c.

Formal Concept Analysis (FCA) [9] is a mathematical theory of data analysis using formal contexts and concept lattices. It was introduced by Rudolf Wille in 1984, and builds on applied lattice and order theory that were developed by Birkhoff et al. [3].

Definition 1. *A formal context (or an extraction context) is a triplet* $K = (X, Y, \mathcal{R})$*, where* X *represents a finite set of objects,* Y *a finite set of attributes (or properties) and* $\mathcal{R}$ *is a binary (incidence) relation (i.e., as a subset of the Cartesian product of the set of documents and the set of words). Each couple* $(x, a) \in \mathcal{R}$ *expresses that the object* $x \in X$ *has the attribute* $a \in Y$*.*

Definition 2. *The image of* x *by the relation* $\mathcal{R}$ *is defined as* $x.\mathcal{R} = \{a \in Y | (x, a) \in \mathcal{R}\}$*.*

The image of a set X *by the relation* $\mathcal{R}$ *is defined as* $X.\mathcal{R} = \bigcup_{x \in X} \{a \in Y | (x, a) \in \mathcal{R}\}$

Definition 3. *The relative product or composition of two binary relations* $\mathcal{R}$ *and* $\mathcal{R}'$ *is* $\mathcal{R} \circ \mathcal{R}' = \{(e, e') | \exists t | ((e, t) \in \mathcal{R})$ *and* $((t, e') \in \mathcal{R}')\}$*.*

Definition 4. *The converse of the relation* $\mathcal{R}$ *is* $\mathcal{R}^{-1} = \{(e, e') | (e', e) \in \mathcal{R}\}$*.*

Definition 5. *The identity relation denoted $\mathcal{I}(A)$ is a binary relation on a set A (i.e., it is a subset of the Cartesian product $A \times A$), such that $\forall a \in A, a.\mathcal{I}(A) = \{a\}$.*

Definition 6. *The cardinality of $\mathcal{R}$ is defined by: $Card(\mathcal{R})$ = the number of pairs $(e, e') \in \mathcal{R}$.*

Definition 7. *Let $\mathcal{R}$ be a binary relation defined between X and Y: A rectangle of $\mathcal{R}$ is a Cartesian product of two non empty sets $A \subseteq X$ and $B \subseteq Y$ and $A \times B \subseteq \mathcal{R}$, where A is the domain (also called objects), and B is the range (also called attributes) of the rectangle. The rectangular closure of a binary relation is: $\mathcal{R}^* = X.\mathcal{R} \times \mathcal{R}.Y$.*
A Rectangle $A \times B \subseteq \mathcal{R}$ is called non enlargeable if: $A \times B \subseteq A' \times B' \subseteq \mathcal{R} \Rightarrow (A = A') \wedge (B = B')$. In terms of formal concept analysis, a non enlargeable rectangle is called a formal concept.

Definition 8. *Let $K = (X, Y, \mathcal{R})$ be a formal context. If A and B are non-empty sets, such that $A \subseteq X$ and $B \subseteq Y$. The pair (A, B) is a formal concept if and only if $A \times B$ is a non-enlargeable rectangle [10].*

Definition 9. *Let $(X, Y, \mathcal{R})$ be a formal context and $a \in Y$ an arbitrary attribute. Let v be a vector such that $v = a.\mathcal{R}^{-1} \times S$, where S is the universal set. The hyper rectangle denoted by $H_a(\mathcal{R})$ is a sub-relation of $\mathcal{R}$ such that $H_a(\mathcal{R}) = \mathcal{R} \cap (v \circ L)$.*
Where L is the universal relation ($L = S \times S$).
$H_a(\mathcal{R})$ may be expressed differently as: $H_a(\mathcal{R}) = \mathcal{I}(a.\mathcal{R}^{-1}) \circ \mathcal{R}$.

The hyper rectangle associated to an element a is the union of all non-enlargeable rectangles containing a. In formal concept analysis, a non-enlargeable rectangle is also called a concept. Therefore a hyper rectangle is also called hyper concept.

Definition 10. *The weight of the hyper rectangle $H_a(\mathcal{R})$ denoted $W(H_a(\mathcal{R}))$ is defined by a generalization of a metric introduced in [11]:*

$$W(H_a(\mathcal{R})) = \frac{r}{d*c} * (r - (d + c)).$$

Wherein r is the cardinality of $H_a(\mathcal{R})$ (i.e. the number of pairs in the binary relation $H_a(\mathcal{R})$), d is the cardinality of its domain, and c is the cardinality of its codomain.

The quantity $\frac{r}{d*c}$ provides a measure of the density of the hyper rectangle, whether the quantity $(r - (d + c))$ is a measure of the economy of information.

The idea behind this metric is that, in the case of a concept having d objects and c attributes, there are $r = d * c$ links (because every object is linked to every attribute). However, instead of storing the whole $d * c$ links, one can store $d + c$ links (d links between the concept and the objects and c links between the concept and the attributes). Therefore, the quantity $r - (d + c)$ represents the

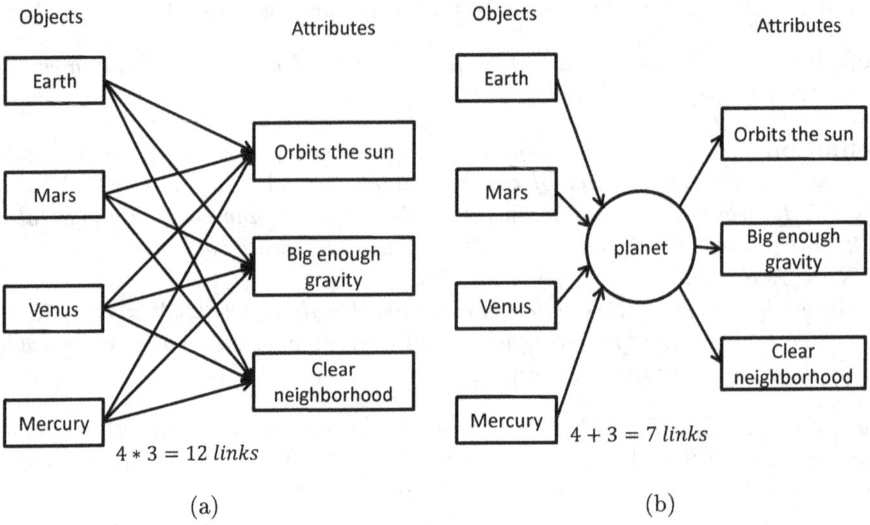

Fig. 1. Illustrating economy of information in terms of number of links. (a) Without considering the concept. (b) After considering a concept.

economy of information or storage space. Figure 1 illustrates this notion for a concept of 4 objects and 3 attributes.

A concept has a density of 1 but a hyper concept has a lower density. The economy of information corresponding to a hyper concept can be approximated by multiplying the quantity $r - (d + c)$ by the density $\frac{r}{d*c}$.

In this proposed metric, generally higher weights reflect more generic words.

Definition 11. *The Optimal Hyper Rectangle denoted* $maxHa(\mathcal{R})$ *is the hyper rectangle which has the maximum weight. That is* $W_a(H_a(\mathcal{R})) \geq W_b(H_b(\mathcal{R})) \forall b \neq a, b \in Cod(\mathcal{R})$.

Figure 2 shows the computation of hyper rectangles associated with each attribute of an illustrative relation of eight documents and eight words. It shows as well the corresponding weight of each hyper rectangle. Notice that the sixth word has the maximum weight.

Definition 12. *The Remaining Binary Relation is the relation* $\mathcal{R}$ *minus the optimal Hyper Rectangle:* $R_m(\mathcal{R}) = \mathcal{R} - maxH_a(\mathcal{R})$. *The remaining relation is useful for splitting a relation into a hierarchy of hyper rectangles.*

Figure 3 illustrates the computation of the remaining relation of the example shown in figure 2.

c=card(codomain)=8

d=card(domain)=8

Original relation

	police	officer	arrest	prison	guilty	man	united	states
a0	1	0	0	0	0	1	0	0
a1	1	0	0	0	0	1	1	0
a2	1	1	0	0	0	1	1	0
a3	0	1	0	0	0	1	1	1
a4	1	0	0	1	1	0	0	0
a5	1	1	1	0	1	0	0	0
a6	0	1	1	1	0	0	0	0
a7	0	1	1	0	1	0	0	0

r=card(R)=26

Hyper rectangle weights

$$W(H(R)) = \frac{r}{d * c} * (r - (d + c))$$

	police	officer	arrest	prison	guilty	man	united
a0	1	0	0	0	0	1	0
a1	1	0	0	0	0	1	1
a2	1	1	0	0	0	1	1
a4	1	0	0	1	1	0	0
a5	1	1	1	0	1	0	0

$$W(H(R)) = \frac{16}{7 * 5} * (16 - (7 + 5)) = 1.83$$

	police	**officer**	arrest	prison	guilty	man	united	states
a2	1	1	0	0	0	1	1	0
a3	0	1	0	0	0	1	1	1
a5	1	1	1	0	1	0	0	0
a6	0	1	1	1	0	0	0	0
a7	0	1	1	0	1	0	0	0

$$W(H(R)) = \frac{18}{8 * 5} * (18 - (8 + 5)) = 2.25$$

	police	officer	**arrest**	prison	guilty
a5	1	1	1	0	1
a6	0	1	1	1	0
a7	0	1	1	0	1

$$W(H(R)) = \frac{10}{5*3} * (10 - (5 + 3)) = 1.33$$

	police	officer	arrest	**prison**	guilty
a4	1	0	0	1	1
a6	0	1	1	1	0

$$W(H(R)) = \frac{6}{5*2} * (6 - (5 + 2)) = -0.6$$

	police	officer	arrest	prison	**guilty**
a4	1	0	0	1	1
a5	1	1	1	0	1
a7	0	1	1	0	1

$$W(H(R)) = \frac{10}{5*3} * (10 - (5 + 3)) = 1.33$$

	police	officer	**man**	united	states
a0	1	0	1	0	0
a1	1	0	1	1	0
a2	1	1	1	1	0
a3	0	1	1	1	1

$$W(H(R)) = \frac{13}{5*4} * (13 - (5 + 4)) = \mathbf{2.6}$$

	police	officer	man	**united**	states
a1	1	0	1	1	0
a2	1	1	1	1	0
a3	0	1	1	1	1

$$W(H(R)) = \frac{11}{5 * 3} * (11 - (5 + 3)) = 2.2$$

	officer	man	united	**states**
a3	1	1	1	1

$$W(H(R)) = \frac{4}{4*1} * (4 - (4 + 1)) = -1$$

Hyper rectangles

Fig. 2. Computing Hyper Rectangles associated with each attribute and their corresponding weight.

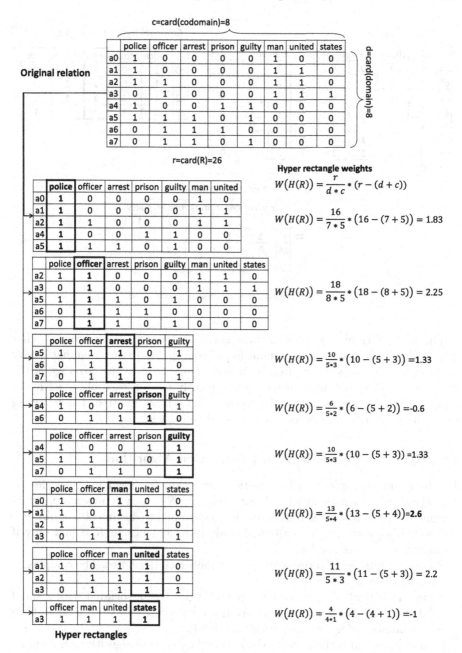

Original relation

	police	officer	arrest	prison	guilty	man	united	states
a0	1	0	0	0	0	1	0	0
a1	1	0	0	0	0	1	1	0
a2	1	1	0	0	0	1	1	0
a3	0	1	0	0	0	1	1	1
a4	1	0	0	1	1	0	0	0
a5	1	1	1	0	1	0	0	0
a6	0	1	1	1	0	0	0	0
a7	0	1	1	0	1	0	0	0

\-

Optimal hyper rectangle

	police	officer	**man**	united	states
a0	1	0	**1**	0	0
a1	1	0	**1**	1	0
a2	1	1	**1**	1	0
a3	0	1	**1**	1	1

=

Remaining relation

	police	officer	arrest	prison	guilty
a4	1	0	0	1	1
a5	1	1	1	0	1
a6	0	1	1	1	0
a7	0	1	1	0	1

Fig. 3. Computing the remaining relation.

The remaining binary relation undergoes the same splitting process. That is, it will be split into the maximum hyper concept and the remaining binary relation. This process is repeated in an iterative way until the coverage of the full original relation. The process is subsequently run on each obtained hyper concept in an iterative way in order to construct a browsing tree. The following subsections describe this process in a more detailed way.

3.1 Hyper Rectangle Keyword Extraction Algorithm

The hyper rectangle keyword extraction algorithm is shown in Algorithm 1. The algorithm takes as input a binary relation $\mathcal{R}$ which represents a list of documents. The lines of the binary relation correspond to documents (or objects) and columns to words (or attributes). Pair (o, a) belongs to $\mathcal{R}$ if and only if document o contains word a.

If m is the number of objects and n the number of attributes. The complexity of this algorithm is $O((mn)^3)$.

Figure 4 illustrates a part of a hyper rectangles tree associated with a small set of documents. Each depth (or level) in the browsing tree will therefore have a certain number of keywords associated with it.

Figure 5 illustrates the steps of the keyword extraction algorithm on an illustrative example. Note that the hyper rectangles keyword extraction algorithm is run on each document category as it will be shown in the next section.

Algorithm 1. Hyper Rectangles Keywords Extraction

- Input: Binary relation $\mathcal{R}$ representing a set of preprocessed files.
- Parameter: max_tree_depth maximum level of the hyper rectangles tree.
- Output: List of keywords.

1: **procedure** ComputeHyperRectangleKeywords($\mathcal{R}$, $list\_of\_keywords$)
2:     $list\_of\_keywords \leftarrow \{\}$
3:     $relations\_queue \leftarrow empty\_queue$
4:     $relations\_queue.Enqueue(\mathcal{R})$
5:     **while** $relations\_queue \neq empty\_queue$ **do**
6:        $par\_relation \leftarrow relations\_queue.Dequeue()$ ▷ parent relation
7:        $current\_relation \leftarrow par\_relation$
8:        **do**
9:           Split $current\_relation$ into $maxHa$ and R_m
10:          $best\_attribute \leftarrow attribute(maxHa)$
11:          $list\_of\_keywords.Add(best\_attribute)$
12:          $child\_relations(par\_relation).Add(maxHa)$
13:          $current\_relation \leftarrow R_m$
14:       **while** $current\_relation$ is not empty
15:       **if** $tree\_depth(par\_relation) < max\_tree\_depth$ **then**
16:          **for all** r in $child\_relations(par\_relation)$ **do**
17:             $relations\_queue.Enqueue(e)$
18:          **end for**
19:       **end if**
20:    **end while**
21:    **return** $list\_of\_keywords$
22: **end procedure**

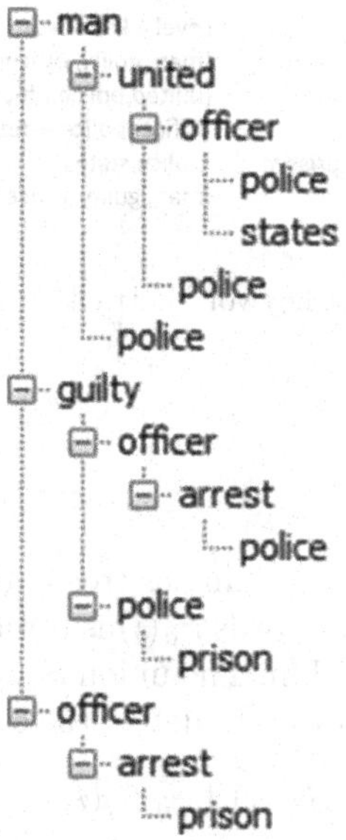

Fig. 4. Hyper Rectangles Tree.

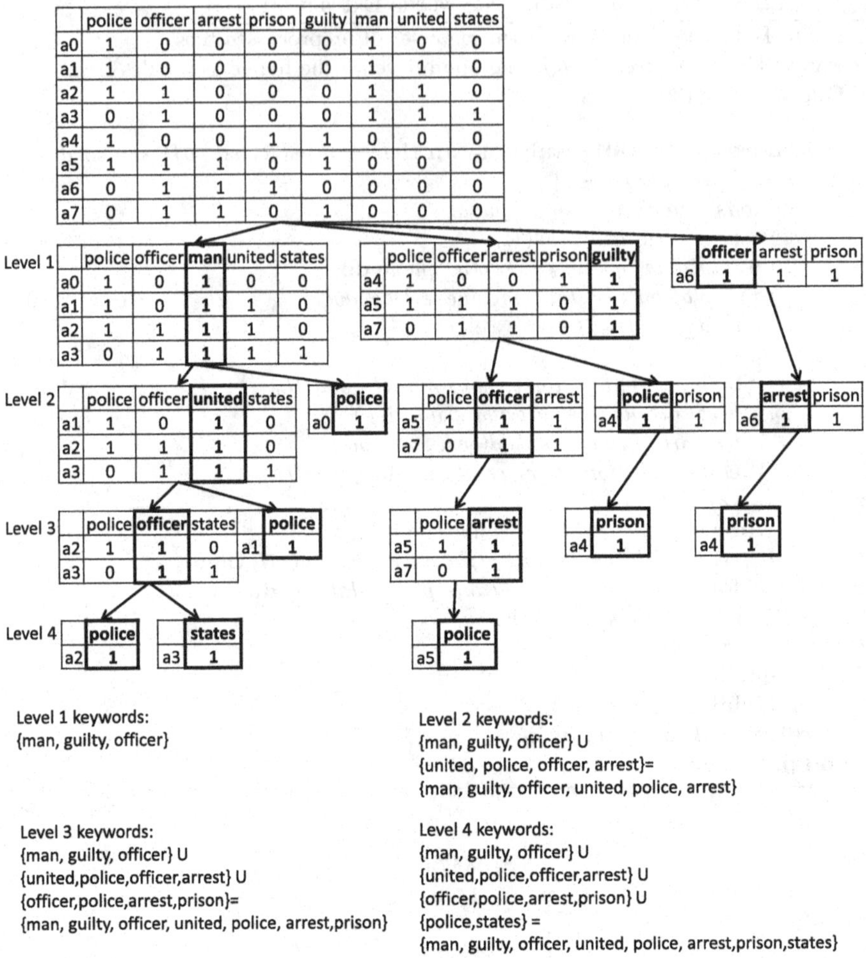

Fig. 5. Running the hyper concept keyword extraction algorithm of a small illustrative relation.

4 Classification

For a given depth d of the hyper rectangles tree and for each category i of documents in a corpus A, a set of keywords $S_d(i)$ is obtained using the hyper rectangular keyword extraction algorithm. The union S_d of those sets is the total set of keywords which will be used in order to detect the category of unseen documents.

$$S_d = \bigcup_{i \in A} S_d(i).$$

The size of S_d increases with the considered depth d of the hyper rectangles tree. Note also that there might be some intersections between the sets corresponding to different categories.

In order to be fed to a classifier, each document o in A is represented as a binary feature vector v in which each feature represents whether or not the document o contains the word w (for all words $w \in S_d$).

Several classification algorithms exist. Some classifiers are known to be efficient but very sensitive to outliers. Some other classifiers are less sensitive to outliers but their performance is very weak. In this study, we used the random forest classifier which is a strong classifier and not very sensitive to outliers [4].

Random forests is an ensemble learning method for classification that operates by constructing a multitude of decision trees at training time and outputting the class that is the mode of the classes output by individual trees. Each decision tree is constructed as follows:

- If the number of cases in the training set is N, sample n cases such as $n < N$ at random but with replacement from the original data. This sample will be the training set for growing the tree.
- If there are M input variables, a number $m < M$ is specified such that at each node, m variables are selected at random from M and the best split on these m variables is used to split the node. Value m is held constant during the forest growing.
- Each tree is grown to the largest extent possible. There is no pruning.

The random forest classification algorithm outputs for each feature vector the most probable category to which it belongs.

5 Experiments

5.1 Database Description

In this paper, we used the Reuters-21578 dataset which is a widely used dataset for text categorization [16]. In order to ensure reproducibility of the results, we used the same experimental setup introduced by Ana Cardoso-Cachopo [5,6]. This database contains 7674 news articles ranging into 8 different categories. 5485 articles are used in the training set and the remaining 2189 articles are used in the test set.

5.2 Results

The keywords extracted from each category in the training set are used as features and fed to a random forest classifier with 1000 random trees. Figure 6 illustrates the correct classification rates for increasing depth of the hyper rectangles tree.

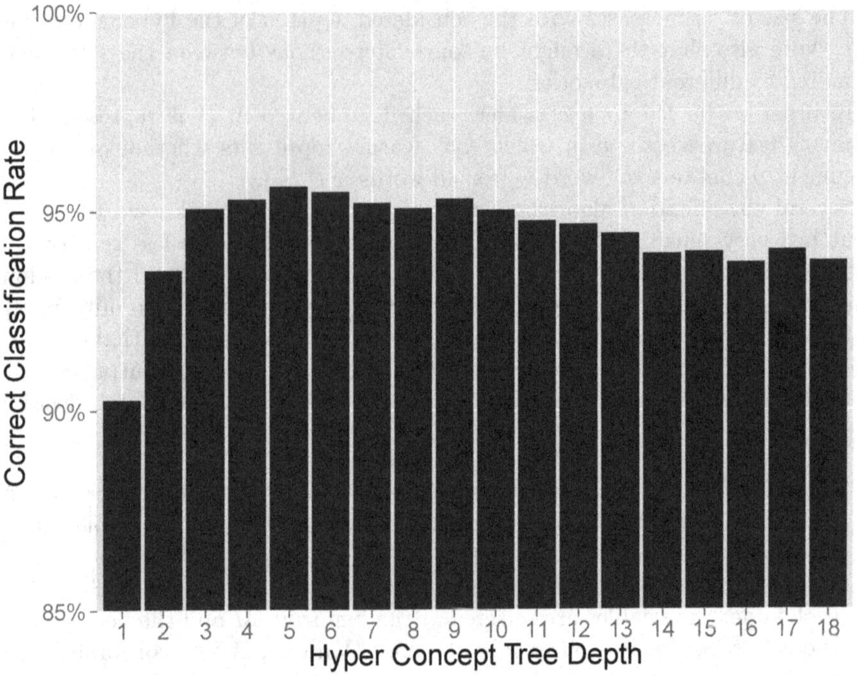

Fig. 6. Classification results for varying hyper rectangles tree depth.

5.3 Discussion

The correct classification rates reaches its maximum 95.61% at depth 5 and then slowly drops because of the large increase of the number of irrelevant keywords introduced at the deep levels of the hyper rectangles tree. Those keywords are not useful for classification because they are not enough generic to represent any document category.

Depth 5 of the hyper rectangules tree is therefore used in the subsequent results of this paper.

Figure 7 illustrates the number of keywords obtained for each depth of the hyper rectangles tree. Generally speaking, the number of keywords becomes stable after a certain depth of the hyper rectangles tree (when reaching the total number of words). This specific depth greatly depends on the size of the corpus.

It is interesting to note that the accuracy depends on the number of training examples. Table 1 shows the accuracy for each document category. It is clear that the categories with high number of training examples are correctly classified with high accuracy.

Finally, our method achieves comparable results with state-of-the-art methods in terms of text categorization as illustrated in table 2.

Table 1. Classification performance for each separate document category

Category	# occurences	Accuracy	Precision	Recall	F-Measure
acq	2292	97.53%	94.09%	98.42%	97.97%
crude	374	99.54%	98.26%	93.39%	96.37%
earn	3923	98.77%	98.44%	99.08%	98.92%
grain	51	99.91%	100.00%	80.00%	88.85%
interest	271	98.58%	87.88%	71.60%	82.96%
money-fx	293	98.13%	81.08%	68.97%	81.00%
ship	144	99.41%	92.59%	69.44%	81.77%
trade	326	99.36%	87.65%	94.67%	96.96%
All	7674	95.61 %	92.50%	84.45%	90.60%

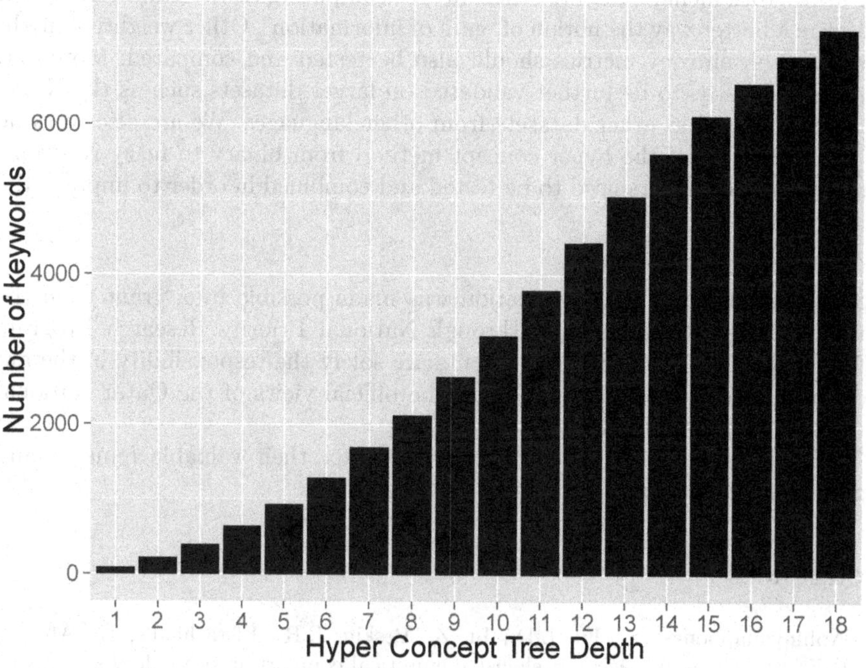

Fig. 7. Classification results for varying hyper rectangles tree depth.

Table 2. Classification accuracy compared with state-of-the-art methods

Method	Accuracy
Our Method	95.61 %
Yoshikawa et al. [21]	94 %
Jia et al. [12]	96.12 %
Kurian et al. [14]	92.37 %
Cardoso-Cachopo et al. [7]	96.98 %
Lee et al. [15]	94.97 %

6 Conclusion

We have proposed a new method for document categorization based on keyword extraction using the hyper rectangular method. Keywords are then fed into a random forest classification algorithm. Categorization results suggest that this method successfully extracts relevant keywords that can be used to determine the category of documents successfully.

Comparison shows that our conceptual approach can lead to accurate results which are comparable with state-of-the-art methods in document categorization.

Future work includes testing other ways of computing the weight metric which reflect in a better way the notion of "gain of information". Other weighting methods based on entropy metrics should also be tested and compared. Moreover, our method needs to be further validated on larger datasets such as the News-20 corpus as well as other datasets from other languages. We are also working on the expansion of the hyper concept method from binary to fuzzy relations. Finally, other classifiers need to be tested and combined in order to improve the performance.

Acknowledgment. This publication was made possible by a grant from the Qatar National Research Fund through National Priority Research Program (NPRP) No. 06-1220-1-233. Its contents are solely the responsibility of the authors and do not necessarily represent the official views of the Qatar National Research Fund or Qatar University.

We would also like to thank the reviewers for their valuable remarks and suggestions which helped improving this paper.

References

1. Aphinyanaphongs, Y., Fu, L.D., Li, Z., Peskin, E.R., Efstathiadis, E., Aliferis, C.F., Statnikov, A.: A comprehensive empirical comparison of modern supervised classification and feature selection methods for text categorization. Journal of the Association for Information Science and Technology 65(10), 1964–1987 (2014)
2. Azam, N., Yao, J.: Comparison of term frequency and document frequency based feature selection metrics in text categorization. Expert Systems with Applications 39(5), 4760–4768 (2012)

3. Birkhoff, G.: Lattice theory, vol. 25. American Mathematical Soc. (1967)
4. Breiman, L.: Random forests. Machine Learning 45(1), 5–32 (2001)
5. Cardoso-Cachopo, A.: Datasets for single label text categorization. artificial Intelligence Group, Department of Information Systems and Computer Science Instituto Superior Tecnico, Portugal (2009) http://web.ist.utl.pt/~acardoso/datasets/
6. Cardoso-Cachopo, A.: Improving Methods for Single-label Text Categorization. Ph.D. thesis, Instituto Superior Tecnico, Universidade Tecnica de Lisboa (2007)
7. Cardoso-Cachopo, A., Oliveira, A.: Combining lsi with other classifiers to improve accuracy of single-label text categorization. In: First European Workshop on Latent Semantic Analysis in Technology Enhanced Learning-EWLSATEL, vol. (2007)
8. Ferjani, F., Jaoua, A., Elloumi, S., Yahia, S.B.: Hyper-rectangular relation decomposition and dimensionality reduction. In: 13th International Conference on Relational and Algebraic Methods in Computer Science, RAMiCS 2013 (2012)
9. Ganter, B.: Two basic algorithms in concept analysis. In: Kwuida, L., Sertkaya, B. (eds.) ICFCA 2010. LNCS, vol. 5986, pp. 312–340. Springer, Heidelberg (2010)
10. Ganter, B., Wille, R.: Formal concept analysis: mathematical foundations. Springer Science & Business Media (2012)
11. Jaoua, A.: Pseudo-conceptual text and web Structuring. In: 16th International Conference on Conceptual Structures (ICCS 2008) (2008)
12. Jia, S., Liang, J., Xie, Y., Deng, L.: A novel feature voting model for text classification. In: 2014 11th International Conference on Fuzzy Systems and Knowledge Discovery (FSKD), pp. 306–311. IEEE (2014)
13. Jiang, S., Pang, G., Wu, M., Kuang, L.: An improved k-nearest-neighbor algorithm for text categorization. Expert Systems with Applications 39(1), 1503–1509 (2012)
14. Kurian, A., Josephine, M., Jeyabalaraja, V.: Scaling down dimensions and feature extraction in document repository classification. International Journal of Data Mining Techniques and Applications (2014)
15. Lee, L.H., Wan, C.H., Rajkumar, R., Isa, D.: An enhanced support vector machine classification framework by using euclidean distance function for text document categorization. Applied Intelligence 37(1), 80–99 (2012)
16. Lewis, D.D.: Reuters-21578 text categorization test collection, distribution 1.0 (1997). http://www.research.att.com/~lewis/reuters21578.html
17. Li, C.H., Yang, J.C., Park, S.C.: Text categorization algorithms using semantic approaches, corpus-based thesaurus and wordnet. Expert Systems with Applications 39(1), 765–772 (2012)
18. Llc, B.: Relational Model: Relational Algebra, Relational Database Management System, Object-Relational Impedance Mismatch, Synonym, Codd's Theorem. General Books LLC (2010). https://books.google.com.qa/books?id=JgDFbwAACAAJ
19. Uğuz, H.: A two-stage feature selection method for text categorization by using information gain, principal component analysis and genetic algorithm. Knowledge-Based Systems 24(7), 1024–1032 (2011)
20. Yang, J., Liu, Y., Zhu, X., Liu, Z., Zhang, X.: A new feature selection based on comprehensive measurement both in inter-category and intra-category for text categorization. Information Processing & Management 48(4), 741–754 (2012)
21. Yoshikawa, Y., Iwata, T., Sawada, H.: Latent support measure machines for bag-of-words data classification. In: Advances in Neural Information Processing Systems, pp. 1961–1969 (2014)

Solving a Tropical Optimization Problem via Matrix Sparsification

Nikolai Krivulin

Saint Petersburg State University, Faculty of Mathematics and Mechanics, Universitetsky Ave. 28, 198504, St. Petersburg, Russia
nkk@math.spbu.ru

Abstract. An optimization problem, which arises in various applications as that of minimizing the span seminorm, is considered in the framework of tropical mathematics. The problem is to minimize a nonlinear function defined on vectors over an idempotent semifield, and calculated by means of multiplicative conjugate transposition. We find the minimum of the function, and give a partial solution which explicitly represents a subset of solution vectors. We characterize all solutions by a system of simultaneous equation and inequality, and exploit this characterization to investigate properties of the solutions. A matrix sparsification technique is developed to extend the partial solution to a wider solution subset, and then to a complete solution described as a family of subsets. We offer a backtracking procedure that generates all members of the family, and derive an explicit representation for the complete solution. Numerical examples and graphical illustrations of the results are presented.

Keywords: Tropical algebra, Idempotent semifield, Optimization problem, Span seminorm, Sparse matrix, Backtracking, Complete solution.

1 Introduction

Tropical (idempotent) mathematics focuses on the theory and applications of semirings with idempotent addition, and had its origin in the seminal works published in the 1960s by Pandit [Pan61], Cuninghame-Green [CG62], Giffler [Gif63], Hoffman [Hof63], Vorob'ev [Vor63], Romanovskiĭ [Rom64], Korbut [Kor65], and Peteanu [Pet67]. An extensive study of tropical mathematics was motivated by real-world problems in various areas of operations research and computer science, including path analysis in graphs and networks [Pan61, Pet67], machine scheduling [CG62, Gif63], production planning and control [Vor63, Rom64]. The significant progress achieved in the field over the past few decades is reported in several research monographs, such as ones by Kolokoltsov and Maslov [KM97], Golan [Gol03], Heidergott et al. [HOvdW06], Gondran and Minoux [GM08], Butkovič [But10], as well as in a wide range of contributed papers.

Since the early studies [Gif63, Hof63, Rom64, Pet67], optimization problems that can be examined in the framework of tropical mathematics have formed a notable research domain in the field. These problems are formulated to minimize

W. Kahl et al. (Eds.): RAMiCS 2015, LNCS 9348, pp. 326–343, 2015.
DOI: 10.1007/978-3-319-24704-5_20

or maximize functions defined on vectors over idempotent semifields (semirings with multiplicative inverses), and may involve constraints in the form of tropical linear equations and inequalities. The objective functions can be both linear and nonlinear in the tropical mathematics setting.

The span (range) vector seminorm, which is defined as the maximum deviation between components of a vector, presents one of the objective functions encountered in practice. Specifically, this seminorm can serve as the optimization criterion for just-in-time scheduling (see, e.g., T'kindt and Billaut [TB06]), and finds applications in real-world problems that involve time synchronization in manufacturing, transportation networks, and parallel data processing.

In the context of tropical mathematics, the span seminorm has been put by Cuninghame-Green [CG79], and Cuninghame-Green and Butkovič [CGB04]. The seminorm was used by Butkovič and Tam [BT09] and Tam [Tam10] in a tropical optimization problem drawn from machine scheduling. A manufacturing system was considered, in which machines start and finish under some precedence constraints to make components for final products. The problem was to find the starting time for each machine to provide the completion times that are spread over a shortest time interval. A solution was given within a combined framework that involves two reciprocally dual idempotent semifields. A similar problem in the general setting of tropical mathematics was examined by Krivulin in [Kri13], where a direct, explicit solution was suggested. However, the results obtained present partial solutions, rather than a complete solution to the problems.

Consider the tropical optimization problem formulated in [Kri13] as an extension of the problem of minimizing the span seminorm, and represent it in a slightly different form to

$$\text{minimize} \quad \boldsymbol{q}^{-}\boldsymbol{x}(\boldsymbol{A}\boldsymbol{x})^{-}\boldsymbol{p},$$

where $\boldsymbol{p}$ and $\boldsymbol{q}$ are given vectors, $\boldsymbol{A}$ is a given matrix, $\boldsymbol{x}$ is the unknown vector, the minus in the superscript indicates conjugate transposition of vectors, and the matrix-vector multiplications are thought of in the sense of tropical algebra.

The purpose of this paper is to extend the partial solution of the problem, which is obtained in [Kri13] in the form of an explicit representation of a subset of solution vectors, to a complete solution describing all vectors that solve the problem. We combine the approach developed in [Kri04, Kri09, Kri12, Kri13, Kri14, Kri15b, Kri15a] to reduce the problem to a system of simultaneous equation and inequality, with a new matrix sparsification technique to obtain all solutions to the system in a direct, compact vector form.

We start with a brief overview of basic definitions, notation, and preliminary results of tropical mathematics in Section 2 to provide a general framework for the solutions in the later sections. Specifically, a lemma that offers two equivalent representations for a vector set is presented, which is of independent interest. In Section 3, we formulate the minimization problem to be solved, find the minimum in the problem, and give a partial solution in the form of an explicit representation of a subset of solution vectors. We characterize all solutions to the problem by a system of simultaneous equation and inequality, and exploit this characterization to investigate properties of the solutions.

In Section 4, we develop a matrix sparsification technique, which consists in dropping entries below a prescribed threshold in the matrix $\boldsymbol{A}$ without affecting the solution of the problem. By combining this technique with the above characterization, the partial solution obtained in Section 3 is extended to a wider solution subset, which includes the partial solution as a particular case.

Section 5 focuses on the derivation of a complete solution to the problem. We describe all solutions of the problem as a family of subsets of solution vectors, and propose a backtracking procedure that allows one to generate all members in the family. The section concludes with our main result, which offers an explicit representation for the complete solution in a compact vector form.

Numerical examples and graphical illustrations are also included in the text to provide additional insights into the results obtained.

2 Preliminary Results

In this section, we give a brief overview of the main definitions, notation, and preliminary results used in the subsequent solution to the tropical optimization problem under study. Concise introductions to and thorough discussion of tropical mathematics are presented in various forms in a range of works, including [KM97, Gol03, HOvdW06, ABG07, Lit07, GM08, SS09, But10]. In the overview below, we mainly follow the results in [Kri09, Kri14, Kri15a, Kri15b], which offer a unified framework to obtain explicit solutions in a compact form. For further details, one can consult the publications listed before.

2.1 Idempotent Semifield

Let $\mathbb{X}$ be a nonempty set that is closed under two associative and commutative operations, addition $\oplus$ and multiplication $\otimes$, which have their neutral elements, zero $\mathbb{0}$ and identity $\mathbb{1}$. Addition is idempotent to yield $x \oplus x = x$ for all $x \in \mathbb{X}$. Multiplication is invertible, which implies that each nonzero $x \in \mathbb{X}$ has an inverse x^{-1} to satisfy the equality $x \otimes x^{-1} = \mathbb{1}$. Moreover, multiplication distributes over addition, and has $\mathbb{0}$ as the absorbing element. Under these conditions, the system $\langle \mathbb{X}, \mathbb{0}, \mathbb{1}, \oplus, \otimes \rangle$ is commonly referred to as the idempotent semifield.

The idempotent addition produces a partial order, by which $x \leq y$ if and only if $x \oplus y = y$. With respect to this order, the inequality $x \oplus y \leq z$ is equivalent to two inequalities $x \leq z$ and $y \leq z$. Moreover, addition and multiplication are isotone in each argument, whereas the multiplicative inversion is antitone.

The partial order is assumed to extend to a consistent total order over $\mathbb{X}$.

The power notation with integer exponents is used for iterated multiplication to define $x^0 = \mathbb{1}$, $x^p = x \otimes x^{p-1}$, $x^{-p} = (x^{-1})^p$ for any nonzero x and positive integer p. Moreover, the equation $x^p = a$ is assumed to have a solution $x = a^{1/p}$ for all a, which extends this notation to rational exponents, and thereby makes the semifield algebraically closed (radicable).

In what follows, the multiplication sign $\otimes$ is dropped for simplicity. The relation symbols and the minimization problems are thought of in terms of the above order, which is induced by idempotent addition.

As examples of the general semifield under consideration, one can take

$$\mathbb{R}_{\max,+} = \langle \mathbb{R} \cup \{-\infty\}, -\infty, 0, \max, + \rangle, \quad \mathbb{R}_{\min,+} = \langle \mathbb{R} \cup \{+\infty\}, +\infty, 0, \min, + \rangle,$$
$$\mathbb{R}_{\max,\times} = \langle \mathbb{R}_{+} \cup \{0\}, 0, 1, \max, \times \rangle, \quad \mathbb{R}_{\min,\times} = \langle \mathbb{R}_{+} \cup \{+\infty\}, +\infty, 1, \min, \times \rangle,$$

where $\mathbb{R}$ is the set of real numbers and $\mathbb{R}_{+} = \{x \in \mathbb{R} | x > 0\}$.

Specifically, the semifield $\mathbb{R}_{\max,+}$ has addition $\oplus$ given by the maximum, and multiplication $\otimes$ by the ordinary addition, with the null $\mathbb{0} = -\infty$ and identity $\mathbb{1} = 0$. Each $x \in \mathbb{R}$ has its inverse x^{-1} equal to $-x$ in standard notation. The power x^y is defined for any $x, y \in \mathbb{R}$ and coincides with the arithmetic product xy. The order induced by addition corresponds to the natural linear order on $\mathbb{R}$.

2.2 Matrix and Vector Algebra

We now consider matrices over $\mathbb{X}$ and denote the set of matrices with m rows and n columns by $\mathbb{X}^{m \times n}$. A matrix with all entries equal to $\mathbb{0}$ is called the zero matrix. A matrix is row- (column-) regular, if it has no zero rows (columns).

For any matrices $\boldsymbol{A} = (a_{ij})$, $\boldsymbol{B} = (b_{ij})$, and $\boldsymbol{C} = (c_{ij})$ of appropriate size, and a scalar x, matrix addition, matrix and scalar multiplication are routinely defined entry-wise by the formulae

$$\{\boldsymbol{A} \oplus \boldsymbol{B}\}_{ij} = a_{ij} \oplus b_{ij}, \qquad \{\boldsymbol{B}\boldsymbol{C}\}_{ij} = \bigoplus_{k} b_{ik} c_{kj}, \qquad \{x\boldsymbol{A}\}_{ij} = x a_{ij}.$$

For any matrix $\boldsymbol{A} \in \mathbb{X}^{m \times n}$, its transpose is the matrix $\boldsymbol{A}^T \in \mathbb{X}^{n \times m}$.

For a nonzero matrix $\boldsymbol{A} = (a_{ij}) \in \mathbb{X}^{m \times n}$, the multiplicative conjugate transpose is the matrix $\boldsymbol{A}^{-} = (a_{ij}^{-}) \in \mathbb{X}^{n \times m}$ with the elements $a_{ij}^{-} = a_{ji}^{-1}$ if $a_{ji} \neq \mathbb{0}$, and $a_{ij}^{-} = \mathbb{0}$ otherwise.

Consider square matrices in the set $\mathbb{X}^{n \times n}$. A matrix is diagonal if it has all off-diagonal entries equal to $\mathbb{0}$. A diagonal matrix whose diagonal entries are all equal to $\mathbb{1}$ is the identity matrix represented by $\boldsymbol{I}$.

Suppose that a matrix $\boldsymbol{A}$ is row-regular. Clearly, the inequality $\boldsymbol{A}\boldsymbol{A}^{-} \geq \boldsymbol{I}$ is then valid. Moreover, if the row-regular matrix $\boldsymbol{A}$ has exactly one nonzero entry in every row, then the inequality $\boldsymbol{A}^{-}\boldsymbol{A} \leq \boldsymbol{I}$ holds as well.

The matrices with only one column (row) are routinely referred to as the column (row) vectors. Unless otherwise indicated, the vectors are considered below as column vectors. The set of column vectors of order n is denoted by $\mathbb{X}^{n}$.

A vector that has all components equal to $\mathbb{0}$ is the zero vector denoted $\boldsymbol{0}$. If a vector has no zero components, it is called regular.

For any vectors $\boldsymbol{a} = (a_i)$ and $\boldsymbol{b} = (b_i)$ of the same order, and a scalar x, addition and scalar multiplication are performed component-wise by the rules

$$\{\boldsymbol{a} \oplus \boldsymbol{b}\}_i = a_i \oplus b_i, \qquad \{x\boldsymbol{a}\}_i = x a_i.$$

In the context of $\mathbb{R}^2_{\max,+}$, these vector operations are illustrated in the Cartesian coordinate system on the plane in Fig. 1.

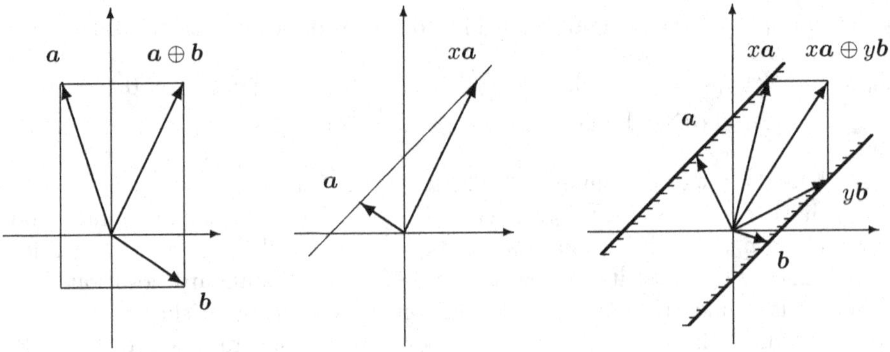

Fig. 1. Addition (left), scalar multiplication (middle), and a linear span (right) of vectors in $\mathbb{R}^2_{\max,+}$.

The left picture shows that, in terms of $\mathbb{R}^2_{\max,+}$, vector addition uses a rectangle rule. The sum of two vectors is the upper right vertex of the rectangle formed by the lines that are drawn through the end points of the vectors parallel to the coordinate axes. Scalar multiplication is given in the middle by the shift of the end point of a vector along the line at 45° to the axes.

Let $\boldsymbol{x}$ be a regular vector and $\boldsymbol{A}$ be a square matrix of the same order. It is clear that the vector $\boldsymbol{Ax}$ is regular only when the matrix $\boldsymbol{A}$ is row-regular. Similarly, the row vector $\boldsymbol{x}^T\boldsymbol{A}$ is regular provided that $\boldsymbol{A}$ is column-regular.

For any nonzero vector $\boldsymbol{x} = (x_i) \in \mathbb{X}^n$, the multiplicative conjugate transpose is the row vector $\boldsymbol{x}^- = (x_i^-)$, where $x_i^- = x_i^{-1}$ if $x_i \neq \mathbb{0}$, and $x_i^- = \mathbb{0}$ otherwise. The following properties of the conjugate transposition are easy to verify.

For any nonzero vectors $\boldsymbol{x}$ and $\boldsymbol{y}$, the equality $(\boldsymbol{x}\boldsymbol{y}^-)^- = \boldsymbol{y}\boldsymbol{x}^-$ is valid. When the vectors $\boldsymbol{x}$ and $\boldsymbol{y}$ are regular and have the same size, the component-wise inequality $\boldsymbol{x} \leq \boldsymbol{y}$ implies that $\boldsymbol{x}^- \geq \boldsymbol{y}^-$ and vice versa.

For any nonzero column vector $\boldsymbol{x}$, the equality $\boldsymbol{x}^-\boldsymbol{x} = \mathbb{1}$ holds. Moreover, if the vector $\boldsymbol{x}$ is regular, then the matrix inequality $\boldsymbol{x}\boldsymbol{x}^- \geq \boldsymbol{I}$ is valid as well.

2.3 Linear Dependence

A vector $\boldsymbol{b} \in \mathbb{X}^m$ is linearly dependent on vectors $\boldsymbol{a}_1, \ldots, \boldsymbol{a}_n \in \mathbb{X}^m$ if there exist scalars $x_1, \ldots, x_n \in \mathbb{X}$ such that the vector $\boldsymbol{b}$ can be represented by a linear combination of these vectors as $\boldsymbol{b} = x_1\boldsymbol{a}_1 \oplus \cdots \oplus x_n\boldsymbol{a}_n$. Specifically, the vector $\boldsymbol{b}$ is collinear with a vector $\boldsymbol{a}$ if $\boldsymbol{b} = x\boldsymbol{a}$ for some scalar x.

To describe a formal criterion for a vector $\boldsymbol{b}$ to be linearly dependent on vectors $\boldsymbol{a}_1, \ldots, \boldsymbol{a}_n$, we take the latter vectors to form the matrix $\boldsymbol{A} = (\boldsymbol{a}_1, \ldots, \boldsymbol{a}_n)$, and then introduce a function that maps the pair $(\boldsymbol{A}, \boldsymbol{b})$ to the scalar

$$\delta(\boldsymbol{A}, \boldsymbol{b}) = (\boldsymbol{A}(\boldsymbol{b}^-\boldsymbol{A})^-)^-\boldsymbol{b}.$$

The following result was obtained in [Kri04] (see also [Kri09, Kri12]).

Lemma 1. *A vector $\boldsymbol{b}$ is linearly dependent on vectors $\boldsymbol{a}_1, \dots, \boldsymbol{a}_n$ if and only if the condition $\delta(\boldsymbol{A}, \boldsymbol{b}) = \mathbb{1}$ holds, where $\boldsymbol{A} = (\boldsymbol{a}_1, \dots, \boldsymbol{a}_n)$.*

The set of all linear combinations of vectors $\boldsymbol{a}_1, \dots, \boldsymbol{a}_n \in \mathbb{X}^m$ form a linear span of the vectors, which is closed under vector addition and scalar multiplication. A linear span of two vectors in $\mathbb{R}^2_{\max,+}$ is displayed in Fig. 1 (right) as a strip between two thick hatched lines drawn at 45° to the axes.

A system of vectors $\boldsymbol{a}_1, \dots, \boldsymbol{a}_n$ is linearly dependent if at least one vector in the system is linearly dependent on others, and linearly independent otherwise.

Two systems of vectors are considered equivalent if each vector of one system is a linear combination of vectors of the other system. Equivalent systems of vectors obviously have a common linear span.

Let $\boldsymbol{a}_1, \dots, \boldsymbol{a}_n$ be a system that may include linearly dependent vectors. To construct an equivalent linearly independent system, we use a procedure that sequentially reduces the system until it becomes linearly independent. The procedure applies the criterion provided by Lemma 1 to examine the vectors one by one to remove a vector if it is linearly dependent on others, or to leave the vector in the system otherwise. It is not difficult to see that the procedure results in a linearly independent system that is equivalent to the original one.

2.4 Representation Lemma

We apply properties of the conjugate transposition to obtain a useful result that offers an equivalent representation for a set of vectors $\boldsymbol{x} \in \mathbb{X}^n$, which is defined by boundaries given by a double inequality with vectors $\boldsymbol{g}, \boldsymbol{h} \in \mathbb{X}^n$.

Lemma 2. *Let $\boldsymbol{g}$ be a vector and $\boldsymbol{h}$ a regular vector such that $\boldsymbol{g} \leq \boldsymbol{h}$. Then, the following statements are equivalent:*

1. *The vector $\boldsymbol{x}$ satisfies the double inequality*

$$\alpha \boldsymbol{g} \leq \boldsymbol{x} \leq \alpha \boldsymbol{h}, \qquad \alpha > \mathbb{0}. \tag{1}$$

2. *The vector $\boldsymbol{x}$ is given by the equality*

$$\boldsymbol{x} = (\boldsymbol{I} \oplus \boldsymbol{g}\boldsymbol{h}^-)\boldsymbol{u}, \qquad \boldsymbol{u} > \boldsymbol{0}. \tag{2}$$

Proof. We verify that both representations follow from each other. First, suppose that a vector $\boldsymbol{x}$ satisfies double inequality (1). Left multiplication of the right inequality at (1) by $\boldsymbol{g}\boldsymbol{h}^-$ yields $\boldsymbol{g}\boldsymbol{h}^-\boldsymbol{x} \leq \alpha \boldsymbol{g}\boldsymbol{h}^-\boldsymbol{h} = \alpha \boldsymbol{g}$. Considering the left inequality, we see that $\boldsymbol{x} \geq \alpha \boldsymbol{g} \geq \boldsymbol{g}\boldsymbol{h}^-\boldsymbol{x}$, and hence write $\boldsymbol{x} = \boldsymbol{x} \oplus \boldsymbol{g}\boldsymbol{h}^-\boldsymbol{x}$. With $\boldsymbol{u} = \boldsymbol{x}$, we obtain $\boldsymbol{x} = \boldsymbol{u} \oplus \boldsymbol{g}\boldsymbol{h}^-\boldsymbol{u} = (\boldsymbol{I} \oplus \boldsymbol{g}\boldsymbol{h}^-)\boldsymbol{u}$, which gives (2).

Now assume that $\boldsymbol{x}$ is a vector given by (2). Take a scalar $\alpha = \boldsymbol{h}^-\boldsymbol{u}$ and write $\boldsymbol{x} = (\boldsymbol{I} \oplus \boldsymbol{g}\boldsymbol{h}^-)\boldsymbol{u} \geq \boldsymbol{g}\boldsymbol{h}^-\boldsymbol{u} = \alpha \boldsymbol{g}$, which provides the left inequality in (1). Furthermore, it follows from the inequalities $\boldsymbol{h} \geq \boldsymbol{g}$ and $\boldsymbol{h}\boldsymbol{h}^- \geq \boldsymbol{I}$ that $\boldsymbol{x} = (\boldsymbol{I} \oplus \boldsymbol{g}\boldsymbol{h}^-)\boldsymbol{u} \leq (\boldsymbol{h}\boldsymbol{h}^- \oplus \boldsymbol{g}\boldsymbol{h}^-)\boldsymbol{u} = (\boldsymbol{h} \oplus \boldsymbol{g})\boldsymbol{h}^-\boldsymbol{u} = \boldsymbol{h}\boldsymbol{h}^-\boldsymbol{u} = \alpha \boldsymbol{h}$, and therefore, the right inequality is valid as well. □

Fig. 2 offers a graphical illustration in terms of $\mathbb{R}^2_{\max,+}$ for the representation lemma. An example set defined by inequality (1) is depicted on the left. The rectangle formed by horizontal and vertical lines drawn through the end points of the vectors $\boldsymbol{g} = (g_1, g_2)^T$ and $\boldsymbol{h} = (h_1, h_2)^T$ shows the boundaries of the set given by (1) with $\alpha = 0$. The whole set is then represented as the strip area between thick hatched lines, which is covered when the rectangle shifts at 45° to the axes in response to the variation of α.

According to representation (2), the same area is shown on the right as the linear span of the columns in the matrix $\boldsymbol{I} \oplus \boldsymbol{g}\boldsymbol{h}^-$, where $\boldsymbol{g}\boldsymbol{h}^- = (h_1^{-1}\boldsymbol{g}, h_2^{-1}\boldsymbol{g})$.

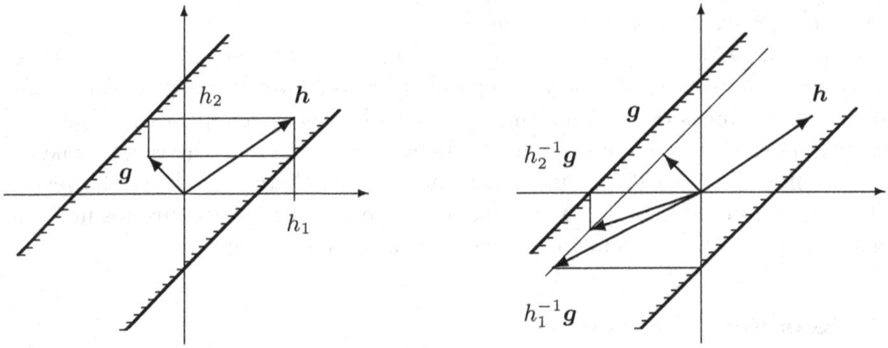

Fig. 2. An example set defined in $\mathbb{R}^2_{\max,+}$ by conditions (1) (left) and (2) (right).

3 Tropical Optimization Problem

We start this section with the formulation of a general tropical optimization problem, which arises in constrained approximation in the sense of the span seminorm, and finds applications in optimal scheduling in just-in-time manufacturing [Kri13]. Below, we find the minimum value, and offer a partial solution of the problem. Then, we reduce the problem to the solution of simultaneous equation and inequality, and investigate properties of the solution set.

Given a matrix $\boldsymbol{A} \in \mathbb{X}^{m\times n}$ and vectors $\boldsymbol{p} \in \mathbb{X}^m$, $\boldsymbol{q} \in \mathbb{X}^n$, the problem is to find regular vectors $\boldsymbol{x} \in \mathbb{X}^n$ that

$$\text{minimize} \quad \boldsymbol{q}^-\boldsymbol{x}(\boldsymbol{A}\boldsymbol{x})^-\boldsymbol{p}. \tag{3}$$

First, we note that substitution of $\alpha\boldsymbol{x}$, where $\alpha \neq \mathbb{0}$, for the vector $\boldsymbol{x}$ does not affect the objective function, and thus all solutions of (3) are scale-invariant.

A partial solution to the problem formulated in a slightly different form was given in [Kri13]. We include the proof of this result into the next lemma for the sake of completeness, and to provide a starting point for further examination.

Lemma 3. *Let* $\boldsymbol{A}$ *be a row-regular matrix,* $\boldsymbol{p}$ *be nonzero and* $\boldsymbol{q}$ *regular vectors. Then, the minimum value in problem* (3) *is equal to*

$$\Delta = (\boldsymbol{A}\boldsymbol{q})^-\boldsymbol{p}, \tag{4}$$

and all regular vectors $\boldsymbol{x}$ *that produce this minimum are defined by the system*

$$\boldsymbol{q}^{-}\boldsymbol{x} = \alpha, \qquad \boldsymbol{A}\boldsymbol{x} \geq \alpha\Delta^{-1}\boldsymbol{p}, \qquad \alpha > \mathbb{0}. \tag{5}$$

Specifically, the minimum is attained at any vector $\boldsymbol{x} = \alpha\boldsymbol{q}$, *where* $\alpha > \mathbb{0}$.

Proof. To obtain the minimum value of the objective function in problem (3), we derive a lower bound for the function, and then show that this bound is strict.

Suppose that $\boldsymbol{x}$ is a regular solution of the problem. Since $\boldsymbol{x}\boldsymbol{x}^{-} \geq \boldsymbol{I}$, we have $(\boldsymbol{q}^{-}\boldsymbol{x})^{-1}\boldsymbol{x} = (\boldsymbol{q}^{-}\boldsymbol{x}\boldsymbol{x}^{-})^{-} \leq \boldsymbol{q}$. Next, left multiplication by the matrix $\boldsymbol{A}$ gives the inequality $(\boldsymbol{q}^{-}\boldsymbol{x})^{-1}\boldsymbol{A}\boldsymbol{x} \leq \boldsymbol{A}\boldsymbol{q}$, where both sides are regular vectors. Finally, conjugate transposition followed by right multiplication by the vector $\boldsymbol{p}$ yields the lower bound $\boldsymbol{q}^{-}\boldsymbol{x}(\boldsymbol{A}\boldsymbol{x})^{-}\boldsymbol{p} \geq (\boldsymbol{A}\boldsymbol{q})^{-}\boldsymbol{p} = \Delta > \mathbb{0}$.

With $\boldsymbol{x} = \boldsymbol{q}$, the objective function becomes $\boldsymbol{q}^{-}\boldsymbol{x}(\boldsymbol{A}\boldsymbol{x})^{-}\boldsymbol{p} = (\boldsymbol{A}\boldsymbol{q})^{-}\boldsymbol{p} = \Delta$, and therefore, Δ is the minimum value of the problem.

Considering that all solutions are scale-invariant, we see that not only the vector $\boldsymbol{q}$, but also any vector $\boldsymbol{x} = \alpha\boldsymbol{q}$ with nonzero α solves the problem.

Furthermore, all vectors $\boldsymbol{x}$ that yield the minimum must satisfy the equation

$$\boldsymbol{q}^{-}\boldsymbol{x}(\boldsymbol{A}\boldsymbol{x})^{-}\boldsymbol{p} = \Delta.$$

To examine the equation, we put $\alpha = \boldsymbol{q}^{-}\boldsymbol{x} > \mathbb{0}$, and rewrite it in an equivalent form as the system

$$\boldsymbol{q}^{-}\boldsymbol{x} = \alpha, \qquad (\boldsymbol{A}\boldsymbol{x})^{-}\boldsymbol{p} = \alpha^{-1}\Delta.$$

It is easy to see from the first equation that each solution $\boldsymbol{x}$ satisfies the condition $\boldsymbol{x} \leq \alpha\boldsymbol{q}$. Indeed, after left multiplication of this equation by the vector $\boldsymbol{q}$, which is regular and hence $\boldsymbol{q}\boldsymbol{q}^{-} \geq \boldsymbol{I}$, we immediately obtain $\boldsymbol{x} \leq \boldsymbol{q}\boldsymbol{q}^{-}\boldsymbol{x} = \alpha\boldsymbol{q}$.

Furthermore, the second equation can be written as two opposite inequalities $(\boldsymbol{A}\boldsymbol{x})^{-}\boldsymbol{p} \leq \alpha^{-1}\Delta$ and $(\boldsymbol{A}\boldsymbol{x})^{-}\boldsymbol{p} \geq \alpha^{-1}\Delta$. However, the condition $\boldsymbol{x} \leq \alpha\boldsymbol{q}$ leads to $(\boldsymbol{A}\boldsymbol{x})^{-}\boldsymbol{p} \geq \alpha^{-1}(\boldsymbol{A}\boldsymbol{q})^{-}\boldsymbol{p} = \alpha^{-1}\Delta$, which makes the second inequality superfluous.

Consider the first inequality $(\boldsymbol{A}\boldsymbol{x})^{-}\boldsymbol{p} \leq \alpha^{-1}\Delta$, and verify that it is equivalent to $\boldsymbol{A}\boldsymbol{x} \geq \alpha\Delta^{-1}\boldsymbol{p}$. Left multiplication of the former inequality by the regular vector $\alpha\Delta^{-1}\boldsymbol{A}\boldsymbol{x}$ yields $\alpha\Delta^{-1}\boldsymbol{p} \leq \alpha\Delta^{-1}\boldsymbol{A}\boldsymbol{x}(\boldsymbol{A}\boldsymbol{x})^{-}\boldsymbol{p} \leq \boldsymbol{A}\boldsymbol{x}$. At the same time, left multiplication of the latter inequality by $\alpha^{-1}\Delta(\boldsymbol{A}\boldsymbol{x})^{-}$ gives the former one.

As a result, the system under investigation reduces to the form of (5). □

The following statement is an important consequence of Lemma 3.

Corollary 4. *Let* $\boldsymbol{A}$ *be a row-regular matrix,* $\boldsymbol{p}$ *be nonzero and* $\boldsymbol{q}$ *regular vectors. Then, the set of regular solutions of problem* (3) *is closed under addition.*

Proof. Suppose vectors $\boldsymbol{x}$ and $\boldsymbol{y}$ are regular solutions of problem (3) such that the vector $\boldsymbol{x}$ satisfies system (5), whereas $\boldsymbol{y}$ solves the system

$$\boldsymbol{q}^{-}\boldsymbol{y} = \beta, \qquad \boldsymbol{A}\boldsymbol{x} \geq \beta\Delta^{-1}\boldsymbol{p}, \qquad \beta > \mathbb{0}.$$

Furthermore, we immediately verify that $\boldsymbol{q}^{-}(\boldsymbol{x} \oplus \boldsymbol{y}) = \boldsymbol{q}^{-}\boldsymbol{x} \oplus \boldsymbol{q}^{-}\boldsymbol{y} = \alpha \oplus \beta$ and $\boldsymbol{A}(\boldsymbol{x} \oplus \boldsymbol{y}) = \boldsymbol{A}\boldsymbol{x} \oplus \boldsymbol{A}\boldsymbol{y} \geq (\alpha \oplus \beta)\Delta^{-1}\boldsymbol{p}$, which shows that the sum $\boldsymbol{x} \oplus \boldsymbol{y}$ also obeys system (5), where α is replaced by $\alpha \oplus \beta$. □

Note that an application of Lemma 2 provides problem (3) with another representation of the solution $\boldsymbol{x} = \alpha\boldsymbol{q}$ in the form

$$\boldsymbol{x} = (\boldsymbol{I} \oplus \boldsymbol{q}\boldsymbol{q}^{-})\boldsymbol{u}, \qquad \boldsymbol{u} > \boldsymbol{0}.$$

However, this representation is not sufficiently different from that offered by Lemma 3. Indeed, considering that the vector $\boldsymbol{q}$ is regular, we immediately obtain $\boldsymbol{x} = (\boldsymbol{I} \oplus \boldsymbol{q}\boldsymbol{q}^{-})\boldsymbol{u} = \boldsymbol{q}\boldsymbol{q}^{-}\boldsymbol{u} = \alpha\boldsymbol{q}$, where we take $\alpha = \boldsymbol{q}^{-}\boldsymbol{u}$.

4 Extended Solution via Matrix Sparsification

To extend the partial solution obtained in the previous section, we first suggest an entry-wise thresholding (dropping) procedure to sparsify the matrix in the problem. Then, we apply the sparsified matrix to find new solutions, and illustrate the result with an example, followed by a graphical representation.

4.1 Matrix Sparsification

As the first step to derive an extended solution of problem (3), we use a procedure that sets each entry of the matrix $\boldsymbol{A}$ to $\mathbb{0}$ if it is below a threshold value determined by both this matrix and the vectors $\boldsymbol{p}$ and $\boldsymbol{q}$, and leaves the entry unchanged otherwise. The next result introduces the sparsified matrix, and shows that the sparsification does not affect the solution of the problem.

Lemma 5. *Let $\boldsymbol{A} = (a_{ij})$ be a row-regular matrix, $\boldsymbol{p} = (p_i)$ be a nonzero vector, $\boldsymbol{q} = (q_j)$ be a regular vector, and $\Delta = (\boldsymbol{A}\boldsymbol{q})^{-}\boldsymbol{p}$. Define the sparsified matrix $\widehat{\boldsymbol{A}} = (\widehat{a}_{ij})$ with the entries*

$$\widehat{a}_{ij} = \begin{cases} a_{ij}, & \text{if } a_{ij} \geq \Delta^{-1}p_iq_j^{-1}; \\ \mathbb{0}, & \text{otherwise}. \end{cases} \tag{6}$$

Then, replacing the matrix $\boldsymbol{A}$ by $\widehat{\boldsymbol{A}}$ in problem (3) does not change the solutions of the problem.

Proof. We first verify that the sparsification retains the minimum value given by Lemma 3 in the form $\Delta = (\boldsymbol{A}\boldsymbol{q})^{-}\boldsymbol{p}$. We define indices k and s by the conditions

$$k = \arg\max_{1 \leq i \leq m} (a_{i1}q_1 \oplus \cdots \oplus a_{in}q_n)^{-1}p_i, \qquad s = \arg\max_{1 \leq j \leq n} a_{kj}q_j,$$

and then represent Δ by using the scalar equality

$$\Delta = \bigoplus_{i=1}^{m}(a_{i1}q_1 \oplus \cdots \oplus a_{in}q_n)^{-1}p_i = (a_{k1}q_1 \oplus \cdots \oplus a_{kn}q_n)^{-1}p_k = (a_{ks}q_s)^{-1}p_k.$$

The regularity of $\boldsymbol{A}$ and $\boldsymbol{q}$ guarantees that $a_{i1}q_1 \oplus \cdots \oplus a_{in}q_n > \mathbb{0}$ for all i. Since $\boldsymbol{p}$ is nonzero, we see that $\Delta > \mathbb{0}$ as well as that $a_{ks} > \mathbb{0}$ and $p_k > \mathbb{0}$.

Let us examine an arbitrary row i in the matrix $\boldsymbol{A}$. The above equality for Δ yields the inequality $\Delta \geq (a_{i1}q_1 \oplus \cdots \oplus a_{in}q_n)^{-1}p_i$, which is equivalent to the inequality $a_{i1}q_1 \oplus \cdots \oplus a_{in}q_n \geq \Delta^{-1}p_i$. Because the order defined by the relation $\leq$ is assumed total, the last inequality is valid if and only if the condition $a_{ij}q_j \geq \Delta^{-1}p_i$ holds for some j.

Thus, we conclude that each row i of $\boldsymbol{A}$ has at least one entry a_{ij} to satisfy the inequality

$$a_{ij} \geq \Delta^{-1}p_iq_j^{-1}. \tag{7}$$

Now consider row k in the matrix $\boldsymbol{A}$ to verify the inequality $a_{kj} \leq \Delta^{-1}p_kq_j^{-1}$ for all j. Indeed, provided that $a_{kj} = \mathbb{0}$, the inequality is trivially true. If $a_{kj} > \mathbb{0}$, then we have $(a_{kj}q_j)^{-1}p_k \geq (a_{k1}q_1 \oplus \cdots \oplus a_{kn}q_n)^{-1}p_k = \Delta$, which gives the desired inequality. Since $\Delta = (a_{ks}q_s)^{-1}p_k$, we see that row k has entries which turns inequality (7) into an equality, but no entries for which (7) becomes strict.

Suppose that inequality (7) fails for some i and j. Provided that $p_i > \mathbb{0}$, we write $a_{ij} < \Delta^{-1}p_iq_j^{-1} \leq (a_{i1}q_1 \oplus \cdots \oplus a_{in}q_n)q_j^{-1}$, which gives the inequality $a_{ij}q_j < a_{i1}q_1 \oplus \cdots \oplus a_{in}q_n$. The last inequality means that decreasing $a_{ij}q_j$ through lowering of a_{ij} down to $\mathbb{0}$ does not affect the value of $a_{i1}q_1 \oplus \cdots \oplus a_{in}q_n$, and hence the value of $\Delta \geq (a_{i1}q_1 \oplus \cdots \oplus a_{in}q_n)^{-1}p_i$. Note that if $p_i = \mathbb{0}$, then Δ does not depend at all on the entries in row i, including, certainly, a_{ij}.

We now verify that all entries a_{ij} that do not satisfy inequality (7) can be set to $\mathbb{0}$ without affecting not only the minimum value Δ, but also the regular solutions of problem (3). First, note that all vectors $\boldsymbol{x} = (x_j)$ providing the minimum in the problem are determined by the equation $\boldsymbol{q}^{-}\boldsymbol{x}(\boldsymbol{A}\boldsymbol{x})^{-}\boldsymbol{p} = \Delta$.

We represent this equation in the scalar form

$$(q_1^{-1}x_1 \oplus \cdots \oplus q_n^{-1}x_n) \bigoplus_{i=1}^{m} (a_{i1}x_1 \oplus \cdots \oplus a_{in}x_n)^{-1}p_i = \Delta,$$

which yields that $a_{i1}x_1 \oplus \cdots \oplus a_{in}x_n \geq \Delta^{-1}(q_1^{-1}x_1 \oplus \cdots \oplus q_n^{-1}x_n)p_i$ for all i.

Assume the matrix $\boldsymbol{A}$ to have an entry, say a_{ij}, that satisfies the condition $a_{ij} < \Delta^{-1}p_iq_j^{-1}$, and thereby violates inequality (7). Provided that $p_i = \mathbb{0}$, the condition leads to the equality $a_{ij} = \mathbb{0}$. Suppose that $p_i > \mathbb{0}$, and write

$$a_{ij}x_j < \Delta^{-1}p_iq_j^{-1}x_j \leq \Delta^{-1}(q_1^{-1}x_1 \oplus \cdots \oplus q_n^{-1}x_n)p_i \leq a_{i1}x_1 \oplus \cdots \oplus a_{in}x_n.$$

This inequality implies that, for each solution of the above equation, the term $a_{ij}x_j$ does not contribute to the value of the entire sum $a_{i1}x_1 \oplus \cdots \oplus a_{in}x_n$ involved in the calculation of the left-hand side of the equation. Therefore, we can set a_{ij} to $\mathbb{0}$ without altering the solutions of this equation.

It remains to see that setting the entries a_{ij}, which do not satisfy inequality (7), to $\mathbb{0}$ is equivalent to the replacement of the matrix $\boldsymbol{A}$ by the matrix $\widehat{\boldsymbol{A}}$. □

The matrix obtained after the sparsification procedure for problem (3) is referred to below as the sparsified matrix of the problem.

Note that the sparsification of the matrix $\boldsymbol{A}$ according to definition (6) is actually determined by the threshold matrix $\Delta^{-1}\boldsymbol{p}\boldsymbol{q}^{-}$, which contains the threshold values for corresponding entries of $\boldsymbol{A}$.

Let $\widehat{\boldsymbol{A}}$ be the sparsified matrix for $\boldsymbol{A}$, based on the threshold matrix $\Delta^{-1}\boldsymbol{p}\boldsymbol{q}^{-}$. Then, it follows directly from (6) that the inequality $\widehat{\boldsymbol{A}}^{-} \leq \Delta\boldsymbol{q}\boldsymbol{p}^{-}$ is valid.

4.2 Extended Solution Set

We now assume problem (3) already has a sparsified matrix. Under this assumption, we use the characterization of solutions given by Lemma 3 to improve the partial solution provided by this lemma by further extending the solution set.

Theorem 6. *Let $\boldsymbol{A}$ be a row-regular sparsified matrix of problem* (3) *with a nonzero vector $\boldsymbol{p}$ and a regular vector $\boldsymbol{q}$.*

Then, the minimum value in the problem is equal to $\Delta = (\boldsymbol{A}\boldsymbol{q})^{-}\boldsymbol{p}$, and attained at any vector $\boldsymbol{x}$ given by the conditions

$$\alpha\Delta^{-1}\boldsymbol{A}^{-}\boldsymbol{p} \leq \boldsymbol{x} \leq \alpha\boldsymbol{q}, \qquad \alpha > \mathbb{0}; \tag{8}$$

or, equivalently, by the conditions

$$\boldsymbol{x} = (\boldsymbol{I} \oplus \Delta^{-1}\boldsymbol{A}^{-}\boldsymbol{p}\boldsymbol{q}^{-})\boldsymbol{u}, \qquad \boldsymbol{u} > \boldsymbol{0}. \tag{9}$$

Proof. It follows from Lemma 3 and Lemma 5 that the minimum value, given by $\Delta = (\boldsymbol{A}\boldsymbol{q})^{-}\boldsymbol{p}$, and the regular solutions do not change after sparsification.

Considering that, by Lemma 3, all regular solutions are defined by system (5), we need to show that each vector $\boldsymbol{x}$, which satisfies (8), also solves (5).

Note that the set of vectors given by inequality (8) is not empty. Indeed, as the matrix $\boldsymbol{A}$ is sparsified, the inequality $\boldsymbol{A}^{-} \leq \Delta\boldsymbol{q}\boldsymbol{p}^{-}$ holds. Consequently, we obtain $\Delta^{-1}\boldsymbol{A}^{-}\boldsymbol{p} \leq \Delta^{-1}\Delta\boldsymbol{q}\boldsymbol{p}^{-}\boldsymbol{p} = \boldsymbol{q}$, which results in $\alpha\Delta^{-1}\boldsymbol{A}^{-}\boldsymbol{p} \leq \alpha\boldsymbol{q}$.

By using properties of conjugate transposition, we have $(\boldsymbol{A}\boldsymbol{q}\boldsymbol{q}^{-})^{-} = \boldsymbol{q}(\boldsymbol{A}\boldsymbol{q})^{-}$ and $\boldsymbol{q}\boldsymbol{q}^{-} \geq \boldsymbol{I}$. Then, we write $\boldsymbol{q}^{-}\boldsymbol{A}^{-} \geq \boldsymbol{q}^{-}(\boldsymbol{A}\boldsymbol{q}\boldsymbol{q}^{-})^{-} = \boldsymbol{q}^{-}\boldsymbol{q}(\boldsymbol{A}\boldsymbol{q})^{-} = (\boldsymbol{A}\boldsymbol{q})^{-}$. After left multiplication of (8) by $\boldsymbol{q}^{-}$, we obtain

$$\alpha = \alpha\Delta^{-1}(\boldsymbol{A}\boldsymbol{q})^{-}\boldsymbol{p} \leq \alpha\Delta^{-1}\boldsymbol{q}^{-}\boldsymbol{A}^{-}\boldsymbol{p} \leq \boldsymbol{q}^{-}\boldsymbol{x} \leq \alpha\boldsymbol{q}^{-}\boldsymbol{q} = \alpha,$$

and thus arrive at the first equality at (5).

In addition, it follows from the row regularity of $\boldsymbol{A}$ and the left inequality in (8) that $\boldsymbol{A}\boldsymbol{x} \geq \alpha\Delta^{-1}\boldsymbol{A}\boldsymbol{A}^{-}\boldsymbol{p} \geq \alpha\Delta^{-1}\boldsymbol{p}$, which gives the second inequality at (5).

Finally, application of Lemma 2 provides the representation of the solution in the form of (9), which completes the proof. □

Example 7. As an illustration, we examine problem (3), where $m = n = 2$, in the framework of the semifield $\mathbb{R}_{\max,+}$ with the matrix and vectors given by

$$\boldsymbol{A} = \begin{pmatrix} 2 & 0 \\ 4 & 1 \end{pmatrix}, \qquad \boldsymbol{p} = \begin{pmatrix} 5 \\ 2 \end{pmatrix}, \qquad \boldsymbol{q} = \begin{pmatrix} 1 \\ 2 \end{pmatrix}.$$

We start with the evaluation of the minimum value by calculating

$$\boldsymbol{A}\boldsymbol{q} = \begin{pmatrix} 3 \\ 5 \end{pmatrix}, \qquad \Delta = (\boldsymbol{A}\boldsymbol{q})^{-}\boldsymbol{p} = 2.$$

Next, we find the threshold and sparsified matrices. With $\mathbb{0} = -\infty$, we write

$$\Delta^{-1}\boldsymbol{p}\boldsymbol{q}^{-} = \begin{pmatrix} 2 & 1 \\ -1 & -2 \end{pmatrix}, \qquad \widehat{\boldsymbol{A}} = \begin{pmatrix} 2 & \mathbb{0} \\ 4 & 1 \end{pmatrix}, \qquad \Delta^{-1}\widehat{\boldsymbol{A}}^{-}\boldsymbol{p}\boldsymbol{q}^{-} = \begin{pmatrix} 0 & -1 \\ -2 & -3 \end{pmatrix}.$$

The solution given by (8) is represented as follows:

$$\alpha\boldsymbol{x}' \leq \boldsymbol{x} \leq \alpha\boldsymbol{x}'', \qquad \boldsymbol{x}' = \Delta^{-1}\widehat{\boldsymbol{A}}^{-}\boldsymbol{p} = \begin{pmatrix} 1 \\ -1 \end{pmatrix}, \qquad \boldsymbol{x}'' = \boldsymbol{q} = \begin{pmatrix} 1 \\ 2 \end{pmatrix}, \qquad \alpha \in \mathbb{R}.$$

By applying (9), we obtain the solution in the alternative form

$$\boldsymbol{x} = \boldsymbol{B}\boldsymbol{u}, \qquad \boldsymbol{B} = \boldsymbol{I} \oplus \Delta^{-1}\widehat{\boldsymbol{A}}^{-}\boldsymbol{p}\boldsymbol{q}^{-} = \begin{pmatrix} 0 & -1 \\ -2 & 0 \end{pmatrix}, \qquad \boldsymbol{u} \in \mathbb{R}^2.$$

A graphical illustration of the solution is given in Fig. 3, which shows both the known partial solution by Lemma 3 (left), and the new extended solution provided by Theorem 6 (middle). In the left picture, the solution is depicted as a thick line drawn through the end point of the vector $\boldsymbol{q}$ at 45° to the axes.

The extended solution in the middle is represented by a strip between two hatched thick lines, which includes the previous solution as the upper boundary. Due to (8), this strip is drawn as the area covered when the vertical segment between the ends of the vectors $\boldsymbol{x}'$ and $\boldsymbol{x}''$ shifts at 45° to the axes. Solution (9) is depicted as the linear span of columns in the matrix $\boldsymbol{B} = (\boldsymbol{b}_1, \boldsymbol{b}_2)$.

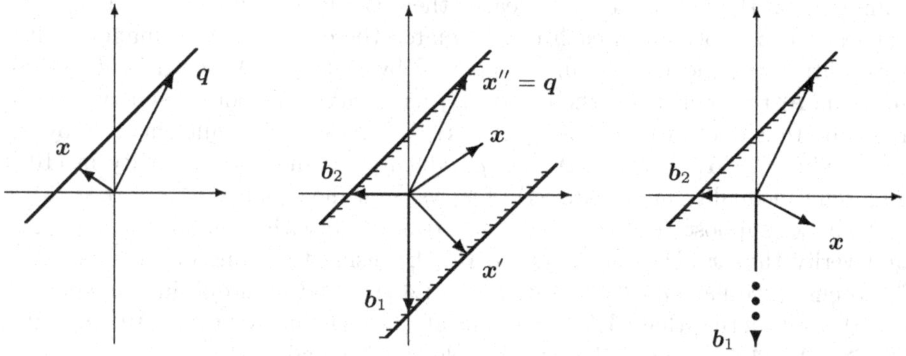

Fig. 3. Partial (left), extended (middle), and complete (right) solutions.

5 Complete Solution

We are now in a position to derive a complete solution to the problem. We start with the description of all solutions as a family of sets, each defined by a matrix obtained from the sparsified matrix of the problem. We discuss a backtracking procedure that generates all members in the family of solutions. Finally, we combine these solutions to provide a direct representation of a complete solution that describes, in a compact closed form, all solutions to the problem.

5.1 Derivation of All Solutions

The next result offers a simple way to describe all solutions to problem (3).

Theorem 8. *Let* $\boldsymbol{A}$ *be a row-regular sparsified matrix for problem* (3) *with a nonzero vector* $\boldsymbol{p}$ *and a regular vector* $\boldsymbol{q}$*, and* $\mathcal{A}$ *be the set of matrices obtained from* $\boldsymbol{A}$ *by fixing one nonzero entry in each row and setting the other ones to* $\mathbb{0}$.

Then, the minimum value in (3) *is equal to* $\Delta = (\boldsymbol{A}\boldsymbol{q})^{-}\boldsymbol{p}$*, and all regular solutions* $\boldsymbol{x}$ *are given by the conditions*

$$\alpha\Delta^{-1}\boldsymbol{A}_1^{-}\boldsymbol{p} \leq \boldsymbol{x} \leq \alpha\boldsymbol{q}, \qquad \alpha > \mathbb{0}, \qquad \boldsymbol{A}_1 \in \mathcal{A}; \tag{10}$$

or, equivalently, by the conditions

$$\boldsymbol{x} = (\boldsymbol{I} \oplus \Delta^{-1}\boldsymbol{A}_1^{-}\boldsymbol{p}\boldsymbol{q}^{-})\boldsymbol{u}, \qquad \boldsymbol{u} > \boldsymbol{0}, \qquad \boldsymbol{A}_1 \in \mathcal{A}. \tag{11}$$

Proof. It follows from Lemma 3 that all solutions of problem (3) are defined by system (5). Therefore, to prove the theorem, we need to show that each solution of system (5) is a solution of (10) with some matrix $\boldsymbol{A}_1 \in \mathcal{A}$, and vice versa.

Consider any matrix $\boldsymbol{A}_1 \in \mathcal{A}$, and note that it is row-regular. Moreover, the inequalities $\boldsymbol{A}_1 \leq \boldsymbol{A}$ and $\boldsymbol{A}_1^{-} \leq \boldsymbol{A}^{-}$ hold. In the same way as in Theorem 6, we see that since $\boldsymbol{A}_1^{-} \leq \boldsymbol{A}^{-} \leq \Delta\boldsymbol{q}\boldsymbol{p}^{-}$, the double inequality at (10) has solutions.

Let $\boldsymbol{x}$ be a solution to system (5). First, we take the inequality $\boldsymbol{A}\boldsymbol{x} \geq \alpha\Delta^{-1}\boldsymbol{p}$, and examine every corresponding scalar inequality to determine the maximal summand on the left-hand side. Clearly, there is a matrix $\boldsymbol{A}_1 \in \mathcal{A}$ with nonzero entries that are located in each row to match these maximal summands. With this matrix, the inequality can be replaced by $\boldsymbol{A}_1\boldsymbol{x} \geq \alpha\Delta^{-1}\boldsymbol{p}$ without loss of solution. At the same time, the matrix $\boldsymbol{A}_1$ has exactly one nonzero entry in each row, and thus obeys the inequality $\boldsymbol{A}_1^{-}\boldsymbol{A}_1 \leq \boldsymbol{I}$. After right multiplication by $\boldsymbol{x}$, we obtain $\boldsymbol{x} \geq \boldsymbol{A}_1^{-}\boldsymbol{A}_1\boldsymbol{x} \geq \alpha\Delta^{-1}\boldsymbol{A}_1^{-}\boldsymbol{p}$, which gives the left inequality in (10). The right inequality in (10) directly follows from the equality $\boldsymbol{q}^{-}\boldsymbol{x} = \alpha$ at (5).

Next, we suppose that the vector $\boldsymbol{x}$ satisfies (10) with some matrix $\boldsymbol{A}_1 \in \mathcal{A}$, and verify that $\boldsymbol{x}$ also solves system (5). By using the same arguments as in Theorem 6, we have $\boldsymbol{q}^{-}\boldsymbol{A}_1^{-} \geq (\boldsymbol{A}_1\boldsymbol{q})^{-} \geq (\boldsymbol{A}\boldsymbol{q})^{-}$, and then obtain the equality at (5). Considering that $\boldsymbol{A}\boldsymbol{A}_1^{-} \geq \boldsymbol{I}$, we take the left inequality at (10) to write $\boldsymbol{A}\boldsymbol{x} \geq \alpha\Delta^{-1}\boldsymbol{A}\boldsymbol{A}_1^{-}\boldsymbol{p} \geq \alpha\Delta^{-1}\boldsymbol{p}$, which yields the inequality at (5).

An application of Lemma 2 completes the proof. □

Note that the solution sets defined by different matrices from the set $\mathcal{A}$ in Theorem 8 can have nonempty intersection, as shown in the next example.

Example 9. Suppose that the matrix in Example 7 is replaced by its sparsified matrix, and consider the problem with

$$\boldsymbol{A} = \begin{pmatrix} 2 & \mathbb{0} \\ 4 & 1 \end{pmatrix}, \qquad \boldsymbol{p} = \begin{pmatrix} 5 \\ 2 \end{pmatrix}, \qquad \boldsymbol{q} = \begin{pmatrix} 1 \\ 2 \end{pmatrix}.$$

Since the sparsification of the matrix does not change the minimum in the problem, we still have $\Delta = (\boldsymbol{A}\boldsymbol{q})^{-}\boldsymbol{p} = 2$.

Consider the set $\mathcal{A}$, which is formed of the matrices obtained from $\boldsymbol{A}$ by keeping only one nonzero entry in each row. This set consists of two matrices

$$\boldsymbol{A}_1 = \begin{pmatrix} 2 & \mathbb{0} \\ 4 & \mathbb{0} \end{pmatrix}, \qquad \boldsymbol{A}_2 = \begin{pmatrix} 2 & \mathbb{0} \\ \mathbb{0} & 1 \end{pmatrix}.$$

Let us write the solutions defined by these matrices in the form of (11). First, we calculate the matrices

$$\Delta^{-1}\boldsymbol{A}_1^{-}\boldsymbol{p}\boldsymbol{q}^{-} = \begin{pmatrix} 0 & -1 \\ \mathbb{0} & \mathbb{0} \end{pmatrix}, \qquad \Delta^{-1}\boldsymbol{A}_2^{-}\boldsymbol{p}\boldsymbol{q}^{-} = \begin{pmatrix} 0 & -1 \\ -2 & -3 \end{pmatrix}.$$

Using the first matrix yields the solution

$$\boldsymbol{x} = \boldsymbol{B}_1\boldsymbol{u}, \qquad \boldsymbol{B}_1 = \boldsymbol{I} \oplus \Delta^{-1}\boldsymbol{A}_1^{-}\boldsymbol{p}\boldsymbol{q}^{-} = \begin{pmatrix} 0 & -1 \\ \mathbb{0} & 0 \end{pmatrix}, \qquad \boldsymbol{u} \in \mathbb{R}^2.$$

The second solution coincides with that obtained in Example 7 in the form

$$\boldsymbol{x} = \boldsymbol{B}_2\boldsymbol{u}, \qquad \boldsymbol{B}_2 = \boldsymbol{I} \oplus \Delta^{-1}\boldsymbol{A}_2^{-}\boldsymbol{p}\boldsymbol{q}^{-} = \begin{pmatrix} 0 & -1 \\ -2 & 0 \end{pmatrix}, \qquad \boldsymbol{u} \in \mathbb{R}^2.$$

The first solution is displayed in Fig. 3 (right) as the half-plane below the thick hatched line. Clearly, this area completely covers the strip region in Fig. 3 (middle), offered by the second solution.

5.2 Backtracking Procedure for Generating Solutions

Consider a backtracking search procedure that finds all solutions to problem (3) with the sparsified matrix $\boldsymbol{A}$ in an economical way. To generate all matrices in $\mathcal{A}$, the procedure examines each row in the matrix $\boldsymbol{A}$ to fix one nonzero entry in the row and to set the other entries to zeros. After selecting a nonzero entry in the current row, the subsequent rows are modified to reduce the number of remaining alternatives. Then, a nonzero entry in the next row of the modified matrix is fixed if any exists, and the procedure continues repeatedly.

Suppose that every row of the modified matrix has exactly one nonzero entry. This matrix is considered as a solution matrix $\boldsymbol{A}_1 \in \mathcal{A}$, and stored in a solution list. Furthermore, if the modified matrix has zero rows, it does not provide a solution. In either case, the procedure returns to roll back all last modifications, and to fix the next nonzero entry in the current row if there is any, or goes back to the previous row otherwise. The procedure is completed when no more nonzero entries in the first row of the matrix $\boldsymbol{A}$ can be selected.

To describe the technique used to reduce search, suppose that the procedure, which has fixed one nonzero entry in each of the rows $1, \ldots, i-1$, currently selects a nonzero entry in row i of the modified matrix $\widetilde{\boldsymbol{A}}$, say the entry $\widetilde{a}_{ij}$ in column j, whereas the other entries in the row are set to zero.

Any solution vector $\boldsymbol{x}$ must satisfy the inequality $\widetilde{\boldsymbol{A}}\boldsymbol{x} \geq \alpha\Delta^{-1}\boldsymbol{p}$ in system (5). Specifically, the scalar inequality for row i, where only the entry $\widetilde{a}_{ij}$ is nonzero, reads $\widetilde{a}_{ij}x_j \geq \alpha\Delta^{-1}p_i$, or, equivalently, $x_j \geq \alpha\Delta^{-1}\widetilde{a}_{ij}^{-1}p_i$. If $p_i > \mathbb{0}$, then the inequality determines a lower bound for x_j in the solution under construction.

Assuming $p_i > \mathbb{0}$, consider the entries of column j in rows $k = i+1, \ldots, n$. Provided that the condition $\widetilde{a}_{kj} \geq \widetilde{a}_{ij}p_i^{-1}p_k$ is satisfied for row k, we write $\widetilde{a}_{kj}x_j \geq \alpha\widetilde{a}_{ij}p_i^{-1}p_k\Delta^{-1}\widetilde{a}_{ij}^{-1}p_i \geq \alpha\Delta^{-1}p_k$, which means that the inequality at (5) for this row is valid regardless of x_l for $l \neq j$. In this case, further examination of nonzero entries $\widetilde{a}_{kl}$ in row k cannot impose new constraints on the element x_l in the vector $\boldsymbol{x}$, and thus is not needed. These entries can be set to zeros without affecting the inequality, which may decrease the number of search alternatives.

Example 10. As a simple illustration of the technique, we return to Example 9, where the initial sparsified matrix and its further sparsifications are given by

$$\boldsymbol{A} = \begin{pmatrix} 2 & \mathbb{0} \\ 4 & 1 \end{pmatrix}, \qquad \boldsymbol{A}_1 = \begin{pmatrix} 2 & \mathbb{0} \\ 4 & \mathbb{0} \end{pmatrix}, \qquad \boldsymbol{A}_2 = \begin{pmatrix} 2 & \mathbb{0} \\ \mathbb{0} & 1 \end{pmatrix}.$$

The procedure first fixes the entry $a_{11} = 2$. Since $a_{21} = 4$ is greater than $a_{11}p_1^{-1}p_2 = -1$, the procedure sets a_{22} to $\mathbb{0}$, which immediately excludes the matrix $\boldsymbol{A}_2$ from further consideration, and hence reduces the analysis to $\boldsymbol{A}_1$.

5.3 Representation of Complete Solution in Closed Form

A complete solution to problem (3) can be expressed in a closed form as follows.

Theorem 11. *Let $\boldsymbol{A}$ be a row-regular sparsified matrix for problem* (3) *with a nonzero vector $\boldsymbol{p}$ and a regular vector $\boldsymbol{q}$, and $\mathcal{A}$ be the set of matrices obtained from $\boldsymbol{A}$ by fixing one nonzero entry in each row and setting the other ones to $\mathbb{0}$.*

Let $\boldsymbol{B}$ be the matrix, which is formed by putting together all columns of the matrices $\boldsymbol{B}_1 = \boldsymbol{I} \oplus \Delta^{-1}\boldsymbol{A}_1^{-}\boldsymbol{p}\boldsymbol{q}^{-}$ for every $\boldsymbol{A}_1 \in \mathcal{A}$, and $\boldsymbol{B}_0$ be a matrix whose columns comprise a maximal linear independent system of the columns in $\boldsymbol{B}$.

Then, the minimum value in (3) *is equal to $\Delta = (\boldsymbol{A}\boldsymbol{q})^{-}\boldsymbol{p}$, and all regular solutions are given by*

$$\boldsymbol{x} = \boldsymbol{B}_0\boldsymbol{v}, \qquad \boldsymbol{v} > \mathbf{0}.$$

Proof. Suppose that the set $\mathcal{A}$ consists of k elements, which can be enumerated as $\boldsymbol{A}_1, \ldots, \boldsymbol{A}_k$. For each $\boldsymbol{A}_i \in \mathcal{A}$, we define the matrix $\boldsymbol{B}_i = \boldsymbol{I} \oplus \Delta^{-1}\boldsymbol{A}_i^{-}\boldsymbol{p}\boldsymbol{q}^{-}$.

First, note that by Theorem 8, the set of vectors $\boldsymbol{x}$ that solve problem (3) is the union of subsets, each of which corresponds to one index $i = 1, \ldots, k$, and contains the vectors given by $\boldsymbol{x} = \boldsymbol{B}_i\boldsymbol{u}_i$, where $\boldsymbol{u}_i > \mathbf{0}$ is a vector.

We now verify that all solutions to the problem can also be represented as

$$\boldsymbol{x} = \boldsymbol{B}_1\boldsymbol{u}_1 \oplus \cdots \oplus \boldsymbol{B}_k\boldsymbol{u}_k, \qquad \boldsymbol{u}_1, \ldots, \boldsymbol{u}_k > \mathbf{0}. \tag{12}$$

Indeed, any solution provided by Theorem 8 can be written in the form of (12). At the same time, since the solution set is closed under addition by Corollary 4, any vector $\boldsymbol{x}$ given by representation (12) solves the problem. Therefore, this representation describes all solutions to the problem.

With the matrix $\boldsymbol{B} = (\boldsymbol{B}_1, \ldots, \boldsymbol{B}_k)$ and the vector $\boldsymbol{u} = (\boldsymbol{u}_1^T, \ldots, \boldsymbol{u}_k^T)^T$, we rewrite (12) in the form

$$\boldsymbol{x} = \boldsymbol{B}\boldsymbol{u}, \qquad \boldsymbol{u} > \boldsymbol{0},$$

which specifies each solution to be a linear combination of columns in $\boldsymbol{B}$.

Clearly, elimination of a column that linearly depends on some others leaves the linear span of the columns unchanged. By eliminating all dependent columns, we reduce the matrix $\boldsymbol{B}$ to a matrix $\boldsymbol{B}_0$ to express any solution to the problem by a linear combination of columns in $\boldsymbol{B}_0$ as $\boldsymbol{x} = \boldsymbol{B}_0\boldsymbol{v}$, where $\boldsymbol{v} > \boldsymbol{0}$ is a vector, and thus complete the proof. □

Example 12. We again consider results of Example 9 to examine the matrices

$$\boldsymbol{B}_1 = \begin{pmatrix} 0 & -1 \\ \mathbb{0} & 0 \end{pmatrix}, \qquad \boldsymbol{B}_2 = \begin{pmatrix} 0 & -1 \\ -2 & 0 \end{pmatrix}.$$

We take the dissimilar columns from $\boldsymbol{B}_1$ and $\boldsymbol{B}_2$, and denote them by

$$\boldsymbol{b}_1 = \begin{pmatrix} 0 \\ \mathbb{0} \end{pmatrix}, \qquad \boldsymbol{b}_2 = \begin{pmatrix} -1 \\ 0 \end{pmatrix}, \qquad \boldsymbol{b}_3 = \begin{pmatrix} 0 \\ -2 \end{pmatrix}.$$

Next, we put these columns together to form the matrix

$$\boldsymbol{B} = (\boldsymbol{b}_1\ \boldsymbol{b}_2\ \boldsymbol{b}_3) = \begin{pmatrix} 0 & -1 & 0 \\ \mathbb{0} & 0 & -2 \end{pmatrix}.$$

Furthermore, we examine the matrix $\boldsymbol{B}_1 = (\boldsymbol{b}_1, \boldsymbol{b}_2)$ to calculate $\delta(\boldsymbol{B}_1, \boldsymbol{b}_3)$, and then to apply Lemma 1. Since we have

$$(\boldsymbol{b}_3^- \boldsymbol{B}_1)^- = \boldsymbol{B}_1(\boldsymbol{b}_3^- \boldsymbol{B}_1)^- = \begin{pmatrix} 0 \\ -2 \end{pmatrix}, \qquad \delta(\boldsymbol{B}_1, \boldsymbol{b}_3) = (\boldsymbol{B}_1(\boldsymbol{b}_3^- \boldsymbol{B}_1)^-)^- \boldsymbol{b}_3 = 0 = \mathbb{1},$$

the column $\boldsymbol{b}_3$ is linearly dependent on the others, and thus can be removed.

Considering that the columns $\boldsymbol{b}_1$ and $\boldsymbol{b}_2$ are obviously not collinear, none of them can be further eliminated. With $\boldsymbol{B}_0 = \boldsymbol{B}_1$, a complete solution to the problem is given by $\boldsymbol{x} = \boldsymbol{B}_0\boldsymbol{v}$, where $\boldsymbol{v} > \boldsymbol{0}$, and depicted in Fig. 3 (right).

6 Conclusions

In many tropical optimization problems encountered in real-world applications, it is not too difficult to obtain a particular solution in an explicit form, whereas finding all solutions may be a hard problem. This paper was concerned with a multidimensional optimization problem that arises in various applications as the problem of minimizing the span seminorm, and is formulated to minimize a nonlinear function defined on vectors over an idempotent semifield by a given matrix. To obtain a complete solution of the problem, we first characterized all solutions by a system of simultaneous vector equation and inequality, and then developed a new matrix sparsification technique. This technique was applied to the description of all solutions to the problem in an explicit compact vector form.

The extension of the characterization of solutions and sparsification technique proposed in the paper to other tropical optimization problems may be of particular interest and present important directions for future work.

Acknowledgments. The author is very grateful to three reviewers for their valuable comments and useful suggestions that have been incorporated in the final version.

References

[ABG07] Akian, M., Bapat, R., Gaubert, S.: Max-plus algebra. In: Hogben, L. (ed.) Handbook of Linear Algebra. Discrete Mathematics and Its Applications, pp. 25–1–25–17. Taylor and Francis, Boca Raton (2007)

[BT09] Butkovič, P., Tam, K.P.: On some properties of the image set of a max-linear mapping. In: Litvinov, G.L., Sergeev, S.N. (eds.) Tropical and Idempotent Mathematics, Contemp. Math., vol. 495, pp. 115–126. AMS (2009)

[But10] Butkovič, P.: Max-linear Systems: Theory and Algorithms. Springer Monographs in Mathematics. Springer, London (2010)

[CG62] Cuninghame-Green, R.A.: Describing industrial processes with interference and approximating their steady-state behaviour. Oper. Res. Quart. 13, 95–100 (1962)

[CGB04] Cuninghame-Green, R.A., Butkovič, P.: Bases in max-algebra. Linear Algebra Appl. 389, 107–120 (2004)

[CG79] Cuninghame-Green, R.: Minimax Algebra. Lecture Notes in Economics and Mathematical Systems, vol. 166. Springer, Berlin (1979)

[Gif63] Giffler, B.: Scheduling general production systems using schedule algebra. Naval Res. Logist. Quart. 10, 237–255 (1963)

[Gol03] Golan, J.S.: Semirings and Affine Equations Over Them: Theory and Applications, Mathematics and Its Applications, vol. 556. Springer, New York (2003)

[GM08] Gondran, M., Minoux, M.: Graphs, Dioids and Semirings: New Models and Algorithms, Operations Research/Computer Science Interfaces, vol. 41. Springer, New York (2008)

[HOvdW06] Heidergott, B., Olsder, G.J., van der Woude, J.: Max-plus at Work: Modeling and Analysis of Synchronized Systems. Princeton Series in Applied Mathematics. Princeton Univ. Press, Princeton (2006)

[Hof63] Hoffman, A.J.: On abstract dual linear programs. Naval Res. Logist. Quart. 10, 369–373 (1963)

[KM97] Kolokoltsov, V.N., Maslov, V.P.: Idempotent Analysis and Its Applications, Mathematics and Its Applications, vol. 401. Kluwer Acad. Publ, Dordrecht (1997)

[Kor65] Korbut, A.A.: Extremal spaces. Soviet Math. Dokl. 6, 1358–1361 (1965)

[Kri04] Krivulin, N.K.: On solution of linear vector equations in idempotent algebra. In: Chirkov, M.K. (ed.) Mathematical Models. Theory and Applications, Issue 5, pp. 105–113. Saint Petersburg State Univ., St. Petersburg (2004) (in Russian)

[Kri09] Krivulin, N.K.: Methods of Idempotent Algebra for Problems in Modeling and Analysis of Complex Systems. Saint Petersburg Univ. Press, St. Petersburg (2009) (in Russian)

[Kri12] Krivulin, N.: A solution of a tropical linear vector equation. In: Yenuri, S. (ed.) Advances in Computer Science. Recent Advances in Computer Engineering Series, vol. 5, pp. 244–249. WSEAS Press (2012)

[Kri13] Krivulin, N.: Explicit solution of a tropical optimization problem with application to project scheduling. In: Biolek, D., Walter, H., Utu, I., von Lucken, C. (eds.) Mathematical Methods and Optimization Techniques in Engineering, pp. 39–45. WSEAS Press (2013)

[Kri14] Krivulin, N.: Complete solution of a constrained tropical optimization problem with application to location analysis. In: Höfner, P., Jipsen, P., Kahl, W., Müller, M.E. (eds.) RAMiCS 2014. LNCS, vol. 8428, pp. 362–378. Springer, Cham (2014)

[Kri15a] Krivulin, N.: Extremal properties of tropical eigenvalues and solutions to tropical optimization problems. Linear Algebra Appl. 468, 211–232 (2015)

[Kri15b] Krivulin, N.: A multidimensional tropical optimization problem with nonlinear objective function and linear constraints. Optimization 64, 1107–1129 (2015)

[Lit07] Litvinov, G.: Maslov dequantization, idempotent and tropical mathematics: A brief introduction. J. Math. Sci (N. Y.) 140, 426–444 (2007)

[Pan61] Pandit, S.N.N.: A new matrix calculus. J. SIAM 9, 632–639 (1961)

[Pet67] Peteanu, V.: An algebra of the optimal path in networks. Mathematica 9(2), 335–342 (1967)

[Rom64] Romanovskiĭ, I.V.: Asymptotic behavior of dynamic programming processes with a continuous set of states. Soviet Math. Dokl. 5, 1684–1687 (1964)

[SS09] Speyer, D., Sturmfels, B.: Tropical mathematics. Math. Mag. 82, 163–173 (2009)

[Tam10] Tam, K.P.: Optimizing and Approximating Eigenvectors in Max-Algebra. PhD thesis, The University of Birmingham, Birmingham (2010)

[TB06] T'kindt, V., Billaut, J.-C.: Multicriteria Scheduling: Theory, Models and Algorithms. Springer, Berlin (2006)

[Vor63] Vorob'ev, N.N.: The extremal matrix algebra. Soviet Math. Dokl. 4, 1220–1223 (1963)

Towards Antichain Algebra

Bernhard Möller

Institut für Informatik, Universität Augsburg, D-86135 Augsburg, Germany
bernhard.moeller@informatik.uni-augsburg.de

Abstract. We use an algebra of preference strict-orders to give a formal derivation of the standard Block-Nested Loop (BNL) algorithm for computing the best or maximal objects w.r.t. such an order. This derivation is presented in terms of antichains, i.e., sets of mutually incomparable objects. We define an approximation relation between antichains that reflects the steps taken by the BNL algorithm. This induces a semilattice and the operator computing the maximal objects of a subset can be viewed as a closure operator in an associated pre-ordered set and hence yields a characterisation of antichains in terms of a Galois connection.

Keywords: Preference relations, Maximal objects, Block-nested loop algorithm, Lattice theory, Galois connections.

1 Introduction

The motivation for this work arose in the area of preference databases (see [7]). Classical databases had supported only queries with so-called *hard constraints*, by which the objects sought in the database are clearly and sharply characterised. Hence, if there are no exact matches the empty result set is returned, which is often very frustrating for users. As a remedy, over the last decades queries with *soft constraints* have been studied. These constraints arise from a formalisation of the *user's preferences* in the form of partial strict-orders.

For instance, a person wanting to have a vacation may prefer inexpensive hotels closer to the beach over expensive ones further off. This could be formalised as the following preference relation $\prec$ between tuples s, t:

$$\begin{aligned} s \prec t \Leftrightarrow_{df} \; & (t.prize < s.prize \wedge t.dist \leq s.dist) \vee \\ & (t.prize \leq s.prize \wedge t.dist < s.dist) \end{aligned}$$

A query with such a preference order may then return the set of "best" or maximal objects found in the search space. As usual in partial orders, the maximal objects are pairwise incomparable, i.e., form an *antichain*.

If the search space has two dimensions, like in the above example, it can be depicted in a 2D rectangular coordinate system. The maxima then are the end points of a stair-case like shape, a.k.a. the "skyline" [1], see Fig. 1.

In earlier papers [8, 7] we have developed an algebraic calculus for reasoning about the set $a \rhd p$ of maximal objects in a set p w.r.t. a preference relation a, independent of the special application area of databases. In the present paper

W. Kahl et al. (Eds.): RAMiCS 2015, LNCS 9348, pp. 344–361, 2015.
DOI: 10.1007/978-3-319-24704-5_21

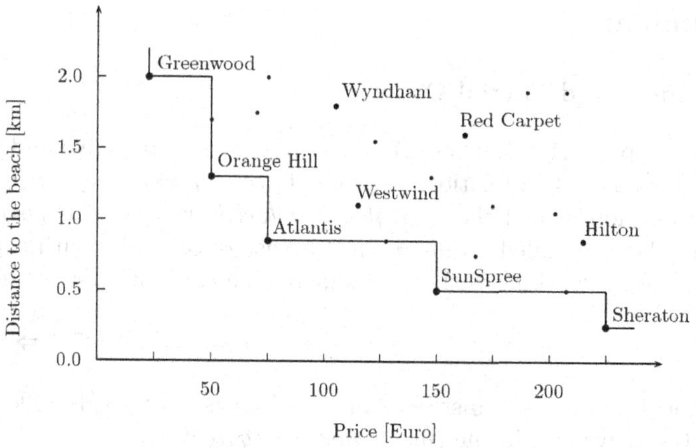

Fig. 1. A skyline diagram

we extend these results by a number of additional ones. As a test case we give a derivation of the standard *Block-Nested Loop (BNL)* algorithm (e.g. [1]) for computing the maximal objects. To the best of our knowledge this is the first calculational treatment of that algorithm. A closer analysis exhibits that there is an approximation order between antichains underlying that algorithm which even induces a semilattice structure. While, w.r.t. the inclusion order, the maxima operator $a \triangleright p$ is antitone (i.e., monotonically decreasing) in a, it is neither isotone (i.e., monotonically increasing) nor antitone in p. Fortunately, isotony can be recovered by passing to the approximation order. Last, the maxima operator can be viewed as a closure operator in an associated preordered set and hence yields a characterisation of antichains in terms of a Galois connection.

The paper is structured as follows. In Sect. 2 we recapitulate basic notions about preorders and orders as well as the algebraic notions in terms of semirings that underlie our calculus. Sect. 3 presents basic results about the algebraic representation of the maxima operator. Next to new properties concerning the relation between what we call normality of a strict-order and its noetherity, we show a couple of auxiliary results for the following sections. Sect. 4 provides various characterisations of antichains and properties concerning the maxima of a union of sets. In Sect. 5 we give the announced calculational derivation of the BNL algorithm. Sect. 6 presents an approximation order between antichains and shows that it induces a semilattice as well as some results on isotony and suprema preservation of the maxima operator. The BNL algorithm is shown to construct an ascending chain of antichains w.r.t. that order. In Sect. 7 we then prove that a modified version of the approximation order exhibits the maxima operator as a closure operator in a preordered set and hence as an adjoint in a Galois connection. Since both closures and Galois connections are usually only dealt with in partial orders, we provide the necessary results on the preorder case, partially in an Appendix. The paper finishes with a brief conclusion and outlook in Sect. 8.

2 Preliminaries

2.1 Preorders and Partial Orders

A *preorder* is a pair $(A, \leq)$, where A is a set and $\leq$ is a reflexive and transitive binary relation on A. The relation $\sim$ defined by $x \sim y \Leftrightarrow_{df} x \leq y \wedge y \leq x$ is an equivalence relation, called the *equivalence induced by* $\leq$. If $\leq$ is also antisymmetric then $(A, \leq)$ is called an *order*; in this case $\sim$ coincides with equality.

A useful tool for working with preorders are the rules of *indirect inequality*:

$$x \leq y \;\Leftrightarrow\; (\forall\, z : z \leq x \Rightarrow z \leq y)\,, \qquad x \leq y \;\Leftrightarrow\; (\forall\, z : y \leq z \Rightarrow x \leq z)\,.$$

The direction $(\Rightarrow)$ needs transitivity of $\leq$, whereas $(\Leftarrow)$ needs reflexivity. By combining these, we obtain the rule of *indirect equivalence*:

$$x \sim y \;\Leftrightarrow\; (\forall\, z : z \leq x \Leftrightarrow z \leq y) \;\Leftrightarrow\; (\forall\, z : x \leq z \Leftrightarrow y \leq z)\,.$$

2.2 Algebraic Notions

Throughout we assume an *idempotent semiring* $(S, +, 0, \cdot, 1)$. This means that $+$ and $\cdot$ are associative operators on set S, with neutral elements 0 and 1, resp.; moreover, $+$ is assumed to be commutative and idempotent, i.e., to satisfy $a+a = a$ for all $a \in S$. Finally, $\cdot$ is assumed to distribute through $+$ in both arguments and to preserve 0, i.e., $0 \cdot a = 0 = a \cdot 0$.

Because of the properties of $+$ one can define a partial order $\leq$ on S by $a \leq b \Leftrightarrow_{df} a + b = b$. It is called the *natural order* or *subsumption order*. It induces an upper semilattice in which $+$ is the binary supremum operator. If that semilattice is even a complete lattice and $\cdot$ distributes through arbitrary suprema then S is called a *quantale*.

A prominent example of an idempotent semiring, that is even a quantale, is provided by the set of all binary relations over a set M, with union as $+$ and relational composition as $\cdot$. The roles of 0 and 1 are played by the empty relation $\emptyset$ and the identity relation I. The natural order coincides with relational inclusion. *Partial identity relations* $I_N =_{df} \{(x, x) \,|\, x \in N\} \subseteq I$, a.k.a. *coreflexives* or *monotypes*, can be used to encode subsets $N \subseteq M$ as relations.

Inspired by that, we model preference relations between database tuples abstractly by general semiring elements $a \in S$ and sets of database tuples by *tests* $p \leq 1$, analogous to the above partial identity relations. Tests p are required to have a complement $\neg p$ relative to 1, uniquely characterised by the conditions $p + \neg p = 1$ and $p \cdot \neg p = 0 = \neg p \cdot p$.

The set of all tests of S is denoted by $\mathsf{test}(S)$; it forms a Boolean algebra with $+$ as supremum and $\cdot$ as infimum, least element 0 and greatest element 1. Between tests the order $\leq$ is the abstract counterpart of set inclusion. We define the difference operator for $p, q \in \mathsf{test}(S)$ as $p - q =_{df} p \cdot \neg q$ and assume that it associates to the left. We also note that it is right-commutative, i.e., satisfies $p - q - r = p - r - q$.

An element $p \in \mathsf{test}(S)$ is called *atomic* if $p \neq 0$ and $\forall q \in \mathsf{test}(S) : q \leq p \Rightarrow q = 0 \lor q = p$. While general tests stand for sets of database tuples, atomic tests correspond to single database tuples. Because of that we will frequently use set-theoretic terminology when talking about them, such as "all objects in p" and the like. Finally, we note that an atom x in a Boolean algebra satisfies $x \leq p+q \Rightarrow x \leq p \lor x \leq q$. In particular,

$$x \not\leq p \Rightarrow x \leq \neg p\ . \tag{1}$$

Tests are used to define the central operators of a *modal semiring*, namely box and diamond which can be defined in a forward and backward form. In the present note we will only use the forward diamond $|a\rangle : \mathsf{test}(S) \to \mathsf{test}(S)$, which can be axiomatised by

$$|a\rangle q \leq p \Leftrightarrow \neg p \cdot a \cdot q \leq 0\ , \qquad |a \cdot b\rangle q = |a\rangle|b\rangle q =_{df} |a\rangle(|b\rangle q)\ .$$

Informally, the test $|a\rangle q$ represents all database tuples that are a-related to (or dominated by) some tuple in the set represented by q. Hence $|a\rangle$ can be viewed as an algebraic form of the inverse image operator on binary relations. In particular, the *domain* of element a can be defined as the inverse image of the largest test 1 as $\ulcorner a =_{df} |a\rangle 1$.

A corresponding forward box operator $|a]$ is defined as the De Morgan dual of $|a\rangle$ by $|a]q =_{df} \neg|a\rangle\neg q$. It is an algebraic counterpart of Dijkstra's wlp operator and can be used to define an algebraic version of Hoare triples.

Diamond and box satisfy many useful laws (e.g. [3]). The most important ones for diamond are additivity (and hence isotony) in both arguments:

$$|a+b\rangle p = |a\rangle p + |b\rangle p\ , \qquad |a\rangle(p+q) = |a\rangle p + |a\rangle q\ .$$

In fact, $|a\rangle$ preserves arbitrary suprema; if S is a quantale then $|\ \rangle$ preserves abitrary suprema in both arguments.

If the underlying semiring is a *Kleene algebra*, i.e., has an operation $*$ for finite iteration with the standard axioms (e.g. [6]), we have the *unfold* and *induction rules* for the diamond of a starred element:

$$|a^*\rangle p = p + |a\rangle|a^*\rangle p\ , \tag{star-dia-unfold}$$

$$p \leq q \land |a\rangle q \leq q \Rightarrow |a^*\rangle p \leq q\ . \tag{star-dia-induct}$$

3 Strict-Orders and Maxima

Definition 3.1. An element a is called *d-transitive* ("d" standing for "diamond") if all tests p satisfy $|a\rangle|a\rangle p \leq |a\rangle p$. By the second diamond axiom this is equivalent to $|a \cdot a\rangle p \leq |a\rangle p$. It is, however, more liberal than stipulating $a \cdot a \leq a$; for the case of relations both formulations coincide. An element a is called *d-irreflexive* if for all atomic tests x we have $x \cdot |a\rangle x \leq 0$. A d-transitive and d-irreflexive element is called a *strict-order*.

Corollary 3.2. *For d-transitive a and test p, $|a\rangle|a+1\rangle p = |a\rangle p = |a+1\rangle|a\rangle p$.*

Proof. We only show the first equation; the second one is symmetric. By distributivity of diamond, $|1\rangle$ being the identity, the assumption and the definition of $\leq$ we have $|a\rangle|a+1\rangle p = |a\rangle|a\rangle p + |a\rangle|1\rangle p = |a\rangle|a\rangle p + |a\rangle p = |a\rangle p$. □

In a Kleene algebra, for any d-transitive element a and test p,

$$|a^*\rangle p = |a+1\rangle p = |a\rangle p + |1\rangle p = |a\rangle p + p \ . \tag{2}$$

Definition 3.3. The *best* or *maximal* objects w.r.t. an element a and a test p are represented by the test

$$a \rhd p =_{df} p - |a\rangle p \ .$$

This can be understood as follows. The expression $|a\rangle p$, the inverse image of p under preference element a, denotes the set of objects that are dominated by some object in p. Hence $p - |a\rangle p$ consists of the non-dominated and hence maximal objects in p.

The following lemma collects useful properties of the maximality operator; proofs can be found in [7].

Lemma 3.4. *The following holds for arbitrary elements a, b and test p:*

1. $a \rhd 0 = 0$.
2. $a \rhd 1 = \neg\ulcorner a$.
3. $\ulcorner b \leq \ulcorner a \Leftrightarrow a \rhd 1 \leq b \rhd 1$.
4. $a \rhd p \leq p$.
5. $a \rhd (a \rhd p) = a \rhd p$.
6. $(a+b) \rhd p = (a \rhd p) \cdot (b \rhd p)$.
7. $b \leq a \Rightarrow a \rhd p \leq b \rhd p$, *i.e.,* $\rhd$ *is antitone in its first argument.*
8. $1 \leq a \Rightarrow a \rhd p = 0$.

So far, we have not required any special properties of the elements a that represent, e.g., preference relations. Instead of d-transitivity or d-irreflexivity we need an assumption that such elements admit "enough" maximal objects. This is expressed by requiring every non-maximal object to be dominated by some maximal one. In a setting with finitely many objects, such as a database, and a preference relation on them this property is always satisfied. We will treat the case of infinite sets in Theorem 3.9 and Cor. 3.11 where we establish a connection between the notions of normality and being noetherian.

Definition 3.5. An element a is called *normal* [7] if $\forall p : |a\rangle p \leq |a\rangle(a \rhd p)$, meaning that every object dominated by some object of p is also dominated by a maximal object of p. By $a \rhd p \leq p$ and isotony of diamond this is equivalent to

$$\forall p : |a\rangle p = |a\rangle(a \rhd p) \ . \tag{3}$$

One of the most important applications of normality is the following law.

Theorem 3.6. *Let a be normal. Then* $a \rhd (p+q) = a \rhd (a \rhd p + a \rhd q)$.

This theorem paves the way for a distributed computation of maxima, as for disjoint p and q the calculations $a \rhd p$ and $a \rhd q$ are independent. Early examples are found in [2, 10], further ones again in [1]. For a proof of the theorem see [7]; it generalises from $+$ to arbitrary existing suprema in $\mathsf{test}(S)$.

Next we present the announced connection between noetherity and the existence of maximal elements which will be used in Sect. 6.

Definition 3.7. An object a is called *noetherian* if, for all tests p,

$$a \rhd p \leq 0 \Rightarrow p \leq 0 \ .$$

This definition can be understood as follows. By contraposition and leastness of 0 it is equivalent to

$$p \neq 0 \Rightarrow a \rhd p \neq 0 \ ,$$

which means that every non-empty p contains at least one maximal object (which is the dual of the usual well-foundedness condition). In the relational case it is therefore also equivalent to the absence of infinitely ascending chains. For details see [4]. The following properties are straightforward by Boolean algebra.

Corollary 3.8. *Assume an element a.*

1. *For arbitrary test p we have $a \rhd p \leq 0$ iff $p \leq |a\rangle p$.*
2. *a is Noetherian iff for all tests p we have $p \leq |a\rangle p \Rightarrow p \leq 0$.*

Theorem 3.9. *Let $a \in S$ be noetherian and let a^* be its reflexive and transitive closure. Then for any $q \in \mathsf{test}(S)$ we have $q \leq |a^*\rangle(a \rhd q)$. Informally, this means that any point in the set abstractly represented by q is dominated w.r.t. a^* by some point in $a \rhd q$.*

Proof.

$$
\begin{array}{ll}
& q \leq |a^*\rangle(a \rhd q) \\
\Leftrightarrow & \quad \{\!\mid \text{Boolean algebra} \mid\!\} \\
& q - |a^*\rangle(a \rhd q) \leq 0 \\
\Leftarrow & \quad \{\!\mid \text{noetherity of } a \text{ and Corollary 3.8.2} \mid\!\} \\
& q - |a^*\rangle(a \rhd q) \leq |a\rangle(q - |a^*\rangle(a \rhd q)) \\
\Leftrightarrow & \quad \{\!\mid \text{Boolean algebra} \mid\!\} \\
& q \leq |a^*\rangle(a \rhd q) + |a\rangle(q - |a^*\rangle(a \rhd q)) \\
\Leftrightarrow & \quad \{\!\mid \text{(star-dia-unfold) and distributivity} \mid\!\} \\
& q \leq a \rhd q + |a\rangle|a^*\rangle(a \rhd q) + |a\rangle(q - |a^*\rangle(a \rhd q)) \\
\Leftrightarrow & \quad \{\!\mid \text{Boolean algebra and distributivity} \mid\!\} \\
& q - a \rhd q \leq |a\rangle(|a^*\rangle(a \rhd q) + (q - |a^*\rangle(a \rhd q)) \\
\Leftrightarrow & \quad \{\!\mid \text{Boolean algebra} \mid\!\} \\
& q \cdot |a\rangle q \leq |a\rangle(|a^*\rangle(a \rhd q) + q) \\
\Leftarrow & \quad \{\!\mid \text{lattice algebra} \mid\!\} \\
& |a\rangle q \leq |a\rangle(|a^*\rangle(a \rhd q) + q)
\end{array}
$$

$\Leftarrow \quad \{\![$ isotony of diamond $]\!\}$
$q \leq |a^*\rangle(a \triangleright q) + q$
$\Leftrightarrow \quad \{\![$ lattice algebra $]\!\}$
TRUE .

□

Corollary 3.10. *If a is noetherian and d-transitive then for all tests p, q we have*

$$q \leq |a+1\rangle(a \triangleright q) \ ,$$
$$p \leq |a+1\rangle q \Leftrightarrow a \triangleright p \leq |a+1\rangle q \ .$$

Proof. The first claim follows from Th. 3.9 and (2).

For the second claim, ($\Rightarrow$) follows from $a \triangleright p \leq p$. For ($\Leftarrow$) we have, by the first claim, the assumption with isotony of diamond and finally d-transitivity of a and hence of $a+1$ that $p \leq |a+1\rangle(a \triangleright p) \leq |a+1\rangle|a+1\rangle q \leq |a+1\rangle q$. □

This allows a much shorter proof of the following property from [7].

Corollary 3.11. *A noetherian and d-transitive element is normal.*

Proof. For arbitrary test q we obtain by Cor. 3.10, isotony of diamond and Cor. 3.2 that $q \leq |a+1\rangle(a \triangleright q) \Rightarrow |a\rangle q \leq |a\rangle|a+1\rangle(a \triangleright q) \Leftrightarrow |a\rangle q \leq |a\rangle(a \triangleright q)$. □

In [7] it is also proved that every normal element is noetherian and d-transitive. We conclude with a further property of d-transitive elements.

Lemma 3.12. *If a is d-transitive then for all p we have* $a \triangleright (|a+1\rangle p) = a \triangleright p$.

Proof. $a \triangleright (|a+1\rangle p)$
$= \quad \{\![$ definition of $a \triangleright$ $]\!\}$
$|a+1\rangle p - |a\rangle|a+1\rangle p$
$= \quad \{\![$ Cor. 3.2 $]\!\}$
$(|a\rangle p + p) - |a\rangle p$
$= \quad \{\![$ right distributivity of $-$ $]\!\}$
$(|a\rangle p - |a\rangle p) + (p - |a\rangle p)$
$= \quad \{\![$ Boolean algebra $]\!\}$
$p - |a\rangle p$
$= \quad \{\![$ definition of $a \triangleright$ $]\!\}$
$a \triangleright p$.

□

4 Antichains

An antichain is a set M of objects of a partially ordered set such that any two objects of M are incomparable. Equivalently, M is an antichain if it coincides with its set of maximal elements, characterised algebraically as follows.

Definition 4.1. Given a semiring object a, a test p is an a-*antichain* if $p = a \triangleright p$, i.e., if p is a fixed point of the operator $a \triangleright$. The set of all a-antichains is denoted by $\mathrm{AC}(a)$. By Lm. 3.4.1, $0 \in \mathrm{AC}(a)$ for every a. When a is clear from the context we will just write "antichain" instead of "a-antichain".

Lemma 4.2. *p is an antichain $\Leftrightarrow p \leq \neg|a\rangle p \Leftrightarrow p \cdot |a\rangle p \leq 0$. In particular, if a is d-irreflexive then every atomic test is an antichain.*

Proof. By the definition of $\triangleright$, order theory, definition of $-$, $\cdot$ coinciding with infimum on tests, reflexivity of $\leq$ and Boolean algebra,

$$\begin{aligned} &p = a \triangleright p \Leftrightarrow p = p - |a\rangle p \Leftrightarrow p \leq p - |a\rangle p \Leftrightarrow \\ &p \leq p \wedge p \leq \neg|a\rangle p \Leftrightarrow p \leq \neg|a\rangle p \Leftrightarrow p \cdot |a\rangle p \leq 0\ . \end{aligned}$$

□

Corollary 4.3. *$\mathrm{AC}(a)$ is downward closed, i.e., $p \in \mathrm{AC}(a) \wedge q \leq p \Rightarrow q \in \mathrm{AC}(a)$.*

Proof. By isotony we have $q \cdot |a\rangle q \leq p \cdot |a\rangle p \leq 0$. □

Lemma 4.4. *Consider tests p, q. Then $p + q$ is an antichain iff p and q are antichains and $p \cdot |a\rangle q \leq 0 \wedge q \cdot |a\rangle p \leq 0$.*

Proof.

$$\begin{aligned} & p + q \text{ antichain} \\ \Leftrightarrow\quad & \{\!\!\{ \text{ by Lm. 4.2 } \}\!\!\} \\ & (p+q) \cdot |a\rangle(p+q) \leq 0 \\ \Leftrightarrow\quad & \{\!\!\{ \text{ distributivity } \}\!\!\} \\ & p \cdot |a\rangle p \leq 0 \wedge p \cdot |a\rangle q \leq 0 \wedge q \cdot |a\rangle p \leq 0 \wedge q \cdot |a\rangle q \leq 0 \\ \Leftrightarrow\quad & \{\!\!\{ \text{ by Lm. 4.2 } \}\!\!\} \\ & p, q \text{ antichains} \wedge p \cdot |a\rangle q \leq 0 \wedge q \cdot |a\rangle p \leq 0\ . \end{aligned}$$

□

Lemma 4.5. *For $p, q \in \mathrm{AC}(a)$ we have $a \triangleright (p+q) = (p - |a\rangle q) + (q - |a\rangle p)$.*

Proof.

$$\begin{aligned} & a \triangleright (p+q) \\ =\quad & \{\!\!\{ \text{ definition of } \triangleright \}\!\!\} \\ & (p+q) - |a\rangle(p+q) \\ =\quad & \{\!\!\{ \text{ distributivity } \}\!\!\} \\ & (p+q) - (|a\rangle p + |a\rangle q) \\ =\quad & \{\!\!\{ \text{ De Morgan } \}\!\!\} \\ & (p+q) - |a\rangle p - |a\rangle q \\ =\quad & \{\!\!\{ \text{ distributivity and right-commutativity of } - \}\!\!\} \\ & (p - |a\rangle p - |a\rangle q) + (q - |a\rangle q - |a\rangle p) \\ =\quad & \{\!\!\{ p, q \text{ antichains and Lm. 4.2 } \}\!\!\} \\ & (p - |a\rangle q) + (q - |a\rangle p)\ . \end{aligned}$$

□

5 Deriving the BNL Algorithm

We now give an algebraic, calculational derivation of the BNL algorithm in [1] for computing the maximal objects of a set. For this, we assume that the test algebra of the underlying semiring is finite and hence *atomic*, i.e., every test is the sum of the atoms below it.

Consider a test r that represents all available tuples in a database and let a be a fixed strict-order representing a preference relation. The task is to compute $a \rhd r$, i.e., a test representing the set of all a-maximal objects in r.

A common technique for deriving an algorithmic solution of a specification is to make a constant of the specification into a parameter and then calculate an inductive or recursive version of the generalised specification.

Here we make r into a parameter called u. So for test u we define the function $ma(u)$ that computes the maxima of u w.r.t. preference a as

$$ma(u) =_{df} a \rhd u \ .$$

The aim is now to develop a recursive version of the function ma by induction on the size of the parameter u. By the assumptions of finiteness and atomicity the size $|u|$ of u can be defined as the cardinality of the set of atoms below u.

Base Case $|u| = 0$. Then $u = 0$ and we have $ma(0) = 0 - |a\rangle 0 = 0$.

Inductive Case. Choose an atomic test $x \leq u$ and set $v =_{df} u - x$. By the definitions, Th. 3.6, d-irreflexivity of a, atomicity of x, and the definition of ma:

$$ma(u) = a \rhd (x + v) = a \rhd (a \rhd x + a \rhd v) = a \rhd (x + a \rhd v) = a \rhd (x + ma(v)) \ .$$

Now we observe that $a \rhd v$ is an antichain and define an auxiliary function

$$inc(x, p) =_{df} a \rhd (x + p) \ ,$$

where x is an atomic test and p an antichain. Then we can continue the above derivation to obtain $ma(u) = inc(x, ma(v))$.

Altogether, we have derived the recursion

$$\begin{aligned} ma(u) = \ & \mathsf{if}\ u = 0\ \mathsf{then}\ 0 \\ & \quad\quad\quad\quad \mathsf{else}\ \ \mathsf{choose\ atom}\ x \leq u\ \mathsf{in} \\ & \quad\quad\quad\quad\quad\quad inc(x, ma(u - x)) \ . \end{aligned}$$

Our original task is now solved using the call $ma(r)$. We will transform this recursion into a simpler one in Sect. 6.

Next we derive a recursive version of the function $inc(x, p)$. The parameter p is frequently called the *(working) window*. It contains candidates for objects of the overall maxima set and is incrementally adapted as the single tuples x are inspected in turn.

Base Case $|p| = 0$ and hence $p = 0$: we have $inc(x, 0) = a \rhd (x + 0) = a \rhd x = x$.

Inductive Case: choose an atomic test $y \leq p$ and set $q =_{df} p - y$.

Subcase 1: $x \le |a\rangle y$, i.e., x is dominated by y. Therefore x cannot be maximal in r and can be discarded. Let us show this formally. First, by isotony of diamond, $x \le |a\rangle p$, since $y \le p$, and hence $x - |a\rangle p \le 0$ by Boolean algebra. Moreover, again by isotony of diamond, d-transitivity of a and p being an antichain,

$$p \cdot |a\rangle x \le p \cdot |a\rangle|a\rangle p \le p \cdot |a\rangle p \le 0 \ .$$

By Boolean algebra therefore $p \le \neg|a\rangle x$ and hence $p - |a\rangle x = p$. Now Lm. 4.5 with q specialised to x shows $inc(x,p) = p$.

Subcase 2: $x \not\le |a\rangle y$. Then x is not dominated by y and cannot be discarded immediately. Rather, two subcases arise. If x dominates y then y can be discarded from the window p. Otherwise y still remains a candidate for a maximal object, while x needs to be compared with the remainder q of the window p. Again, we do the formal calculations.

First, since x is an atomic test, (1) implies $x \le \neg|a\rangle y$ and $x - |a\rangle y = x$.

Subcase 2.1: $y \le |a\rangle x$ and hence $y - |a\rangle x = 0$. By Lm. 4.5, distributivity, Boolean algebra and Lm. 4.5 again we obtain

$$\begin{aligned} inc(x,p) &= a \rhd (x+y+q) = (x - |a\rangle(y+q)) + ((y+q) - |a\rangle x) \\ &= (x - |a\rangle y - |a\rangle q) + (y - |a\rangle x) + (q - |a\rangle x) \\ &= (x - |a\rangle q) + (q - |a\rangle x) = inc(x,q) \ . \end{aligned}$$

Subcase 2.2: $y \not\le |a\rangle x$, hence $y \le \neg|a\rangle x$ and therefore $y \cdot |a\rangle x \le 0$ and $y - |a\rangle x = y$ by atomicity of y. Since $y \in p$ and p is an antichain, we know that also $y \cdot |a\rangle q \le 0$, hence $y \cdot |a\rangle(x+q) \le 0$. Moreover, since we are in a case where $x \cdot |a\rangle y = 0$, we know that also $(x+q) \cdot |a\rangle y \le 0$. Now

$$\begin{aligned} & inc(x,p) \\ = & \quad \{\!\!\{ \text{ by Lm. 4.5 } \}\!\!\} \\ & (x - |a\rangle p) + (p - |a\rangle x) \\ = & \quad \{\!\!\{ \text{ above decomposition } p = y + q \text{, additivity of diamond and Boolean algebra } \}\!\!\} \\ & (x - |a\rangle y - |a\rangle q) + (y - |a\rangle x) + (q - |a\rangle x) \\ = & \quad \{\!\!\{ \text{ by } x - |a\rangle y = x \text{ and } y - |a\rangle x = y \text{, as remarked above } \}\!\!\} \\ & (x - |a\rangle q) + y + (q - |a\rangle x) \\ = & \quad \{\!\!\{ \text{ rearrangement and Lm. 4.5, since } q \le p \text{ by Cor. 4.3 is an antichain and } x \cdot q \le x \cdot p \le 0 \ \}\!\!\} \\ & y + inc(x,q) \ . \end{aligned}$$

With this, the recursive version of inc is complete:

$$\begin{aligned} inc(x,p) = \ & \mathsf{if}\ p = 0\ \mathsf{then}\ x \\ & \mathsf{else}\ \ \text{choose atom } y \le p \ \mathsf{in} \\ & \qquad \mathsf{if}\ x \le |a\rangle y\ \mathsf{then}\ p \\ & \qquad \mathsf{else}\ \ \mathsf{if}\ y \le |a\rangle x\ \mathsf{then}\ inc(x, p-y) \\ & \qquad\qquad \mathsf{else}\ \ y + inc(x, p-y) \ . \end{aligned}$$

We show the algorithm at work in our example from Fig. 1 in Sect. 1. The test r represents the set of hotels (with abbreviated names) $\{\mathsf{GW}, \mathsf{WH}, \mathsf{OH}, \mathsf{A}, \mathsf{WW}, \mathsf{SSp}, \mathsf{RC}, \mathsf{H}, \mathsf{SH}\}$. The strict-order a is the relation $\prec$. We show the values of p, x and y, all in set notation, during the evaluation of the recursion for $inc(x, p)$ with initial values $x = \{\mathsf{SSp}\}$ and $p = \{\mathsf{GW}, \mathsf{H}, \mathsf{WH}, \mathsf{WW}\}$.

step	1	2	3	4	5
p	$\{\mathsf{GW}, \mathsf{H}, \mathsf{WH}, \mathsf{WW}\}$	$\{\mathsf{GW}, \mathsf{H}, \mathsf{WH}\}$	$\{\mathsf{H}, \mathsf{WH}\}$	$\{\mathsf{WH}\}$	$\emptyset$
y	$\{\mathsf{WW}\}$	$\{\mathsf{GW}\}$	$\{\mathsf{H}\}$	$\{\mathsf{WH}\}$	–
partial result	$\{\mathsf{WW}\}$	$\{\mathsf{GW}\}$	$\emptyset$	$\{\mathsf{WH}\}$	x

In the first step we choose $y = \{\mathsf{WW}\}$. Then $x \not\leq |\prec\rangle y$ and $y \not\leq |\prec\rangle x$. Therefore y is preserved as a partial result and the recursion continues with the remainder of the window. The second and fourth steps are analogous. In step 3 y is dominated by x and hence discarded. Altogether, $\{\mathsf{GW}, \mathsf{SSp}, \mathsf{WH}, \mathsf{WW}\}$ is returned.

6 The Lattice Structure of Antichains

In this section we will exhibit a lattice structure on the set of antichains w.r.t. a strict-order. To this end we first define an approximation relation.

Definition 6.1. Test p is *improved by* test q, in symbols $p \sqsubseteq q$, if q results from removing some objects of p that are dominated by q-objects and possibly adding others that are not dominated by p-objects. Formally,

$$p \sqsubseteq q \Leftrightarrow_{df} p - |a\rangle q \leq q \wedge q \cdot |a\rangle p \leq 0 \ .$$

By Boolean algebra and distributivity we equivalently have

$$p \sqsubseteq q \Leftrightarrow p \leq |a + 1\rangle q \wedge q \cdot |a\rangle p \leq 0 \ .$$

Lemma 6.2.
1. *$\forall p \in \mathsf{test}(S) : 0 \sqsubseteq p$.*
2. *$\sqsubseteq$ is reflexive precisely on $\mathrm{AC}(a)$, i.e., $p \sqsubseteq p \Leftrightarrow p \in \mathrm{AC}(a)$.*
3. *$\sqsubseteq$ is antisymmetric.*
4. *If a is d-transitive, then for antichains the second conjunct in the definition of $\sqsubseteq$ is implied by the first one, i.e., for $p, q \in \mathrm{AC}(a)$, $p \sqsubseteq q \Leftrightarrow p \leq |a+1\rangle q$.*
5. *If a is d-transitive then $\sqsubseteq$ is transitive and hence a partial order on $\mathrm{AC}(a)$.*
6. *If a is normal then $p \sqsubseteq a \rhd p$.*

Proof.
1. Immediate from the definition and Lm. 4.2.
2. $\quad p \sqsubseteq q \wedge q \sqsubseteq p$

$\Leftrightarrow \quad$ {[definition]}

$\quad p - |a\rangle q \leq q \wedge q \cdot |a\rangle p \leq 0 \wedge q - |a\rangle p \leq p \wedge p \cdot |a\rangle q \leq 0$

$\Leftrightarrow \quad$ {[commutativity of $\wedge$ and Boolean algebra]}

$$
\begin{array}{ll}
& p - |a\rangle q \leq q \wedge p \leq \neg|a\rangle q \wedge q - |a\rangle p \leq p \wedge q \leq \neg|a\rangle p \\
\Leftrightarrow & \{\!\!\{ \text{ since } p \leq \neg|a\rangle q \wedge q \leq \neg|a\rangle p \text{ imply} \\
& \quad p - |a\rangle q = p \text{ and } q - |a\rangle p = q \}\!\!\} \\
& p \leq q \wedge q \leq p \\
\Rightarrow & \{\!\!\{ \text{ antisymmetry of } \leq \}\!\!\} \\
& p = q \ .
\end{array}
$$

3. Assume $p \leq |a+1\rangle q$, which is equivalent to $p \leq |a\rangle q + q$. Then by isotony and distributivity of diamond, d-transitivity of a and Lm. 4.2,

$$q \cdot |a\rangle p \leq q \cdot (|a\rangle|a\rangle q + |a\rangle q) = q \cdot |a\rangle q = 0 \ .$$

4. By Part 4, isotony, d-transitivity of a and hence of $a+1$, distributivity, and Part 4 again:

$$
\begin{array}{l}
p \sqsubseteq q \wedge q \sqsubseteq s \Leftrightarrow p \leq |a+1\rangle q \wedge q \leq |a+1\rangle s \\
\quad \Rightarrow p \leq |a+1\rangle|a+1\rangle s \Rightarrow p \leq |a+1\rangle s \Leftrightarrow p \sqsubseteq s \ .
\end{array}
$$

5. By definition of $\sqsubseteq$, normality of a (3), definition of $a \rhd$ and Boolean algebra,

$$
\begin{array}{l}
p \sqsubseteq a \rhd p \Leftrightarrow p - |a\rangle(a \rhd p) \leq a \rhd p \wedge (a \rhd p) \cdot |a\rangle p \leq 0 \Leftrightarrow p - |a\rangle p \leq \\
\quad a \rhd p \wedge (a \rhd p) \cdot |a\rangle p \leq 0 \Leftrightarrow \mathsf{TRUE} \ .
\end{array}
$$

□

We show now that the order $\sqsubseteq$ holds, in particular, between p and $inc(x,p)$. Therefore the BNL algorithm produces a $\sqsubseteq$-ascending chain of antichains. It ends with the $\sqsubseteq$-largest antichain $a \rhd r$, where r is again the set of all tuples under consideration.

Theorem 6.3. *Assume a to be a noetherian strict-order.*

1. *The operator $a \rhd$ transforms all $\leq$-suprema existing in* $\mathsf{test}(S)$ *into $\sqsubseteq$-suprema in* $\mathrm{AC}(a)$.
2. *The operator $a \rhd$ is isotone w.r.t. $\leq$ and $\sqsubseteq$, i.e.,*

$$\forall p, q \in \mathsf{test}(S) : p \leq q \Rightarrow a \rhd p \sqsubseteq a \rhd q \ .$$

3. $\mathrm{AC}(a)$ *is an upper semilattice with $p \sqcup q = a \rhd (p+q)$ and hence $inc(x,p) = p \sqcup x$ and $0 \sqcup p = p$.*
4. *If $(S, \leq)$ is a quantale then* $\mathrm{AC}(a)$ *is a complete lattice with $\bigsqcup_{\sqsubseteq} A = a \rhd (\Sigma A)$, where Σ is the supremum operator on $(S, \leq)$.*
5. *For atomic test x with $x \cdot p = 0$ and $p \in \mathrm{AC}(a)$ we have $p \sqsubseteq inc(x,p)$.*
6. *The operator $a \rhd$ preserves $\sqcup$ on* $\mathrm{AC}(a)$.
7. *The operator $a \rhd$ is also isotone w.r.t. $\sqsubseteq$ and $\sqsubseteq$ on arbitrary tests, i.e.,*

$$\forall p, q \in \mathsf{test}(S) : p \sqsubseteq q \Rightarrow a \rhd p \sqsubseteq a \rhd q \ .$$

Proof. We recall the following characterisation of the supremum s of a subset X of a partially ordered set M (provided it exists):

$$\forall y \in M : s \leq y \Leftrightarrow (\forall x \in X : x \leq y) \ . \qquad (*)$$

1. Let $T \subseteq \textsf{test}(S)$ have $\leq$-supremum z. Then

$$\begin{array}{ll} & \forall q \in a \triangleright T : q \sqsubseteq y \\ \Leftrightarrow & \quad \{\!\![\text{ definition of } \sqsubseteq \,]\!\!\} \\ & \forall q \in a \triangleright T : q \leq |a+1\rangle y \\ \Leftrightarrow & \quad \{\!\![\text{ definition of } a \triangleright T \,]\!\!\} \\ & \forall p \in T : a \triangleright p \leq |a+1\rangle y \\ \Leftrightarrow & \quad \{\!\![\text{ by Cor. 3.10 }]\!\!\} \\ & \forall p \in T : p \leq |a+1\rangle y \\ \Leftrightarrow & \quad \{\!\![\text{ definition of } z \,]\!\!\} \\ & z \leq |a+1\rangle y \\ \Leftrightarrow & \quad \{\!\![\text{ by Cor. 3.10 }]\!\!\} \\ & a \triangleright z \leq |a+1\rangle y \\ \Leftrightarrow & \quad \{\!\![\text{ definition of } \sqsubseteq \,]\!\!\} \\ & a \triangleright z \sqsubseteq y \,. \end{array}$$

   Hence, by ($*$), $a \triangleright z$ is the $\sqsubseteq$-supremum of the image set $a \triangleright T$ of T under $a \triangleright$.
2. Immediate from Part 1.
3. Immediate from Part 1.
4. Immediate from Part 1.
5. By d-irreflexivity of a we have $x \in \mathrm{AC}(a)$. Hence Part 3 entails $p \sqsubseteq x \sqcup p = a \triangleright (x+p) = inc(x,p)$.
6. For $p, q \in \mathrm{AC}(a)$, by Part 3, idempotence of $\triangleright$ (Lm. 3.4.5), Th. 3.6, Part 3,
$$a \triangleright (p \sqcup q) = a \triangleright (a \triangleright (p+q)) = a \triangleright (p+q) = a \triangleright (a \triangleright p + a \triangleright q) = a \triangleright p \sqcup a \triangleright q \,.$$
7. For $p, q \in \textsf{test}(S)$, by definition of $\sqsubseteq$, Part 2 and Lm. 3.12
$$p \sqsubseteq q \Leftrightarrow p \leq |a+1\rangle q \Rightarrow a \triangleright p \sqsubseteq a \triangleright (|a+1\rangle q) \Leftrightarrow a \triangleright p \sqsubseteq a \triangleright q \,. \qquad \square$$

It should be noted that noetherity is essential for these results. As a counterexample to $\leq$-isotony of $a \triangleright$, consider the semiring of binary relations on the set $\mathbb{N}$ of natural numbers. Take a to be the usual strict-order on $\mathbb{N}$ so that $a+1$ is the standard order on $\mathbb{N}$. Choose as p and q the tests encoding $\{0\}$ and $\mathbb{N}$. Then $p \leq q$, but $a \triangleright p = p \not\sqsubseteq \emptyset = a \triangleright q$, since $p \not\leq \emptyset = |a+1\rangle \emptyset$.

We conclude with an application of the algebra for bringing the function ma from Sect. 5 into tail-recursive form, as a preparation for transliterating it into loop form (see e.g. [9] for details of that). The essential observation is that $\sqcup$ as a supremum operator is associative and has the $\sqsubseteq$-least element 0 as its neutral element. We define an auxiliary function $mat(p,u) =_{df} p \sqcup ma(u)$ with an additional parameter p that will accumulate the end result during the recursion. By neutrality of 0 we can solve the original task as $ma(u) = mat(0,u)$. Now we calculate a recursive version of mat based on the one for ma. In the termination case $u = 0$ we obtain $mat(p,0) = p \sqcup 0 = p$. In the recursive case for $u \neq 0$ we get by the definitions, Th. 6.3.3, associativity of $\sqcup$ and the definitions again

$$\begin{array}{c} mat(p,u) = p \sqcup inc(x, ma(u-x)) = p \sqcup (x \sqcup ma(u-x)) = \\ (p \sqcup x) \sqcup ma(u-x) = mat(p \sqcup x, u-x) \,, \end{array}$$

which is a tail-recursive call.

7 Maxima as a Closure Operator

7.1 Closure Operators

We recall the definition of a closure operator.

Definition 7.1. A *closure operator* on a partially ordered set $(L, \leq)$ is a total function $f : L \to L$ with the following properties:

- $x \leq f(x)$ (extensivity)
- $x \leq y \Rightarrow f(x) \leq f(y)$ (isotony)
- $f(f(x)) = f(x)$ (idempotence)

Consider now a noetherian strict-order a. By Lm.6.2.6, Thm. 6.3.7 and Lm. 3.4.5 $a\triangleright$ satisfies all three of these properties w.r.t. $\sqsubseteq$. Unfortunately, however, $\sqsubseteq$ is not even a preorder on $\mathsf{test}(S)$, since by Lm. 6.2.2 reflexivity holds exactly on AC(a). To remedy this, we define another comparison relation on $\mathsf{test}(S)$.

Definition 7.2. For a given a we set $p \preceq_a q \Leftrightarrow_{df} a \triangleright p \sqsubseteq a \triangleright q$.

Lemma 7.3. *$\preceq$ is a preorder, but not a partial order. We have $p \preceq q \wedge q \preceq p \Leftrightarrow a \triangleright p = a \triangleright q$. Finally, $p \leq q \Rightarrow p \preceq q$.*

Proof. Reflexivity and transitivity are immediate from reflexivity and transitivity of $\sqsubseteq$. The second claim follows from the antisymmetry of $\sqsubseteq$; it also shows that in general $\prec_a$ is not antisymmetric. The final claim is immediate from Th. 6.3.2 and the definitions. □

With this definition we can now actually view $a\triangleright$ as a closure operator if we carry over that notion to the case of preorders.

Definition 7.4. Consider a preorder $(L, \leq)$ with induced equivalence relation $\sim$. An endofunction $H : L \to L$ on a is called *weakly idempotent* if $H(H(x)) \sim H(x)$ for all $x \in L$. We call H a *kernel operator* if it is isotone, weakly idempotent and *contractive*; by the latter property we mean $H(x) \leq x$ for all $x \in L$. Symmetrically, we call H a *closure operator* if it is isotone, weakly idempotent and *extensive*; by the latter property we mean $x \leq H(x)$ for all $x \in L$.

In each case, the image set $H(A)$ coincides with the sets of *weak fixed points* of H, i.e., with the set $\{x \in A \mid H(x) \sim x\}$. Now we have the following result.

Lemma 7.5. *$a\triangleright$ is a closure operator w.r.t. $\preceq$.*

Proof. Since we already know that $a\triangleright$ is idempotent, it suffices to show extensivity and isotony w.r.t. $\preceq$.
Extensivity: by the definition of $\preceq$, idempotence of $a\triangleright$ (Lm. 3.4.5) and reflexivity of $\sqsubseteq$ we have $p \preceq a \triangleright p \Leftrightarrow a \triangleright p \sqsubseteq a \triangleright (a \triangleright p) \Leftrightarrow a \triangleright p \sqsubseteq a \triangleright p \Leftrightarrow \mathsf{TRUE}$.
Isotony: by the definition of $\preceq$, idempotence of $a\triangleright$ (Lm. 3.4.5) and the definition of $\preceq$ again we obtain

$$p \preceq q \Leftrightarrow a \triangleright p \sqsubseteq a \triangleright q \Leftrightarrow a \triangleright (a \triangleright p) \sqsubseteq a \triangleright (a \triangleright q) \Leftrightarrow a \triangleright p \preceq a \triangleright q\,.$$

□

7.2 A Galois Connection for the Maxima Operator

Since we have established the maxima operator as a closure operator, we can use a well-known result concerning Galois connections, again adapted to the case of preorders rather than partial orders.

Definition 7.6. Consider two preorders $(A, \leq_A)$ and $(B, \leq_B)$ and total functions $F : A \to B$ and $G : B \to A$. Then the pair (F, G) is called a *Galois connection (GC)* between A and B iff

$$\forall x \in A : \forall y \in B : F(x) \leq_B y \Leftrightarrow x \leq_A G(y) .$$

Then F is called the *lower*, G the *upper adjoint* of the GC.

Details of the theory of Galois connections for the preorder case can be found in the Appendix. The following Lm. is well known (e.g. [5]) for the case of partial orders; we adapt it to preorders.

Lemma 7.7. *Every closure operator $H : L \to L$ induces the following Galois connection between L and $H(L)$:*

$$H(x) \leq y \Leftrightarrow x \leq \iota(y) ,$$

where ι is the embedding of $H(L)$ into L, i.e., $\iota(y) = y$ for $y \in H(L)$.

Proof. ($\Rightarrow$) By extensivity of H and the assumption, $x \leq f(x) \leq y = \iota(y)$.
($\Leftarrow$) First, by weak idempotence of H we have $H(y) \sim y$ for all $y \in H(L)$. Now, by isotony of H we obtain $x \leq \iota(y) \Rightarrow H(x) \leq H(\iota(y)) = H(y) \sim y$. □

Hence for $p \in \mathsf{test}(S)$ and $q \in \mathrm{AC}(a)$ we have the Galois connection

$$a \triangleright p \preceq q \Leftrightarrow p \preceq \iota(q) .$$

As a lower adjoint therefore the $a\triangleright$ operator preserves all existing $\preceq$-suprema (see Th. 9.7 in the Appendix). This nicely rounds off the small collection of preservation results in Th. 6.3.

8 Conclusion

We have presented an algebraic account of an approximation relation between antichains that induces a semilattice and renders the maxima operator isotone in several ways. Moreover, the maxima operator has been shown to be a closure operator in an associated preordered set and hence satisfies a Galois connection. We have shown the calculus at work in the non-trivial example of the BNL algorithm. Therefore we are convinced that the theory developed here will be useful for many further calculational derivations involving the maxima operator and antichains.

Acknowledgment. I am grateful to Peter Höfner, Martin E. Müller, Patrick Roocks, Andreas Zelend and the anonymous referees for careful proofreading and helpful remarks, and to Roland Backhouse for asking the questions that led to Sects. 6 and 7. He also suggested significant simplifications to an earlier version of this paper, in particular the property in Lm. 6.2.4.

References

[1] Börzsönyi, S., Kossmann, D., Stocker, K.: The skyline operator. In: Proceedings of the 17th International Conference on Data Engineering, pp. 421–430 (2001)
[2] Dehne, F.: $o(n^{1/2})$ algorithms for the maximal elements and ECDF searching problem on a mesh-connected parallel computer. Inf. Process. Lett. 22(6), 303–306 (1986)
[3] Desharnais, J., Möller, B., Struth, G.: Kleene algebra with domain. ACM Trans. Comput. Logic 7(4), 798–833 (2006)
[4] Desharnais, J., Möller, B., Struth, G.: Algebraic notions of termination. Logical Methods in Computer Science 7(1) (2011)
[5] Erne, M., Koslowski, J., Melton, A., Strecker, G.: A primer on Galois connections. In: Proc. 1991 Summer Conference on General Topology and Applications in Honor of Mary Ellen Rudin and Her Work. Annals of the New York Academy of Sciences, vol. 704, pp. 103–125. New York Academy of Sciences (1993)
[6] Kozen, D.: A completeness theorem for Kleene algebras and the algebra of regular events. Information and Computation 110(2), 366–390 (1994)
[7] Möller, B., Roocks, P.: An algebra of database preferences. Journal of Logical and Algebraic Methods in Programming 84, 456–481 (2015)
[8] Möller, B., Roocks, P., Endres, M.: An Algebraic Calculus of Database Preferences. In: Gibbons, J., Nogueira, P. (eds.) MPC 2012. LNCS, vol. 7342, pp. 241–262. Springer, Heidelberg (2012)
[9] Partsch, H.: Specification and Transformation of Programs - A Formal Approach to Software Development. Texts and Monographs in Computer Science. Springer (1990)
[10] Stojmenovic, I., Miyakawa, M.: An optimal parallel algorithm for solving the maximal elements problem in the plane. Parallel Computing 7(2), 249–251 (1988)

9 Appendix: Galois Connections between Preorders

We investigate in how far the standard properties of Galois connections between partial orders hold for general preorders as well. For a good summary of the standard case see e.g. [5].

9.1 Definition and Basic Properties

Consider two preorders $(A, \leq_A)$ and $(B, \leq_B)$ and total functions $F : A \to B$ and $G : B \to A$. Then the pair (F, G) is called a *Galois connection (GC)* between A and B iff

$$\forall x \in A : \forall y \in B : F(x) \leq_B y \Leftrightarrow x \leq_A G(y) \ .$$

Then F is called the *lower*, G the *upper adjoint* of the GC.

In the sequel we shall suppress the indices of the preorders involved in a Galois connection.

The functions in a GC are quasi-inverses of each other:

Lemma 9.1 (Quasi-Inverses (QI)). *Assume that $F : A \to B$ and $G : B \to A$ form a GC between A and B. Then*

$$\forall x \in A : x \leq G(F(x)) \;, \qquad \forall y \in B : F(G(y)) \leq y \;.$$

Proof. By the GC and reflexivity $x \leq G(F(x)) \Leftrightarrow F(x) \leq F(x) \Leftrightarrow \mathsf{TRUE}$. □

From (QI) we obtain

Corollary 9.2 (Isotony). *The adjoints of a GC are isotone.*

Proof. By transitivity, since QI entails $z \leq G(F(z))$, and GC:

$$x \leq z \Rightarrow x \leq G(F(z)) \Leftrightarrow F(x) \leq F(z) \;.$$

□

On the other hand, isotony and (QI) imply that we have a GC:

Lemma 9.3 (O. Ore). *(F, G) form a GC iff F and G are isotone and quasi-inverses of each other.*

Proof. We only need to show the if-part. By isotony of G, and (QI):

$$F(x) \leq y \Rightarrow G(F(x)) \leq G(y) \Rightarrow x \leq G(y) \;.$$

Symmetrically one shows $x \leq F(y) \Rightarrow F(x) \leq y$. □

For the following results we lift the equivalence $\sim$ induced by the preorder $\leq$ pointwise to functions by setting

$$F_1 \sim F_2 \Leftrightarrow_{df} \forall x : F_1(x) \sim F_2(x) \;.$$

Then the rule of indirect equivalence immediately entails the following uniqueness property.

Lemma 9.4 (Determination). *Let (F_i, G_i) $(i = 1, 2)$ be GCs between A and B. Then each adjoint determines the other one uniquely up to $\sim$, i.e.,*

$$F_1 \sim F_2 \Leftrightarrow G_1 \sim G_2 \;.$$

Now we deal with iterated application of the adjoints.

Corollary 9.5. $F \circ G \circ F \sim F$ *and* $G \circ F \circ G \sim G$.

Proof. From (QI) we know $x \leq G(F(x))$. Isotony implies $F(x) \leq F(G(F(x)))$. On the other hand, (QI) gives us $F(G(F(x))) \leq F(x)$, so that the claim follows by definition of $\sim$.

The claim on orders is immediate from that. □

Corollary 9.6. *Under the assumptions of Cor. 9.5, $F \circ G$ and $G \circ F$ are weakly idempotent and hence a closure and a kernel operator, respectively.*

9.2 Galois Connections and Extremal Elements

Consider an arbitrary preorder $(M, \leq)$. The sets of lower and upper bounds of a subset $X \subseteq M$ are given by

$$y \in \mathsf{lwb}\,X \Leftrightarrow_{df} \forall x \in X : x \leq y\ , \qquad z \in \mathsf{upb}\,X \Leftrightarrow_{df} \forall x \in X : z \leq x\ .$$

A quick calculation shows that $\mathsf{upb}\,X \supseteq Y \Leftrightarrow X \subseteq \mathsf{lwb}\,Y$. So $(\mathsf{upb}, \mathsf{lwb})$ is a GC between $(\mathcal{P}(M), \supseteq)$ and $(\mathcal{P}(M), \subseteq)$.

Based on these we can define the sets of least and greatest objects of X as

$$\mathsf{lst}\,X =_{df} X \cap \mathsf{lwb}\,X\ , \qquad \mathsf{gst}\,X =_{df} X \cap \mathsf{upb}\,X\ .$$

Finally, the sets of infima and suprema of X are given as

$$\mathsf{inf}\,X =_{df} \mathsf{gst}\,\mathsf{lwb}\,X\ , \qquad \mathsf{sup}\,X =_{df} \mathsf{lst}\,\mathsf{upb}\,X\ .$$

Note that any of these sets may be empty. All objects in a set $\mathsf{lst}\,X$ or $\mathsf{gst}\,X$ are $\sim$-equivalent.

Theorem 9.7. *Let (F, G) form a GC. Then*

1. *F preserves all existing suprema, i.e., $F(\mathsf{sup}\,X) \subseteq \mathsf{sup}\,F(X)$ for all $X \subseteq A$.*
2. *G preserves all existing infima, i.e., $G(\mathsf{inf}\,Y) \subseteq \mathsf{inf}\,F(Y)$ for all $Y \subseteq B$.*

Proof. We only show 1. First,

$$\begin{array}{ll} & \mathsf{TRUE} \\ \Leftrightarrow & \{\!\![\ \mathsf{sup}\,X \subseteq \mathsf{upb}\,X \text{ (definition of } \mathsf{sup}) \ ]\!\!\} \\ & \forall x \in X : \forall s \in \mathsf{sup}\,X : x \leq s \\ \Rightarrow & \{\!\![\ \text{isotony}\ ]\!\!\} \\ & \forall x \in X : \forall s \in \mathsf{sup}\,X : F(x) \leq F(s) \\ \Rightarrow & \{\!\![\ \text{the definitions}\ ]\!\!\} \\ & F(\mathsf{sup}\,X) \subseteq \mathsf{upb}\,F(x)\ . \end{array}$$

Second,

$$\begin{array}{ll} & \forall y \in \mathsf{upb}\,F(X) : \forall x \in X : F(x) \leq y \\ \Leftrightarrow & \{\!\![\ \text{GC}\ ]\!\!\} \\ & \forall y \in \mathsf{upb}\,F(X) : \forall x \in X : x \leq G(y) \\ \Leftrightarrow & \{\!\![\ \text{definition}\ ]\!\!\} \\ & \forall y \in \mathsf{upb}\,F(X) : G(y) \in \mathsf{upb}\,X \\ \Leftrightarrow & \{\!\![\ \text{definition of } \mathsf{sup}\ ]\!\!\} \\ & \forall y \in \mathsf{upb}\,F(X) : \forall s \in \mathsf{sup}\,X : s \leq G(y) \\ \Leftrightarrow & \{\!\![\ \text{GC}\ ]\!\!\} \\ & \forall y \in \mathsf{upb}\,F(X) : \forall s \in \mathsf{sup}\,X : F(s) \leq y \\ \Leftrightarrow & \{\!\![\ \text{definitions}\ ]\!\!\} \\ & F(\mathsf{sup}\,X) \subseteq \mathsf{lwb}\,\mathsf{upb}\,F(X)\ . \end{array}$$

Now the claim is immediate from the definitions. □

Decomposition of Database Preferences on the Power Set of the Domain

Patrick Roocks

Institut für Informatik, Universität Augsburg, D-86135 Augsburg, Germany
roocks@informatik.uni-augsburg.de

Abstract. Database preferences allow defining strict orders on the tuples of a data set and selecting the optimal elements w.r.t. this order. In our prior work we have shown that in common implementations of preferences a small set of preference operators and operands is sufficient to express arbitrary strict orders. We have suggested preference decomposition algorithms to prove this expressiveness. In the present paper we define the induced preference on the power set of the original data set and transfer our decomposition results to this setting. We modify the algorithms of our prior work to reduce the term length and complexity of the resulting decompositions. This optimization turns out to be very efficient especially for power set preferences.

Keywords: Relational algebra, Preference decomposition, Power set.

1 Introduction

Database preferences realise soft constraints in queries for relational databases. They allow selecting highly relevant tuples that are optimal compromises for the user. A popular subclass are *Pareto preferences*, which induce strict orders having *Pareto optimal* tuples as their maxima. This concept was introduced as *Skyline operator* [1] to the database community. This operator is typically used to find optimal objects w.r.t. goals that tend to conflict. As an example, consider Figure 1 where we search for the best cars in a data set when optimizing the dimensions of horse power and fuel consumption simultaneously.

In the present work we consider sets of tuples with a given preference on these tuples. For the application, we are interested to construct a preference on sets starting from the given strict order (i.e., preference). For example, assume that a user wants to rent a car and has the choice between two car rentals offering different choices of cars. The car fleets of these rental agencies are depicted in Figure 1. The user prefers powerful cars with low fuel consumption, hence the cars with id's 6, 8 and 7 are optimal for her. But none of the fleets contains all these maxima. Assume further that both rental agencies do not accept reservations for an individual car; they just guarantee that one of the cars is available. The question arises, which car rental is superior for the user? Obviously none of these fleets is strictly better; some arrows point from fleet A to fleet B, some in the converse direction.

W. Kahl et al. (Eds.): RAMiCS 2015, LNCS 9348, pp. 362–379, 2015.
DOI: 10.1007/978-3-319-24704-5_22

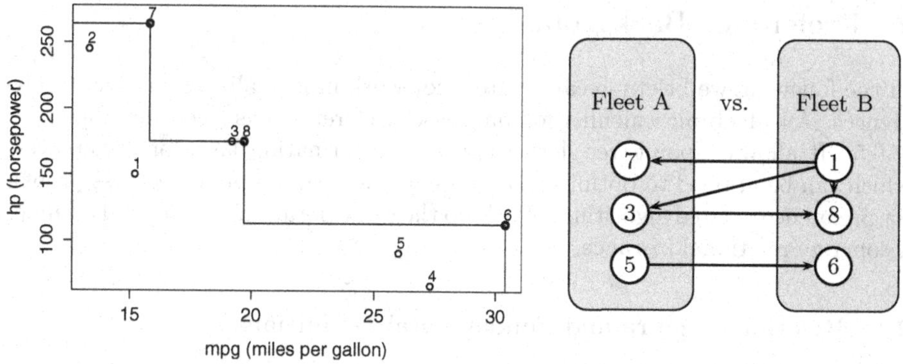

Fig. 1. Left: Pareto optima of cars with low fuel consumption (i.e., high *(mpg)* value) and high horsepower *(hp)*. Right: Comparison of two car fleets with cars from the left diagram. Arrows point from worse to better objects in the sense of the Pareto optima.

To study this in detail, we have to consider preferences on sets, i.e., the extension of a preference to the power set of the domain. In the terminology of [2] this is called a *power construction*. Corresponding to the definitions in [2,13] there are three different possibilities of power constructions for strict orders, which have been used there in contexts other than preferences. In [2] different variants of programming semantics are discussed, where power constructions play an important role throughout. In [13], the focus lies on modelling non-deterministic computations. One of the power constructions for strict orders has already been used in the context of database preferences in [12].

In [10] we have presented a decomposition approach for database preferences. We have shown that every strict order can be expressed with a quite restricted set of simple operators and operands which are available in common frameworks for database preferences. We especially consider the *Skyline* feature in the commercial database *Exasol Exasolution* [4] or our *rPref* package [9]. In the concluding remarks in [10] we have noticed that these decompositions can produce quite lengthy terms which can be minimized with known transformation laws. In the present paper we optimize these decomposition algorithms by introducing equivalence classes of equally good tuples.

We compare the unoptimized and the optimized versions of our algorithms on power set preferences. It turns out that this optimization leads to shorter preference terms especially on the power set preferences. The implementation of our algorithms and the comparative study is available in an R script [8].

The remainder of the paper is structured as follows: in Section 2 we recapitulate the formal foundations from prior work and formally introduce power set preferences. In Section 3 we present the optimized decomposition algorithms. These are applied to power set preferences in Section 4. We end with a conclusion and an outlook.

2 Preference Background

In the following we briefly recapitulate prior work in formalizing database preferences. An algebraic calculus for database preference has been developed in [7,6,5]. It allows a point-free derivation of transformation laws for preferences which can be applied to optimization of preference queries or correctness proofs for preference-related algorithms. We keep the same notation as in [6] and assume a concrete relational instance.

2.1 Relation Algebra and Fundamental Definitions

For a domain D we define a concrete relation algebra of binary homogeneous relations on $D^2 = D \times D$. We define the following special relations: The empty relation $0 =_{df} \emptyset$, the identity $\mathbb{1} =_{df} \{(t,t) \mid t \in D\}$ and the universal relation $\top =_{df} D^2$.

The fundamental operations for relations u, v are the relational union $u + v$, the composition $u \cdot v$, the intersection $u \sqcap v$ and the inclusion order $u \leq v$. We use the following conventions regarding literals in formulas:

- a, b, c, d are binary homogeneous relations.
- p, q, r, s are subidentities, i.e., $p \leq \mathbb{1}$, each representing a set $M_p \subseteq D$ which is related to p via $p = \{(t,t) \leq \mathbb{1} \mid t \in M_p\}$. They are also called *tests*.
- x, y, z are subidentities which are non-zero and atomic w.r.t. $+$, also called *atomic tests*. They represent singletons, i.e., $x = \{(t,t)\}$ for some $t \in D$.

According to the database scenario, which is our field of application, a test is a *data set* or a *set of tuples*, i.e., a (partial) table of a database. An atomic test is a *tuple* and models a single row of a database table. Hence we use the term *test* synonymously with *data set*, and *atomic tests* represent *tuples*. For tests, composition and intersection coincide, i.e., we have $p \cdot q = p \sqcap q$.

An element $t \in D$ from the data set is modelled as $x = \{(t,t)\} \subset D^2$ which allows the left/right restrictions $x \cdot a$ and $a \cdot x$ for a relation a by the usual relational composition. For sake of readability we introduce the following shorthand notations:

- Selection of a tuple $x = \{(t,t')\}$ from a data set p is defined as (with a slight abuse of $\in$, as x is formally a subset of p)

$$x \in p \quad \Leftrightarrow_{df} \quad x \leq p \ \wedge \ x \text{ is atomic} .$$

- To express that x is *a-related* to y we define

$$x\,a\,y \quad \Leftrightarrow_{df} \quad (t,t') \in a \qquad \text{where } x = \{(t,t)\} \wedge y = \{(t',t')\} .$$

Based on the above conventions we define the following operations:

- The converse relation $a^{-1} =_{df} \{(t',t) \in \top \mid (t,t') \in a\}$,
- the general complement $\overline{a} =_{df} \{(t,t') \in \top \mid (t,t') \notin a\}$,

- the complement of tests $\neg p =_{df} \{(t,t) \in \mathbb{1} \mid (t,t) \notin p\}$,
- the difference of tests $p - q =_{df} p \cdot \neg q$,
- the (inverse) image of a relation a w.r.t. a test p:

$$\langle a|p =_{df} \{(t,t) \in \mathbb{1} \mid \exists t' \in D : (t',t) \in (p \cdot a)\} \qquad \text{(image)},$$
$$|a\rangle p =_{df} \{(t,t) \in \mathbb{1} \mid \exists t' \in D : (t,t') \in (a \cdot p)\} \qquad \text{(inverse image)}.$$

Powers are given by $a^0 =_{df} \mathbb{1}$ and $a^i =_{df} a \cdot a^{i-1}$. We will also need the *Hasse diagram* of a relation, i.e., the transitive reduction. For a transitive and irreflexive relation a this is given by $a \sqcap \overline{a^2}$, cf. [11].

Because the focus of this work lies on the algorithmic approach to preference decomposition we think that such a concrete definition of a relation algebra is more appropriate than an abstract axiomatisation of the relational operations. An *abstract relation algebra* uses algebraic axiomatisations of tests, diamonds, etc., and has been used in the context of preferences in [6].

2.2 The Preference Framework

We recapitulate the definitions of preferences and related operations from [10].

Definition 2.1 (Preferences with SV). A relation a is a *preference* if and only if it is irreflexive and transitive, i.e., a *strict partial order*. In relation algebra, this formally corresponds to $a \sqcap \mathbb{1} = 0$ and $a^2 \leq a$.

Every preference a will be associated with an SV-relation s_a. This has to be an equivalence relation on the domain of a, where the equivalence classes contain "equally good" objects (SV is short for "substitutable values"). It must fulfil the compatibility conditions $\mathsf{s}_a \cdot a \leq a$ and $a \cdot \mathsf{s}_a \leq a$.

A preference a is a *layered preference*, also known as *strict weak order*, if and only if additionally *negative transitivity* $(\overline{a})^2 \leq \overline{a}$ holds. □

Compatibility for an SV relation implies $\mathsf{s}_a \sqcap a = 0$, which is shown in [5], Corollary 5.2. For a layered preference a the relation $\mathsf{s}_a = \overline{a + a^{-1}}$ (equivalently $a + \mathsf{s}_a = \overline{a^{-1}}$) is a possible SV-relation, and this is the most intuitive way to define which tuples are equivalent w.r.t. a preference. There are alternative candidates, e.g., $s_a = \mathbb{1}$ is a valid SV-relation for any preference. In general, i.e., for non-layered preferences, the construction $\mathsf{s}_a = \overline{a + a^{-1}}$ does not fulfill the compatibility conditions. A counterexample can be found in [5], Example 5.6.

If a tuple x is a-related to y, formally $(x\,a\,y)$ we say that y *is better than* x *w.r.t. to the preference* a. The compatibility conditions in Definition 2.1 ensure for a tuple y better than x that all tuples from the SV-equivalence class of y are better than those equivalent to x. By convention, the empty preference 0 has the SV-relation $\mathsf{s}_0 =_{df} \top$, i.e., all tuples are equivalent. This satisfies $s_0 = \overline{0 + 0^{-1}}$.

Definition 2.2 (Preference selection). In the scope of database preferences the maximum operator $\triangleright$ selecting the a-maximal elements on r is defined by

$$a \triangleright r =_{df} r - |a\rangle r,$$

which is also known as *preference selection* on the data set r w.r.t. preference a.

We now recapitulate the definitions of the two most important complex preference operators, Pareto composition and prioritisation on a single domain, as given in [10]. These operators are widely used in the context of database preferences and hence they are important building blocks for our decomposition approach. In contrast, + is typically not used as a complex preference operator, as + does not preserve the strict order property in general.

Definition 2.3 (Prioritisation and Pareto on single domain). Let a, b be preferences with associated SV-relations $\mathsf{s}_a, \mathsf{s}_b$. The *prioritisation with SV on a single domain* is given by, where $\sqcap$ binds tighter than +,

$$a \,\&\, b =_{df} a + \mathsf{s}_a \sqcap b \,,$$

whereas the *Pareto composition on a single domain with SV* is defined as

$$a \otimes b =_{df} (a + \mathsf{s}_a) \sqcap b + a \sqcap (b + \mathsf{s}_b) \,.$$

We say that $a \star b$ with $\star \in \{\&, \otimes\}$ is *SV-preserving*, if $\mathsf{s}_{a \star b} = \mathsf{s}_a \sqcap \mathsf{s}_b$ is fulfilled.

Unless otherwise specified, we will assume that the SV-relation of a layered preference is set to the SV-preserving relation.

The intuition behind the prioritisation is the lexicographic order: $a \,\&\, b$ means "better in a or {equal in a and better in b}". The Pareto composition is used to compose equal important wishes, $a \otimes b$ means "equal or better in {a or b} and strictly better in one of them". Both operators are associative and $\otimes$ is even commutative. They both preserve preferences, i.e., for preferences a, b the object $a \star b$ for $\star \in \{\&, \otimes\}$ is a preference again. The & operator even preserves layered preferences, which is shown in [10], Corollary 2.6.

Furthermore, 0 is a neutral element for both operators (note that $\mathsf{s}_0 = \top$), i.e., we have

$$0 \otimes a = a \otimes 0 = a, \quad a \,\&\, 0 = 0 \,\&\, a = a \,.$$

To compare preference relations we introduce a concept of equivalence of preferences w.r.t. a given data set.

Definition 2.4 (r-equivalence). Let a, b be preferences and r a data set. We say that a and b are r-equivalent, if and only if $r \cdot a \cdot r = r \cdot b \cdot r$.

This is equivalent to $(x\, a\, y) \Leftrightarrow (x\, b\, y)$ for all tuples $x, y \in r$. Intuitively, r-equivalence of a and b means that the Hasse diagrams of a and b on the data set r are identical.

An important role is played by the Boolean preference constructor, recapitulated subsequently. This constructor is supported in the most common implementations of database preferences [4,9].

Definition 2.5 (Boolean preference). Let $\rho : D \to \{\mathsf{true}, \mathsf{false}\}$ be a predicate which can be evaluated over all elements in D. Then $\mathsf{is\_true}(\rho)$ is a *Boolean preference* defined by

$$\mathsf{is\_true}(\rho) =_{df} \rho^{-1}(\mathsf{false}) \times \rho^{-1}(\mathsf{true})$$

We define $\mathsf{s}_{\mathsf{is\_true}(\rho)} =_{df} (\rho^{-1}(\mathsf{false}) \times \rho^{-1}(\mathsf{false})) + (\rho^{-1}(\mathsf{true}) \times \rho^{-1}(\mathsf{true}))$. □

Note that this definition fulfills $\mathsf{s}_{\mathsf{is\_true}(\rho)} = \overline{\mathsf{is\_true}(\rho) + \mathsf{is\_true}(\rho)^{-1}}$.

In the *Exasolution* [4] implementation of preferences we can simply write the logical condition ρ in the `PREFERRING` clause of the database query. For example, `select * from mtcars preferring mpg=15` (equivalent to `psel(mtcars, true(mpg==15))` in rPref) selects those cars with an *mpg* value of 15 if such tuples exist. If not, then all cars are returned.

In [10] we have defined *tuple and set preferences* as sub-constructors of *Boolean* preferences. In the following we give an explicit definition of tuple and set preferences in a point-free algebraic fashion.

Definition 2.6 (Set/tuple preference). For a set of tuples $p \leq \mathbb{1}$ we define the *set preference*

$$\mathsf{t}(p) =_{df} \neg p \cdot \top \cdot p \; .$$

If $|p| = 1$, i.e., p is a singe tuple, we also say that $\mathsf{is\_true}(p)$ is a *tuple preference.* The associated SV-relation is given by $\mathsf{s}_{\mathsf{t}(p)} =_{df} p \cdot \top \cdot p + \neg p \cdot \top \cdot \neg p$.

With $\mathsf{t}(x)$ we can express that the tuple x is preferred over all other tuples in $(\mathbb{1} - x)$. Using $\mathsf{t}(\cdot)$ and the $\{\&, \otimes\}$ operators, we can construct arbitrary strict orders, as we will summarize in the following section.

2.3 Preference Decomposition

For a data set r, every preference a can be transformed into a preference term b which is r-equivalent to a and consists of set preferences and the $\{\&, \otimes\}$ operators. We call this transformation *preference decomposition.* The set of necessary operands and operators can be limited in two different ways detailed below. In both cases all strict orders can be decomposed.

1. The resulting preference consists only of a $\otimes$-composition of set preferences, shown in [10], Theorem 4.1. For example, the "N-shaped" preference a on the data set $r = x_1 + ... + x_4$, depicted in Figure 2(1), has the r-equivalent decomposition

$$((\mathsf{t}(x_1) \otimes \mathsf{t}(x_2)) \;\&\; \mathsf{t}(x_3)) \otimes (\mathsf{t}(x_2) \;\&\; \mathsf{t}(x_4)) \; .$$

   Roughly speaking, parallel nodes in the diagram of the preference are $\otimes$-composed and serial connections result in &-chains.
2. Alternatively we can construct a term of tuple preferences connected by both of the operators $\{\&, \otimes\}$, shown in [10], Theorem 4.4. The above preference decomposes into

$$\mathsf{t}(x_1) \otimes \mathsf{t}(x_2) \otimes \mathsf{t}(x_1 + x_2 + x_3) \otimes \mathsf{t}(x_2 + x_4) \; .$$

   Here the better-than-relations of the preference are encoded in the inclusion order of the involved sets, e.g., $(x \, a \, y)$ is realized by a preference $\mathsf{t}(p) \otimes \mathsf{t}(q)$ with $y \in p \leq q$, $x \in q$ and $x \notin p$.

2.4 Power Set Preferences

In the following we define the power construction for preferences on a given data set. According to [2,13] there are three natural extensions of a strict order to the power set of the domain which we will recapitulate in the following definition. We will assume a finite data set in the following, as we need finiteness to show that the power construction yields a preference.

Definition 2.7 (Power set preference). Let a be a preference on a finite data set r. We introduce preferences π_i^a for $i \in \{0, 1, 2\}$ on the power set $\mathcal{P}(r) = \{p \mid p \leq r\}$ by defining for all $u, v \in \mathcal{P}(r)$:

$$\begin{aligned} u\,\pi_0^a\,v \quad &\Leftrightarrow_{df} \quad v \neq 0 \ \wedge \ \forall y \in v : \exists x \in u : x\,a\,y \\ u\,\pi_1^a\,v \quad &\Leftrightarrow_{df} \quad u \neq 0 \ \wedge \ \forall x \in u : \exists y \in v : x\,a\,y \\ u\,\pi_2^a\,v \quad &\Leftrightarrow_{df} \quad u\,(\pi_0^a \sqcap \pi_1^a)\,v \end{aligned}$$

Intuitively π_0 means that a set v is better than a non-empty set u, formally $u\,\pi_0^a\,v$, if every tuple in v dominates some tuple in u. For $u\,\pi_1^a\,v$ we require that every tuple in v is dominated by some tuple in u. Finally for $u\,\pi_2^a\,v$ both of these conditions have to be fulfilled, formally resulting in the intersection of π_0^a and π_1^a. In Figure 2 we show (partial) graphs of all power set preferences π_i^a ($i \in \{0, 1, 2\}$) for the "N-shaped" preference a in Figure 2(1). The power set preference π_1^a has already been used in the context of database preferences in [12], Definition 3.1.

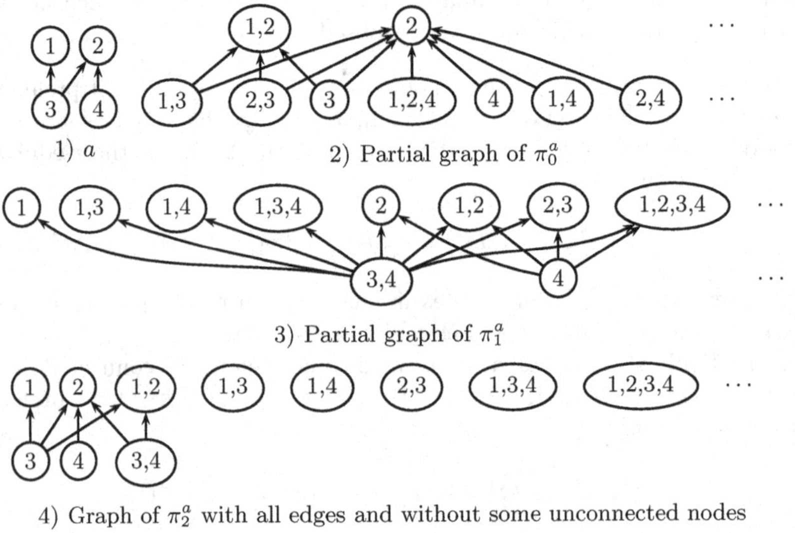

Fig. 2. Preference a and a partial diagram of its induced power set preferences π_j^a for $j \in \{0, 1, 2\}$. A circled i is short for x_i and a circled $i_1, ..., i_k$ short for $x_{i_1} + ... + x_{i_k}$.

Note that, without the second conjunct excluding empty sets in π_1^a and π_2^a, Definition 2.7 would lead to a relation not being irreflexive (and hence not a preference). Formally, for $u\,\pi_3^a\,v \Leftrightarrow \forall y \in v : \exists x \in v : x\,a\,y$, we get $0\,\pi_3^a\,0$.

However, in the following we will exclude the empty set from the considered domain as it is not interesting for our application. For a data set r we define

$$\widehat{r} =_{df} \mathcal{P}(r)\backslash\{0\} = \{p \mid p \leq r \wedge p \neq 0\} ,$$

which will be the domain of power set preferences throughout the remainder of the paper.

Subsequently we introduce a quantifier-free representation of the predicates $u\,\pi_i^a\,v$, following the idea of [2], Theorem 2.31.

Corollary 2.8. *Let a be a preference on a data set r. For all $u, v \in \widehat{r}$ we have*

$$\begin{array}{lll} u\,\pi_0^a\,v & \Leftrightarrow_{df} & v \leq \langle a|u \ , \\ u\,\pi_1^a\,v & \Leftrightarrow_{df} & u \leq |a\rangle v \ . \end{array}$$

Proof. Immediately from the definition of $\widehat{r}$, $|a\rangle(\cdot)$ and $\langle a|(\cdot)$. □

Lemma 2.9. *Definition 2.7 is well-formed, i.e., π_i^a are indeed preferences.*

Proof. We will show this just for π_1^a. For π_0^a analogous arguments hold and for $\pi_2^a = \pi_0^a \sqcap \pi_1^a$ we exploit that preferences are preserved under intersection. We have to show that π_1^a is irreflexive and transitive. For $u, v, w \in \widehat{r}$ we have, using Corollary 2.8, the isotony of the diamond and the transitivity of a,

$$u\,\pi_1^a\,v \ \wedge\ v\,\pi_1^a\,w \quad \Rightarrow \quad u \leq |a\rangle(|a\rangle w) \leq |a^2\rangle w \leq |a\rangle w \quad \Rightarrow \quad u\,\pi_1^a\,w \ ,$$

showing the transitivity of π_1^a. Next, we show irreflexivity by contradiction, i.e., we assume $u\,\pi_1^a\,u$ for $u \in \widehat{r}$. We calculate:

$$\begin{array}{ll} & u\,\pi_1^a\,u \\ \Leftrightarrow & \{\!\![\ \text{Corollary 2.8}\ ]\!\!\} \\ & u \leq |a\rangle u \\ \Leftrightarrow & \{\!\![\ \text{shunting } (p \cdot q \leq s \Leftrightarrow p \leq \neg q + s \text{ with } p = u, q = \neg|a\rangle u, s = 0)\ ]\!\!\} \\ & u - |a\rangle u \leq 0 \\ \Leftrightarrow & \{\!\![\ \text{definition of } \rhd\ ]\!\!\} \\ & a \rhd u = 0 \end{array}$$

As $u \in \widehat{r}$ we have $u \neq 0$, hence this implies that the maxima set of a non-empty set w.r.t. a preference (strict order) a is empty. By the assumption of a finite r in Definition 2.7, we also have that $u \leq r$ is finite. Thus an empty maxima set $a \rhd u$ is a contradiction. Hence we have shown the irreflexivity of π_1^a. □

Now we know that the preference property is preserved under extending the preference to its power set. But for a layered preference a the power set preference π_2^a is not a layered preference in general. For example, let $r = x_1 + x_2$ and $(x_1 \, a \, x_2)$ the only better-than-relation in a. By the definition of π_2^a we have $(x_1 \, \pi_2^a \, x_2)$. The set $p = x_1 + x_2$ is incomparable to both x_1 and x_2, i.e., formally $(x_1 \, \overline{\pi_2^a} \, p)$ and $(p \, \overline{\pi_2^a} \, x_2)$. Hence π_2^a is not a layered preference, as negative transitivity is violated.

Note that on $\mathcal{P}(r)$ the empty set 0 is incomparable to all sets in $\widehat{r}$ w.r.t. π_i^a for all $i \in \{0, 1, 2\}$. On $\widehat{r}$, the power set preferences π_0^a and π_1^a preserve layered preferences, as we show subsequently.

Lemma 2.10. *Let a be a layered preference and r a data set. Then, the preferences $\widehat{r} \cdot \pi_0^a \cdot \widehat{r}$ and $\widehat{r} \cdot \pi_1^a \cdot \widehat{r}$ are layered preferences.*

Proof. Let a be negative transitive, i.e., $(\overline{a})^2 \leq \overline{a}$. We have to show that the preferences $\widehat{r} \cdot \pi_i^a \cdot \widehat{r}$ are also negative transitive. We show this for $b = \widehat{r} \cdot \pi_1^a \cdot \widehat{r}$ and $u, v, w \in \widehat{r}$:

$$u \, \overline{b} \, v \wedge v \, \overline{b} \, w \quad \Leftrightarrow \quad \neg(u \, (\widehat{r} \cdot \pi_1^a \cdot \widehat{r}) \, v) \wedge \neg(v \, (\widehat{r} \cdot \pi_1^a \cdot \widehat{r}) \, w)$$

$$\Leftrightarrow \quad \{\!\!\{ \; u, v, w \in \widehat{r}, \text{ definition of } \pi_a^1 \text{ and moving negation inside } \}\!\!\}$$

$$u = 0 \vee \exists x \in u : \forall y \in v : x \, \overline{a} \, y \; \wedge \; v = 0 \vee \exists x' \in v : \forall y' \in w : x' \, \overline{a} \, y'$$

$$\Leftrightarrow \quad \{\!\!\{ \; u, v \in \widehat{r}, \text{ hence } u \neq 0 \text{ and } v \neq 0 \; \}\!\!\}$$

$$\exists x \in u : \forall y \in v : x \, \overline{a} \, y \; \wedge \; \exists x' \in v : \forall y' \in w : x' \, \overline{a} \, y'$$

$$\Rightarrow \quad \{\!\!\{ \text{ specialization } y = x', \text{ reorganization of quantifiers } \}\!\!\}$$

$$\exists x \in u : \exists x' \in v : \forall y' \in w : x \, \overline{a} \, x' \; \wedge \; x' \, \overline{a} \, y'$$

$$\Leftrightarrow \quad \{\!\!\{ \text{ definition of } (\cdot)^2 \; \}\!\!\}$$

$$\exists x \in u : \forall y' \in w : x \, (\overline{a})^2 \, y'$$

$$\Rightarrow \quad \{\!\!\{ \text{ logic and negative transitivity of } a \; \}\!\!\}$$

$$u = 0 \vee \exists x \in u : \forall y' \in w : x \, \overline{a} \, y'$$

$$\Leftrightarrow \quad \{\!\!\{ \text{ moving negation outside, definition of } \pi_a^1 \text{ and } u, w \in \widehat{r} \; \}\!\!\}$$

$$\neg(u \, (\widehat{r} \cdot \pi_1^a \cdot \widehat{r}) \, w) \quad \Leftrightarrow \quad u \, \overline{b} \, w$$

For $\widehat{r} \cdot \pi_0^a \cdot \widehat{r}$ an analogous argument holds. Thus for $i \in \{0, 1\}$, the preference $\widehat{r} \cdot \pi_i^a \cdot \widehat{r}$ is negatively transitive and hence a layered preference. □

3 Optimized Decomposition Algorithms

The decomposition algorithms from [10] allow expressing any preference (strict order) with Boolean preferences and the $\{\&, \otimes\}$ operators, as exemplified in Section 2.3.

The obvious problem of that approach, which we have discussed in the outlook of [10], is that these algorithms generate much redundancy. For example the empty preference 0 on a data set $r = x_1 + ... + x_n$ is inflated to $\mathtt{t}(x_1) \otimes ... \otimes \mathtt{t}(x_n)$ with both constructions. Especially, layered preferences with many tuples per layer result in lengthy terms as we see in the following example.

Example 3.1. Let $b = \mathtt{t}(x_1 + x_2)$ on $r = x_1 + ... + x_5$. We apply the decomposition into $\{\&, \otimes\}$ and tuple preferences from [10]. This results in the r-equivalent preference

$$(\mathtt{t}(x_1) \mathbin{\&} (\mathtt{t}(x_3) \otimes \mathtt{t}(x_4) \otimes \mathtt{t}(x_5))) \otimes (\mathtt{t}(x_2) \mathbin{\&} (\mathtt{t}(x_3) \otimes \mathtt{t}(x_4) \otimes \mathtt{t}(x_5))) \ .$$

Considering the Hasse diagram of b, given in Figure 3, we see that x_1 is equivalent to x_2 in the sense that their sets of a-predecessors and a-successors are identical, and the same holds for x_3, x_4 and x_5. Our idea is to identify all nodes which are equivalent in this sense and then to apply our decomposition algorithms to these simpler graphs. We exemplify this idea in Figure 3.

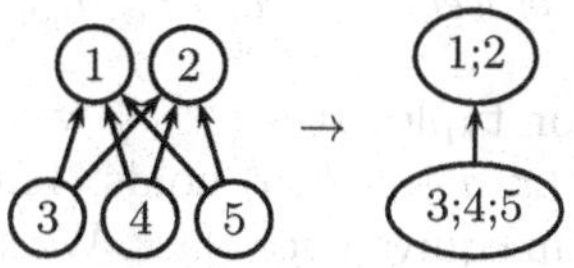

Fig. 3. Preference b (left) and its simplified graph (right). A circled $i_1; ...; i_k$ is short for the equivalence class $[\![x_{i_1}]\!] = ... = [\![x_{i_k}]\!]$.

3.1 Elimination of Equivalent Nodes

In the following we give a formal definition of the elimination of equivalent nodes. The induced preference on the quotient set is called the *minimized preference.*

Definition 3.2 (Minimized preference). Let a be a preference and r a data set. We define an equivalence relation $\sim_{a,r}$ where all tuples having the same predecessors and successors are in one equivalence class. Formally we define for all $u, v \in r$:

$$u \sim_{a,r} v \quad \Leftrightarrow_{df} \quad r \cdot |a\rangle u = r \cdot |a\rangle v \ \wedge \ r \cdot \langle a| u = r \cdot \langle a| v \ .$$

Let $r_{\min} =_{df} r/{\sim_{a,r}}$ be the quotient set of $\sim_{a,r}$. For $u, v \in r$ we define the minimized preference $a_{\min}$ on the equivalence classes $[\![u]\!], [\![v]\!] \in r_{\min}$ by

$$[\![u]\!] \, a_{\min} \, [\![v]\!] \quad \Leftrightarrow_{df} \quad u \, a \, v \ .$$

Further we define $\mathsf{s}_{a_{\min}} = \sim_{a,r}$ as the SV-relation of this preference.

All other relational operations are lifted to the set of equivalence classes in the canonical way, especially we define $0_{\min} =_{df} 0$ and $1\!\!1_{\min} =_{df} 1\!\!1/{\sim_{a,r}}$. Relational composition with some other relation b on the original data set r (and not the quotient set) is canonically defined for $[\![u]\!] \in r_{\min}, v \in r$ by

$$[\![u]\!] \, (a_{\min} \cdot b) \, v \quad \Leftrightarrow_{df} \quad \exists w \in \mathbb{1} : [\![u]\!] \, a_{\min} \, [\![w]\!] \ \wedge \ w \, b \, v \; ,$$

and symmetrically we define $b \cdot a_{\min}$. This means that w and its associated equivalence class $[\![w]\!]$ establishes the connection between common relations and relations on equivalence classes w.r.t. $\sim_{a,r}$.

The set preference having an equivalence class as argument, formally $\mathrm{t}([\![x]\!])$, is defined on the original domain r. The equivalence class $[\![x]\!] \leq r$ is considered as the set of all tuples equivalent to x w.r.t. $\sim_{a,r}$. □

This construction is very similar to constructing minimal automata by identifying equivalent states, cf. [3].

The definition of $\sim_{a,r}$ is directly connected to the compatibility conditions for SV-relations in Definition 2.1. This implies that $\sim_{a,r}$ is the maximal SV-relation for a.

Lemma 3.3. *Definition 3.2 is well-formed, i.e.,* $a_{\min}$ *is indeed a preference.*

Proof. First we show that for tuples $x, y \in [\![u]\!]$ with $u \in r_{\min}$, i.e., tuples in the same equivalence class, $\neg(x \, (a + a^{-1}) \, y)$ holds. This means that there are no better-than-relations within an equivalence class. We show this by contradiction. Assume that $(x \, (a + a^{-1}) \, y)$ holds. W.l.o.g. we presume $(x \, a \, y)$. This implies $x \in |a\rangle y$ by definition and $y \notin |a\rangle y$ by irreflexivity of a. This is a contradiction to $x, y \in [\![u]\!]$ by the definition of $[\![u]\!]$. With this, the definition of $a_{\min}$ and the property that a is a preference we can simply verify that $a_{\min}$ is also a preference. By definition it is clear that $a_{\min} \cdot \sim_{a,r} = \sim_{a,r} \cdot a_{\min} = a_{\min}$ holds, hence $\sim_{a,r}$ is a valid SV-relation for $a_{\min}$. □

3.2 Minimized Decomposition

We apply the elimination of equivalent nodes to the decomposition algorithms.

Definition 3.4. *Let a be a preference and $r \leq \mathbb{1}$ a finite data set. We define a $\otimes$-composition of set preferences where each set is upward closed w.r.t. $a_{\min} + \mathbb{1}_{\min}$, using the definitions for $r_{\min}$, $a_{\min}$ and $\mathbb{1}_{\min}$ from Definition 3.2, by*

$$\textsc{Dec\_Min}_1(a, r) =_{df} \bigotimes_{[\![x]\!] \in r_{\min}} \mathrm{t}(r \cdot \langle a_{\min} + \mathbb{1}_{\min} | [\![x]\!]) \; .$$

We define $\textsc{Dec\_Min}_2(a, r)$ *as given in Algorithm 1.*

Algorithm 1 is a decomposition into set preferences and $\{\&, \otimes\}$ where each set corresponds to the equivalence class of a node in the preference graph, i.e., a tuple of the data set, or, in the case of minimized preferences, a subset of the data set.

Algorithm 1. Preference decomposition into set preferences and $\{\&, \otimes\}$ using the definitions for $r_{\min}$ and $a_{\min}$ from Definition 3.2

Input: Preference to decompose a, data set r

Output: r-equivalent decomposition b_{res}

function $\text{DEC\_MIN}_2(a, r)$
    $a_h \leftarrow r_{\min} \cdot (a_{\min} \sqcap \overline{(a_{\min})^2}) \cdot r_{\min}$ // Hasse diagram of $a_{\min}$ on $r_{\min}$
    $b[r_{\min}] \leftarrow 0$ // initialization of array b of preferences
    $m \leftarrow (a_{\min} \rhd r_{\min})$ // start traversing with eq. classes of maxima
    while $m \neq 0_{\min}$ **do**
        for all $[\![y]\!] \in m$ **do** // pref. for y, collect and $\otimes$-compose successors,
            $b[[\![y]\!]] \leftarrow \left(\bigotimes_{[\![x]\!] \in \langle a_h | [\![y]\!]} b[[\![x]\!]]\right) \,\&\, \mathsf{t}([\![y]\!])$ // and add pref. on $[\![y]\!]$
        end for
        $b[\langle a_h | m] \leftarrow 0$ // delete preferences of m-successors
        $m \leftarrow (a_{\min} \rhd |a_h\rangle m))$ // find a-maximal predecessors of m
    end while
    $b_{\text{res}} \leftarrow \bigotimes_{[\![x]\!] \in r_{\min}} b[[\![x]\!]]$; **return** b_{res} // $\otimes$-compose final preference
end function

The indices of the array $b[\cdot]$ in Algorithm 1 are the equivalence classes $r_{\min}$. The values of $b[\cdot]$ are preferences. The assignment $b[m] \leftarrow c$ for a non-empty set $m \subseteq r_{\min}$ is used as a shorthand notation for simultaneous assignments $b[[\![x]\!]] \leftarrow c$ for all $[\![x]\!] \in m$. We also assume that in all assignments the neutrality of 0 is used, implying that $b[[\![x]\!]] \leftarrow 0 \star c$ is executed as $b[[\![x]\!]] \leftarrow c$ for $\star \in \{\&, \otimes\}$.

Subsequently we will show the correctness of these decomposition algorithms. Both are canonical transformations of the algorithms presented and proven to be correct in [10], where the domain changes from r to the quotient set $r/\sim_{a,r}$. We will use the correctness proofs of that paper in the following arguments.

Theorem 3.5. *Let a be a preference and r a data set. Then both* $\text{DEC\_MIN}_1(a, r)$ *and* $\text{DEC\_MIN}_2(a, r)$ *are r-equivalent to a.*

Proof. In Lemma 3.3 we have shown that $a_{\min}$ is indeed a preference. Hence the correctness of the decomposition algorithms in [10] implies that $a_{\min}$ is $r_{\min}$-equivalent to $\text{DEC\_MIN}_1(a, r)$ and $\text{DEC\_MIN}_2(a, r)$. Additionally this implies that $\text{DEC\_MIN}_1(a, r)$ and $\text{DEC\_MIN}_2(a, r)$ are also well-defined preferences on $r_{\min}$. Immediately by Definition 3.2 we get that

$$r \cdot b \cdot r = r \cdot b_{\min} \cdot r \quad \text{for} \quad b \in \{a, \text{DEC\_MIN}_1(a, r), \text{DEC\_MIN}_2(a, r)\}$$

holds, i.e., the minimized preference is r-equivalent to non-minimized one. Finally we calculate, using the $r_{\min}$-equivalence of a and $\text{DEC\_MIN}_i(a, r)$,

$$\begin{aligned} r \cdot a \cdot r &= r \cdot a_{\min} \cdot r = r \cdot r_{\min} \cdot a_{\min} \cdot r_{\min} \cdot r \\ &= r \cdot r_{\min} \cdot \text{DEC\_MIN}_i(a, r) \cdot r_{\min} \cdot r = r \cdot \text{DEC\_MIN}_i(a, r) \cdot r \ , \end{aligned}$$

for $i \in \{1, 2\}$. For the composition of the subidentities $r_{\min}$ and r we apply Definition 3.2. The above calculation shows the claim. □

As a first example of these algorithms we apply the optimized decompositions to the simple layered preference from Example 3.1.

Example 3.6. Let $b = \mathsf{t}(x_1 + x_2)$ on $r = x_1 + ... + x_5$, as illustrated in Figure 3. For the optimized decompositions we get:

$$\begin{aligned}
\text{DEC_MIN}_1(b, r) &= \mathsf{t}([\![x_1]\!]) \otimes \mathsf{t}([\![x_1]\!] + [\![x_3]\!]) = \mathsf{t}(x_1 + x_2) \otimes \mathsf{t}(r) \ , \\
\text{DEC_MIN}_2(b, r) &= \mathsf{t}([\![x_1]\!]) \,\&\, \mathsf{t}([\![x_3]\!]) = \mathsf{t}(x_1 + x_2) \,\&\, \mathsf{t}(x_1 + x_2 + x_3) \ .
\end{aligned}$$

The final term for $\text{DEC\_MIN}_2(b, r)$ is much shorter than the decomposition from Example 3.1.

4 Minimized Decomposition of Power Set Preferences

In the following we consider the decomposition of the minimized power set preferences $(\pi_i^a)_{\min}$ for a given preference a on the data set r. Throughout we will use $\widehat{r}$ as domain for the power set preferences and their Hasse diagrams.

4.1 Examples

Example 4.1. Reconsider the N-shaped preference a from Figure 2(1). We apply the elimination of equivalent nodes from Definition 3.2 to the power set preferences π_i^a for $i \in \{0, 1, 2\}$.

Consider π_1^a and its minimized variant $(\pi_1^a)_{\min}$, illustrated in Figure 4. The diagram of $(\pi_1^a)_{\min}$ is isomorphic to the original preference a. For the equivalence classes of $\sim_{\pi_1^a, r}$ we get:

$$\begin{aligned}
[\![x_1]\!] &= \{x_1 + y \mid y \le x_3 + x_4\} \\
[\![x_2]\!] &= \{x_2 + y \mid y \le x_1 + x_3 + x_4\} \\
[\![x_3]\!] &= \{x_3\} \\
[\![x_4]\!] &= \{x_3, x_3 + x_4\}
\end{aligned}$$

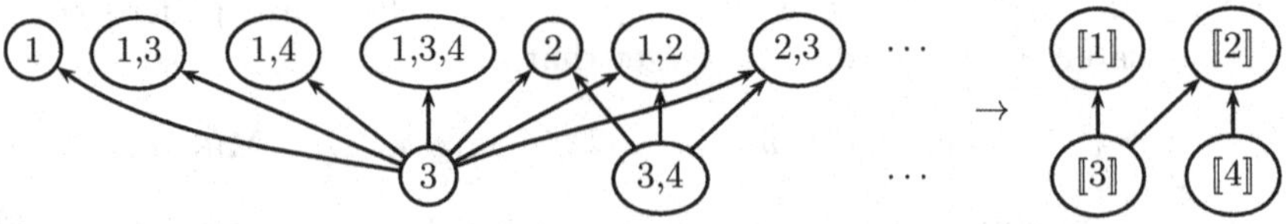

Fig. 4. Partial diagram of Preference π_1^a (left) and diagram of $(\pi_i^a)_{\min}$ for $i \in \{0, 1\}$ (right), where a circled $[\![i]\!]$ is short for $[\![x_i]\!]$.

By definition of π_1^a, only the existence of dominating elements (here x_1 and x_2) in a set v is of interest to determine if $(u \, \pi_1^a \, v)$ holds. For example, x_1 is dominating x_3 w.r.t. a. Hence all sets v with $x_1 \in v$ are better than x_3.

Consequently, sets like x_1, $x_1 + x_4$ and $x_1 + x_3$ belong to the same equivalence class w.r.t. $\sim_{\pi_1^a,r}$.

According to decomposition methods from Definition 3.4 and Algorithm 1 we get for the decomposed preference terms

$$\text{DEC\_MIN}_1(\pi_1^a, r) = \mathsf{t}([\![x_1]\!]) \otimes \mathsf{t}([\![x_2]\!]) \otimes \mathsf{t}([\![x_1]\!] + [\![x_2]\!] + [\![x_3]\!]) \otimes \mathsf{t}([\![x_1]\!] + [\![x_4]\!]) ,$$
$$\text{DEC\_MIN}_2(\pi_1^a, r) = ((\mathsf{t}([\![x_1]\!]) \otimes \mathsf{t}([\![x_2]\!])) \,\&\, \mathsf{t}([\![x_3]\!])) \otimes (\mathsf{t}([\![x_2]\!]) \,\&\, \mathsf{t}([\![x_4]\!])) .$$

Next we consider the power set preference π_0^a. The Hasse diagram of $(\pi_0^a)_{\min}$ has the same structure as that of $(\pi_1^a)_{\min}$ but the evaluation of the equivalence classes w.r.t. $\sim_{\pi_0^a,r}$ yields a different result:

$$\begin{aligned} [\![x_1]\!] &= \{x_1,\ x_1 + x_2\} \\ [\![x_2]\!] &= \{x_2\} \\ [\![x_3]\!] &= \{x_3 + y \mid y \leq x_1 + x_2 + x_3\} \\ [\![x_4]\!] &= \{x_4 + y \mid y \leq x_1 + x_2\} \end{aligned}$$

The equivalence relation for π_0^a is isomorphic to that of π_1^a and the isomorphism is given by the mapping $\phi : r \to r$ with

$$\phi(x_1) = x_4,\ \phi(x_4) = x_1,\ \phi(x_2) = x_3,\ \phi(x_3) = x_2 .$$

Formally it holds that $\sim_{\pi_1^a,r} = \sim_{\pi_0^a,\phi(r)}$. But note that this isomorphism does not change the better-than-relations, the graph of π_0^a is still the same as in Figure 4 (right). Only the existence of dominated tuples in a set is relevant to determine better sets w.r.t. π_0^a. For example, we can add any tuple to x_3 and this set is still worse than x_2, formally $(u\,\pi_0^a\,x_2)$ holds if $x_3 \in u$.

Finally we consider $\pi_2^a = \pi_0^a \sqcap \pi_1^a$. Now for $u\,\pi_2^a\,v$ with $u, v \in \widehat{r}$ we require that all tuples in u are dominated by tuples in v, and simultaneously the tuples in v have to dominate those of u.

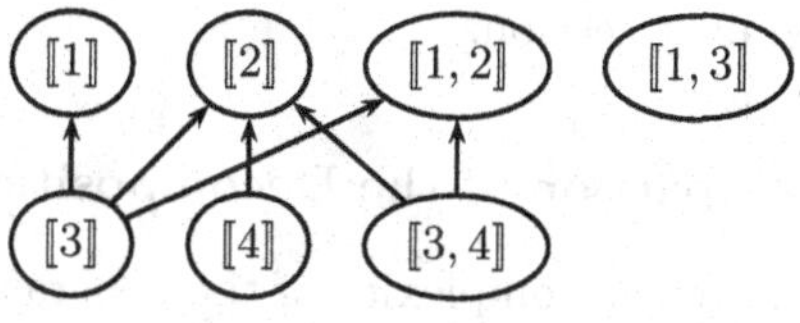

Fig. 5. Diagram of preference $(\pi_2^a)_{\min}$. A circled $[\![i_1, ..., i_k]\!]$ is short for $[\![x_{i_1} + ... + x_{i_k}]\!]$.

This changes the Hasse diagram of π_2^a, depicted in Figure 5, compared to that of π_0^a and π_1^a (Figure 4). Each equivalences class in $r_{\min} - [\![x_1 + x_3]\!]$ forms an equivalence class of its own w.r.t. $\sim_{\pi_2^a,r}$. Their better-than-relations are identical to the non-minimized variant depicted in Figure 2(4). The class $[\![x_1 + x_3]\!]$, collecting the incomparable tuples w.r.t. π_2^a, can be formally characterized by

$$u \in [\![x_1 + x_3]\!] \quad \Leftrightarrow \quad u \sqcap (x_1 + x_2) \neq 0 \ \wedge\ u \sqcap (x_3 + x_4) \neq 0 .$$

This formalizes that all sets containing dominated as well as dominating nodes are incomparable to all other sets w.r.t. π_2^a.

In the example above the Hasse diagrams of π_i^a for $i = 0$ and $i = 1$ are isomorphic to that of a. This is not always the case as we will show in the following example.

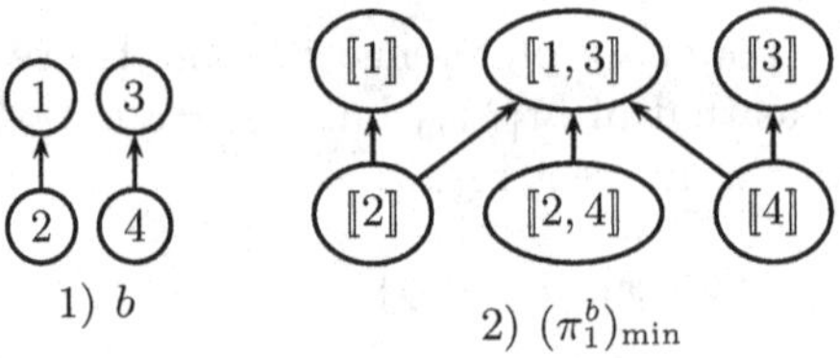

Fig. 6. Preference b and its minimized induced power set preference $(\pi_1^b)_{\min}$.

Example 4.2. Consider the preference $b = (\mathsf{t}(x_1)\ \&\ \mathsf{t}(x_2)) \otimes (\mathsf{t}(x_3)\ \&\ \mathsf{t}(x_4))$ on $r = x_1 + ... + x_4$ and its minimized power set preference $(\pi_1^b)_{\min}$, depicted in Figure 6. For the preference π_1^b we get the following decompositions:

$$\begin{aligned}
\text{DEC_MIN}_1(\pi_1^b, r) = {} & \mathsf{t}([\![x_1]\!]) \otimes \mathsf{t}([\![x_1 + x_3]\!]) \otimes \mathsf{t}([\![x_3]\!]) \otimes \mathsf{t}([\![x_1]\!] + [\![x_2]\!]) \otimes \\
& \mathsf{t}([\![x_1 + x_3]\!] + [\![x_2]\!]) \otimes \mathsf{t}([\![x_1 + x_3]\!] + [\![x_2 + x_4]\!]) \otimes \\
& \mathsf{t}([\![x_1 + x_3]\!] + [\![x_4]\!]) \otimes \mathsf{t}([\![x_3]\!] + [\![x_4]\!]) \\
\text{DEC_MIN}_2(\pi_1^b, r) = {} & (\mathsf{t}([\![x_1]\!])\ \&\ \mathsf{t}([\![x_2]\!])) \otimes (\mathsf{t}([\![x_3]\!])\ \&\ \mathsf{t}([\![x_3 + x_4]\!])) \otimes \\
& (\mathsf{t}([\![x_1 + x_3]\!])\ \&\ (\mathsf{t}([\![x_2]\!]) \otimes \mathsf{t}([\![x_2 + x_4]\!]) \otimes \mathsf{t}([\![x_4]\!])))
\end{aligned}$$

The resulting terms in this example may still look quite complex. Still, the minimization is an advantage in many cases as we will underline in the quantitative comparison in the next section.

4.2 Quantitative Comparison of the Decomposition Approaches

In the following we compare the complexity of the generated terms by counting operands and operators. For a preference a, we define $|a|_{\mathsf{t}}$ to be the number of $\mathsf{t}(\cdot)$-operands in a and $|a|_{\otimes}$ the number of $\otimes$-operators in a. For example consider $a = \mathsf{t}(x_1) \otimes (\mathsf{t}(x_2)\ \&\ \mathsf{t}(x_3)) \otimes \mathsf{t}(x_4)$. We get $|a|_{\mathsf{t}} = 4$ and $|a|_{\otimes} = 2$.

The number of $\otimes$-operators is the main factor for the computational complexity of the preference evaluation, i.e., the determination of the maxima $a \triangleright r$ for a given data set r. As discussed in [10], the costs to evaluate a $\otimes$-chain $a_1 \otimes ... \otimes a_n$ of layered preferences a_i quickly increases with the length n. The & operator preserves layered preferences whereas $\otimes$ does not.

Next to the preference a from Figure 2 and b from Figure 6 we consider c, d, e as depicted in Figure 7. The preference c is similar to the N-shaped preference

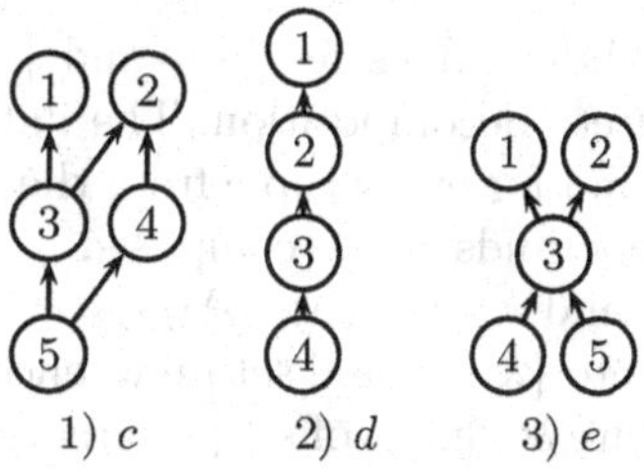

Fig. 7. Hasse diagrams of preferences c, d, e used in the experiments.

a, but contains an additional tuple x_5 which is worse than all other tuples. Like a and b the preference c is also not a layered preference. The preferences d and e are layered preferences. We have chosen these examples, as they contain typical sub-graphs of larger preferences. They allow us to study the principal effects, e.g., term complexity depending on the number of tuples, for some typical parallel and serial constructions of preferences.

Table 1. Quantitative comparison of different preference decompositions of power set preferences. We abbreviate $(\cdot)_P = \text{DECOMP\_PARETO}(\cdot, r)$, $(\cdot)_T = \text{DECOMP\_TUPLE}(\cdot, r)$, $(\cdot)_{M1} = \text{DEC\_MIN}_1(\cdot, r)$, $(\cdot)_{M2} = \text{DEC\_MIN}_2(\cdot, r)$ and $r_4 = x_1 + ... + x_4$, $r_5 = r_4 + x_5$.

Pref.	r	$\lvert(\cdot)_P\rvert_t$	$\lvert(\cdot)_P\rvert_\otimes$	$\lvert(\cdot)_T\rvert_t$	$\lvert(\cdot)_T\rvert_\otimes$	$\lvert(\cdot)_{M1}\rvert_t$	$\lvert(\cdot)_{M1}\rvert_\otimes$	$\lvert(\cdot)_{M2}\rvert_t$	$\lvert(\cdot)_{M2}\rvert_\otimes$
π_1^a	r_4	15	14	31	27	4	3	5	2
π_2^a	r_4	15	14	18	14	7	6	10	6
π_1^b	r_4	15	14	23	19	6	5	8	4
π_2^b	r_4	15	14	15	11	7	6	7	3
π_1^c	r_5	31	30	119	111	5	4	6	2
π_2^c	r_5	31	30	53	45	12	11	21	14
π_1^d	r_4	15	14	75	63	4	3	4	0
π_2^d	r_4	15	14	21	17	6	5	7	4
π_1^e	r_5	31	30	303	287	3	2	3	0
π_2^e	r_5	31	30	54	44	6	5	7	3

In Table 1 we summarize the term complexity for each decomposition method, each preference ($a, ..., e$ from Figures 2, 6 and 7) and the power set preferences $\pi_i^{(\cdot)}$ for $i = 1$ and $i = 2$. We omitted $i = 0$ as this is very similar to $i = 1$ and hence the complexity is nearly the same. For the preferences a, b, d, e it is even exactly the same, because their Hasse diagrams are symmetric, cf. the isomorphism shown in Example 4.1. All results have been retrieved with the help of the R script [8] available on the web.

4.3 Discussion of the Results

For the non-minimized Pareto decompositions $(\cdot)_P$ in Table 1 we get in all cases $\lvert(\cdot)_P\rvert_t = 2^i - 1$ and $\lvert(\cdot)_P\rvert_\otimes = 2^i - 2$ for the data set r_i, $i \in \{4, 5\}$ having the cardinality $|r_i| = i$. These values for $|\cdot|_t$ and $|\cdot|_\otimes$ follow directly from the construction of the Pareto decomposition. The data set has the cardinality $|\widehat{r_i}| = 2^i - 1$ for $i \in \{4, 5\}$, and for every tuple from the data set one $\mathsf{t}(\cdot)$ operand is generated and then all operands are $\otimes$-composed.

For the preferences π_1^d and π_1^e we get $|(\cdot)_{M2}|_\otimes = 0$. This means that the minimized decomposition into $\{\&, \otimes\}$ and set preferences does not contain any $\otimes$ operator. Hence π_1^d and π_1^e are &-chains of set preferences. This is what we expect from Lemma 2.10, i.e., the power construction preserves layered preferences, which can be expressed as &-chains of set preferences, as shown in Lemma 3.8 in [10].

For all the other quantitative results we have no such "obvious" explanation. We consider this quantitative summary as an empirical result, giving some evidence that the minimization according to the $\sim_{(\cdot),r}$ equivalence relation is quite useful when decomposing power set preferences.

5 Conclusion and Outlook

In the present paper we have worked on the problem, mentioned in the outlook of [10], that the preference decompositions from that paper generate quite lengthy terms containing much redundancy.

This paper can be considered as a first step to find *minimal decompositions*, i.e., the search for a minimal complex preference term for a given strict order. Again we have restricted ourselves to the simple case of Boolean preferences, as they have pleasant theoretical properties. For future research it would be interesting to obtain preference terms like "high mpg value $\otimes$ high hp value" (as sketched in the example from the introduction) from a given strict order. Currently, this looks like a NP-hard combinatorial problem to us.

Moreover we have considered the power construction of preferences, where the term complexity of the Boolean decompositions can be reduced. The minimized decomposition also helped us to see some interesting theoretic properties of power set preferences. For example, in some cases the diagram of a preference is isomorphic to its power set extension. With the presented results we have lifted the expressiveness result from [10] to the power construction.

Power set preferences occur naturally when a user has to decide between given sets of alternatives where she searches for Pareto optima of the elements in the sets. This concept has already been applied in the context of database preferences, cf. [12]. With our approach we offer a way for a compact description of power set preferences in common preference query languages [9,4], only requiring set preferences and the $\{\&, \otimes\}$ operators. For the decomposition DEC_MIN_1, set preferences and just the $\otimes$ operator are sufficient to express arbitrary power set preferences.

For future research we are also interested in making more general statements about the term complexity of (minimized) decompositions. For example, a short formula to get the results from Table 1 would be a nice theoretical result. Currently, these numbers have been obtained experimentally by an R script. Based on such a formula, one could search for upper limits of the term complexity and more general connections between properties of preferences and their term complexity.

Acknowledgement. I am grateful to Bernhard Möller and the anonymous referees for their helpful remarks.

References

1. Borzsony, S., Kossmann, D., Stocker, K.: The skyline operator. In: 17th International Conference on Data Engineering, pp. 421–430 (2001)
2. Brink, C., Rewitsky, I.: A paradigm for program semantics: power structures and duality. CSLI Publications, Stanford (2001)
3. Eilenberg, S.: Automata, languages, and machines, vol. 59. Academic press (1974)
4. Mandl, S., Kozachuk, O., Endres, M., Kießling, W.: Preference Analytics in EXASolution. In: 16th Conference on Database Systems for Business, Technology, and Web (2015). http://tinyurl.com/pxco8d4
5. Möller, B., Roocks, P.: An algebra of database preferences. Journal of Logical and Algebraic Methods in Programming 84(3), 456–481 (2015). http://www.sciencedirect.com/science/article/pii/S2352220815000188
6. Möller, B., Roocks, P.: An algebra of layered complex preferences. In: Kahl, W., Griffin, T.G. (eds.) RAMICS 2012. LNCS, vol. 7560, pp. 294–309. Springer, Heidelberg (2012)
7. Möller, B., Roocks, P., Endres, M.: An Algebraic Calculus of Database Preferences. In: Gibbons, J., Nogueira, P. (eds.) MPC 2012. LNCS, vol. 7342, pp. 241–262. Springer, Heidelberg (2012)
8. Roocks, P.: R Script containing the experiments from Section 4 (2015). http://www.p-roocks.de/powerset-prefs-ramics.r
9. Roocks, P.: The rPref Package: Preferences and Skyline Computation in R (2015). http://www.p-roocks.de/rpref
10. Roocks, P.: Preference decomposition and the expressiveness of preference query languages. In: Hinze, R., Voigtländer, J. (eds.) MPC 2015. LNCS, vol. 9129, pp. 71–92. Springer, Heidelberg (2015). http://dx.doi.org/10.1007/978-3-319-19797-5_4
11. Schmidt, G., Ströhlein, T.: Relations and Graphs: Discrete Mathematics for Computer Scientists. EATCS Monographs on Theoretical Computer Science (1993)
12. Wang, Q., Balke, W.T., Kießling, W., Huhn, A.: P-news: Deeply personalized news dissemination for mpeg-7 based digital libraries. In: Heery, R., Lyon, L. (eds.) ECDL 2004. LNCS, vol. 3232, pp. 256–268. Springer, Heidelberg (2004). http://dx.doi.org/10.1007/978-3-540-30230-8_24
13. Winskel, G.: On powerdomains and modality. Theor. Comput. Sci. 36, 127–137 (1985)

Roughness by Residuals

Algebraic Description of Rough Sets and an Algorithm for Finding Core Relations

Martin E. Müller

University of Augsburg, Department of Computer Science
m.e.mueller@informatik.uni-augsburg.de

Abstract. Rough set theory (RST) focuses on forming posets of equivalence relations to describe sets with increasing accuracy. The connection between modal logics and RST is well known and has been extensively studied in their relation algebraic (RA) formalisation. RST has also been interpreted as a variant of intuitionistic or multi-valued logics and has even been studied in the context of logic programming.

This paper presents a detailed formalisation of RST in RA by way of residuals, motivates its generalisation and shows how results can be used to prove many RST properties in a simple algebraic manner (as opposed to many tedious and error-prone set-theoretic proofs). A further abstraction to an entirely point-free representation shows the correspondence to Kleene algebras with domain.

Finally, we show how an RA-perspective on RST allows to derive an abstract algorithm for finding reducts from a mere analysis of the properties of the RA-construction rather than by a data-driven approach.

1 Introduction

Rough set theory (RST) as introduced by [9] is a method developed for relational data analysis. Originally it was presented as a purely set-theoretic approach by abstracting from feature based information systems to systems of equivalence relations. It has been studied in many different contexts; to name just a few there are (modal) logics, e.g. [3,7,8,20], logic programming, e.g. [5], fuzzy set theory, e.g. [2] and formal concept analysis, e.g. [19,18].

In this paper, we first give a short set-theoretic introduction to RST. We then characterise RST in terms of residuals (section 3) using characteristic functions of sets. Section 4 briefly sketches how to further generalise the characterisation and reveals common properties to Kleene algebra with domain operators. In section 5 the formalism from section 3 is used to derive an efficient description of so-called core relations. This description turns out to be a concise specification of a common core discovery algorithm, [14].

2 Rough Set Theory by Set Theory

Notation. $\mathcal{U}$ denotes the domain of discourse. Sets are denoted by lowercase letters r, s, t, relations by uppercase letters P, Q, R and object variables are x, y, z.

W. Kahl et al. (Eds.): RAMiCS 2015, LNCS 9348, pp. 380–394, 2015.
DOI: 10.1007/978-3-319-24704-5_23

Functions are written f, g, h; the characteristic function of a set s is $\dot{s} : \mathcal{U} \to \mathbf{2}$ and the kernel relation induced by $f : s \to t$ is indicated by $\widetilde{f} : s \rightharpoondown s$. Sets of relations are typeset in boldface letters $\mathbf{P}, \mathbf{Q}, \mathbf{R}$. Complements, converse, duality and composition are written $\overline{s}$ or $\overline{R}$, $R^{\smile}$, $R^{\mathsf{d}} = \overline{R}^{\smile}$ and $P \mathbin{;} Q$. The quotient or partition induced by an equivalence R on a set s is written s/R and $[x]_R$ is the R-equivalence class of x. When clear from context, we drop stacked operators (i.e. $\widetilde{s}$ for $\widetilde{\dot{s}}$, s/f for $s/\widetilde{f}$, etc).

(Pre-) images are denoted by $P.s$ and $s.P$; if $s = \{x\}$ is a singleton, we may write $x.P$ and $P.x$. The universal and null relation are $\mathbb{T}_{(s,t)} = s \times t$ and $\mathbb{L}_{(s,t)} = \emptyset$ and $\mathit{1}$ denotes the identity (we implicitly assume type consistency). To indicate the interpretation of a subidenty as a set we also write $s_{\mathit{1}} = \mathit{1} \cap (s \times s)$.

2.1 Information Systems

An information system consists of a domain set $\mathcal{U}$ with a set $\mathbf{F}$ of total functions:

$$\mathfrak{I} = \langle \mathcal{U}, \mathbf{F} \rangle \text{ where } \mathbf{F} = \{f_i : \mathcal{U} \to V_i : i \in \mathbf{n}\} \tag{1}$$

and $\mathbf{F}$, V_i, and $\mathcal{U}$ are finite. Such systems are usually represented as tables with a row for each element $x \in \mathcal{U}$ and a column for each feature $f \in \mathbf{F}$ and the value $f(x)$ in the x-row and f-column. An example for an information system containing knowledge about geometric figures over the domain $\mathcal{U} = \{□, ■, ▪, ●, △, ◆, ○\}$ is shown in figure 1.[1] Readers familiar with formal concept analysis will recognise $\langle \mathcal{U}, \mathbf{F}, I \rangle$ as a formal context with all $f \in \mathbf{F}$ being attributes and $I : \mathcal{U} \rightharpoondown \mathbf{F}$ with $xIf :\Longleftrightarrow f(x) = \mathbf{1}$; see, e.g., [4,12,13].

2.2 Rough Set Data Analysis: Objects and Definability

RST explores the knowledge encoded in kernel relations, [9,11,6]. Let $\mathbf{R}$ be the set of kernel relations induced by $\mathbf{F}$. The *indiscernability relation* generated by $\mathbf{R}$ is defined as

$$\overset{\approx}{\mathbf{R}} := \bigcap_{R \in \mathbf{R}} R. \tag{2}$$

Trivially, $\overset{\approx}{\mathbf{R}} \subseteq R$ for all $R \in \mathbf{R}$ and $[x]_{\overset{\approx}{\mathbf{R}}} \subseteq [x]_R$ for any $x \in \mathcal{U}$ and $R \in \mathbf{R}$. Finally, $\langle \mathbf{R}, \subseteq \rangle$ forms a complete lattice on all equivalences on $\mathcal{U}$ with least upper bound operation $P \sqcup R = (P \cup R)^*$, $\overset{\approx}{\mathbf{R}}$ being the smallest element and $\bigsqcup \mathbf{R}$ the largest. Hence, the *knowledge base*

$$\left\langle \mathcal{U}, \left\{ \overset{\approx}{\mathbf{P}} : \mathbf{P} \subseteq \mathbf{R} \right\} \right\rangle \tag{3}$$

contains all information available to group elements of $\mathcal{U}$ into sets defined in terms of unions and intersections of equivalence classes (e.g. in terms of feature value assignments in CNF-formulae).

---

[1] We assume all feature value sets V_i to be pairwise disjoint (by renaming).

$\mathfrak{I}$	**F** col	shp	edg	siz	$\widetilde{col}$ □ ■ ▪ ● △ ◆ ○	$\widetilde{shp}$ □ ■ ▪ ● △ ◆ ○	$\widetilde{\widetilde{\mathbf{R}}} = \bigcap\{\widetilde{col}, \widetilde{shp}\}$ □ ■ ▪ ● △ ◆ ○
□	w	square	4	S	1 0 0 0 1 0 1	1 1 1 0 0 0 0	1 0 0 0 0 0 0
■	b	square	4	B	0 1 1 0 0 1 0	1 1 1 0 0 0 0	0 1 1 0 0 0 0
▪	b	square	4	S	0 1 1 0 0 1 0	1 1 1 0 0 0 0	0 1 1 0 0 0 0
●	g	circle	1	S	0 0 0 1 0 0 0	0 0 0 1 0 0 1	0 0 0 1 0 0 0
△	w	triangle	3	B	1 0 0 0 1 0 1	0 0 0 0 1 0 0	0 0 0 0 1 0 0
◆	b	diamond	4	S	0 1 1 0 0 1 0	0 0 0 0 0 1 0	0 0 0 0 0 1 0
○	w	circle	1	S	1 0 0 0 1 0 1	0 0 0 1 0 0 1	0 0 0 0 0 0 1

Fig. 1. An information system, two kernel relations and their indiscernability relation.

The *lower R-approximation of s* is defined as the union of all R-classes that are contained in s; the corresponding *upper approximation* is the union of all classes containing at least one element of s:

$$[\![R]\!]s := \{x \in \mathcal{U} : [x]_R \subseteq s\} \tag{4}$$
$$\langle\!|R|\!\rangle s := \{x \in \mathcal{U} : [x]_R \cap s \neq \emptyset\}\,. \tag{5}$$

Both approximation operators are isotone in their set arguments, but $[\![\ ]\!]$ is antitone in its relation argument:

$$P \subseteq R \Longrightarrow \langle\!|P|\!\rangle s \subseteq \langle\!|R|\!\rangle s \text{ but } P \subseteq R \Longrightarrow [\![R]\!]s \subseteq [\![P]\!]s. \tag{6}$$

The proof is deferred to section 4. A set $s \subseteq \mathcal{U}$ is *roughly R-definable*, if $[\![R]\!]s \neq \emptyset$ or $\langle\!|R|\!\rangle s \neq \mathcal{U}$. If $[\![R]\!]s = \langle\!|R|\!\rangle s$ we also have $[\![R]\!]s = \langle\!|R|\!\rangle s = s$ and s is called *(exactly) R-definable* (for a proof, see equations (17,18) and section 4). As already suggested by notation, upper and lower approximations are dual operations in the usual sense of modal logics:

$$[\![R]\!]\overline{s} = \overline{\langle\!|R|\!\rangle s}. \tag{7}$$

For a proof, see 4. In most cases we will examine *sets* of relations; for better readability we write $[\![\mathbf{R}]\!]$ instead of $[\![\widetilde{\widetilde{\mathbf{R}}}]\!]$.A *classification* $\mathfrak{s} = \{s_i : i \in \mathbf{n}\}$ is a collection of classes $s_i \subseteq \mathcal{U}$. Usually, one assumes a classification to be a partitioning $\mathcal{U}/\widetilde{f}$ induced by a *classifier* $f : \mathcal{U} \rightarrow \mathbf{n}$. Rough set approximations can be lifted to arbitrary classifications by building approximations of the classes:

$$[\![R]\!]\mathfrak{s} := \{[\![R]\!]s : s \in \mathfrak{s}\} \text{ and } \langle\!|R|\!\rangle\mathfrak{s} := \{\langle\!|R|\!\rangle s : s \in \mathfrak{s}\}\,. \tag{8}$$

Hence, $[\![R]\!]s$ can be interpreted as a special case of $\mathfrak{s} = \{s\}$ or $[\![R]\!](\mathcal{U}/\widetilde{s}) = [\![R]\!]\{s, \overline{s}\}$ with $\widetilde{s}$ being the kernel relation induced by the characteristic function of s.[2] With rough set theory being a set based approach we will stick to the arbitrary classification based view for the remainder of this section. Given two sets $\mathbf{P}, \mathbf{R}$ of equivalences, one wants to evaluate their expressiveness against

[2] To be precise, $[\![R]\!]s = \bigcup\{s' \in [\![R]\!]\mathcal{U}/\widetilde{s} : s' \cap s \neq \emptyset\}$.

each other and/or with respect to a given classification $\mathfrak{s}$. A relation R is *($\mathfrak{s}$-) dispensable in* $\mathbf{R}$, if $[\![\mathbf{R} - \{R\}]\!]\mathfrak{s} = [\![\mathbf{R}]\!]\mathfrak{s}$. Then, R does not contribute to the knowledge in $\mathbf{R}$. Otherwise it is called *indispensable.* $\mathbf{R}$ is *($\mathfrak{s}$-) irreducible*, iff it does not contain any $\mathfrak{s}$-dispensable relation. Finally, $\mathbf{P}$ is called a *($\mathfrak{s}$-)reduct of* $\mathbf{R}$, iff $\mathbf{P} \subseteq \mathbf{R}$ and $\mathbf{P}$ is $\mathfrak{s}$-irreducible. Reducts are not unique, hence we denote the set of all $\mathfrak{s}$-reducts of $\mathbf{R}$ by $\mathrm{Red}_{\mathfrak{s}}(\mathbf{R})$; their intersection $\mathrm{Cor}_{\mathfrak{s}}(\mathbf{R}) := \bigcap \mathrm{Red}_{\mathfrak{s}}(\mathbf{R})$ is called the *core* and its elements are called *essential.*[3] To compare the descriptive power of equivalences $\mathbf{P}$ to that of another set of equivalences $\mathbf{R}$, we define the $\mathbf{P}$*-positive set of* $\mathbf{R}$ *against* $\mathfrak{s}$ to be the union set of all $\mathbf{P}$-lower approximations of $\widetilde{\widetilde{\mathbf{R}}}$ classes:

$$[\![\mathbf{P} \lessdot \mathbf{R}]\!]\mathfrak{s} := \bigcup_{s \in \mathfrak{s}} [\![\mathbf{P} \lessdot \mathbf{R}]\!]s := \bigcup_{s \in \mathfrak{s}} \bigcup_{t \in s/\widetilde{\widetilde{\mathbf{R}}}} [\![\mathbf{P}]\!]t\,. \tag{9}$$

$[\![\lessdot]\!]$ is well defined for arbitrary classifications, partitions and single sets $\mathfrak{s} = \{s\}$ but the result of $[\![\mathbf{P} \lessdot \mathbf{R}]\!]\mathfrak{s}$ is always a flat set. If the $\mathbf{P}$-positive set of $\mathbf{R}$ includes the $\mathbf{Q}$-positive set of $\mathbf{R}$ (w.r.t. $\mathfrak{s}$), then $\mathbf{P}$ obviously contains more $\mathbf{R}$-knowledge (w.r.t. $\mathfrak{s}$) than $\mathbf{Q}$. We say that $\mathbf{P}$ *($\mathbf{R}$-) refines* $\mathbf{Q}$ (on $\mathfrak{s}$) and write

$$\mathbf{P} \overset{\mathbf{R}}{\succeq}_{\mathfrak{s}} \mathbf{Q} :\Longleftrightarrow [\![\mathbf{P} \lessdot \mathbf{R}]\!]\mathfrak{s} \supseteq [\![\mathbf{Q} \lessdot \mathbf{R}]\!]\mathfrak{s}. \tag{10}$$

In most cases, properties of relation sets are compared with respect to the entire set of objects in the universe such that $\mathfrak{s} = \{\mathcal{U}\} = \mathcal{U}/\mathbb{T}$. We then simply drop the arguments and say that $\mathbf{P} \overset{\mathbf{R}}{\succeq} \mathbf{Q}$ iff $[\![\mathbf{P} \lessdot \mathbf{R}]\!]\mathcal{U} \supseteq [\![\mathbf{Q} \lessdot \mathbf{R}]\!]\mathcal{U}$. If we assume $\mathfrak{s} = \{s, \overline{s}\} = \mathcal{U}/\dot{s}$, the following equivalence shows that "more knowledge" as expressed by $\succeq$ simply means bigger regions of lower approximations for both s and its complement $\overline{s}$:

$$\mathbf{P} \overset{\widetilde{s}}{\succeq} \mathbf{R} \overset{(a)}{\Longleftrightarrow} [\![\mathbf{R}]\!]s \subseteq [\![\mathbf{P}]\!]s \overset{(b)}{\Longleftrightarrow} [\![\mathbf{R}]\!]\overline{s} \subseteq [\![\mathbf{P}]\!]\overline{s}. \tag{11}$$

Proof (Equation 11). By definition ($\succeq$ to $[\![\lessdot]\!]$; $U/\widetilde{s} = \{s, \overline{s}\}$), isotony, and $[\![R]\!]s \subseteq s \subseteq \langle\!|R|\!\rangle s$ we show validity of (11a):

$$\begin{aligned}
\mathbf{P} \overset{\widetilde{s}}{\succeq} \mathbf{R} &\Longleftrightarrow \bigcup_{c \in \{s,\overline{s}\}} [\![\mathbf{P}]\!]c \supseteq \bigcup_{c \in \{s,\overline{s}\}} [\![\mathbf{R}]\!]c \\
&\Longleftrightarrow [\![\mathbf{R}]\!]s \cup [\![\mathbf{R}]\!]\overline{s} \subseteq [\![\mathbf{P}]\!]s \cup [\![\mathbf{P}]\!]\overline{s} \\
&\overset{(*)}{\Longrightarrow} [\![\mathbf{R}]\!]s \cup [\![\mathbf{R}]\!]\overline{s} \cup \overline{s} \subseteq [\![\mathbf{P}]\!]s \cup [\![\mathbf{P}]\!]\overline{s} \cup \overline{s} \\
&\overset{(a)}{\Longleftrightarrow} [\![\mathbf{R}]\!]s \cup \overline{s} \subseteq [\![\mathbf{P}]\!]s \cup \overline{s}
\end{aligned}$$

Adding s instead of $\overline{s}$ in line $(*)$, we obtain $(11b)$. Then,

$$[\![\mathbf{R}]\!]\overline{s} \subseteq [\![\mathbf{P}]\!]\overline{s} \Longrightarrow [\![\mathbf{R} \lessdot \widetilde{s}]\!]\overline{s} \subseteq [\![\mathbf{P} \lessdot \widetilde{s}]\!]\overline{s} \Longrightarrow \mathbf{P} \overset{\widetilde{s}}{\succeq}_{\overline{s}} \mathbf{R}$$

and the same for s which proves the reverse direction for $(*)$.

---

[3] The definitions of dispensability and all following concepts are given on simple sets s in RST literature.

2.3 Relation Algebra

We assume the reader to be familiar with (concrete) relation algebra, [17], and only briefly repeat the definition of *residuals*. For relations P, Q and T with matching (co-)domains, the right and left residuals are defined as follows:

$$\begin{array}{ll} P\backslash\!\backslash Q := \overline{P^{\smile} ; \overline{Q}} \text{ or } R \subseteq P\backslash\!\backslash Q \iff P;R \subseteq Q \\ Q/\!/P := \overline{\overline{Q};P^{\smile}} \text{ or } R \subseteq Q/\!/P \iff R;P \subseteq Q \end{array} \tag{12}$$

with R being the biggest solution of the respective inequalities. Sets $s \subseteq \mathcal{U}$ can be represented relationally as a blocks $s \times s \subseteq \mathcal{U} \times \mathcal{U}$, as a subidentities $s_{1} = 1 \cap (s \times s)$, or characteristic functions $\dot{s} : \mathcal{U} \to \mathbf{2}$. In the homogenous setting, we require $M = \mathcal{U}$. Note that $\dot{s}$ is a heterogenous representation which requires a domain operation for translation into the homogenous representation.

Elements are singleton sets: $x \in s \iff \{x\} \subseteq s$, i.e. $\vec{x}$ has a singleton domain $\{x\}$, $s \times s = s_{1} = \{\langle x, x\rangle\}$ and $\{x\} = \dot{s}.\mathbf{1}$.

3 A Relation Algebraic Approach to Rough Sets

We first consider characteristic functions for a proper algebraisation.

3.1 Basic Constructions

For any set s, let $\underset{\cdot}{s}$ be an arbitrary but fixed element of s. Let there be a partition $\mathcal{U}/\tilde{f} = \mathfrak{s} = \{s_i : i \in \mathbf{n}\}$. We then call $\{\underset{\cdot}{s_i} : i \in \mathbf{n}\}$ a *representation system* of f and we can reconstruct arbitrary $f \in \mathbf{F}$ from $\tilde{f}$ by only knowing $f(\underset{\cdot}{s_i})$ and $\tilde{f}$:

$$f(x) = f(\underset{\cdot}{s_i}) \iff x\tilde{f}\underset{\cdot}{s_i} \iff x.\vec{s}_i = \mathcal{U} \iff \vec{x}^{\smile} ; \vec{s} = \mathbb{T}. \tag{13}$$

Using $\underset{\cdot}{s}$, one can reconstruct s from $\tilde{s}$ with a trick (see footnote 2) exploiting the fact $s = \bigcup\{s' \in \mathcal{U}/\tilde{s} : \underset{\cdot}{s} \in s'\}$.

3.2 Residuals of Characteristic Functions

We first give an equivalent residual based description of lower approximations:

$$[\![R]\!]s \overset{(a)}{=\!=} R\backslash\!\backslash\dot{s}\,.\mathbf{1} \overset{(b)}{=\!=} R\backslash\!\backslash\tilde{s}\,.\underset{\cdot}{s} \overset{(c)}{=\!=} (R\backslash\!\backslash\tilde{s}\,.\mathcal{U}) \cap s. \tag{14}$$

Based on the fact that $x \in [\![R]\!]s$ implies $[x]_R \subseteq s$ the proof idea is as follows:

$$xRy \to y \in s \overset{\text{FOL}}{\iff} \neg(xRy \wedge y\bar{\dot{s}}\mathbf{1}) \overset{;}{\iff} x\,\overline{R^{\smile} ; \bar{\dot{s}}\mathbf{1}} \overset{\backslash\!\backslash}{\iff} x\,R\backslash\!\backslash\dot{s}\,\mathbf{1}. \tag{15}$$

A more detailed version can be found in the appendix. Also, the desired duality of $[\![\ ]\!]$ and $\langle\!|\ |\!\rangle$ in equation (7) can be expressed using the complementation of characteristic functions:

$$\langle\!| R |\!\rangle\bar{s} := R\backslash\!\backslash s\,.\mathbf{0} = R;\bar{s}\,.\mathbf{0} = \overline{R;\bar{s}\,.\mathbf{1}} = \overline{R\backslash\!\backslash s\,.\mathbf{1}} = \overline{[\![R]\!]s}, \tag{16}$$

	2	
$\dot{c}$	**0**	**1**
□	1	0
■	1	0
▪	0	1
●	0	1
△	0	1
◆	1	0
○	1	0

	$\mathcal{U}$						
$\widetilde{shp}$	□	■	▪	●	△	◆	○
□	*1*	*1*	*1*	0	0	0	0
■	*1*	*1*	*1*	0	0	0	0
▪	*1*	*1*	1	0	0	0	0
●	0	0	0	1	0	0	*1*
△	0	0	0	0	1	0	0
◆	0	0	0	0	0	1	0
○	0	0	0	*1*	0	0	*1*

	2	
$\widetilde{shp}\backslash\!\backslash\dot{c}$	**0**	**1**
□	0	0
■	0	0
▪	0	0
●	0	0
△	0	1
◆	1	0
○	0	0

Hence, $\widetilde{shp}\backslash\!\backslash\dot{c}\,.\,\{\mathbf{1}\} = \{\triangle\} = [\![\widetilde{shp}]\!]\,\{$▪, ●, △$\}$.

Fig. 2. Lower approximations by residuals and characteristic functions

i.e. $\langle\!|R|\!\rangle s = R\backslash\!\backslash\dot{\overline{s}}\,.\mathbf{0}$. Next, we note that $R;\dot{\overline{s}} = \overline{R;\dot{s}}$ for any equivalence R. In particular, we have $\widetilde{s};\dot{s} = \overline{\widetilde{s};\dot{\overline{s}}} = \widetilde{s}\backslash\!\backslash\dot{s}$ which gives $s = [s]_{\widetilde{s}} = \widetilde{s}\backslash\!\backslash\dot{s}\,.\mathbf{1}$, i.e. image choice determines the preimage to be s or its complement. The fact that $[\![R]\!]s = s \Longleftrightarrow \langle\!|R|\!\rangle s = s$ (R-definability of s) can be shown by

$$R\backslash\!\backslash\dot{s} = \dot{s} \Longrightarrow R;\dot{s} \subseteq \dot{s} \wedge \dot{\overline{s}} \subseteq R;\dot{\overline{s}} \tag{17}$$

$$\overline{R\backslash\!\backslash\dot{\overline{s}}} = \dot{s} \Longrightarrow R;\dot{s} \subseteq \dot{s} \wedge R;\dot{\overline{s}} \subseteq \dot{\overline{s}} \tag{18}$$

which, together, means $R;\dot{\overline{s}} = \dot{\overline{s}}$, then $R;\dot{s} = \dot{s}$ by Schröder, then $R;\dot{\overline{s}} = \overline{R;\dot{s}}$ by complementation and therefore $R\backslash\!\backslash\dot{s} = \dot{s} \Longleftrightarrow \dot{s} = \overline{R\backslash\!\backslash\dot{\overline{s}}}$. An even nicer proof is given in section 4

3.3 Classifications, Positive Regions, Implication

Restricting classifications to quotients in (8) and using the residual based representation in (14), $[\![\mathbf{R}]\!]\mathfrak{c}$ can be characterised as follows:

$$[\![\mathbf{R}]\!]\mathfrak{c} = \left\{ \tilde{\tilde{R}}\backslash\!\backslash\dot{c_i}\,.\mathbf{1} : c_i \in \mathcal{U}/Q \right\} = \left\{ \tilde{\tilde{R}}\backslash\!\backslash f\,.i : \{i\} \in \mathbf{n} \right\} \tag{19}$$

Similarly, the definition of positive regions can be rewritten as

$$[\![\mathbf{R} \lessdot Q]\!]\mathcal{U} = \bigcup_{i \in \mathbf{n}} \tilde{\tilde{\mathbf{R}}}\backslash\!\backslash Q\,.\underset{\cdot}{c_i} = \tilde{\tilde{\mathbf{R}}}\backslash\!\backslash Q\,.\mathcal{U}. \tag{20}$$

In other words, the $\mathbf{R}$-positive region with respect to Q coincides with the preimage set of the right residual; and it does so for arbitrary equivalences Q and subsets $s \subseteq \mathcal{U}$:

$$[\![\mathbf{R} \lessdot Q]\!]s = (\,\tilde{\tilde{\mathbf{R}}}\backslash\!\backslash Q\,.\mathcal{U}) \cap s. \tag{21}$$

Equations (20) and (21) follow immediately from the proof of equation (14) by generalising $\mathfrak{s} = \{s, \overline{s}\}$ to $\mathfrak{c} = \mathcal{U}/Q$. Another conclusion for more data-driven applications is that the positive region under a classification is the same as the lower approximation of the entire data set (i.e. we still know which data points can be correctly classified but we forget into which class it has been classified). From this, we can conclude for any $\mathfrak{c} = \mathcal{U}/Q$

$$\mathbf{P} \overset{\mathbf{R}}{\succeq}_{\mathfrak{c}} \mathbf{Q} \overset{(10)}{\Longleftrightarrow} [\![\mathbf{P} \lessdot \mathbf{R}]\!]\mathfrak{c} \supseteq [\![\mathbf{Q} \lessdot \mathbf{R}]\!]\mathfrak{c} \overset{(20)}{\Longleftrightarrow} \tilde{\tilde{\mathbf{Q}}}\backslash\!\backslash\tilde{\tilde{\mathbf{R}}} \subseteq \tilde{\tilde{\mathbf{P}}}\backslash\!\backslash\tilde{\tilde{\mathbf{R}}} \tag{22}$$

which shows the irrelevance of Q for the comparison of $\mathbf{P}$ to $\mathbf{Q}$ under $\mathbf{R}$ and finally leads to a point-free characterisation. From equation (22) we can also derive[4]

$$\tilde{\tilde{\mathbf{Q}}}\backslash\!\backslash\tilde{\tilde{\mathbf{R}}} \subseteq \tilde{\tilde{\mathbf{P}}}\backslash\!\backslash\tilde{\tilde{\mathbf{R}}} \Longleftrightarrow \tilde{\tilde{\mathbf{P}}};\overline{\tilde{\tilde{\mathbf{R}}}} \subseteq \tilde{\tilde{\mathbf{Q}}};\overline{\tilde{\tilde{\mathbf{R}}}} \overset{!}{\Longleftrightarrow} \tilde{\tilde{\mathbf{P}}} \subseteq \tilde{\tilde{\mathbf{Q}}}. \tag{23}$$

Also, if $\mathbf{Q} \subseteq \mathbf{P}$, it follows that $\tilde{\tilde{\mathbf{P}}} \subseteq \tilde{\tilde{\mathbf{Q}}}$, which expresses the fact that loss of knowledge results in loss of information (or "predictive power"):

$$\mathbf{Q} \subseteq \mathbf{P} \Longrightarrow \mathbf{P} \overset{\mathbf{R}}{\succeq}_{\mathfrak{c}} \mathbf{Q} \Longleftrightarrow [\![\mathbf{Q} \lessdot \mathbf{R}]\!]\mathfrak{c} \subseteq [\![\mathbf{P} \lessdot \mathbf{R}]\!]\mathfrak{c} \tag{24}$$

for arbitrary equivalences $\tilde{\tilde{\mathbf{R}}}$ and Q with $\mathfrak{c} = \mathcal{U}/Q$.

Finally, once we have found an algebraic definition of positive regions, we can reversely define $[\![\ ]\!]$ as a simplified version of $[\![\ \lessdot\ ]\!]$:

$$[\![\mathbf{R}]\!]s \overset{(14a)}{=\!=} \tilde{\tilde{\mathbf{R}}}\backslash\!\backslash\dot{s}.\mathbf{1} \overset{(14c)}{=\!=} \tilde{\tilde{\mathbf{R}}}\backslash\!\backslash\tilde{\dot{s}}.\mathcal{U} \cap s \overset{(20)}{=\!=} [\![\mathbf{R} \lessdot \tilde{\dot{s}}]\!]s. \tag{25}$$

Then it is clear why refinement is also referred to as *implication*:

$$\mathbf{P} \succeq \mathbf{Q} \overset{\text{Def}}{\Longleftrightarrow} \forall x \in \mathcal{U} : x \in [\![\mathbf{P} \lessdot \mathbf{Q}]\!]\mathcal{U} \tag{26}$$

$$\overset{(20)}{\Longleftrightarrow} \forall x \in \mathcal{U} : x \in \tilde{\tilde{\mathbf{P}}}\backslash\!\backslash\tilde{\tilde{\mathbf{Q}}}.\mathcal{U} \tag{27}$$

$$\Longleftrightarrow \forall x, y \in \mathcal{U} : y\tilde{\tilde{\mathbf{P}}}x \longrightarrow y\tilde{\tilde{\mathbf{Q}}}x. \tag{28}$$

For a proof, see appendix. This way we have reformulated the idea of "logic implication" trough the intuitive meaning of residuals rather than by a long-winded set theoretic treatment as in [10]. As an example, see figure 3.

Finally, $\overset{R}{\succeq}$ forms a preorder on $\mathbb{R}$ for any equivalence R on $\mathcal{U}$: Reflexivity is trivial and $\mathbf{P} \succeq \mathbf{Q} \succeq \mathbf{R}$ implies $\mathbf{P} \succeq \mathbf{R}$ by transitivity of $\subseteq$ in equation (22). Sadly, $\succeq$ does *not* form a poset since it is *not* antisymmetric:

$$\mathbf{P} \succeq \mathbf{Q} \wedge \mathbf{Q} \succeq \mathbf{P} \not\Longrightarrow \mathbf{P} = \mathbf{Q} \tag{29}$$

(just consider $\mathbf{P} = \{\mathit{1}\}$ and $\mathbf{Q} = \{\widetilde{\{x\}} : x \in \mathcal{U}\}$ which does imply $\tilde{\tilde{\mathbf{P}}} = \tilde{\tilde{\mathbf{Q}}}$, but $\mathbf{P} \neq \mathbf{Q}$). Hence, reducts are not unique and dispensability of relations is always relative to the set of equivalences under consideration which again demonstrates the conciseness of the relation algebraic formalisation of RST.

---

[4] The equivalence marked "!" only holds for $\tilde{\tilde{\mathbf{R}}} \neq \mathbb{T}$. However, $\tilde{\tilde{\mathbf{R}}} = \mathbb{T}$ only if all features in $\mathbf{F}$ are constant which we can safely assume to be not the case.

$[\![\tilde{\dot{edg}} \lessdot \tilde{\dot{shp}}]\!]$	□	■	▪	◆	△	●	○
□	0	0	0	0	0	0	0
■	0	0	0	0	0	0	0
▪	0	0	0	0	0	0	0
◆	0	0	0	0	0	0	0
△	0	0	0	0	1	0	0
●	0	0	0	0	0	1	1
○	0	0	0	0	0	1	1

$[\![\tilde{\dot{shp}} \lessdot \tilde{\dot{edg}}]\!]$	□	■	▪	◆	△	●	○
□	1	1	1	1	0	0	0
■	1	1	1	1	0	0	0
▪	1	1	1	1	0	0	0
◆	1	1	1	1	0	0	0
△	0	0	0	0	1	0	0
●	0	0	0	0	0	1	1
○	0	0	0	0	0	1	1

Fig. 3. Refinement, inclusion, implication

4 Pointfree Rough Sets

We now treat relations R and sets s in a unified way by assuming all sets to be represented by subidentities. Let $\mathfrak{U} := \wp(\mathcal{U} \times \mathcal{U})$ and $\mathrm{set}_\mathfrak{U} := \{s \in \mathfrak{U} : s \subseteq 1_\mathcal{U}\}$. Using this notation we can finally prove equation (6) pointfree:

Proof. Let $s \in \mathrm{set}_\mathfrak{U}$ and $\mathbf{P} \subseteq \mathbf{R} \subset \mathfrak{U}$ are sets of equivalences. Then $R := \overset{\approx}{\mathbf{R}} \subseteq \overset{\approx}{\mathbf{P}} =: P$.

1. $\mathbf{P} \subseteq \mathbf{R} \Longrightarrow [\![\mathbf{P}]\!]s \subseteq [\![\mathbf{R}]\!]s$.
   By isotony, $\overset{\approx}{\mathbf{R}}\mathbin{;}s^- \subseteq \overset{\approx}{\mathbf{P}}\mathbin{;}s^-$. By definition of residuals, $\overline{\overset{\approx}{\mathbf{R}}\backslash\!\backslash s} \subseteq \overline{\overset{\approx}{\mathbf{P}}\backslash\!\backslash s}$, and by complementation and definition of $[\![\ ]\!]$ through $\backslash\!\backslash$ one obtains $[\![\mathbf{P}]\!]s \subseteq [\![\mathbf{R}]\!]s$.
2. $\mathbf{P} \subseteq \mathbf{R} \Longrightarrow (\!|\mathbf{R}|\!\rangle s \subseteq (\!|\mathbf{P}|\!\rangle s$.
   Again, $\mathbf{P} \subseteq \mathbf{R}$ implies $\overset{\approx}{\mathbf{R}}\mathbin{;}s \subseteq \overset{\approx}{\mathbf{P}}\mathbin{;}s$. By definition of $\backslash\!\backslash$ and symmetry of indiscernibility relations, $\overline{\overset{\approx}{\mathbf{R}}\backslash\!\backslash s^-} \subseteq \overline{\overset{\approx}{\mathbf{P}}\backslash\!\backslash s^-}$. This is equivalent to $\overline{[\![\mathbf{R}]\!]\overline{s}} \subseteq \overline{[\![\mathbf{P}]\!]\overline{s}}$ and, by equation (7), $(\!|\mathbf{R}|\!\rangle s \subseteq (\!|\mathbf{P}|\!\rangle s$.

In a next step, we interpret subidentities as restrictions or conditions to be satisfied when classified by equivalences. [1] introduce domain operators to reason over specifications of state transitions and tests in Kleenealgebra (KAD). RST is a very simple instance: Sets $\mathbf{R}$ of equivalences are the "actions" we perform to approximate classes, and membership correspond to the tests. To lift set complementation to subidentities, one defines for $s \in \mathrm{set}_\mathfrak{U}$ a complementation operator $s^- := \overline{s} \cap 1$. As a result, $\langle \mathfrak{U}, \cup, \mathbin{;}, \emptyset, 1, {}^*, {}^-, \mathrm{set}_\mathfrak{U}\rangle$ is a Kleene algebra with tests $\mathrm{set}_\mathfrak{U}$. To define the binary approximation operator $[\![\cdot]\!] : \mathfrak{U} \times \mathrm{set}_\mathfrak{U} \to \mathrm{set}_\mathfrak{U}$ we make use of a domain operator $\langle R| : \mathfrak{U} \to \mathrm{set}_\mathfrak{U}$:

$$\langle R| := \min\{X \in \mathrm{set}_\mathfrak{U} : R \subseteq X\mathbin{;}R\} \tag{30}$$

Then, by domain laws, by R being an equivalence and by $s \in \mathrm{set}_\mathfrak{U}$,

$$\langle R|\, s = \langle R\mathbin{;}s\,| = \langle\, \overline{R\backslash\!\backslash s^-}\,| = (\!|R|\!\rangle s. \tag{31}$$

Building the complement of $\langle\ |$ relative to $C \in \mathfrak{U}$, we have

$$-\langle R| = \max\left\{X \in \mathrm{set}_\mathfrak{U} :\ X\mathbin{;}R \subseteq \overline{C}\right\} = \overline{C\mathbin{;}R^\smile}\,. \tag{32}$$

Hence,

$$[R|\,t := -\langle R|\,t^- = \overline{t^- ; R} = R\backslash t = [\![R]\!]t. \tag{33}$$

Whereas the above proof of equation (6a) demonstrates the isotony $\mathbf{P} \subseteq \mathbf{R} \Longrightarrow [\![\mathbf{P}]\!]s \subseteq [\![\mathbf{R}]\!]s$, antitony with respect to $\tilde{\tilde{\mathbf{R}}}$ and $\tilde{\tilde{\mathbf{P}}}$ directly follows from he fact that $R \subseteq P \Longrightarrow [P|\,s \subseteq [R|\,s$. Also, by a single application of the Schröder rule and contraposition,

$$\langle R|\,s \subseteq t \Longleftrightarrow \langle R|\,t^- \subseteq s^- \Longleftrightarrow s \subseteq \overline{\langle R|\,t^-} \Longleftrightarrow s \subseteq [R|\,t \tag{34}$$

we have shown the Galois-connection (GC) between $[\;|/[\![\;]\!]$ and $\langle\;|/\langle\!|\;|\!\rangle$. From this observation many (tedious) proofs become obsolete since it implies that, e.g. $[\![R]\!]\overline{s} = \overline{\langle\!| R |\!\rangle s}$ (proving equation (7)) or $\langle\!| R|\!\rangle s = s \Longleftrightarrow [\![R]\!]s = s$ (replacing equations (17,18)).[5]

5 Dispensability and Irreducibility

Having reformulated all basic RST operations in RA, it is straightforward to formalise advanced concepts.

5.1 Redundancy and Reducts

Using the reflexivity of $\succeq$, relative redundancy of equivalences can be redefined as follows: R is Q-redundant in $\mathbf{R}$ on s iff

$$\mathbf{R} \overset{Q}{\succeq}_s \mathbf{R} - \{R\} \;\wedge\; \mathbf{R} - \{R\} \overset{Q}{\succeq}_s \mathbf{R}. \tag{35}$$

Equivalently, for $\mathbf{R} = \mathbf{P} \dot{\cup} \{R\}$ (i.e. $R \notin \mathbf{P}$),

$$\tilde{\tilde{\mathbf{R}}}\backslash\!\backslash Q\,.s = \tilde{\tilde{\mathbf{P}}}\backslash\!\backslash Q\,.s \text{ or } \tilde{\tilde{\mathbf{R}}}\backslash\!\backslash\tilde{\tilde{\mathbf{P}}} = \tilde{\tilde{\mathbf{P}}}\backslash\!\backslash\tilde{\tilde{\mathbf{R}}}. \tag{36}$$

Hence,

$$\begin{array}{lll} \mathbf{Q} \in \mathrm{Red}_{s/Q}(\mathbf{R}) :\Longleftrightarrow & & \\ \text{(a)} \quad \mathbf{Q} \subseteq \mathbf{R} & |\text{: Reduction} & \\ \text{(b)} \wedge \mathbf{Q} \overset{Q}{\succeq}_s \mathbf{R} & |\text{: } \tilde{\tilde{\mathbf{Q}}} = \tilde{\tilde{\mathbf{R}}} & \\ \text{(c)} \wedge \forall \mathbf{P} \subset \mathbf{Q} : \mathbf{Q} \overset{Q}{\succeq}_s \mathbf{P} \wedge \mathbf{P} \overset{Q}{\not\succeq}_s \mathbf{Q} & |\text{: Irreducibility} & \end{array} \tag{37}$$

From (a) it follows that $\tilde{\tilde{\mathbf{R}}} \subseteq \tilde{\tilde{\mathbf{Q}}}$, (b) translates to $\tilde{\tilde{\mathbf{Q}}} \subseteq \tilde{\tilde{\mathbf{R}}}$ by equations (22-24) such that, together, $\tilde{\tilde{\mathbf{Q}}} = \tilde{\tilde{\mathbf{R}}}$. Then, in particular, for $\mathbf{P} = \mathbf{Q} - \{R\}$, part (c) means that $\mathbf{Q}$ can be a Q-reduct of $\mathbf{R}$ on s only if (note the implication):

$$\begin{array}{rl} \forall \mathbf{P} \subset \mathbf{Q} : & \mathbf{Q} \overset{Q}{\succeq}_s \mathbf{P} \wedge \mathbf{P} \overset{Q}{\not\succeq}_s \mathbf{Q} \\ \Longrightarrow \neg\exists R \in \mathbf{Q} : & [\![\mathbf{Q}]\!]s/Q = [\![\mathbf{Q} - \{R\}]\!]s/Q \\ \Longleftrightarrow \forall R \in \mathbf{Q} : & [\![\mathbf{Q} \lessdot Q]\!]s \supset [\![\mathbf{Q} - \{R\} \lessdot Q]\!]s \end{array} \tag{38}$$

[5] Had we started with the GC stated above, $[\![R]\!]u \subseteq u \subseteq \langle\!| R |\!\rangle u$ and $[\![1]\!]u = u = \langle\!| 1 |\!\rangle u$, and setting $s = t = u$ immediately delivers the desired result.

where the last inequation shows that (c) can be rewritten as a strict implication: $\mathbf{Q}$ is irreducible, iff for any $\mathbf{P} \subset \mathbf{Q}$, $\mathbf{Q} \overset{Q}{\succ}_s \mathbf{P}$. Reformulating equation (37 a-c) algebraically, we have

$$\mathbf{Q} \subset \mathbf{R}\,,\ \tilde{\tilde{\mathbf{R}}}\backslash\!\!\backslash Q = \tilde{\tilde{\mathbf{Q}}}\backslash\!\!\backslash Q \text{ and } \forall \mathbf{P} \subset \mathbf{Q} : \tilde{\tilde{\mathbf{P}}}\backslash\!\!\backslash Q \subset \tilde{\tilde{\mathbf{Q}}}\backslash\!\!\backslash Q\,. \tag{39}$$

This characterisation of irreducibility is composed of one equation and two strict inequalities which has two big advantages: First, with a suitable $\{R\} \subseteq \mathbf{Q} \cap \overline{\mathbf{P}}$, we can disprove $\mathbf{Q}$ being a reduct by showing that R is dispensable. Vice versa, R is *essential* if (i.e. $\tilde{\tilde{\mathbf{R}}}$-indispensable) if it is indispensable in *all* reducts. Second, instead of checking all $x \in \mathcal{U}$ (as the definition in equation (37) would demand), it suffices to inspect only a (suitable) set of representatives of $\mathcal{U}/Q$ to efficiently identify essential relations and then (non-deterministically) build possible reducts by adding relations.[6]

5.2 Identification of Essential Relations

Algebraically, essential relations can be characterised very concisely:

$$\begin{array}{c} R \in \mathbf{Q} \text{ is essential} \\ \text{w.r.t. } Q \text{ on } s, \end{array} \iff (\mathbf{Q} - \{R\})\backslash\!\!\backslash Q \subset \mathbf{Q}\backslash\!\!\backslash Q\,. \tag{40}$$

But in real life, a proof of the strict inclusion in equation (40) requires a look into the data. We start by translating the residual notation back to a pointwise perspective using positive regions: Assume that $\mathbf{P} \subset \mathbf{Q} \subseteq \mathbf{R}$ and $\mathbf{P} = \mathbf{R} - \{R\}$. R is essential, if

$$\forall \mathbf{P} \subseteq \mathbf{R} : [\![\mathbf{P} \lessdot Q]\!]s \subset [\![\mathbf{P} \cup \{R\} \lessdot Q]\!]s, \tag{41}$$

which is a necessary precondition for R being an element of all reducts (its removal implies loss of discernibility (equation 37 (c)). Pointwise speaking, there are at least two different objects that can be distinguished by R only:

$$\begin{aligned} R \text{ is essential} &\iff \exists x, y \in s : \forall P \in \mathbf{P} : x\overline{R}y \wedge xPy & (42)\\ &\iff \exists x, y \in s : x(\tilde{\tilde{\mathbf{P}}} - R)y. & (43)\end{aligned}$$

In terms of relation matrices, $(P)_{x,y} = \mathbf{1}$ for all $P \in \mathbf{P}$ and $(R)_{x,y} = \mathbf{0}$.

This gives rise to defining a simple exhaustive test procedure that happens to coincide with what in RST is known as analysis by *discernibility matrices*: We define

$$\Delta\mathbf{Q} : \mathcal{U} \times \mathcal{U} \to \wp(\mathbf{R}') \text{ with } \Delta\mathbf{Q}(x, y) = \{R' \in \mathbf{R}' : x\overline{R}y\} \tag{44}$$

[6] "Efficiency" crucially depends on: The effort of finding r, the fact whether r contains the "right" representatives, the existence of a non-empty core $\mathbf{P}$ and the relative size of the core to all reducts—with the worst case being an empty core and two reducts $\mathbf{Q}$ and $\mathbf{R} - \mathbf{Q}$.

Let $s = \{\square, \blacksquare, \bullet, \triangle, \blacklozenge, \circ\} \subseteq \mathcal{U}$ (leaving out $\blacksquare$). For this example, we choose $Q = \mathit{1}$.

$$\begin{array}{llll}
\Delta\mathbf{Q}(\square,\blacksquare) = \{\tilde{col}, \tilde{edg}, \tilde{siz}\} & \Delta\mathbf{Q}(\blacksquare,\circ) = \{\tilde{col}, \tilde{edg}, \tilde{shp}, \tilde{siz}\} \\
\Delta\mathbf{Q}(\square,\bullet) = \{\tilde{col}, \tilde{edg}, \tilde{shp}\} & \\
\Delta\mathbf{Q}(\square,\triangle) = \{\tilde{edg}, \tilde{shp}\} & \Delta\mathbf{Q}(\bullet,\triangle) = \{\tilde{col}, \tilde{edg}, \tilde{shp}, \tilde{siz}\} \\
\Delta\mathbf{Q}(\square,\blacklozenge) = \{\tilde{col}, \tilde{edg}, \tilde{shp}\} & \Delta\mathbf{Q}(\bullet,\blacklozenge) = \{\tilde{col}, \tilde{edg}, \tilde{shp}\} \\
\Delta\mathbf{Q}(\square,\circ) = \{\tilde{edg}, \tilde{shp}\} & \Delta\mathbf{Q}(\bullet,\circ) = \{\tilde{col}\} \\
\Delta\mathbf{Q}(\blacksquare,\bullet) = \{\tilde{col}, \tilde{edg}, \tilde{shp}, \tilde{siz}\} & \Delta\mathbf{Q}(\triangle,\blacklozenge) = \{\tilde{col}, \tilde{edg}, \tilde{shp}, \tilde{siz}\} \\
\Delta\mathbf{Q}(\blacksquare,\triangle) = \{\tilde{col}, \tilde{edg}, \tilde{shp}\} & \Delta\mathbf{Q}(\triangle,\circ) = \{\tilde{edg}, \tilde{shp}, \tilde{siz}\} \\
\Delta\mathbf{Q}(\blacksquare,\blacklozenge) = \{\tilde{shp}, \tilde{siz}\} & \Delta\mathbf{Q}(\blacklozenge,\circ) = \{\tilde{col}, \tilde{edg}, \tilde{shp}\}
\end{array}$$

Because $\circ R \bullet$ for any $R \in \mathbf{R}$ other than $\tilde{col}$, $\mathrm{Cor}_{s/\mathit{1}}(\mathbf{R}) = \left\{\tilde{col}\right\}$ is essential. There are two reducts in $\mathrm{Red}_{s/\mathit{1}}(\mathbf{R})$, namely $\mathbf{Q} = \left\{\tilde{col}, \tilde{shp}\right\}$ and $\mathbf{Q}' = \left\{\tilde{col}, \tilde{edg}, \tilde{siz}\right\}$.

Fig. 4. $\Delta\mathbf{Q}$ for core relation identification

where the R' denotes the name of R and find that

$$R \text{ is essential} \iff \exists x, y \in s : \Delta\mathbf{Q}(x, y) = \{R'\}. \tag{45}$$

which requires $|\mathbf{Q}|(\frac{1}{2}n^2)$ tests (with $n = |s|$). Since one hardly wants to identify every single object $\{x\} \in s/\mathit{1}$ but only classes $c \in s/Q$, the problem above can be reduced to comparing pairs of elements from different classes; in the best case by comparing only two representatives $\underset{\cdot}{c}$ and $\underset{\cdot}{c}'$ for every pair of classes—and, in the ideal case, comparing the *same* pair of representatives for *every* relation. Then, the efficiency gain increases polynomially in the increasing coarseness Q on s, reducing the number of tests to a maximum of $|\mathbf{Q}|(\frac{1}{2}m^2)$ where $m = |s/Q| < n$.

Suppose $r_P := \{\underset{\cdot}{c} : c \in \mathfrak{c}\} \subseteq s$ being a "suitable" representation of $\mathfrak{c} = \{[x]_Q : x \in s\}$ (the indexing r is required because the existence of a globally suitable representation is yet unclear). Then, we can reformulate indispensability of R as

$$\forall P \in \mathbf{Q} : \exists \underset{\cdot}{c} \in r_P : [\underset{\cdot}{c}]_{\approx_{\mathbf{Q}}} \subseteq c \wedge \underset{\cdot}{c} \notin [\![\mathbf{P}]\!]c \iff (42). \tag{46}$$

Hence, given a suitable representation system $r \subseteq s$, we can efficiently deduce that R is essential on s by testing wether it is indispensable on r:
First, recall that $\mathbf{Q} = \mathbf{P} \dot{\cup} \{R\}$. Suppose

$$\forall P \in \mathbf{Q} : \exists \underset{\cdot}{c} \in r_P : [\underset{\cdot}{c}]_{\approx_{\mathbf{Q}}} \subseteq c \wedge \underset{\cdot}{c} \notin [\![\mathbf{P}]\!]c$$

is true. Then,

$$\begin{aligned}
[\underset{\cdot}{c}]_{\approx_{\mathbf{Q}}} \subseteq c &\overset{[\![\,]\!]}{\Longrightarrow} \underset{\cdot}{c} \in [\![\mathbf{Q}]\!]c \\
&\overset{\approx}{\iff} \underset{\cdot}{c} \in [\![\mathbf{P} \cup \{R\}]\!]c \overset{\approx}{\iff} \underset{\cdot}{c} \in [\![\tilde{\tilde{\mathbf{P}}} \cap R]\!]c \\
&\overset{<}{\iff} \underset{\cdot}{c} \in [\![\mathbf{Q} < Q]\!]c \overset{\subseteq}{\iff} \underset{\cdot}{c} \in [\![\mathbf{Q} < Q]\!]s
\end{aligned} \tag{47}$$

$$\begin{aligned}\underset{\cdot}{c} \notin \llbracket \mathbf{P} \rrbracket c &\overset{\llbracket\,\rrbracket}{\Longleftrightarrow} [\underset{\cdot}{c}]_{\overset{\approx}{\mathbf{P}}} \not\subseteq c \\ &\overset{[\,]}{\Longleftrightarrow} \exists y : \underset{\cdot}{c}\overset{\approx}{\mathbf{P}} y \notin c \overset{\approx}{\Longleftrightarrow} \exists y : \forall P \in \mathbf{P} : \underset{\cdot}{c} P y \notin c \\ &\overset{\lessdot}{\Longleftrightarrow} \underset{\cdot}{c} \notin \llbracket \mathbf{P} \lessdot Q \rrbracket c \overset{\subseteq}{\Longleftrightarrow} \underset{\cdot}{c} \notin \llbracket \mathbf{P} \lessdot Q \rrbracket s\end{aligned} \tag{48}$$

Note that $\underset{\cdot}{c}$ depends on P whereas $c = [\underset{\cdot}{c}]_Q \in s/Q$ does not. With $y \notin c$, we know $y \in c' = [\underset{\cdot}{c}']_Q$. Assuming a "suitable" r, we can choose $y = \underset{\cdot}{c}'$:

$$(48) \overset{y=\underset{\cdot}{c}'}{\Longleftrightarrow} \forall P \in \mathbf{P} : \underset{\cdot}{c} P \underset{\cdot}{c}'. \tag{49}$$

But since (a) $[\underset{\cdot}{c}]_{\overset{\approx}{\mathbf{Q}}} \subseteq c$, (b) $c \cap c' = \emptyset$, (c) $R \in \mathbf{Q} - \mathbf{P}$ and (d) for all $P \in \mathbf{P}$, $\underset{\cdot}{c} P \underset{\cdot}{c}'$, it follows that R is the only relation in $\mathbf{Q}$ that discerns $x = \underset{\cdot}{c}$ and $y = \underset{\cdot}{c}'$. Hence, it follows that

$$\exists x, y \in s : \forall P \in \mathbf{P} : x\overline{R}y \wedge xPy \Longleftrightarrow R \text{ is essential} . \tag{50}$$

As a result, if some R is a core relation, we can find r_P for every $P \neq R$ to show that $x\overline{R}y$ and xPy for $x, y \in r_P$. Formally, this is not a problem because we can assume "$\underset{\cdot}{\ }$" to always pick the "right" element by a perfect guess. For an implementation, non-determinism cannot be resolved that easily. A working algorithm therefore would have to make use of a suitable heuristically guided search/selection procedure. But even then, it is unclear whether there is a *single* $r = r_P$ for all P. Actually, the proof above implies that r_P are not unique in general: From $\underset{\cdot}{c}\overset{\approx}{\mathbf{P}}\underset{\cdot}{c}'$ as in (40) follows that $[\underset{\cdot}{c}]_{\overset{\approx}{\mathbf{P}}} = [\underset{\cdot}{c}']_{\overset{\approx}{\mathbf{P}}}$ such that one might choose any two elements $x \in [\underset{\cdot}{c}]_{\overset{\approx}{\mathbf{P}}} \cap c$ and $y \in [\underset{\cdot}{c}']_{\overset{\approx}{\mathbf{P}}} \cap c'$.

Is $\widetilde{col}$ a core relation in $\mathbf{Q}$ or $\mathbf{Q}'$ w.r.t. *1* on s as in figure 4?

$\mathbf{Q}$: $\widetilde{col}$ is indispensable in $\mathbf{Q}$, if $\mathbf{P} = \left\{\widetilde{shp}\right\}$ fails on a suitable $\widetilde{shp}$-representation:
Let $r_{\overset{\approx}{\mathbf{P}}} = \{\blacksquare, \blacklozenge, \circ, \triangle\}$ and $\underset{\cdot}{c} = \blacksquare$. Then, $[\blacksquare]_{\overset{\approx}{\mathbf{Q}}} = \{\blacksquare\}$ but $\blacksquare \notin \{\} = \llbracket \widetilde{shp} \rrbracket \{\blacksquare\}$.
To demonstrate the importance of right choice, consider the counterexample:
Let $r_{\overset{\approx}{\mathbf{P}}} = \{\blacksquare, \blacklozenge, \circ, \triangle\}$ and $\underset{\cdot}{c} = \blacklozenge$. Then, $[\blacklozenge]_{\overset{\approx}{\mathbf{Q}}} = \{\blacklozenge\}$ and $\blacklozenge \in \{\blacklozenge\} = \llbracket \widetilde{shp} \rrbracket \{\blacklozenge\}$.

$\mathbf{Q}'$: $\widetilde{col}$ is indispensable in $\mathbf{Q}'$, if $\mathbf{P} = \left\{\widetilde{edg}, \widetilde{siz}\right\}$ fails on a suitable $\overset{\approx}{\mathbf{P}}$-representation:
Let $r_{\overset{\approx}{\mathbf{P}}} = \{\blacksquare, \square, \triangle, \bullet\}$ and $\underset{\cdot}{c} = \square$. Then, $[\square]_{\overset{\approx}{\mathbf{Q}'}} = \{\square\}$ but $\square \notin \{\} = \llbracket \mathbf{P} \rrbracket \{\square\}$.
Also, different choices may deliver the desired result:
Let $r_{\overset{\approx}{\mathbf{P}}} = \{\blacksquare, \square, \triangle, \bullet\}$ and $\underset{\cdot}{c} = \bullet$. Then, $[\bullet]_{\overset{\approx}{\mathbf{Q}'}} = \{\bullet\}$ and $\bullet \notin \{\} = \llbracket \mathbf{P} \rrbracket \{\bullet\}$.
Let $r_{\overset{\approx}{\mathbf{P}}} = \{\blacksquare, \blacklozenge, \triangle, \circ\}$ and $\underset{\cdot}{c} = \blacklozenge$. Then, $[\blacklozenge]_{\overset{\approx}{\mathbf{Q}'}} = \{\blacklozenge\}$ and $\blacklozenge\overset{\approx}{\mathbf{P}}\square \notin \{\blacklozenge\}$.

Fig. 5. Checking for core relations by representation systems

5.3 Finding Core Relations

Supposing a perfect choice of $\d{c}$,

$$\begin{array}{l} R \text{ is indispensable} \\ \text{in } \mathbf{Q} \text{ w.r.t. } Q \text{ on } s, \end{array} \text{ iff } \d{c} \notin [\![\mathbf{P}]\!]c \not\subseteq [\d{c}]_{\widetilde{\widetilde{\mathbf{Q}}}} \subseteq [\![\mathbf{Q}]\!]c \subseteq [\d{c}]_Q. \tag{51}$$

Hence, the following "algorithm" for finding essential relations could be constructed along these lines:

1. Compute $r := \{\d{c} : c \in ([\![\mathbf{R} \lessdot Q]\!]s)/Q\}$ (if not done yet).
2. For every $x, y \in r$, compute $(\mathbf{d})_{x,y} = \{R' : R \in \mathbf{R} \wedge x\overline{R}y\}$ (where R' denotes the mere name of R rather than R itself).
3. $\mathrm{Cor}_{\mathfrak{c}}(\mathbf{R}) := \{R : \exists x, y \colon (\mathbf{d})_{x,y} = \{R'\}\}$
4. For every $\mathbf{P} \in \wp(\mathbf{R} - \mathrm{Cor}_{\mathfrak{c}}(\mathbf{R}))$ in order of cardinality:
   $\mathbf{Q} := \mathbf{P} \cup \mathrm{Cor}_{\mathfrak{c}}(\mathbf{R})s$ is a Q-reduct of $\mathbf{R}$ w.r.t. s, iff:
   $\mathbf{Q} \overset{Q}{\succeq}_s \mathbf{R}$ and $\mathbf{Q}$ is not a superset of any other reduct.

This method, derived from our algebraic analysis, exactly corresponds to Skowron's approach in [14] for finding Q-reducts of $\mathbf{R}$ on s by discernibility matrices.

6 Conclusion

The main contribution of this paper is an algebraic formalisation of rough set theory using residual operations and characteristic relations. The emphasis on residuals results from the connection between RST operators to modal operators and weakest/strongest pre-/ postconditions. However, the formalisation in section 3 is still not point-free as it heavily depends on set representations as characteristic functions and their preimages. The reason for this approach is related ongoing work: Equation (25) shows that RST works on arbitrary partitions just as well as on simple sets; i.e. RST can handle information systems with any kind of total functions. Formal concept analysis (FCA), on the other hand, describes sets of domain elements by *attributes*; i.e. binary features that create only partitions $\{s, \overline{s}\}$. Any information system can be transformed into such an attribute system ("formal context") and by using methods similar to those presented in section 3 one can simulate FCA through RST (and vice versa). With formal contexts being heterogenous relations, domain and codomain operations (extent and intent) are not symmetric as they are in RST which motivates an analysis of KAD in this context. Section 4 demonstrates how this approach can be used to drastically simplify both RST and FCA formalisations.

A second contribution is the algebraic formalisation of core relations in section 5. It is shown how the specification of an algorithm to discover core relations naturally evolves from its mere description.

All results presented in this article are part of ongoing work: The basic approach presented in section 4 requires a complete entirely point-free formalisation of RST. In a next step, FCA should be treated within KAD. The residual based representation of approximation operators offers connections to dilation and erosion operators in mathematical morphology (e.g. [15]). Finally, we want to examine modal logics as another interlink: modal and multivalued logics for RST have been studied extensively and [16] present a bi-intuitionistic logic that which appears a very promising approach towards generalisation or specialisation of theories by reasoning about hypotheses.

Acknowledgements. The author wishes to thank Bernhard Möller and the anonymous referees for valuable comments.

References

1. Desharnais, J., Möller, B., Struth, G.: Kleene algebra with domain. ACM Transactions on Computational Logic 7(4) (2006)
2. Dubois, D., Prade, H.: Rough fuzzy sets and fuzzy rough sets. International Journal of General Systems 12(2–3) (1990)
3. Düntsch, I.: A logic for rough sets. Theoretical Computer Science 179(1–2), 427–436 (1997)
4. Ganter, B., Wille, R.: Formal Concept Analysis. Springer (1999)
5. Małuszyński, J., Szałas, A., Vitória, A.: Paraconsistent logic programs with four-valued rough sets. In: Chan, C.-C., Grzymala-Busse, J.W., Ziarko, W.P. (eds.) RSCTC 2008. LNCS (LNAI), vol. 5306, pp. 41–51. Springer, Heidelberg (2008)
6. Müller, M.E.: Relational Knowledge Discovery. Cambridge University Press (2012)
7. Nakamura, A.: A rough logic based on incomplete information and its application a rough logic based on incomplete information and its application. International Journal of Approximate Reasoning 15, 367–378 (1996)
8. Parsons, S., Kubat, M.: A first-order logic for reasoning under uncertainty using rough sets. Journal of Intelligent Manufacturing 5, 211–232 (1994)
9. Pawlak, Z.: On Rough Sets. Bulletin of the EATCS 24, 94–184 (1984)
10. Pawlak, Z.: Rough Sets - Theoretical Aspects of reasoning about Data. D: System Theory, Knowledge Engineering and Problem Solving, vol. 9. Kluwer Academic Publishers (1991)
11. Polkowski, L.: Rough Sets - Mathematical Foundations. Advances in Soft Computing. Physica (2002)
12. Priss, U.: An FCA interpretation of relation algebra. In: Missaoui, R., Schmidt, J. (eds.) ICFCA 2006. LNCS (LNAI), vol. 3874, pp. 248–263. Springer, Heidelberg (2006)
13. Priss, U.: Relation algebra operations on formal contexts. In: Rudolph, S., Dau, F., Kuznetsov, S.O. (eds.) ICCS 2009. LNCS(LNAI), vol. 5662, pp. 257–269. Springer, Heidelberg (2009)
14. Skowron, A., Rauszer, C.: The discernibility matrices and functions in information systems. In: The Discernibility Matrices and Functions in Information Systems. Theory and Decision Library, vol. 11. Springer, Netherlands (1992)
15. Stell, J.G.: Relations in mathematical morphology with applications to graphs and rough sets. In: Winter, S., Duckham, M., Kulik, L., Kuipers, B. (eds.) COSIT 2007. LNCS, vol. 4736, pp. 438–454. Springer, Heidelberg (2007)
16. Stell, J.G., Schmidt, R.A., Rydeheard, D.: Tableau development for a bi-intuitionistic tense logic. In: Höfner, P., Jipsen, P., Kahl, W., Müller, M.E. (eds.) RAMiCS 2014. LNCS, vol. 8428, pp. 412–428. Springer, Heidelberg (2014)

17. Tarski, A.: On the calculus of relations. Journal of Symbolic Logic 6(3), 73–89 (1941)
18. Xu, F., Yao, Y., Miao, D.: Rough set approximations in formal concept analysis and knowledge spaces. In: An, A., Matwin, S., Raś, Z.W., Ślęzak, D. (eds.) ISMIS 2008. LNCS (LNAI), vol. 4994, pp. 319–328. Springer, Heidelberg (2008)
19. Yao, Y., Chen, Y.: Rough set approximations in formal concept analysis. In: Peters, J.F., Skowron, A. (eds.) Transactions on Rough Sets V. LNCS, vol. 4100, pp. 285–305. Springer, Heidelberg (2006)
20. Yao, Y.Y., Lin, T.Y.: Generalization of rough sets using modal logics. Intelligent Automation and Soft Computing 2(2) (1996)

Proofs

Proof (Equation 14).

$$\begin{aligned}
[\![R]\!]s &= \{x : [x]_R \subseteq s\} = \{x : \forall y : xRy \rightarrow y\dot{s}\mathbf{1}\} \\
&= \{x : \neg\exists y : \neg(\neg xRy \vee y\dot{s}\mathbf{1})\} = \{x : \neg\exists y : xRy \wedge y\bar{\dot{s}}\mathbf{1}\} \\
&= \overline{\{x : \exists y : xRy \wedge y\bar{\dot{s}}\mathbf{1}\}} = \overline{\{x : x\, R ; \bar{\dot{s}}\, \mathbf{1}\}} = \left\{x : x\, \overline{R ; \bar{\dot{s}}}\, \mathbf{1}\right\} \\
&\overset{(a)}{=} R \backslash\!\backslash \dot{s} . \mathbf{1} = \overline{R ; \bar{\dot{s}}} . \mathbf{1} = \overline{R ; \bar{\dot{s}} . \mathbf{1}} = \overline{R . \overline{[\underset{\cdot}{s}]_{\tilde{s}}}} = \overline{R ; \tilde{\bar{s}} . \underset{\cdot}{s}} \\
&\overset{(b)}{=} R \backslash\!\backslash \tilde{\dot{s}} . \underset{\cdot}{s} .
\end{aligned}$$

By (a) and since $[\![R]\!]\overline{s} \cap s = \emptyset$ and $[\![R]\!]s \subseteq s$,

$$\begin{aligned}
[\![R]\!]s &= s \cap [\![R]\!]s = s \cap ([\![R]\!]s \cup [\![R]\!]\overline{s}) \\
&= s \cap (R \backslash\!\backslash \dot{s} . \mathbf{1} \cup R \backslash\!\backslash \dot{s} . \mathbf{0}) = s \cap (R \backslash\!\backslash \tilde{\dot{s}} . s \cup R \backslash\!\backslash \tilde{s} . \overline{s}) \\
&\overset{(c)}{=} (R \backslash\!\backslash \tilde{s} . \mathcal{U}) \cap s .
\end{aligned}$$

Proof (Equation 26).

$$\begin{aligned}
\forall x : x \in P \backslash\!\backslash Q . \mathcal{U} &\Longleftrightarrow \forall x, z : x\, P \backslash\!\backslash Q\, z \\
&\Longleftrightarrow \forall x, z : x\, \overline{P ; \overline{Q}}\, z \\
&\Longleftrightarrow \forall x, z : \neg\exists y : (xPy \wedge y\overline{Q}z) \\
&\Longleftrightarrow \forall x, y, z : (x\overline{P}y \vee yQz) \\
&\Longleftrightarrow xPy \longrightarrow yQz \\
&\Longleftrightarrow \forall y : [y]_P \subseteq [y]_Q .
\end{aligned}$$

Author Index